GUIDE TO
PAIRING-BASED
CRYPTOGRAPHY

CHAPMAN & HALL/CRC
CRYPTOGRAPHY AND NETWORK SECURITY

Series Editors
Douglas R. Stinson and Jonathan Katz

Published Titles

Lidong Chen and Guang Gong, Communication System Security

Shiu-Kai Chin and Susan Older, Access Control, Security, and Trust: A Logical Approach

M. Jason Hinek, Cryptanalysis of RSA and Its Variants

Antoine Joux, Algorithmic Cryptanalysis

Jonathan Katz and Yehuda Lindell, Introduction to Modern Cryptography, Second Edition

Nadia El Mrabet and Marc Joye, Guide to Pairing-Based Cryptography

Sankar K. Pal, Alfredo Petrosino, and Lucia Maddalena, Handbook on Soft Computing for Video Surveillance

Burton Rosenberg, Handbook of Financial Cryptography and Security

María Isabel González Vasco and Rainer Steinwandt, Group Theoretic Cryptography

CHAPMAN & HALL/CRC
CRYPTOGRAPHY AND NETWORK SECURITY

GUIDE TO PAIRING-BASED CRYPTOGRAPHY

Edited by

Nadia El Mrabet

SAS - Ecole des Mines de Saint Etienne

Gardanne, France

Marc Joye

NXP Semiconductors

San Jose, USA

CRC Press
Taylor & Francis Group
Boca Raton London New York

CRC Press is an imprint of the
Taylor & Francis Group an **informa** business

A CHAPMAN & HALL BOOK

CRC Press
Taylor & Francis Group
6000 Broken Sound Parkway NW, Suite 300
Boca Raton, FL 33487-2742

© 2017 by Taylor & Francis Group, LLC
CRC Press is an imprint of Taylor & Francis Group, an Informa business

No claim to original U.S. Government works

Printed at CPI on sustainably sourced paper
Version Date: 20161102

International Standard Book Number-13: 978-1-4987-2950-5 (Hardback)

Visit the Taylor & Francis Web site at
http://www.taylorandfrancis.com

and the CRC Press Web site at
http://www.crcpress.com

Contents

Foreword

The field of elliptic curves is an old and well-studied part of number theory. Until the 1980s it was rather arcane outside of a small community of specialists. This changed, abruptly, after Hendrik Lenstra showed how elliptic curves could be used in an efficient algorithm to factor large integers, and Miller and Koblitz showed how they could be used in a version of the Diffie-Hellman key exchange protocol, replacing the multiplicative group of non-zero elements of a finite field. This version was argued to be invulnerable to the so-called "Factor Base" attacks against the original system. Elliptic curves also have an extra feature that the multiplicative groups lack — a pairing.

Pairings are ubiquitous in mathematics. They range from the ordinary inner product in vector spaces, to integration of cycles against differential forms, and to many fundamental uses in algebraic topology. A particular analytic pairing, defined on points of finite order in elliptic curves and abelian varieties over the complex numbers, was given an algebraic definition in 1948 by André Weil. This algebraic version proved to be important in the arithmetic theory of elliptic curves and abelian varieties.

In 1985, Miller and Koblitz independently proposed that the group of an elliptic curve over a finite field be used as a replacement for the group of invertible residues in a finite field in the Diffie-Hellman protocol. Motivated by an attempt to relate the discrete logarithm problem in the two different kinds of groups, Miller recalled that the Weil pairing (if it could be efficiently calculated) would relate the arithmetic in an elliptic curve with multiplication in a finite field. This led him to devise an efficient algorithm to calculate the Weil pairing. This was later shown to give an attack against the discrete logarithm problem for supersingular elliptic curves over a finite field by Menezes, Okamoto, and Vanstone, and over a small set of other classes of curves by Frey and Rück.

Meanwhile, in 1984, Adi Shamir had proposed an interesting thought experiment he called "Identity-Based Cryptosystems." If it could be realized it could allow, for example, using a person's email address as their public key in a public key cryptosystem. However, it was not clear how this could be practically implemented. In 1999, Antoine Joux took the first steps in work that was completed the next year by by Dan Boneh and Matt Franklin. These three authors used the Weil pairing algorithm of Miller as an essential building block. Thus the burgeoning field of Pairing-Based Cryptography was born.

The security of various cryptosystems is usually predicated on the presumed difficulty of specific computational problems. As cryptographic protocols have grown more sophisticated, the number of such assumptions has multiplied.

If G is a finite cyclic group (written additively) with generator P, the *Computational Diffie-Hellman* (CDH) problem is to recover abP when given P, aP and bP. The *discrete logarithm* (DL) problem is to recover a when given P and aP. The presumed difficulty of the Diffie-Hellman problem is the basis for the security of the Diffie-Hellman cryptosystem. If one can solve DL one can solve CDH, and, to date, this has been the only approach. The closely related *distinguishing Diffie-Hellman problem* (DDH) is to distinguish a triple of the form (aP, bP, abP) from a triple of the form (aP, bP, cP) where c is random.

In its abstract form, a pairing is a map $e : M \times N \to R$, where M, N, R are abelian groups, and e is bilinear and non-degenerate (meaning that if $0 \neq P \in M$ there is a $Q \in N$ such that $e(P, Q) \neq 0$). By bilinearity we have $e(aP, bQ) = abe(P, Q)$. One can see that in the case where $M = N$, if CDL is easy in R, then we can solve the DDH in M.

Boneh and Franklin realized that if we were in the above situation (along with a slightly more technical assumption called the *bilinear Diffie-Hellman assumption* — BDH), we could leverage this disparity into creating a viable IBE. Indeed, this was the case for elliptic curves defined over certain fields (assuming standard conjectures). Since then many other applications of pairings have been found, many of which are detailed in this volume. As with any new field there are many subsidiary technical problems that arise and practicalities concerning implementation. A number of these are detailed here.

In conclusion, it is a pleasure to see a comprehensive survey of this important field in the present volume.

Victor S. Miller
Princeton, New Jersey

Symbol Description

δ	A non-zero integer	d	The degree of the twist between E and E'
p	A prime number		
q	A power of a prime number	$\mathbb{G}_1$	A subgroup of order r of $E(\mathbb{F}_p)$
$\mathbb{F}_p$	The finite field of characteristic p	$\mathbb{G}_2$	A subgroup of order r of $E(\mathbb{F}_{p^k}) \setminus E(\mathbb{F}_p)$
π_p	The Frobenius automorphism		
$\mathbb{F}_{p^\delta}$	The extension of degree δ of $\mathbb{F}_p$	$\mathbb{G}_3$	A subgroup of order r of $\mathbb{F}_{p^k}$
$\overline{\mathbb{F}_p}$	The algebraic closure of $\mathbb{F}_p$	e	A pairing $e : \mathbb{G}_1 \times \mathbb{G}_2 \to \mathbb{G}_3$
$E(\mathbb{F}_p)$	An elliptic curve E defined over $\mathbb{F}_p$	e_T	The Tate pairing
P_∞	The point at infinity of an elliptic curve E	e_A	Ate pairing
		e_{tA}	twisted Ate pairing
$\#E(\mathbb{F}_p)$	The order of $E(\mathbb{F}_p)$ (also denoted n)	Ker	The kernel of a morphism
r	A prime number dividing $\#E(\mathbb{F}_p)$	Im	The image of a morphism
$E(\mathbb{F}_p)[r]$	The subgroup of $E(\mathbb{F}_p)$ with order r	card($\mathbb{G}$)	The cardinal of the set $\mathbb{G}$
k	The embedding degree of $E(\mathbb{F}_p)$ with respect to r	$\mathcal{O}$	An order in an imaginary quadratic field
E'	A twisted elliptic curve of E		

1

Pairing-Based Cryptography

Sébastien Canard
Orange

Jacques Traoré
Orange

1.1 Introduction

Initially used in cryptography to break the discrete logarithm problem in the group of points of some elliptic curves, by providing a reduction to the discrete logarithm in finite fields [37] (where subexponential attacks are known), pairings are now considered to be one of the most suitable mathematical tools to design secure and efficient cryptographic protocols. The cryptography relying on pairings is known under the generic term of "pairing-based cryptography."

In this chapter, we review the cryptographic building blocks where the use of a bilinear pairing is definitely an advantage, compared to the historical use of finite fields as in traditional constructions. For each of these blocks, we also explain how they can be used in practice for real-world applications.

1.2   Preliminaries on Pairings

We first give some preliminaries on pairings. More details will be given throughout this book, and here we only sketch some properties and notation that we will need in order to introduce the necessary notions.

1.2.1   Pairing and Bilinear Environment

We first define three groups: $\mathbb{G}_1, \mathbb{G}_2$ (each with additive notation), and $\mathbb{G}_T$ (with multiplicative notation), all of prime order denoted r. In this chapter, a pairing e is then defined as a map $e : \mathbb{G}_1 \times \mathbb{G}_2 \to \mathbb{G}_T$ having the following properties:

1. **bilinearity**, i.e., for all $P_1 \in \mathbb{G}_1, P_2 \in \mathbb{G}_2$ and $a, b \in \mathbb{Z}_r$ we have:

$$e([a]P_1, [b]P_2) = e(P_1, P_2)^{ab} \text{ ; and}$$

2. **non-degeneracy**, i.e., for $P_1 \neq 0_{\mathbb{G}_1}$ and $P_2 \neq 0_{\mathbb{G}_2}$, $e(P_1, P_2) \neq 1_{\mathbb{G}_T}$;

where $0_{\mathbb{G}_1}$ (resp. $0_{\mathbb{G}_2}$ and $1_{\mathbb{G}_T}$) is the neutral element of the group $\mathbb{G}_1$ (resp. $\mathbb{G}_2$ and $\mathbb{G}_T$). The notation $[a]P_1$ here corresponds to the scalar multiplication (in an additive group) of a generator $P_1 \in \mathbb{G}_1$ by a scalar $a \in \mathbb{Z}_r$ (i.e., $P + P + \cdots + P$, a times when $a > 0$ or $-P - P - \ldots - P$, a times when $a < 0$). Additionally to these mathematical properties, cryptographers most of the time want a pairing to be **efficiently computable** and **hard to inverse**, so that these two notions sometimes directly appear in the definition of a pairing.

The above bilinear property can be derived into a lot of equalities, permitting us to play with the scalars and moving them from one "group" to another. We have, for example:

$$e([a]X_1, [b]X_2) = e([b]X_1, [a]X_2) = e([ab]X_1, X_2) = e([a]X_1, X_2)^b = e([b]X_1, X_2)^a = \cdots$$

In the sequel, we then consider a *bilinear environment* as a tuple

$$(r, \mathbb{G}_1, \mathbb{G}_2, \mathbb{G}_T, P_1, P_2, e),$$

where r, $\mathbb{G}_1$, $\mathbb{G}_2$, $\mathbb{G}_T$, and e are defined as above, and where P_1 (resp. P_2) is a generator of $\mathbb{G}_1$ (resp. $\mathbb{G}_2$).

1.2.2   Types of Pairings

There are several ways to describe a pairing, but today the most efficient ones are defined when the groups $\mathbb{G}_1$ and $\mathbb{G}_2$ are elliptic curves, and the group $\mathbb{G}_T$ is the multiplicative group of a finite field. Galbraith, Patterson and Smart [32] have defined three types of pairings:

- **type 1**, when $\mathbb{G}_1 = \mathbb{G}_2$;
- **type 2**, when $\mathbb{G}_1 \neq \mathbb{G}_2$ but an efficiently computable isomorphism $\phi : \mathbb{G}_2 \to \mathbb{G}_1$ is known, while none is known in the other direction;
- **type 3**, when $\mathbb{G}_1 \neq \mathbb{G}_2$ and no efficiently computable isomorphism is known between $\mathbb{G}_1$ and $\mathbb{G}_2$, in either direction.

Although type 1 pairings were mostly used in the early age of pairing-based cryptography, they have gradually been discarded in favor of type 3 pairings. Indeed, today type 1 is not attractive enough from an efficiency point of view, since it involves very large curves.

The constructions given in this chapter all require the use of asymmetric pairings (i.e., of type 2 or type 3). For simplicity, we will only consider pairings of type 3, which is not a strong restriction (see [22]) since these pairings offer the best efficiency. They are moreover compatible

TABLE 1.1 Comparison of the size of parameters, in bits, for different types of cryptography.

	Security level	96	128	256
RSA-based	factorization module	1776	3248	15424
Discrete logarithm-based	key	192	256	512
	group	1776	3248	15424
	group (elliptic curve)	192	256	512
Pairing-based	scalar	192	256	512
	group $\mathbb{G}_1$	192	256	512
	group $\mathbb{G}_2$	384	512	1024
	group $\mathbb{G}_T$	1776	3248	15424

with several computational assumptions, such as the Decision Diffie-Hellman (DDH) in $\mathbb{G}_1$ or $\mathbb{G}_2$, also known as the XDH assumption [14], which does not hold in type 1 pairings (see Remark 1.1 in Section 1.3.3 for some details).

1.2.3 Choice of Parameters

We now consider the size of the parameters that need to be manipulated when using a pairing (some details will also be given throughout the book). Generally speaking, the size of the security parameters is related to the infeasibility of executing best-known attacks on the studied cryptographic system. By convention, a *security level* corresponds to the minimum size for an exhaustive search among the set of all possible secret keys. The figures given for "RSA-based," "discrete logarithm-based," and "pairing-based" are computed by using, in particular, the best algorithms to solve the related problems. One can refer to Chapter 10 for more details on the choice of parameters.

Focusing on pairing-based cryptography, for a given security level, minimal sizes for r (the size of the elliptic curve subgroup) and q^k (the size of the finite field underlying $\mathbb{G}_T$) have to be chosen. The integers r and q are prime numbers, and k is called the *embedding degree*. For 128-bit security, the most relevant choice for an elliptic curve today is the Barreto-Naerhig [8], of equation $Y^2 = X^3 + 5$ over $\mathbb{F}_q$. With such a curve, the most suitable pairing seems to be the optimal Ate [41], which can be written

$$e : E(\mathbb{F}_q) \times E(\mathbb{F}_{q^2}) \longrightarrow \mathbb{F}_{q^{12}}^*$$

where $E(\mathbb{F}_q)$ (resp. $E(\mathbb{F}_{q^2})$) denotes the elliptic curve defined over the finite field $\mathbb{F}_q$ (resp. $\mathbb{F}_{q^2}$).

For a 128-bit security level, this then gives us the following parameters:

- size of the order $q = 256$ bits (for scalars);
- size of $\mathbb{G}_1 = 256$ bits;
- size of $\mathbb{G}_2 = 512$ bits; and
- size of $\mathbb{G}_T = 3248$ bits.

Table 1.1 then compares the size of the different parameters for different types of cryptography. The scores are given based on a work by the ECRYPT II [1].

1.3   Preliminaries on Cryptography

1.3.1   Cryptography in a Nutshell

Cryptography, the art of secret writing, has historically been used to hide the communication between parties. It is today known to address the needs of several important security services:

- **confidentiality**, to be convinced that only the intended receiver of a message can read it;
- **authentication**, to be convinced that a person with whom one is communicating is the right person;
- **integrity**, to be convinced that a received message has not been modified; and
- **non-repudiation**, to be convinced that the origin of a message cannot repudiate it.

But nowadays, cryptographers, most of the time by using techniques from public key cryptography, design much more sophisticated tools that permit them to solve modern (and more complex) problems arising from our digital society:

- perform computations over encrypted data;
- share confidential data;
- guarantee both anonymity and accountability of customers;
- ensure both privacy protection and profiling...

The construction of these new cryptographic solutions requires the use of new mathematical tools, and, in this context, *pairings* are the most important and most often used ones in the current literature.

Before giving some details about pairings, we start by introducing more formally the notions of hash function, encryption, and signature that will be used in this chapter.

1.3.2   Hash Function

Algorithms. Generally speaking, a hash function is a mathematical function defined by

$$\mathcal{H} : \{0,1\}^* \longrightarrow \{0,1\}^k$$

where k is a "small" parameter. It is then a function that permits us to transform data of arbitrary size into a representative data with a fixed size. In cryptography, a hash function mechanism Hash only needs to define the above mathematical function and, in the sequel, we most of the time only use the mathematical notation $\mathcal{H}$. Sometimes, the output of $\mathcal{H}$ is not an element in $\{0,1\}^k$ but a group element, and the hash function is then defined as, e.g., $\mathcal{H} : \{0,1\}^* \longrightarrow \mathbb{G}$.

Security. Regarding security, a cryptographically secure hash function should verify the following properties.

- **Pre-image resistance**, which means that for a given output $h \in \{0,1\}^k$, it is computationally infeasible to find a value $m \in \{0,1\}^*$ such that $\mathcal{H}(m) = h$.
- **2nd pre-image resistance**, saying that for a given input $m \in \{0,1\}^*$, it is computationally infeasible to find a value $m' \in \{0,1\}^*$ such that $\mathcal{H}(m) = \mathcal{H}(m')$.
- **Collision resistance**, which says that it is computationally infeasible to find two values $m, m' \in \{0,1\}^*$ such that $\mathcal{H}(m) = \mathcal{H}(m')$.

Based on these security properties, and due to the birthday paradox [43], the value k should be taken to be equal to 256 for 128-bit security. One can refer to Chapter 8 for more details on hash functions.

1.3.3 Encryption

An encryption scheme aims at protecting the confidentiality of some messages.

Algorithms. In the public key cryptography setting, such a scheme is composed of four main algorithms:

- the Setup algorithm, which, on input of a security parameter λ, outputs a set of parameters param, which depends on the schemes used;
- the key generation KeyGen, which, on input of the parameters param, outputs a key pair composed of a *public key* pk and a *secret key* (also called *private key*) sk;
- the encryption algorithm *per se* Enc, which, on input of the parameters param, the public key pk, and a message m (also called a *plaintext*), outputs a *ciphertext* c;
- the decryption algorithm Dec, which, on input of the parameters param, the secret key sk, and the ciphertext c, outputs the corresponding message m.

Security. Regarding security, the expected property that an encryption scheme should verify is the following one.

- **Indistinguishability**, which states that an adversary against the scheme is not able, given two messages m_0 and m_1 that (s)he may have chosen, and a ciphertext of either m_0 or m_1, to determine which message this ciphertext encrypts. For details, we refer the reader to, e.g., [9].

Example. We give an example of encryption scheme that will be used in the sequel: ElGamal [27]. The key generation is executed as follows:

- let $\mathbb{G}$ be a group (for example, the set of points of an elliptic curve) of prime order r and let P be a generator of $\mathbb{G}$;
- the secret key is $x \in \mathbb{Z}_r^*$ and the public key is $Y = [x]P$.

The encryption and decryption steps are then given by Algorithms 1.1 and 1.2, respectively.

ALGORITHM 1.1   ElGamal encryption algorithm Enc.

Input : message $m \in \mathbb{G}$, public key $pk = (P, Y) \in \mathbb{G}^2$.
Output: ciphertext $c = (T_1, T_2) \in \mathbb{G}^2$.

1 choose $\rho r \in \mathbb{Z}_r^*$ at random
2 compute $T_1 = m + [\rho]Y$ and $T_2 = [\rho]P$
3 output $c = (T_1, T_2)$.

ALGORITHM 1.2   ElGamal decryption algorithm Dec.

Input : ciphertext $c = (T_1, T_2) \in \mathbb{G}^2$, secret key $s_k = x \in \mathbb{Z}_r^*$.
Output: message $m \in \mathbb{G}$.

1 output $m = T_1 - [x]T_2$.

Remark 1.1   The ElGamal encryption scheme is secure if the Decisional Diffie-Hellman (DDH) assumption holds. This assumption states that it is hard to distinguish between the following two tuples:

$$(P, [a]P, [b]P, [c]P) \text{ and } (P, [a]P, [b]P, [ab]P);$$

where P is the generator of a group $\mathbb{G}$ of prime order r and where a, b, c are random integers in $\mathbb{Z}_r^*$.

In fact, this problem is easy (and the ElGamal encryption scheme is not secure in this setting) if the group $\mathbb{G}$ is related to a type 1 pairing (see Section 1.2.2) since one can easily test whether

$$e([a]P, [b]P) = e(P, [c]P)$$

or not, which is equivalent to $ab = c$ or not. For similar reasons this problem is also easy in $\mathbb{G}_1$ or in $\mathbb{G}_2$ in a type 2 pairing (since we can advantageously use the isomorphism $\phi : \mathbb{G}_2 \to \mathbb{G}_1$). As said previously, in a type 3 pairing, when the DDH assumption holds in $\mathbb{G}_1$ or $\mathbb{G}_2$, we then talk about the XDH assumption [14].

1.3.4 Signature

A digital signature is the basic cryptographic tool to prove, at the same time, the authenticity of a message (it comes from the right person if we plug in, e.g., a PKI), its integrity, and its non-repudiation.

Algorithms. A signature scheme is composed of four main algorithms:

- the Setup algorithm, which, on inputting a security parameter λ, outputs a set of parameters param, which is specific to the schemes used;
- the key generation KeyGen, which, on input of the parameters param, outputs a key pair composed of a public key pk and a secret (or private) key sk;
- the signature algorithm Sign, which, on input of the parameters param, the secret key sk, and a message m, outputs a digital signature σ on m;
- the verification algorithm Verif, which, on input of the parameters param, the public key pk, a message m, and a putative signature σ, outputs either 1 if the signature is correct under pk, or 0 otherwise.

Security. Regarding security, a signature scheme should verify the following property.

- **Unforgeability**, which states that it should be infeasible for an attacker of the system to provide a message and its signature that have never been generated by the true owner of the private key. There are then several possibilities for such a defrauder to attack a digital signature scheme, and we refer the interested reader to [33] for more (formal) details.

Example. We give an example of a signature scheme that will be used in the sequel: RSA-PSS [38, 11]. The key generation is executed as follows:

- let $n = pq$ with p and q two distinct prime numbers of sufficient size (see Table 1.1);
- let e be an integer that is prime to $\varphi(n)$ and d such that $ed \equiv 1 \pmod{\varphi(n)}$;
- let $\mathcal{H} : \{0,1\}^* \to \{0,1\}^{\lambda_2}$, $\mathcal{F} : \{0,1\}^* \to \{0,1\}^{\lambda_0}$, $\mathcal{G} : \{0,1\}^* \to \{0,1\}^{\lambda_1}$ be cryptographic hash functions with $\lambda = \lambda_0 + \lambda_1 + \lambda_2 + 1$;
- sk $= (p, q, d)$ and pk $= (n, e)$

The signature and verification algorithms are then given by Algorithms 1.3 and 1.4, respectively, where $\|$ denotes the concatenation operation.

ALGORITHM 1.3   RSA-PSS signature algorithm Sign.

Input : message $m \in \mathbb{Z}_n$, public key $pk = (n, e) \in \mathbb{Z}^2$, secret key $sk = (p, q, d) \in \mathbb{Z}^3$.
Output: signature $(\sigma, y) \in \mathbb{Z}_n^2$.

1 choose ρ at random
2 compute $w = \mathcal{H}(m\|\rho)$, $s = \mathcal{G}(w) \oplus \rho$, and $t = \mathcal{F}(w)$
3 compute $y = 0\|w\|s\|t$
4 compute $\sigma = y^d \pmod{n}$
5 output (σ, y).

ALGORITHM 1.4   RSA-PSS verification algorithm Verif.

Input : message $m \in \mathbb{Z}_n$, public key $pk = (n, e) \in \mathbb{Z}^2$, signature $(\sigma, y) \in \mathbb{Z}_n^2$.
Output: verification ok or not.

1 parse $y = 0\|w\|s\|t$
2 compute $\rho = s \oplus \mathcal{G}(w)$
3 check that $y = \sigma^e \pmod{n}$
4 check that $w = \mathcal{H}(m\|\rho)$, and $t = \mathcal{F}(w)$

1.3.5   Security Model for Cryptography and Pairings

In public key cryptography, there are most of the time three steps to provide evidence that a proposed scheme is secure.

The first step consists of giving a model that idealizes the studied cryptographic protocol. This is exactly what has been done above for encryption and signature schemes with the formal definition of algorithms and parameters. It also consists of clearly defining what an adversary can do to break the security of an encryption or signature scheme.

The next step is to define assumptions stating the difficulty to break some mathematical problems, such as the computation of discrete logarithms or the factorization of integers.

The final step is to prove that breaking the security of the scheme is as hard as solving a mathematical problem supposed to be intractable. During this step, there are some cases where the cryptographer idealizes some cryptographic functions. For example, for a hash function, we talk about the "random oracle model" [10] when we assume that the outputs of a cryptographic hash function are indistinguishable from random values, except that two equal inputs will always give the same output (to keep the deterministic property of a hash function). When there is no such idealization, then we talk about the "standard model."

Most of today's constructions that are proved to be secure in the standard model are based on pairings. This is for example the case for non-interactive zero-knowledge proofs of knowledge, secure in the random oracle using the Fiat-Shamir heuristic [29], and secure in the standard model using the Groth-Sahai technique [35] for pairing equations. Such proofs are for example used to design group signature schemes, as we will see in Section 1.5.

1.4   Short Signatures

Digital signatures are today one of the most famous cryptographic tools used in real-world applications. Since the introduction of public key cryptography, and the publication of RSA cryptosystem [38], many new signature schemes have been proposed, and one of the most challenging tasks for cryptographers has been to design such tools tailored for lightweight devices.

In particular, the way to shorten the size of a digital signature is of primary importance when communication time between two devices is sensitive. As we will see, digital signatures based on the use of a pairing are much more appropriate to this context than traditional ones.

1.4.1   Use Cases

The size of a signed message is an important matter in several use cases. We here describe two of them, based on real-life constraints.

Signed SMS. It is commonly known that an SMS (Short Message Service) is a short message that is sent from one mobile phone to another via the OTA (Over The Air) communication layer. Its payload length is limited by the constraints of the signalling protocol to precisely 140 bytes, which is very small. Although different techniques to obtain a MAC (Message Authentication Code) on an SMS (see for example the ETSI 102 225 standard on secured packet structure for UICC-based applications [2]) exist, they do not permit us to achieve the non-repudiation property that necessitates the use of a true digital signature. Yet non-repudiation can be useful in this context.

   For example, such a signed SMS could be used for payment transactions. Peer-to-peer authenticated payments from one mobile phone to another could be made possible using such a short digital signature, without increasing the number of necessary SMSs to be sent. Another use is the possibility for a consumer to digitally sign a contract for a new phone package or for a new option.

   The size of an SMS can be seen as an important constraint in these cases. For a 128-bit security, an RSA signature necessitates (see below for a more detailed comparison) 4096 bits, and thus 4 SMS to be sent (without taking into account the signed message and an additional potential PKI certificate). By contrast, a BLS signature (see [17] and below) only necessitates 256 bits, that is, roughly 1/4 of an SMS!

Sensor node on-line authentication. Sensor nodes (or RFID tags) are among the less powerful devices used in real-world applications and the communication rate with a reader is very slow. As a typical example, the communication rate between an RFID tag and a reader is only from 9.6 to 19.2 kbits per second. In several practical use cases, such devices have to be authenticated by their environment. As a consequence, the communication rate is an important factor that must be taken into account: the less information the sensor node has to send, the fastest the communications will be. Then a small signature for authentication purposes would be a great advantage to optimize the exchanges.

1.4.2   Related Work: Pairing or Not Pairing

Regarding the state-of-the-art on signature schemes, we can here compare the traditionally implemented ones (namely RSA [38] and EC-DSA [3]) and the most interesting pairing-based signature schemes. The resulting comparison is given in Table 1.2. For the pairing-based signature scheme, we have chosen the BLS one [17] since it is presumably the best suitable choice for our purpose. The size of the parameters are guided by ECRYPT II recommendations on cryptography [1]. The recommended parameters show that there is currently a little advantage to using a pairing-based signature instead of a traditional one. Thus, replacing a current RSA or EC-DSA implementation with a BLS one is relevant only in some particular cases, where the size of a signature is a stringent constraint.

   Apart from the space complexity, time performances are also very good for such schemes. For example, the signature generation of the BLS [17] scheme (see below) is approximately twice

TABLE 1.2 Comparison of the size of digital signatures.

Signature scheme	Security level	Size of signature
RSA		1776 bits
EC-DSA	96 bits	384 bits
BLS		192 bits
RSA		3248 bits
EC-DSA	128 bits	512 bits
BLS		256 bits
RSA		15424 bits
EC-DSA	256 bits	1024 bits
BLS		512 bits

more efficient than the one for EC-DSA [3]. Regarding the verification process, recent results [41] show that this step can now be executed in nearly 200 ms in a smart card.

1.4.3 An Example

In this section, we give the construction of the BLS signature scheme [17], with the slight modification that we describe this scheme in a type 3 pairing setting, as argued above.

Setup and key generation. We consider a bilinear environment $(r, \mathbb{G}_1, \mathbb{G}_2, \mathbb{G}_T, P_1, P_2, e)$, which is chosen during the execution of the Setup algorithm. We also consider a cryptographic hash function $\mathcal{H} : \{0,1\}^* \longrightarrow \mathbb{G}_1$ (modeled as a random oracle [10]).

The key generation algorithm KeyGen consists, for the signer, of choosing at random his/her secret key $x \in \mathbb{Z}_r^*$ and computing the corresponding verification public key pk as $Y = [x]P_2$.

Signature and verification. The signature generation and verification algorithms are then given by Algorithms 1.5 and 1.6, respectively. The verification is correct since

$$
\begin{aligned}
e(\sigma, P_2) &= e([x]\mathcal{H}(m), P_2); \\
&= e(\mathcal{H}(m), [x]P_2) \text{ by using the bilinear property of } e; \\
&= e(\mathcal{H}(m), Y).
\end{aligned}
$$

ALGORITHM 1.5 BLS signature algorithm Sign.

Input : message $m \in \{0,1\}^*$, public key $pk = Y \in \mathbb{G}_2$, secret key $sk = x \in \mathbb{Z}_r^*$.
Output: signature $\sigma \in \mathbb{G}_1$.

1 output $\sigma = [x]\mathcal{H}(m)$.

ALGORITHM 1.6 BLS verification algorithm Verif.

Input : message $m \in \{0,1\}^*$, public key $pk = Y \in \mathbb{G}_2$, signature $\sigma \in \mathbb{G}_1$.
Output: verification ok or not.

1 check that $e(\sigma, P_2) = e(\mathcal{H}(m), Y)$.

1.5   Group Signature Schemes

In a nutshell, a group signature scheme permits members of a group to produce signatures on behalf of the group, in such a way that any verifier of the resulting signature can be convinced that the signature has been produced by a group member, but without being able to determine which one. Group signatures are anonymous and untraceable for everybody, except a designated authority who can, if necessary, "open" a signature to identify the actual signer. They are then used to provide at the same time user anonymity and user accountability.

There exist several variants of group signature, and one of the most promising ones, in the context of mobile services, is called an anonymous credential system. Such a system allows a user to obtain a credential (a driving license, a student card, etc.) from one organization and then later prove possession of this credential to another organization without revealing anything more than the fact that (s)he owns such a credential. Anonymous credentials systems also allow selective disclosure by permitting the user to reveal only some credential attributes or to prove that they satisfy some properties (e.g., age < 25) while hiding all the other credential attribute information (this is in contrast to classical credentials, which only allow the release of all the contained attributes).

1.5.1   General Presentation and Short Model

Algorithms. A group signature scheme is composed of the following algorithms:

- the Setup algorithm, which, on input of the security parameter λ, outputs the issuer's secret key ik, the opener's secret key ok, and the group public key gpk;

- the user key generation UKGn, which, on input gpk, outputs the key pair $(\mathsf{upk}[i], \mathsf{usk}[i])$ for the i-th user;

- the joining interactive protocol Join, which is divided into two interactive algorithms. The issuer plays Iss, which takes on input gpk, ik, and $\mathsf{upk}[i]$. If the issuer accepts, (s)he makes a new entry for i in its registration table reg. The user i executes UJoin, which, on inputs gpk, $\mathsf{usk}[i]$, outputs a private membership signing key denoted $\mathsf{msk}[i]$;

- the group signature algorithm GSign, which, on inputs gpk, a message m, and a private signing key $\mathsf{msk}[i]$, outputs a group signature σ on m;

- the verification algorithm GVerif, which, on inputs gpk, a message m, and a group signature σ, outputs 1 if the signature is valid, and 0 otherwise;

- the opening algorithm Open which, on inputs gpk, a message m, a group signature σ, and the opener key ok, outputs, in a deterministic way, an integer $i \geq 0$ (corresponding to the identity of the group member) and, in a probabilistic way, a proof τ that the i-th user has produced the signature σ on m;

- the Judge algorithm, which, on inputs gpk, a message m, a group signature σ, the public key $\mathsf{upk}[j]$ of j-th user, and a proof τ, outputs 1 if τ is valid and 0 otherwise.

Security. As explained in [12], it is possible to summarize the set of security properties one group signature scheme should satisfy into the four following points.

- **Correctness**, which means that, when executed by honest players, the verification of a group signature outputs 1 with overwhelming probability, and the opening necessarily outputs the right group member's identity and a proof τ that will be accepted by the Judge algorithm with overwhelming probability.

- **Anonymity**, which means that an adversary, given a signature produced by a user (among two users that (s)he may have chosen), is not able to guess (with significant probability) which user among the two generated the signature. This property also includes the user unlinkability, in the sense that it is infeasible to determine whether two different valid group signatures come from the same group member or not.

- **Traceability**, which means that an adversary is not able to output a valid group signature which, once opened, does not identify him/her. This is different from the correctness property since here, the group member is not honest.

- **Non-frameability**, which means that an adversary is not able to falsely accuse an honest user of having produced a valid group signature (s)he did not produce.

1.5.2 Use Cases

There are multiple use cases where group signatures, and their variants, are relevant. We here sketch some of them.

Electronic voting. The aim of electronic voting is to replace traditional paper-ballot voting. Electronic voting possesses numerous advantages: counting is faster, it is more convenient for voters as they do not need to go to a polling station in order to vote, voter turnout can be increased, etc.

Several cryptographic-based electronic voting systems [31, 6, 30] have been proposed in the literature. A possible solution to achieve both the anonymity of the ballot and the verifiability of the outcome of the election is to use an anonymous signature, such as a group signature, or a close variant called list signature [21].

Generally speaking, a voter (i) has to prove that (s)he has the right to vote and (ii) wants to protect the secrecy of his vote (the anonymity of his ballot). Both points are verified if the user can prove that (s)he belongs to the set of eligible voters, while being anonymous. This is exactly the aim of a group signature.

Anonymous electronic cash. The aim of electronic cash (or e-cash) is to replace traditional (material) cash. In electronic cash, a user can withdraw coins in the bank, and then spend them in any shop. Later on, the merchant can deposit the received coins to his own bank. One crucial point with traditional cash is that the user is anonymous and non-traceable when using his money, even with respect to the bank. The latter has in fact no way to make the link between a withdrawal and a spending. One desirable aim of e-cash is to emulate a system satisfying such a property: This is what is called anonymous e-cash.

Several attempts have been done and since the so-called compact e-cash system [18], almost all constructions follow the same principle: When a user withdraws $1, (s)he becomes a member of the group of users who have withdrawn $1. A spending corresponds to a transfer of this coin to a merchant. The transfer is signed, using a group signature, to avoid non-repudiation of this transaction.

Again, a group signature is a good basis to construct a secure e-cash system. There remains, however, some additional work to obtain all the desired functionalities of a truly secure and efficient e-cash system, such as the way to efficiently make non-traceable consecutive payments by the same user [20], to manage double spendings [18], to make a coin transferable [19], etc.

Complex anonymous access control. Generally speaking, a group signature scheme permits anonymous access control: a voter can anonymously access the polling place, a payer can anonymously access the right to pay, a user can anonymously access a building... However, a group signature scheme permits us to manage only with a single group, and then very simple access control policies. As introduced at the beginning of this section, an anonymous credential scheme [23], on the contrary, allows us to manage several groups at once. Let us imagine, for example, a service provider wanting to offer some special rights to certain categories of its customers, such as students under 25 or seniors over 60. A group signature would not be relevant in this case.

This service provider does not necessarily need to obtain all the information about the user (name, address, exact age, etc.) but needs to be sure that these pieces of information (age and student status), called attributes, are correct and that they have been certified by a trusted entity (e.g., the university or the country's authority).

In an anonymous credential system, the trusted entity can certify the attributes (e.g., name, date of birth, nationality, address, etc.) by signing them. Then, each user in possession of such credentials can prove to a third party that (s)he has the correct attributes, and then access the service, while minimizing the information given to the service provider. A user could, for example, prove that (s)he is less than 25, in a certified way, without giving his exact age.

With anonymous credentials systems, a user needs to register only once in order to belong to several groups (such as the group of persons being less than 25, the group of French people, etc.).

1.5.3 Related Work: Pairing or Not Pairing

A lot of work has been done on group signature schemes, and there exist numerous constructions. Among all of them, the most efficient ones follow the same principle [7]:

- a user is the i-th member of the group if (s)he obtained from the issuer a digital signature (denoted msk[i]) on his/her secret value usk[i];

- the anonymity is obtained by not revealing the signature or the secret value during the group signature generation, but only proving the knowledge of the secret usk[i] and its signature upk[i] (valid under the issuer's public key) without revealing the secret or the signature. This can be done by using a well-suited non-interactive zero-knowledge (NIZK) proof of knowledge [39], a cryptographic tool dedicated for such a purpose;

- the anonymity revocation is obtained by adding to the group signature (and to the NIZK proof) the encryption of a part of the issuer's signature msk[i], such that the resulting ciphertext can be decrypted by the opener in order to identify the group member. Since the issuer knows the link between upk[i] and the identity of the i-th group member (stored during the Join protocol), the integer i can be output.

One of the main components for a group signature scheme is then a digital signature scheme that is used by the issuer. Such a signature scheme should (i) permit the issuer to sign a secret value in a blinded manner and (ii) be suitable for an NIZK proof of knowledge. Several constructions exist, which are today mostly of two kinds.

The first one [34] is secure in the standard model (see Section 1.3.5) and can at that time only be instantiated by using a pairing and Groth-Sahai NIZK proofs [35]. The second one [7, 14] is secure in the random oracle model [10], by using the Fiat-Shamir heuristic [29], and is much more efficient than the previous one. In the latter family, there are two kinds of cryptographic assumptions, either based on the flexible RSA assumption [7], or on pairing-based assumptions [14]. The second case is more relevant for a practical implementation, as it permits us to handle the smallest parameters, as shown in Section 1.2.3, and to use short signatures.

1.5.4 An Example

Boneh, Boyen, and Shacham proposed in 2004 a short group signature scheme [14] based on the strong Diffie-Hellman [13] and the decisional linear assumptions.

Key generation. Let $(r, \mathbb{G}_1, \mathbb{G}_2, \mathbb{G}_T, P_1, P_2, e)$ be a bilinear environment. Let $H \in \mathbb{G}_1$, $\zeta_1, \zeta_2 \in \mathbb{Z}_r^*$ and $U, V \in \mathbb{G}_1$ such that $[\zeta_1]U = [\zeta_2]V = H$. Let $\gamma \in \mathbb{Z}_r^*$ and $W = [\gamma]P_2$. Then, ζ_1 and ζ_2

are the secret values ok to open (i.e., revoke the anonymity of) a signature, γ is the secret key ik to add group members, and $(r, P_1, P_2, H, U, V, W)$ is the whole group public key. All these parameters are generated during the Setup algorithm.

During the Join protocol, each group member obtains from the issuer a pair $(A, x) \in \mathbb{G}_1 \times \mathbb{Z}_r^*$ such that $A = [1/(\gamma + x)]P_1$. This pair verifies $e(A, W + [x]P_2) = e(P_1, P_2)$. The value x is the member's secret usk[i] (generated by the UKGn algorithm) and A is the key used to retrieve the identity of a member in case of opening (i.e., anonymity revocation).

Group signature generation. The group signature algorithm GSign is then given in Algorithm 1.7. Here (T_1, T_2, T_3) is a linear encryption of A (see [14] for such an encryption scheme) and the tuple $(c, s_\alpha, s_\beta, s_x, s_{\delta_1}, s_{\delta_2})$ is an NIZK proof of knowledge of a valid certificate (using the Fiat-Shamir heuristic [29]). More precisely, it corresponds to a proof of knowledge of a couple (A, x) such that:

- (A, x) is a valid signature under the issuer's public key W, as

$$e(A, W + [x]P_2) = e(P_1, P_2);$$

- the tuple (T_1, T_2, T_3) is a valid linear encryption of A.

ALGORITHM 1.7   Group signature algorithm GSign.

Input : message m, public key $(r, P_1, P_2, H, U, V, W)$, secret key (A, x).
Output: signature $\sigma = (T_1, T_2, T_3, \Pi)$.

1 choose at random $\alpha, \beta \in \mathbb{Z}_r$
2 compute $T_1 = [\alpha]U$, $T_2 = [\beta]V$ and $T_3 = A + [\alpha + \beta]H$
3 compute $\delta_1 = x\alpha$ and $\delta_2 = x\beta$
4 compute the NIZK proof $\Pi = (c, s_\alpha, s_\beta, s_x, s_{\delta_1}, s_{\delta_2})$ as given by Algorithm 1.8
5 output $\sigma = (T_1, T_2, T_3, \Pi)$.

ALGORITHM 1.8   Group signature NIZK proof algorithm.

Input : message m, public key $(r, P_1, P_2, H, U, V, W)$, secret key (A, x), ciphertext (T_1, T_2, T_3), scalars (δ_1, δ_2).
Output: NIZK proof $\Pi = (c, s_\alpha, s_\beta, s_x, s_{\delta_1}, s_{\delta_2})$.

1 choose at random $r_\alpha, r_\beta, r_x, r_{\delta_1}, r_{\delta_2} \in \mathbb{Z}_r^*$
2 compute $t_1 = [r_\alpha]U$
3 compute $t_2 = [r_\beta]V$
4 compute $t_3 = [r_x]T_1 - [r_{\delta_1}]U$
5 compute $t_4 = [r_x]T_2 - [r_{\delta_2}]V$
6 compute $t_5 = e(T_3, P_2)^{r_x} e(H, W)^{-r_\alpha - r_\beta} e(H, P_2)^{-r_{\delta_1} - r_{\delta_2}}$
7 compute $c = \mathcal{H}(\mathsf{m} \| T_1 \| T_2 \| T_3 \| t_1 \| t_2 \| t_3 \| t_4 \| t_5)$
8 compute $s_\alpha = r_\alpha - c\alpha$, $s_\beta = r_\beta - c\beta$, $s_x = r_x - cx$, $s_{\delta_1} = r_{\delta_1} - c\delta_1$ and $s_{\delta_2} = r_{\delta_2} - c\delta_2$
9 output $\Pi = (c, s_\alpha, s_\beta, s_x, s_{\delta_1}, s_{\delta_2})$.

Verification and opening. The verification algorithm GVerif consists then in verifying this proof of knowledge Π, as described in Algorithm 1.9, using standard techniques, and the open one Open is the decryption of the linear encryption (T_1, T_2, T_3) as:

$$A = T_3 - [\zeta_1]T1 - [\zeta_2]T_2.$$

ALGORITHM 1.9   Group signature NIZK proof verification algorithm GVerif.

Input : message m, public key $(r, P_1, P_2, H, U, V, W)$, ciphertext (T_1, T_2, T_3), NIZK proof
$\Pi = (c, s_\alpha, s_\beta, s_x, s_{\delta_1}, s_{\delta_2})$.

Output: verification ok or not.

1 compute $\tilde{t}_1 = [c]T_1 + [s_\alpha]U$
2 compute $\tilde{t}_2 = [c]T_2 + [s_\beta]V$
3 compute $\tilde{t}_3 = [s_x]T_1 - [s_{\delta_1}]U$
4 compute $\tilde{t}_4 = [s_x]T_2 - [s_{\delta_2}]V$
5 compute $\tilde{t}_5 = \big(e(P_1, P_2)/e(T_3, W)\big)^c e(T_3, P_2)^{s_x} e(H, W)^{-s_\alpha - s_\beta} e(H, P_2)^{-s_{\delta_1} - s_{\delta_2}}$
6 check that $c = \mathcal{H}(\mathsf{m} \| T_1 \| T_2 \| T_3 \| \tilde{t}_1 \| \tilde{t}_2 \| \tilde{t}_3 \| \tilde{t}_4 \| \tilde{t}_5)$.

1.6   Identity-Based Encryption

In public key encryption schemes, as said previously, everyone can encrypt a message by using the public key and only the entity having access to the related secret key will be able to decrypt and read the initial message. But, most of the time, one does not want to send a message to a "public key" but to an entity. This is the objective of Public Key Infrastructures (PKI), introduced in [42]. In fact, the role of a PKI certificate is to make the link between a public key and the owner of this key, thanks to a signature delivered by one (or several) designated trusted certification authority (CA). But the main problem of PKI is that they are relatively complex to deploy and maintain. An alternative is to use identity-based cryptography [40], for which the public key, needed, for example, to encrypt a message, is replaced by the identity of the receiver. For encryption purposes (even if the equivalent exists for signature), we talk about Identity-Based Encryption, or IBE for short.

1.6.1   General Presentation and Short Model

Algorithms. An IBE is generally composed of the following algorithm, which is close to a standard encryption scheme, given above in Section 1.3.3:

- the Setup algorithm, which, on input of a security parameter λ, outputs a set of parameters param, and additionally a master key pair composed of a secret key msk and a corresponding public key mpk. The role of the master secret key is to compute the secret keys for users, based on their identity;

- the key extraction algorithm Ext, which, on input of the parameters param, the master secret key msk, and the identity id of a user, outputs a user secret key usk;

- the encryption algorithm Enc, which, on input of the parameters param, the master public key mpk, an identity id, and a message m, outputs a ciphertext c;

- the decryption algorithm Dec, which, on input of the parameters param, a user secret key usk, and the ciphertext c, outputs the corresponding message m.

Security. An identity-based encryption scheme should verify the following property.

- A special kind of **indistinguishability**, the main difference with the traditional definition (see Section 1.3.3) being that the adversary has access to several secret keys, based on identities of his/her choice, and that (s)he is challenged on a public key id of his/her choice. We refer to [16] for more details.

1.6.2 Use Cases

An IBE permits us to avoid the use of a PKI, and thus may decrease the workload inherent to PKI. However, the generation of the secret key w.r.t. the identity should be done by a specific authority: There are then some similarities between such an authority and a PKI certification authority. The real deployment of such a cryptographic tool could also lead to some problems that necessitate further work. In particular, the fact that the authority knows the secret keys of the users could be an issue in some use cases. Moreover, how to deal with the compromise of a secret key, in particular if this key is related to the name of the user? How to change it? These concerns need to be clearly taken into account but several cryptographic solutions already exist.

Regarding the use of an IBE in the real world, one possibility would be to design a secure channel between two smartphones in order to make it possible for the underlying users to exchange some information, such as an electronic ticket, in a secure way. Another possibility would be to establish such a secure channel between a TEE (Trusted Execution Environment, that is, an environment hosted by a mobile device that runs in isolation from the device's main operating system) and the SIM card. This is a new way to design a solution related to the GlobalPlatform industry association and its standardization of the TEE interfaces [5]. In fact, a key provisioning service can only be done this way on the SIM card side, while permitting us to securely exchange information between the TEE and the SIM card.

1.6.3 Related Work: Pairing or Not Pairing

The concept of identity-based cryptography was introduced by Shamir in 1984 [40] but until 2001, no proposal was deemed secure enough for real use. In 2001, two different schemes were proposed. The first one, due to Cocks [26] and based on quadratic residues, is not efficient enough. The second one, proposed by Boneh and Franklin [16], was then the first IBE scheme which is both secure and efficient. It mainly exploits the structure of a pairing.

Since then, researchers have proposed several schemes to improve (in terms of security, efficiency, or functionality) the latter scheme, and they are all based on the use of a pairing.

1.6.4 An Example

As an example, we describe in this section the Boneh-Franklin IBE [16], essentially for its simplicity. Again, we describe the scheme in the case where one uses a type 3 pairing (instead of a type 1 as used in [16]).

Setup and key generation. We then consider a bilinear environment $(r, \mathbb{G}_1, \mathbb{G}_2, \mathbb{G}_T, P_1, P_2, e)$, which is chosen during the execution of the Setup of the system. We also consider two cryptographic hash functions $\mathcal{H}_1 : \{0,1\}^* \longrightarrow \mathbb{G}_2^*$ and $\mathcal{H}_2 : \mathbb{G}_T \longrightarrow \{0,1\}^\ell$ where ℓ is the size in bits of a message to be encrypted. The Setup also outputs a master secret key $\mathsf{msk} \in \mathbb{Z}_r^*$. The corresponding master public key is then $\mathsf{mpk} = [\mathsf{msk}]P_1$.

The key extraction algorithm Ext consists of computing $P_{\mathsf{id}} = \mathcal{H}_1(\mathsf{id})$ and $\mathsf{usk} = [\mathsf{msk}]P_{\mathsf{id}} \in \mathbb{G}_2$.

Encryption and decryption. The encryption Enc and decryption Dec algorithms are then given by Algorithms 1.10 and 1.11, respectively. The main idea is to produce a one-time-pad (as for the hash ElGamal encryption [24]), where randomness is obtained by a hash function with an input that can be computed by using the pairing and either a random ρ and the identity, or the user secret key. More precisely, the result is correct, since

$$
\begin{aligned}
e(U, \mathsf{usk}) &= e([\rho]P_1, [\mathsf{msk}]P_{\mathsf{id}}) \\
&= e([\mathsf{msk}]P_1, [\rho]P_{\mathsf{id}}) \text{ by using twice the bilinear property of } e \\
&= e(\mathsf{mpk}, [\rho]P_{\mathsf{id}}).
\end{aligned}
$$

ALGORITHM 1.10   Boneh-Franklin IBE encryption algorithm Enc.

Input   : message $\mathsf{m} \in \{0,1\}^\ell$, identity id, master public key $\mathsf{mpk} = [\mathsf{msk}]P_1 \in \mathbb{G}_1$.
Output: ciphertext $\mathsf{c} = (U, V) \in \mathbb{G}_1 \times \{0,1\}^\ell$.

1 compute $P_{\mathsf{id}} = \mathcal{H}_1(\mathsf{id})$
2 choose at random $\rho \in \mathbb{Z}_r^*$
3 compute $U = [\rho]P_1$
4 compute $T = e(\mathsf{mpk}, [\rho]P_{\mathsf{id}}) \in \mathbb{G}_T$ and $V = \mathsf{m} \oplus \mathcal{H}_2(T)$
5 output $\mathsf{c} = (U, V)$.

ALGORITHM 1.11   Boneh-Franklin IBE decryption algorithm Dec.

Input   : ciphertext $\mathsf{c} = (U, V) \in \mathbb{G}_1 \times \{0,1\}^\ell$, secret key $\mathsf{usk} = [\mathsf{msk}]P_{\mathsf{id}} \in \mathbb{G}_2$.
Output: message $\mathsf{m} \in \{0,1\}^\ell$.

1 compute $T' = e(U, \mathsf{usk})$
2 output $\mathsf{m} = V \oplus \mathcal{H}_2(T')$.

1.7   Broadcast Encryption and Traitor Tracing

The concept of Broadcast Encryption (BE) has been introduced by Fiat and Naor [28], as a variant of standard encryption schemes. It permits a broadcaster to encrypt messages to a group of users who have subscribed to a broadcast channel. In a nutshell, the message is encrypted once with a single public key and each user can independently decrypt it by using his/her own secret key. There are several variants of broadcast encryption schemes, depending on the expected properties. In the following, we will focus on Traitor Tracing (TT) schemes.

1.7.1   General Presentation and Short Model

A traitor tracing scheme permit us to detect and identify cards/subscribers involved in the fraudulent dissemination of a secret key, and has been proposed by Chor, Fiat, and Naor [25]. The idea behind it is that by putting special keys in the entitlement flow sent to the decoder/box, and retrieving the key disseminated by piracy services, one can discriminate the involved box and then identify the defrauder.

There are mainly three actors in a traitor tracing scheme:

- a key manager $\mathcal{KM}$ generating the keys, and then giving the keys to other actors;
- a content manager $\mathcal{CM}$ managing the contents, from the encryption to the distribution;
- a set $\mathcal{U}$ of users reading contents, each user being identified by an integer i.

In practice, a content is most of the time encrypted by using a specific or standard secret key encryption scheme (e.g., AES) with a session key K_{sess}. The main problem related to broadcast and traitor tracing is then to give access to this key K_{sess} in a secure and efficient way. This is done by broadcasting a unique encapsulation key E_{sess} that will be used by each user, using some personal credentials.

Algorithms. More formally, a traitor tracing system is defined by the following algorithms:

- the Setup algorithm executed by $\mathcal{KM}$, which, on input of a security parameter λ, generates the key manager secret key msk and the parameters param.
- the user key generation algorithm UserKG, which, on input of the parameters param, the key manager secret key msk, and the index i of the user, outputs the user secret key usk_i for user i. It also adds a new entry into the list $\mathcal{L}$ of registered users.
- the encoding algorithm Enc, which is executed by $\mathcal{KM}$, and, on input of the public parameters, outputs a session key K_{sess} and an encapsulation key E_{sess}.
- the decoding algorithm Dec, which, on input of the parameters param, a user secret key usk_i and an encapsulation key E_{sess}, outputs, in a deterministic way, the related session key K_{sess}.
- the Identify algorithm, which, on input of the parameters param, the manager secret key msk, and a public decoder $\mathcal{D}$, outputs the index i_0 of a user who participated in the creation of the decoder.

Security. A traitor tracing scheme should first verify a variant of the standard indistinguishability property (see [15] for more details). Additionally, many security models have been proposed for the tracing part, which mostly differ on the assumptions made on $\mathcal{D}$. In the full version of [15], Boneh and Franklin considered the three following models.

- **Non-blackbox**, which assumes that it is possible to extract a secret key sk from $\mathcal{D}$. The security of the construction then implicitly relies on the hardness of producing a new secret key, which cannot be linked to an existing one registered in the list of registered users $\mathcal{L}$;
- **Single-key blackbox**, which considers a decoder $\mathcal{D}$ embedding a secret key sk. $\mathcal{D}$ can then be seen as an oracle which, on input of an encapsulation key E_{sess}, returns the output of the decoding algorithm, as $\mathsf{Dec}(\mathsf{param}, \mathsf{sk}, \mathsf{E}_{sess})$;
- **Blackbox**, for which no assumption on $\mathcal{D}$ should be made, except that it decodes with non-negligible probability.

1.7.2 Use Cases

The main use case related to traitor tracing is Pay-TV service, which is nowadays largely deployed in many countries. In a nutshell, it implies a key manager which manages users, a content manager whose role is to broadcast a protected content (executing the Enc algorithm), and finally users (or consumers) who, after a subscription, can access the protected content. Modern Pay-TV services most of time necessitate the user to have a decoder (a.k.a. a box) and a subscription smart card. The latter is given to consumers by the key manager (executing the UserKG algorithm), and permits them (using the Dec algorithm) to decrypt the content protected by the broadcaster. One can find in the literature several ways to attack such a system: (i) content dump inside the box, which can be prevented by a trusted video path and detected by using a forensics watermarking; (ii) content dump outside the box, which is prevented by increasing the robustness of the link protection algorithms and detected by using a forensics watermarking; (iii) smart card reverse engineering (but such an attack may be hard to perform owing to smart card robustness); and (iv) retrieval and publication of the secret key used to protect the content. Traitor tracing schemes are potential solutions to thwart this latter kind of attack.

In fact, the cryptographic part used in such systems is made of several encapsulations of encrypted keys and other data. More precisely, modern Pay TV services are most of the time based on the following key encapsulation:

- the digital content is encrypted by a secret key called a control word cw and is streamed to the user's decoders. The same control word is used by all users;
- the control word cw is sent encrypted by the content manager $\mathcal{CM}$ to the user's decoders, and is updated every couple of seconds, in an Entitlement Control Message (ECM). Each ECM contains the new control word and some other data, such as access rights and an integrity message. The control word is encrypted by the exploitation key ek, which is specific to each user;
- the exploitation key ek is sent encrypted by the key manager $\mathcal{KM}$ to the user's decoders, and is updated every week/month according to the key manager's policy, in an Entitlement Management Message (EMM). Each EMM contains the new exploitation key ek and some other data, such as new access rights and an integrity message. The exploitation key ek is encrypted by a specific management key mk.

One question that should be answered, to design the most suitable traitor tracing scheme, is where to put the scheme in the whole service. In fact, some time ago, fraudsters marketed new decoders based on the use of one or several valid secret keys. Nowadays, fraudsters most of the time publish, over the Internet, the new control word every couple of seconds. In a traitor tracing scheme, one needs to obtain the output of a traitor decryption execution, so as to identify the underlying fraudsters. Then, it seems better to consider implementing a traitor scheme such that the publication of a control word by a fraudster permits us to identify him. As a consequence, the traitor tracing is used at the compilation of the Entitlement Control Message. It follows that the control word cw corresponds to the session key K_{sess} and that the ECM contains the encapsulation key E_{sess}.

It follows that the control word will be the output of the traitor tracing encapsulation step. An important consequence is that the user decoder has to execute the de-encapsulation algorithm every couple of seconds. Then, we clearly need a very efficient traitor tracing scheme, at least for the de-encapsulation phase. As this step is in practice sometimes done inside a smart card, the efficiency constraint can be very important for a real deployment.

1.7.3 Related Work: Pairing or Not Pairing

In this chapter, we only focus on the *non-blackbox* security model, as existing efficient schemes only comply with this model. There are then currently two kinds of construction for traitor tracing schemes:

- **combinatorial schemes** are based on secret key cryptography. Most of the time a tree is generated, in which the leaves correspond to users, and nodes to subgroups. Each user needs all the secret keys along the path from the root to his/her corresponding leaf to be able to decrypt a message. The resulting schemes are time efficient but very space consuming due to key storage;
- **algebraic schemes** are based on public key cryptography. Their usage implies modifying traditional public key encryption schemes so that several users can decrypt a message, and tracing traitors is possible. They permit us to manage only one key per user, independently of the number of customers and defrauders to manage. But, for a long time, the efficiency was missing. Today, several practical solutions exist and they are all based on the use of a pairing.

1.7.4 An Example

As an example, we describe in this section the traitor tracing system given in [36] as it was the most efficient one at the time. We describe it in the case of a type 3 pairing (instead of a type 1 as used in [36]).

Setup and key generation. We consider a bilinear environment $(r, \mathbb{G}_1, \mathbb{G}_2, \mathbb{G}_T, P_1, P_2, e)$, which is chosen during the execution of the **Setup** of the system. The **Setup** also outputs two generators $R_1 \in \mathbb{G}_1$ and $R_2 \in \mathbb{G}_2$ such that $e(P_1, P_2) = e(R_1, R_2)$.

Remark 1.2 To compute such generators, one obvious solution is, for example, to first choose $P_1 \in \mathbb{G}_1$ and $R_2 \in \mathbb{G}_2$ and then to choose at random a secret $x \in \mathbb{Z}_r^*$ and compute $R_1 = [x]P_1 \in \mathbb{G}_1$ and $P_2 = [x]R_2 \in \mathbb{G}_2$. The secret value x can then be deleted as it is sensitive and no longer useful.

The next step consists of choosing $e, v \in \mathbb{Z}_r^*$, then computing $d = e^{-1} \pmod{r}$ and

$$\widetilde{P} = [d]P_1, \widetilde{R} = [d]R_1, \bar{R} = [v]R_1, \hat{R} = [dv]R_1.$$

Similarly to x, the value d can be deleted, as we do not need it anymore.

We finally define the parameters **param** to be $(P_1, R_1, \widetilde{P}, \widetilde{R}, \bar{R}, \hat{R})$ and the manager secret key as $\mathsf{msk} = (e, v, P_2)$.

Regarding the user key generation algorithm **UserKG**, the key manager first has to generate at random $a_i \in \mathbb{Z}_r^*$ ($a_i \neq r - 1$) and then compute the following values:

$$
\begin{aligned}
A_i &= [a_i + v]P_2, \\
b_i &= \frac{1}{a_i + 1} - e \pmod{r}.
\end{aligned}
$$

The user secret key is then the couple $\mathsf{usk}_i = (A_i, b_i)$.

Encoding and decoding. The encoding algorithm **Enc** is given in Algorithm 1.12. It can be seen as a variant of the **ElGamal** encryption scheme (see [27] and Section 1.3.3) in a pairing setting. The session key $\mathsf{K}_{sess} = e(P_1, P_2)^k$ is then used to encrypt the content, using the appropriate secret key-based cryptographic algorithm. The corresponding encrypted content is then denoted by **EC** and is output by a suitable secret key encryption scheme (e.g., **AES**) and the key K_{sess}.

ALGORITHM 1.12 Traitor tracing encoding algorithm **Enc**.

Input : parameters $(P_1, R_1, \widetilde{P}, \widetilde{R}, \bar{R}, \hat{R})$.
Output: couple $(\mathsf{K}_{sess}, \mathsf{E}_{sess})$.

1 choose at random $\rho \in \mathbb{Z}_r^*$
2 compute $C_1 = [\rho]P_1$
3 compute $C_2 = [\rho]\widetilde{R} \ (= [\rho d]R_1)$
4 compute $C_3 = [\rho]\widetilde{P} \ (= [\rho d]P_1)$
5 compute $C_4 = [\rho]\bar{R} \ (= [\rho v]R_1)$
6 compute $C_5 = [\rho]\hat{R} \ (= [\rho d v]R_1)$
7 compute $\mathsf{K}_{sess} = e(P_1, P_2)^k$
8 compute $\mathsf{E}_{sess} = (C_1, C_2, C_3, C_4, C_5)$
9 output $(\mathsf{K}_{sess}, \mathsf{E}_{sess})$.

The decoding algorithm Dec is given by Algorithm 1.13. Using K_{sess} and the used secret key decryption algorithm (e.g., AES), the user can then retrieve the plain content by decrypting EC.

ALGORITHM 1.13 Traitor tracing decoding algorithm Dec.

Input : encapsulation key E_{sess}, user secret key ($usk_i = (A_i, b_i)$).
Output: session key $K_{sess} \in \mathbb{G}_T$.

1 compute $K_{sess} = e(C_1 + [b_i]C_3, A_i) \cdot e([b_i](C_2 - C_5) - C_2 - C_4, R_2)$.

Tracing. Finally, for the Identify algorithm, we give the way to interact with a fraudulent user $\tilde{i}$ able to decrypt the content key. This can be done by the key manager since it has access to the values $(\tilde{i}, e(P_1, P_2)^{b_{\tilde{i}}})$ for each user $\tilde{i}$ having the decoder secret key $usk_{\tilde{i}} = (A_{\tilde{i}}, b_{\tilde{i}})$.

For this purpose, the key manager computes a probe encapsulation key as shown in Algorithm 1.14.

ALGORITHM 1.14 Traitor tracing probe encoding algorithm Enc.

Input : parameters $(P_1, R_1, \widetilde{P}, \widetilde{R}, \bar{R}, \hat{R})$.
Output: couple $(\widetilde{K}_{sess}, \widetilde{E}_{sess})$.

1 choose at random $\rho_1, \rho_2 \in \mathbb{Z}_r^*$
2 compute $\widetilde{C}_1 = [\rho_1]P_1$
3 compute $\widetilde{C}_2 = [\rho_1]\widetilde{R} + [\rho_2]R_1 \ (= [\rho_1 d + \rho_2]R_1)$
4 compute $\widetilde{C}_3 = [\rho_1]\widetilde{P} \ (= [\rho_1 d]P_1)$
5 compute $\widetilde{C}_4 = [\rho_1]\bar{R} \ (= [\rho_1 v]R_1)$
6 compute $\widetilde{C}_5 = [\rho_1]\hat{R} \ (= [\rho_1 dv]R_1)$
7 compute $\widetilde{K}_{sess} = e(P_1, P_2)^k$
8 compute $\widetilde{E}_{sess} = (\widetilde{C}_1, \widetilde{C}_2, \widetilde{C}_3, \widetilde{C}_4, \widetilde{C}_5)$
9 output $(\widetilde{K}_{sess}, \widetilde{E}_{sess})$.

We then assume the existence of a fraudulous decoder $\mathcal{D}$ embedding a unique user secret key denoted $(\widetilde{A}, \widetilde{b})$. The execution of the decryption algorithm on inputting the probe encapsulation key $\widetilde{E}_{sess}$ will output the value

$$\mathsf{FK} = e(P_1, P_2)^{-k} \cdot e(P_1, P_2)^{\rho_2(\widetilde{b}-1)}.$$

Finally, on inputting the set of $(\tilde{i}, e(P_1, P_2)^{b_{\tilde{i}}})$ and FK, it becomes possible to perform an exhaustive search (in the number of users) on the right $\tilde{i}$ such that $b_{\tilde{i}} = \widetilde{b}$, using FK.

1.8 Conclusion

In this chapter, we have introduced and explained the concept of "pairing-based cryptography." We have shown that a pairing is now an important, and sometimes essential, building block for the design of secure and efficient cryptographic protocols used in real-world applications: sensor node authentication, electronic voting, anonymous access control, Pay-TV, etc. The recent introduction of pairings in the ISO/IEC 15946 standard [4] can be seen as an evidence of their maturity and their timeliness for industrial applications.

References

[1] D.SPA.20. *ECRYPT2 Yearly Report on Algorithms and Keysizes (2011-2012)*. European Network of Excellence in Cryptology II, September 2012.

[2] ETSI TS 102 225. *Smart Cards; Secured packet structure for UICC based applications (Release 11)*. European Telecommunications Standards Institute, March 2012.

[3] FIPS PUB 186-3. *Digital Signature Standard (DSS)*. National Institute of Standards and Technology, U.S. Department of Commerce, June 2009.

[4] ISO/IEC FDIS 15946-1. *Information technology – Security techniques – Cryptographic techniques based on elliptic curves*. International Organization for Standardization/International Electrotechnical Commission, July 2015.

[5] TEE White Paper. *The Trusted Execution Environment: Delivering Enhanced Security at a Lower Cost to the Mobile Market*. GlobalPlatform, June 2015. Revised from February 2011.

[6] Ben Adida. Helios: Web-based open-audit voting. In *17th USENIX Security Symposium*, pp. 335–348. USENIX Association, 2008.

[7] Giuseppe Ateniese, Jan Camenisch, Marc Joye, and Gene Tsudik. A practical and provably secure coalition-resistant group signature scheme. In M. Bellare, editor, *Advances in Cryptology – CRYPTO 2000*, volume 1880 of *Lecture Notes in Computer Science*, pp. 255–270. Springer, Heidelberg, 2000.

[8] Paulo S. L. M. Barreto and Michael Naehrig. Pairing-friendly elliptic curves of prime order. In B. Preneel and S. Tavares, editors, *Selected Areas in Cryptography (SAC 2005)*, volume 3897 of *Lecture Notes in Computer Science*, pp. 319–331. Springer, Heidelberg, 2006.

[9] Mihir Bellare, Anand Desai, David Pointcheval, and Phillip Rogaway. Relations among notions of security for public-key encryption schemes. In H. Krawczyk, editor, *Advances in Cryptology – CRYPTO '98*, volume 1462 of *Lecture Notes in Computer Science*, pp. 26–45. Springer, Heidelberg, 1998.

[10] Mihir Bellare and Phillip Rogaway. Random oracles are practical: A paradigm for designing efficient protocols. In V. Ashby, editor, *1st ACM Conference on Computer and Communications Security*, pp. 62–73. ACM Press, 1993.

[11] Mihir Bellare and Phillip Rogaway. The exact security of digital signatures: How to sign with RSA and Rabin. In U. M. Maurer, editor, *Advances in Cryptology – EUROCRYPT '96*, volume 1070 of *Lecture Notes in Computer Science*, pp. 399–416. Springer, Heidelberg, 1996.

[12] Mihir Bellare, Haixia Shi, and Chong Zhang. Foundations of group signatures: The case of dynamic groups. In A. Menezes, editor, *Topics in Cryptology – CT-RSA 2005*, volume 3376 of *Lecture Notes in Computer Science*, pp. 136–153. Springer, Heidelberg, 2005.

[13] Dan Boneh and Xavier Boyen. Short signatures without random oracles and the SDH assumption in bilinear groups. *Journal of Cryptology*, 21(2):149–177, 2008.

[14] Dan Boneh, Xavier Boyen, and Hovav Shacham. Short group signatures. In M. Franklin, editor, *Advances in Cryptology – CRYPTO 2004*, volume 3152 of *Lecture Notes in Computer Science*, pp. 41–55. Springer, Heidelberg, 2004.

[15] Dan Boneh and Matthew K. Franklin. An efficient public key traitor tracing scheme. In M. J. Wiener, editor, *Advances in Cryptology – CRYPTO '99*, volume 1666 of *Lecture Notes in Computer Science*, pp. 338–353. Springer, Heidelberg, 1999.

[16] Dan Boneh and Matthew K. Franklin. Identity based encryption from the Weil pairing. *SIAM Journal on Computing*, 32(3):586–615, 2003.

[17] Dan Boneh, Ben Lynn, and Hovav Shacham. Short signatures from the Weil pairing. *Journal of Cryptology*, 17(4):297–319, 2004.

[18] Jan Camenisch, Susan Hohenberger, and Anna Lysyanskaya. Compact e-cash. In R. Cramer, editor, *Advances in Cryptology – EUROCRYPT 2005*, volume 3494 of *Lecture Notes in Computer Science*, pp. 302–321. Springer, Heidelberg, 2005.

[19] Sébastien Canard, Aline Gouget, and Jacques Traoré. Improvement of efficiency in (unconditional) anonymous transferable e-cash. In G. Tsudik, editor, *Financial Cryptography and Data Security (FC 2008)*, volume 5143 of *Lecture Notes in Computer Science*, pp. 202–214. Springer, Heidelberg, 2008.

[20] Sébastien Canard, David Pointcheval, Olivier Sanders, and Jacques Traoré. Divisible E-cash made practical. In J. Katz, editor, *Public Key Cryptography – PKC 2015*, volume 9020 of *Lecture Notes in Computer Science*, pp. 77–100. Springer, Heidelberg, 2015.

[21] Sébastien Canard, Berry Schoenmakers, Martijn Stam, and Jacques Traoré. List signature schemes. *Discrete Applied Mathematics*, 154(2):189–201, 2006.

[22] Sanjit Chatterjee and Alfred Menezes. On cryptographic protocols employing asymmetric pairings – The role of Ψ revisited. *Discrete Applied Mathematics*, 159(13):1311–1322, 2011.

[23] David Chaum. Showing credentials without identification: Signatures transferred between unconditionally unlinkable pseudonyms. In F. Pichler, editor, *Advances in Cryptology – EUROCRYPT '85*, volume 219 of *Lecture Notes in Computer Science*, pp. 241–244. Springer, Heidelberg, 1986.

[24] Benoît Chevallier-Mames, Pascal Paillier, and David Pointcheval. Encoding-free ElGamal encryption without random oracles. In M. Yung, Y. Dodis, A. Kiayias, and T. Malkin, editors, *Public Key Cryptography – PKC 2006*, volume 3958 of *Lecture Notes in Computer Science*, pp. 91–104. Springer, Heidelberg, 2006.

[25] Benny Chor, Amos Fiat, and Moni Naor. Tracing traitors. In Y. Desmedt, editor, *Advances in Cryptology – CRYPTO '94*, volume 839 of *Lecture Notes in Computer Science*, pp. 257–270. Springer, Heidelberg, 1994.

[26] Clifford Cocks. An identity based encryption scheme based on quadratic residues. In B. Honary, editor, *Cryptography and Coding*, volume 2260 of *Lecture Notes in Computer Science*, pp. 360–363. Springer, Heidelberg, 2001.

[27] Taher ElGamal. A public key cryptosystem and a signature scheme based on discrete logarithms. In G. R. Blakley and D. Chaum, editors, *Advances in Cryptology, Proceedings of CRYPTO '84*, volume 196 of *Lecture Notes in Computer Science*, pp. 10–18. Springer, Heidelberg, 1984.

[28] Amos Fiat and Moni Naor. Broadcast encryption. In D. R. Stinson, editor, *Advances in Cryptology – CRYPTO '93*, volume 773 of *Lecture Notes in Computer Science*, pp. 480–491. Springer, Heidelberg, 1994.

[29] Amos Fiat and Adi Shamir. How to prove yourself: Practical solutions to identification and signature problems. In A. M. Odlyzko, editor, *Advances in Cryptology – CRYPTO '86*, volume 263 of *Lecture Notes in Computer Science*, pp. 186–194. Springer, Heidelberg, 1987.

[30] Atsushi Fujioka, Tatsuaki Okamoto, and Kazuo Ohta. A practical secret voting scheme for large scale elections. In J. Seberry and Y. Zheng, editors, *Advances in Cryptology – AUSCRYPT '92*, volume 718 of *Lecture Notes in Computer Science*, pp. 244–251. Springer, Heidelberg, 1993.

[31] Jun Furukawa, Kengo Mori, and Kazue Sako. An implementation of a mix-net based network voting scheme and its use in a private organization. In D. Chaum et al., ed-

itors, *Towards Trustworthy Elections, New Directions in Electronic Voting*, volume 6000 of *Lecture Notes in Computer Science*, pp. 141–154. Springer, 2010.

[32] Steven D. Galbraith, Kenneth G. Paterson, and Nigel P. Smart. Pairings for cryptographers. *Discrete Applied Mathematics*, 156(16):3113–3121, 2008.

[33] Shafi Goldwasser, Silvio Micali, and Ronald L. Rivest. A digital signature scheme secure against adaptive chosen-message attacks. *SIAM Journal on Computing*, 17(2):281–308, 1988.

[34] Jens Groth. Fully anonymous group signatures without random oracles. In K. Kurosawa, editor, *Advances in Cryptology – ASIACRYPT 2007*, volume 4833 of *Lecture Notes in Computer Science*, pp. 164–180. Springer, Heidelberg, 2007.

[35] Jens Groth and Amit Sahai. Efficient non-interactive proof systems for bilinear groups. In N. P. Smart, editor, *Advances in Cryptology – EUROCRYPT 2008*, volume 4965 of *Lecture Notes in Computer Science*, pp. 415–432. Springer, Heidelberg, 2008.

[36] Philippe Guillot, Abdelkrim Nimour, Duong Hieu Phan, and Viet Cuong Trinh. Optimal public key traitor tracing scheme in non-black box model. In A. Youssef, A. Nitaj, and A. E. Hassanien, editors, *Progress in Cryptology – AFRICACRYPT 2013*, volume 7918 of *Lecture Notes in Computer Science*, pp. 140–155. Springer, Heidelberg, 2013.

[37] Alfred Menezes, Scott A. Vanstone, and Tatsuaki Okamoto. Reducing elliptic curve logarithms to logarithms in a finite field. In *23rd Annual ACM Symposium on Theory of Computing*, pp. 80–89. ACM Press, 1991.

[38] Ronald L. Rivest, Adi Shamir, and Leonard M. Adleman. A method for obtaining digital signature and public-key cryptosystems. *Communications of the Association for Computing Machinery*, 21(2):120–126, 1978.

[39] Amit Sahai. Non-malleable non-interactive zero knowledge and adaptive chosen-ciphertext security. In *40th Annual Symposium on Foundations of Computer Science*, pp. 543–553. IEEE Computer Society Press, 1999.

[40] Adi Shamir. Identity-based cryptosystems and signature schemes. In G. R. Blakley and D. Chaum, editors, *Advances in Cryptology, Proceedings of CRYPTO '84*, volume 196 of *Lecture Notes in Computer Science*, pp. 47–53. Springer, Heidelberg, 1984.

[41] Thomas Unterluggauer and Erich Wenger. Efficient pairings and ECC for embedded systems. In L. Batina and M. Robshaw, editors, *Cryptographic Hardware and Embedded Systems – CHES 2014*, volume 8731 of *Lecture Notes in Computer Science*, pp. 298–315. Springer, Heidelberg, 2014.

[42] Peter Yee. Updates to the internet X.509 public key infrastructure certificate and certificate revocation list (CRL) profile. Request for Comments RFC 6818, Internet Engineering Task Force (IETF), 2013.

[43] Gideon Yuval. How to swindle rabin. *Cryptologia, Volume 3, Issue 3*, p. 187âĂŞ190, 1979.

2

Mathematical Background

Jean-Luc Beuchat
ELCA Informatique SA

Nadia El Mrabet
EMSE - Paris 8

Laura Fuentes-Castañeda
Intel México

Francisco Rodríguez-Henríquez
CINVESTAV-IPN

In this chapter we will review the essential mathematical concepts, definitions, and properties required for the formal description of bilinear pairings. The chapter is organized into three main parts. First, we give in Section 2.1 basic algebra definitions, which will be used in the remaining part of this book. Then, in Section 2.2, we state several facts on finite fields, extension finite fields, and their arithmetic. Finally, in Section 2.3, we give a basic introduction to elliptic curves and their properties, which are especially relevant for pairing-based cryptography.

A more extensive and formal discussion of the above subjects are studied in the specialized references [3, 5, 6, 10, 16, 17, 18, 19, 20].

2.1 Algebra

In order to make this book accessible to readers not too familiar with the mathematics of pairings, we recall basic elements as far as they are needed. We present elementary mathematical definitions and properties. They can be found, for instance, in [11].

2.1.1 Group

DEFINITION 2.1 (Group) A **group** $(\mathbb{G}, \star)$, also denoted by $\mathbb{G}$ when it causes no troubles, is a non-empty set $\mathbb{G}$ together with a binary operation called a group law $\star$. The group law satisfies the following properties:

- The group $\mathbb{G}$ is closed under the group law, i.e., $\forall\, a \in \mathbb{G}$ and $\forall\, b \in \mathbb{G}$ then $(a \star b) \in \mathbb{G}$.
- There exists an element in $\mathbb{G}$, called its **neutral element** (or **identity**) and denoted by e, such that for all $a \in \mathbb{G}$, $a \star e = a = e \star a$.

- The group law is **associative**, i.e., $a \star (b \star c) = (a \star b) \star c$.
- Each element of $\mathbb{G}$ admits a (unique) **symmetric** (also called its **inverse**), i.e., for all $a \in \mathbb{G}$ there exists $b \in \mathbb{G}$ such that $a \star b = e = b \star a$. When the group law is additive (respectively, multiplicative), the inverse of a is usually denoted by $-a$ (respectively, a^{-1}).

In this book, most groups are **commutative** or **abelian**. This means that for all $a, b \in \mathbb{G}$, $a \star b = b \star a$.

Example 2.1

- The set of positive integers $\mathbb{N}$ is not a group under addition. For instance, 1 does not admit an inverse.
- The set of integers forms a group under addition, denoted by $(\mathbb{Z}, +)$. Its neutral element is 0.
- The set of integers is not a group under multiplication. Indeed, 2 does not admit an inverse for the multiplication over $\mathbb{Z}$.

Remark 2.1 The notation $(\mathbb{G}, \star, e)$ can also be used in order to describe a group $\mathbb{G}$ together with its binary operation and neutral element. Given an element $g \in \mathbb{G}$, in this chapter we use $\star^m(g)$ to denote the application of the operation $\star$ $m - 1$ times over the element g, with $m \in \mathbb{N}$. When the operation $\star$ is the addition $+$ (respectively, the multiplication $\times$), the neutral element is denoted by 0 (respectively, 1).

DEFINITION 2.2 (Group order) The **order** of a group $(\mathbb{G}, \star)$ is defined as the number of elements in $\mathbb{G}$.

Remark 2.2 Groups may have finite or infinite order.

Example 2.2

- The group $(\mathbb{Z}, +)$ has an infinite order.
- The group $(\{0, 1\}, +)$ such that the addition if performed modulo 2 (i.e., $1 + 1 = 0$) is a finite group of order 2.

DEFINITION 2.3 (Order of a group element) Let $(\mathbb{G}, \star)$ be a group. Then, the **order** of $g \in \mathbb{G}$ is the smallest positive integer r, such that $\star^r(g) = e$.

Example 2.3 Given $\mathbb{G} = (\mathbb{G}, \star)$, let $g \in \mathbb{G}$ be an element of order 3, then,

$$\star^3(g) = g \star g \star g = e, \quad \text{and} \quad \star^i(g) \neq e, \text{ for } 0 < i < 3.$$

Remark 2.3 Let $\mathbb{G}$ be a finite group with order $n \in \mathbb{N}$. Then, for all $g \in \mathbb{G}$, the order of g divides the group order n, which implies that $\star^n(g) = e$.

DEFINITION 2.4 (Group generator) Given the group $(\mathbb{G}, \star)$, we say that $g \in \mathbb{G}$ is a group **generator**, if for all $h \in \mathbb{G}$ there exists a unique $i \in \mathbb{N}$, such that, $h = \star^i(g)$.

Example 2.4 The element 1 is a generator of the group $(\mathbb{Z}, +)$.

DEFINITION 2.5 (Cyclic group) A group $(\mathbb{G}, \star)$ is **cyclic**, if there exists at least one generator $g \in \mathbb{G}$. The cyclic group generated by g is denoted by $\mathbb{G} = \langle g \rangle$.

The number of generator elements in a finite cyclic group $(\mathbb{G}, \star)$ of order n is defined as $\varphi(n)$, where $\varphi(\cdot)$ denotes the Euler's totien function.[*] Therefore if $\mathbb{G}$ is a group of prime order p, then $\mathbb{G}$ has $\varphi(p) = p - 1$ generators, i.e., $\forall g \in \mathbb{G}$ such that $g \neq e$, $\mathbb{G} = \langle g \rangle$.

Remark 2.4 A group can be described using either additive or multiplicative notation. We illustrate the construction of additive or multiplicative cyclic groups in the following Examples, 2.5 and 2.6.

Example 2.5 Additive notation If a group is described additively using the symbol $+$ to denote the group operation, then usually the identity element is denoted by 0, and the additive inverse of $a \in \mathbb{G}$ is $-a$. Let $m \in \mathbb{N}$, then the application of the operator $+$ over the element a, $m - 1$ times, is denoted as ma. The finite set $\mathbb{Z}_n = \{0, 1, \ldots, n - 1\}$ forms an abelian group $(\mathbb{Z}_n, +)$ with neutral element 0 and of order n. The operator $+$ is defined under addition modulo n. Specifically, if $n = 4$, then the group operation is applied over the group elements $\mathbb{Z}_4$, as shown in the following **Cayley table**,[†]

$\oplus_4$	0	1	2	3
0	0	1	2	3
1	1	2	3	0
2	2	3	0	1
3	3	0	1	2

One could remark that $(\mathbb{Z}_4, +)$ is cyclic with $\varphi(4) = 2$ generators, which are 3 and 1.

$$
\begin{aligned}
3 &\equiv 3 \mod 4, \\
3 + 3 &\equiv 2 \mod 4, \\
3 + 3 + 3 &\equiv 1 \mod 4, \\
3 + 3 + 3 + 3 &\equiv 0 \mod 4,
\end{aligned}
\qquad
\begin{aligned}
1 &\equiv 1 \mod 4, \\
1 + 1 &\equiv 2 \mod 4, \\
1 + 1 + 1 &\equiv 3 \mod 4, \\
1 + 1 + 1 + 1 &\equiv 0 \mod 4.
\end{aligned}
$$

Example 2.6 Multiplicative notation

When a group is described multiplicatively, the group operation is denoted as $\times$, and 1 (respectively a^{-1}) represents the identity element (respectively, the multiplicative inverse of $a \in \mathbb{G}$). Applying the operator $\times$, $m - 1$ times over the element a, with $m \in \mathbb{N}$, is denoted as a^m.

Given a positive integer n, let $\mathbb{Z}_n^*$ denote the set of elements in $\mathbb{Z}_n$ different than zero, that are relatively prime to n, as

$$\mathbb{Z}_n^* = \{a \in \mathbb{Z}_n \mid \gcd(a, n) = 1\}.$$

It is easy to see that $(\mathbb{Z}_n^*, \times)$ is an abelian group of order $\varphi(n)$, whose group operation is the integer multiplication modulo n. The following Cayley table shows the group structure for the case $n = 10$, where $\mathbb{Z}_{10}^* = \{1, 3, 7, 9\}$:

---

[*]Euler's totient function: Let n be a positive integer, then $\varphi(n)$ gives the number of positive integers smaller than or equal to n, that happen to be co-prime to n.

[†]A Cayley table describes the structure of a finite group by showing all the possible group operations that can be performed among the group elements.

$\odot_{10}$	1	3	7	9
1	1	3	7	9
3	3	9	1	7
7	7	1	9	3
9	9	7	3	1

Moreover, $(\mathbb{Z}_{10}^*, \times)$ is a cyclic group equipped with $\varphi(4) = 2$ generators, namely, 3 and 7. It is easy to verify that the repeated application of the group operation over these two generators produces all the elements in the group:

$$
\begin{aligned}
3 &\equiv 3 \mod 10, & 7 &\equiv 7 \mod 10, \\
3 \cdot 3 &\equiv 9 \mod 10, & 7 \cdot 7 &\equiv 9 \mod 10, \\
3 \cdot 3 \cdot 3 &\equiv 7 \mod 10, & 7 \cdot 7 \cdot 7 &\equiv 3 \mod 10, \\
3 \cdot 3 \cdot 3 \cdot 3 &\equiv 1 \mod 10, & 7 \cdot 7 \cdot 7 \cdot 7 &\equiv 1 \mod 10.
\end{aligned}
$$

Example 2.7   Likewise, if p is a prime number, then the set $\mathbb{Z}_p^* = \mathbb{Z}_p - \{0\}$ of order $p - 1$ forms a cyclic finite group $(\mathbb{Z}_p^*, \times, 1)$.

2.1.2   Subgroup

DEFINITION 2.6 (Subgroup)   Let $(\mathbb{G}, \star)$ be a group. A **subgroup** of $\mathbb{G}$ is a non-empty subset $\mathbb{H}$ of $\mathbb{G}$, such that $(\mathbb{H}, \star)$ is also a group. More precisely, $\mathbb{H}$ is closed under $\star$, for all $a \in \mathbb{H}$ then $a^{-1} \in \mathbb{H}$, and $\mathbb{H}$ contains the same neutral element e of $\mathbb{G}$.

THEOREM 2.1   *Lagrange's theorem [18]* . *Let $(\mathbb{G}, \star)$ be a finite abelian group and let $\mathbb{H} = (\mathbb{H}, \star)$ be a subgroup of $\mathbb{G}$. Then the order of $\mathbb{H}$ divides the order of $\mathbb{G}$.*

THEOREM 2.2   *[18, Theorem 8.6]. Let $(\mathbb{G}, \star)$ be an abelian group, and let m be an integer number. The set*

$$
\mathbb{G}\{m\} = \{a \in \mathbb{G} \mid \star^m (a) = e\},
$$

forms a subgroup in $\mathbb{G}$, which is denoted as $(\mathbb{G}\{m\}, \star)$.

Example 2.8   The set $(\mathbb{Z}_{13}^*, \times)$ is an abelian group of order 12 where $\mathbb{Z}_{13}^* = \{1, 2, 3, 4, 5, 6, 7, 8, 9, 10, 11, 12\}$. The sets

$$
\begin{aligned}
\mathbb{Z}_{13}^*\{2\} &= & \{1, 12\} &= \{a \in \mathbb{Z}_{13}^* \mid a^2 = 1\}, \\
\mathbb{Z}_{13}^*\{3\} &= & \{1, 3, 9\} &= \{a \in \mathbb{Z}_{13}^* \mid a^3 = 1\}, \\
\mathbb{Z}_{13}^*\{4\} &= & \{1, 5, 8, 12\} &= \{a \in \mathbb{Z}_{13}^* \mid a^4 = 1\}, \\
\mathbb{Z}_{13}^*\{5\} &= & \{1, 2, 3, 4, 5, 6, 7, 8, 9, 10, 11, 12\} &= \{a \in \mathbb{Z}_{13}^* \mid a^5 = 1\}, \\
\mathbb{Z}_{13}^*\{6\} &= & \{1, 3, 4, 9, 10, 12\} &= \{a \in \mathbb{Z}_{13}^* \mid a^6 = 1\} \\
&\cdots
\end{aligned}
$$

form subgroups of $(\mathbb{Z}_{13}^*, \times)$. The following Cayley table describes the structure of the subgroup $(\mathbb{Z}_{13}^*\{4\}, \times)$:

$\odot_{13}$	1	5	8	12
1	1	5	8	12
5	5	12	1	8
8	8	1	12	5
12	12	8	5	1

THEOREM 2.3   *[18, Theorem 8.7]. Let $(\mathbb{G}, \star)$ be an abelian group, and let m be an integer number such that,*

$$\star^m(\mathbb{G}) = \{\star^m(a) \mid a \in \mathbb{G}\},$$

then $(\star^m(\mathbb{G}), \star)$ is a subgroup of $\mathbb{G}$.

Example 2.9   Once again, let us consider the abelian group $(\mathbb{Z}_{13}^*, \times)$. Applying Theorem 2.3 with $m = 9$ yields

$$(\mathbb{Z}_{13}^*)^9 = \{a^9 \mid a \in \mathbb{Z}_{13}^*\} = \{1, 5, 8, 12\}.$$

One can easily check that

$$
\begin{aligned}
1^9 &\equiv 1 && \bmod 13, \\
2^9 &\equiv 5 && \bmod 13, \\
3^9 &\equiv 1 && \bmod 13, \\
4^9 &\equiv 12 && \bmod 13, \\
5^9 &\equiv 5 && \bmod 13, \\
6^9 &\equiv 5 && \bmod 13, \\
7^9 &\equiv 8 && \bmod 13, \\
8^9 &\equiv 8 && \bmod 13, \\
9^9 &\equiv 1 && \bmod 13, \\
10^9 &\equiv 12 && \bmod 13, \\
11^9 &\equiv 8 && \bmod 13, \\
12^9 &\equiv 12 && \bmod 13.
\end{aligned}
$$

Hence, $((\mathbb{Z}_{13}^*)^9, \times) \cong (\mathbb{Z}_{13}^*\{4\}, \times)$ is a subgroup of $(\mathbb{Z}_{13}^*, \times)$. Repeating the same procedure for $2 \le m \le 6$, the following sets are obtained,

$$
\begin{aligned}
(\mathbb{Z}_{13}^*)^2 &= \{1, 3, 4, 9, 10, 12\}, \\
(\mathbb{Z}_{13}^*)^3 &= \{1, 5, 8, 12\}, \\
(\mathbb{Z}_{13}^*)^4 &= \{1, 3, 9\}, \\
(\mathbb{Z}_{13}^*)^5 &= \{1, 2, 3, 4, 5, 6, 7, 8, 9, 10, 11, 12\}, \\
(\mathbb{Z}_{13}^*)^6 &= \{1, 12\}.
\end{aligned}
$$

which corresponds with the ones obtained from Example 2.8.

Considering Examples 2.8 and 2.9, it is interesting to note that since the group order of $(\mathbb{Z}_{13}^*, \cdot, 1)$ factorizes as $12 = 4 \times 3$, then,

$$
\begin{aligned}
(\mathbb{Z}_{13}^*)^4 &= \mathbb{Z}_{13}^*\{3\}, \\
(\mathbb{Z}_{13}^*)^3 &= \mathbb{Z}_{13}^*\{4\},
\end{aligned}
$$

and also, since $12 = 6 \times 2$,

$$
\begin{aligned}
(\mathbb{Z}_{13}^*)^6 &= \mathbb{Z}_{13}^*\{2\}, \\
(\mathbb{Z}_{13}^*)^2 &= \mathbb{Z}_{13}^*\{6\}.
\end{aligned}
$$

In general, given the group $\mathbb{G} = (\mathbb{G}, \times)$, if the group order can be factorized as, $c \times r$, then

$$\{a^c \mid a \in \mathbb{G}\} = \{a \in \mathbb{G} \mid a^r = 1\};$$

or if the group is described additively as $\mathbb{G} = (\mathbb{G}, +)$, then,

$$\{ca \mid a \in \mathbb{G}\} = \{a \in \mathbb{G} \mid ra = 0\}.$$

2.1.3   Cosets

DEFINITION 2.7 (Equivalence relation)   Let $\mathbb{H} = (\mathbb{H}, \star)$ be a subgroup of $\mathbb{G} = (\mathbb{G}, \star)$. For all $a, b \in \mathbb{G}$, we say that $a \equiv b \pmod{\mathbb{H}}$, if $a \star \bar{b} \in \mathbb{H}$, where $\bar{b}$ is the inverse of b. The expression $\equiv \pmod{\mathbb{H}}$ is known as the **equivalence relation**.

Remark 2.5   This notion is useful because it allows us to classify all the elements in $\mathbb{G}$ into **equivalence classes**.

Given $a \in \mathbb{G}$, we denote by $[a]_{\mathbb{H}}$ the equivalence class that contains the group element a. This class is defined as

$$[a]_{\mathbb{H}} = a \star \mathbb{H} = \{a \star h \mid h \in \mathbb{H}\},$$

Notice that $x \in [a]_{\mathbb{H}} \iff x \equiv a \pmod{\mathbb{H}}$.

Remark 2.6   The equivalence classes are often called **the cosets of $\mathbb{H}$ in $\mathbb{G}$**.

DEFINITION 2.8 (The quotient group)   The set of all cosets is denoted as $\mathbb{G}/\mathbb{H}$ and it forms a group $(\mathbb{G}/\mathbb{H}, \star)$, where the neutral is denoted $[e]_{\mathbb{H}}$, and

$$[a]_{\mathbb{H}} \star [b]_{\mathbb{H}} = [a \star b]_{\mathbb{H}}.$$

This group is called **the quotient group of $\mathbb{G}$ modulo $\mathbb{H}$**.

Example 2.10   Given the set $\mathbb{Z}_6 = \{0, 1, 2, 3, 4, 5\}$, such that $\mathbb{G} = (\mathbb{Z}_6, +)$ is an abelian group under addition modulo 6, by Theorem 2.3, it follows that $\mathbb{H} = (3\mathbb{Z}_6, +)$ is a subgroup of $\mathbb{G}$, where $3\mathbb{Z}_6 = \{0, 3\}$. The cosets of $\mathbb{H}$ in $\mathbb{G}$ are

$$\begin{array}{rcccl}
[0]_{\mathbb{H}} & = & 0 + 3\mathbb{Z}_6 & = & \{0, 3\} \\
[1]_{\mathbb{H}} & = & 1 + 3\mathbb{Z}_6 & = & \{1, 4\} \\
[2]_{\mathbb{H}} & = & 2 + 3\mathbb{Z}_6 & = & \{2, 5\}
\end{array}$$

since $[3]_{\mathbb{H}} = [0]_{\mathbb{H}}$, $[4]_{\mathbb{H}} = [1]_{\mathbb{H}}$, and $[5]_{\mathbb{H}} = [2]_{\mathbb{H}}$. The structure of the abelian group $\mathbb{G}/\mathbb{H} = (\{[0]_{\mathbb{H}}, [1]_{\mathbb{H}}, [2]_{\mathbb{H}}\}, +, [0]_{\mathbb{H}})$ is shown in the following Cayley table

$+$	$[0]_{\mathbb{H}}$	$[1]_{\mathbb{H}}$	$[2]_{\mathbb{H}}$
$[0]_{\mathbb{H}}$	$[0]_{\mathbb{H}}$	$[1]_{\mathbb{H}}$	$[2]_{\mathbb{H}}$
$[1]_{\mathbb{H}}$	$[1]_{\mathbb{H}}$	$[2]_{\mathbb{H}}$	$[0]_{\mathbb{H}}$
$[2]_{\mathbb{H}}$	$[2]_{\mathbb{H}}$	$[0]_{\mathbb{H}}$	$[1]_{\mathbb{H}}$

Notice that the group $\mathbb{G}/\mathbb{H}$ has the same structure of the abelian group $(\mathbb{Z}_3, +)$,

$\oplus_3$	0	1	2
0	0	1	2
1	1	2	0
2	2	0	1

Hence, one can write $\mathbb{G}/\mathbb{H} \cong (\mathbb{Z}_3, +)$.

2.1.4   Morphism of Groups

In what follows, maps between groups, or more elaborate algebraic structures, are considered. So let us introduce this concept and some of its related notions.

DEFINITION 2.9 (Morphism) Let $(\mathbb{G}, \star)$ and $(\mathbb{G}', \star')$ be two groups, with neutral elements e and e', respectively. A (group) **morphism** (or **homomorphism**) f from $\mathbb{G}$ to $\mathbb{G}'$ is a structure-preserving map, i.e., it verifies the following conditions:

- For all $a \in \mathbb{G}$ and all $b \in \mathbb{G}$, $f(a \star b) = f(a) \star' f(b)$.
- $f(e) = e'$.
- For each $a \in \mathbb{G}$, $f(a^{-1}) = (f(a))^{-1}$.

Actually, it may be shown that a map $f \colon \mathbb{G} \to \mathbb{G}'$ is a group morphism if and only if $f(a \star b) = f(a) \star' f(b)$, for all $a, b \in \mathbb{G}$.

Example 2.11 Let $(\mathbb{G}, \star)$ and $(\mathbb{G}', \star')$ be two groups, with neutral elements e and e', respectively. The following map is a morphism:

$$\begin{aligned} \mathbb{G} &\to \mathbb{G}', \\ x &\to e'. \end{aligned}$$

Let $(\mathbb{G}, +)$ be a commutative group, with neutral element e. Let us define recursively $n \cdot x$ for $x \in \mathbb{G}$ and n an integer by $0 \cdot x = e$, and $(n+1) \cdot x = x + (n \cdot x)$. Whence $n \cdot x = \underbrace{x + \cdots + x}_{n \text{ times } x}$. One now extends this notation to negative integers by setting $n \cdot x = -(|n| \cdot x)$ for each $n < 0$. Then the following map is a morphism:

$$\left\{ \begin{aligned} \mathbb{Z} &\to \mathbb{G}, \\ n &\mapsto n \cdot x. \end{aligned} \right.$$

With each group morphism are associated two subgroups, the kernel and the image of the morphism.

DEFINITION 2.10 (Kernel, Image) Let $\mathbb{G}$ and $\mathbb{G}'$ be two groups, and let f be a morphism from $\mathbb{G}$ to $\mathbb{G}'$.
The **kernel** of f is defined by

$$\operatorname{Ker}(f) = \{g \in \mathbb{G} \colon f(g) = e'\}$$

where e' is the neutral element of $\mathbb{G}'$.
The **image** of $f \colon \mathbb{G} \to \mathbb{G}'$ is defined by

$$\operatorname{Im}(f) = \{g' \in \mathbb{G}' \colon \exists g \in \mathbb{G}, f(g) = g'\}.$$

Property 2.1 Let f be a morphism from $\mathbb{G}$ to $\mathbb{G}'$. The kernel of f is a subgroup of $\mathbb{G}$. The image of f is a subgroup of $\mathbb{G}'$.

DEFINITION 2.11 Let $\mathbb{G}$ and $\mathbb{G}'$ be two groups.

- An **isomorphism** between $\mathbb{G}$ and $\mathbb{G}'$ is a morphism that is a bijection. Then its inverse is an isomorphism from $\mathbb{G}'$ to $\mathbb{G}$. In this case, $\mathbb{G}$ and $\mathbb{G}'$ are said to be **isomorphic**.
- An **endomorphism** is a morphism from $\mathbb{G}$ to itself.
- An **automorphism** is a morphism that is both an isomorphism and an endomorphism.

Example 2.12 For every group $\mathbb{G}$, the identity map is an automorphism.

$$\mathbb{G} \rightarrow \mathbb{G},$$
$$x \mapsto x.$$

The inverse map is also an automorphism:

$$\mathbb{G} \rightarrow \mathbb{G},$$
$$x \mapsto -x.$$

2.2 Finite Fields

2.2.1 Ring

Fields, and, more particularly, finite fields, together with elliptic curves, play a fundamental role in the construction of pairings. Here one summarizes their main definitions and results, as far as they are needed hereafter. For further information, refer to [16], for example.

Before introducing finite fields, we present the structure of rings, which is more general.

DEFINITION 2.12 (Ring) A **ring** R is a set with two binary operations $+, \times$ such that

- $(R, +)$ is a commutative group (with 0 as neutral element).
- R is closed under $\times$, which is an associative operation with a neutral element (denoted by 1).
- $\times$ distributes over addition on the left $a \times (b + c) = a \times b + a \times c$, and on the right, $(a + b) \times c = a \times c + b \times c$.

A ring R is said to be **commutative** if $\times$ is commutative.

Example 2.13 The set $(\mathbb{Z}, +, \times)$ is a ring.

Given two rings, say $(R, +, \times)$ and $(R', +', \times')$, a map $f : R \rightarrow R'$ is a **homomorphism of rings** whenever it is a morphism of groups from $(R, +)$ to $(R', +')$, and for every $a, b \in R$, $f(a \times b) = f(a) \times' f(b)$, and $f(1) = 1'$. The homomorphism f is said to be an **endomorphism** (resp. **isomorphism**, **automorphism**) when the rings R and R' are the same (resp. if furthermore f is a bijection, resp. if f is both an endomorphism and an isomorphism).

DEFINITION 2.13 (Ring ideal) Let $(R, +, \times)$ be a ring and $(R, +)$ be its additive group. An **ideal** I of R is an additive subgroup of R that absorbs multiplication by elements of R. Formally, an ideal I verifies the following properties:

- $(I, +)$ is a subgroup of $(R, +)$,
- $\forall x \in I, \forall r \in R : x \times r, r \times x \in I$.

Example 2.14 In the ring $(\mathbb{Z}, +, \times)$:

- The set of even integers is an ideal of $\mathbb{Z}$.
- Let p be a prime number, then the set of multiples of p is an ideal of $\mathbb{Z}$ denoted by $p\mathbb{Z} = \{a \in \mathbb{Z} \text{ such that } \exists x \in \mathbb{Z}, a = x \times p\}$.

DEFINITION 2.14 (Quotient ring) Let $(R, +, \times)$ be a ring and I be an ideal of R. The **quotient ring** is the ring denoted by R/I whose elements are the cosets of I in R subject to special $+$ and $\times$ operations.

Example 2.15 Let p be a prime number, then $\mathbb{Z}/p\mathbb{Z} = \{0, 1, 2, \ldots, p - 1\}$ with the addition and multiplication inherited from $\mathbb{Z}$ modulo p is a quotient ring.

DEFINITION 2.15 (Equivalence class) Let $(R, +, \times)$ be a ring and I an ideal of R. We define the equivalence relation $\equiv \mod I$ as follows: $a \equiv b \mod I$ if and only if $(a - b) \in I$.

If $a \equiv b \mod I$, then a and b are said to be congruent modulo I. The **equivalence class** of an element $a \in R$ is given by $[a]_I = a + I := \{a + r : r \in I\}$.

2.2.2 Finite Fields

In this subsection, we formally define finite fields and their basic properties. As mentioned above, finite fields are the most important mathematical objects for bilinear pairing constructions.

2.2.3 Definition

A **finite field** is a field with finitely many elements. We first give the definition of a general field.

DEFINITION 2.16 (Field) A **field** $\mathbb{F}$ is a set endowed with two binary operations, usually denoted by $+$ (the addition of the field) and by $\times$ (the multiplication), which satisfy the following properties:

- $(\mathbb{F}, +)$ is a commutative group. One denotes by 0 the neutral element of $(\mathbb{F}, +)$.
- Let $\mathbb{F}^\star$ be the subset of $\mathbb{F}$ consisting of non-zero elements, i.e., $\mathbb{F}^\star = F \setminus \{0\}$. Then, $(\mathbb{F}^\star, \times)$ is a commutative group. The elements of $\mathbb{F}^\star$ are said to be invertible for the multiplication law.
- Finally, $+$ and $\times$ interact properly through the **distributivity property**: for every $a, b, c \in \mathbb{F}$, $a \times (b + c) = a \times b + a \times c$.

Example 2.16 The set of real numbers $(\mathbb{R}, +, \times)$ is a field.

One observes that a field is a commutative ring R for which $R \setminus \{0\}$ is a group under multiplication (and in particular $0 \neq 1$).

A homomorphism of fields is a homomorphism of rings between two fields (endomorphisms, isomorphisms, and automorphisms of fields are defined likewise). Two fields are said to be **isomorphic** whenever there exists an isomorphism of field from one to the other.

DEFINITION 2.17 A **subfield** of a field $\mathbb{F}$ is a subset of $\mathbb{F}$, which is itself a field with respect to the field operations inherited from $\mathbb{F}$. The intersection of all subfields of $\mathbb{F}$ is itself a field, called its **prime subfield**. Of course, it is contained into every subfield of $\mathbb{F}$.

The description of the prime subfield as an intersection of all subfields of $\mathbb{F}$, although synthetically, may not be very useful in practice. Fortunately, prime subfields may be described differently. Let $\mathbb{F}$ be a field. Let us define $\phi \colon \mathbb{Z} \to \mathbb{F}$ be the homomorphism of rings defined by

$\phi(n) = n \cdot 1 = \underbrace{1 + 1 + \ldots + 1}_{n \text{ times } 1}$. Recall that ϕ is a morphism of groups from $(\mathbb{Z}, +)$ to $(\mathbb{F}, +)$. The kernel of ϕ is an ideal of $\mathbb{Z}$. By the way, the ideals of $\mathbb{Z}$ are completely known: these are exactly $p\mathbb{Z} = \{ pn \colon n \in \mathbb{Z} \}$, for p ranging over $\mathbb{Z}$. It turns out that there is a unique integer p, called the **characteristic** of $\mathbb{F}$, such that the kernel of ϕ is $p\mathbb{Z}$ [16]. Moreover, p is either a prime number or is equal to 0.

Fields of characteristic zero are necessarily infinite fields. So a finite field has a prime number characteristic. In this case one may give yet another definition of the characteristic of a finite field: it is the least integer $n > 0$ such that $n \cdot 1 = 0$ (notice that such an integer necessarily exists as otherwise $\{ n \cdot 1 \colon n > 0 \}$ would be infinite).

DEFINITION 2.18 (Field characteristic) Given a field $\mathbb{F}$, we say that n is its field characteristic, if n is the smallest positive integer such that $n \cdot 1 = 0$. If such an integer n cannot be defined, we say that the field $\mathbb{F}$ has characteristic 0.

Example 2.17

- The field $(\mathbb{R}, +, \times)$ is a field of characteristic zero.
- Let p be a prime number. If the characteristic of $\mathbb{F}$ is p, then $\mathbb{F}$ is finite. Therefore, the set $\mathbb{F}_p = \{0, 1, 2, \ldots, p - 2, p - 1\}$ defines a **finite field** $(\mathbb{F}_p, +, \times)$, with respect to the addition and multiplication operations modulo p.

Remark 2.7 In pairing-based cryptography, one can consider finite fields of small characteristic (2 or 3) or finite fields of a large prime characteristic. Nevertheless, as shown in [13, 2], pairings over finite fields of small characteristic are no longer secure with respect to the discrete logarithm problem (see Chapter 9). As a consequence, in what follows, one focuses on finite fields of large prime characteristic. A **large prime** is a prime number $p \geq 5$, in practice, the size in bits of a large prime p is at least 256 bits; see Chapters 3, 4, and 10. We make the following obvious observation: any large prime number is an odd integer.

A large prime characteristic finite field is determined — up to isomorphism — by its characteristic, a prime number p and its cardinal. Recall here that the cardinal of a field $\mathbb{F}$ is just the number of its elements. It is usually denoted by $\mathrm{card}(\mathbb{F})$ or by $\#(\mathbb{F})$.

DEFINITION 2.19 (Prime field) Let p be a prime number. Then, the ring $\mathbb{Z}/p\mathbb{Z}$ of the integers mod p is a finite field of characteristic p, with p elements, which is denoted by $\mathbb{F}_p$ and called the **prime field** of characteristic p.

Remark 2.8 For each prime number p, and each field $\mathbb{F}$ of characteristic p, the prime subfield of $\mathbb{F}$ is isomorphic to $\mathbb{F}_p$.

THEOREM 2.4 *If $\mathbb{F}$ is a finite field of characteristic p, then there exists an integer $n > 0$ such that $\mathrm{card}(\mathbb{F}) = p^n$.*

Example 2.18

- $\mathbb{F}_2$ is a finite field. Its cardinal is 2 and it can be represented as $\mathbb{F}_2 = \{0, 1\}$.
- $\mathbb{F}_3 = \{0, 1, 2\}$, its cardinal is 3.
- There is no finite field of characteristic 6.
- For every prime number p, $\mathbb{F}_p = \{0, 1, 2, \ldots, p - 1\}$.

2.2.4 Properties

The computation of pairings relies on the arithmetic of finite fields. In particular, several important optimizations in a pairing execution are based on Fermat's little theorem (for instance, the so-called *denominator elimination* in a pairing computation; see Chapter 3).

THEOREM 2.5 (Fermat's little theorem) *Let p be a prime number and x a non-zero element of $\mathbb{F}_p$. Then, the following identities are true:*

$$x^p \equiv x \mod p,$$
$$\text{if } x \neq 0 \mod p, \ x^{(p-1)} \equiv 1 \mod p.$$

DEFINITION 2.20 (Order) The **order** of an element $x \in \mathbb{F}_p$ is the smallest integer α such that $x^\alpha \equiv 1 \mod p$.

Another important homomorphism when working over finite fields is the Frobenius map.

PROPOSITION 2.1 (Frobenius automorphism) *Let $\mathbb{F}$ be a finite field of characteristic $p > 0$. The map*

$$\pi_p : \mathbb{F} \to \mathbb{F},$$
$$x \to x^p$$

*is an automorphism of field called the **Frobenius** automorphism. If $\mathbb{F} = \mathbb{F}_p$ then the Frobenius automorphism is the identity.*

Remark 2.9 The Frobenius can also be denoted by π when p is clearly defined.

2.2.5 Extensions of Finite Fields

We will need to define the extension of finite fields $\mathbb{F}_{p^k}$ for p, a prime number, and k, a positive integer. The construction of extension of finite fields can be done using a polynomial representation of the field elements. Before formally introducing the extension of a finite field, we first present the definition and properties of polynomials over $\mathbb{F}_p$.

Polynomials over $\mathbb{F}_p$

Let $\mathbb{F}_p[X]$ denote the set of polynomials in the variable X and with coefficients in $\mathbb{F}_p$. We define the notion of irreducible polynomials below.

DEFINITION 2.21 (Irreducible polynomial) A non-constant polynomial $P(X) \in \mathbb{F}_p[X]$ of degree n is said to be **irreducible** over $\mathbb{F}_p$ if it cannot be factored into the product of two non-constant polynomials in the ring $\mathbb{F}_p[X]$ of degree smaller than n.

Example 2.19 Let $p = 257$ and let $\mathbb{F}_p = \{0, 1, 2, \ldots, 255, 256\}$. The polynomial $P(X) = X^4 - 3$ is an irreducible polynomial over $\mathbb{F}_p$.

As a consequence of the existence of irreducible polynomials over $\mathbb{F}_p$, the roots of a polynomial over $\mathbb{F}_p[X]$ are not all included in $\mathbb{F}_p$. If they were, we could not talk about irreducible polynomials. In a general way, the roots of a polynomial over $\mathbb{F}_p[X]$ are all included in $\overline{\mathbb{F}_p}$. We give here an informal definition of the splitting field of a polynomial and of the algebraic closure of a finite field.

DEFINITION 2.22   Let p be a prime number and $\mathbb{F}_p$ be the finite field of characteristic p.

1. The **splitting field** of a polynomial $P(X) \in \mathbb{F}_p[X]$ is the smallest field containing all the roots of $P(X)$.

2. The **algebraic closure** of $\mathbb{F}_p$, denoted by $\overline{\mathbb{F}_p}$, is the set of all roots of all polynomials with coefficients in $\mathbb{F}_p$. Up to an isomorphism, the algebraic closure of a field is unique.

3. A field $\mathbb{F}$ is **algebraically closed** if every non-constant polynomials in $\mathbb{F}[X]$ presents at least one root in $\mathbb{F}$. To put it in another way, the only irreducible polynomials in the polynomial ring $\mathbb{F}[X]$ are those of degree one.

We can prove the existence and uniqueness of any finite field, given its characteristic and cardinality.

THEOREM 2.6   *Let p be a prime number and n a non-zero positive integer. Let $q = p^n$.*

1. *There exists at least one finite field of characteristic p and cardinality q, the splitting field of the polynomial $X^q - X$ over $\mathbb{F}_p$.*

2. *This finite field is denoted $\mathbb{F}_q$ and it is unique up to a homomorphism.*

Remark 2.10   The uniqueness of the finite field of q elements is in fact stronger. Indeed, let $\overline{\mathbb{F}_p}$ be an algebraic closure of $\mathbb{F}_p$. There is in $\overline{\mathbb{F}_p}$ one and only one finite field of q elements, which is the finite field composed of the roots of the polynomial $X^q - X$.

DEFINITION 2.23   The **multiplicative subgroup** of a finite field $\mathbb{F}_q$ is denoted by $\mathbb{F}_q^{\star}$, for q, a power of a prime number. It is composed of the field elements that have a multiplicative inverse, i.e., they are invertible by the multiplication operation.

Let p be a prime number and $q = p^n$ for a non-zero integer n. The value of a pairing is an element belonging to the multiplicative subgroup in a finite field $\mathbb{F}_q$.

THEOREM 2.7   *The multiplicative subgroup $\mathbb{F}_q^{\star}$ of $\mathbb{F}_q$ is cyclic and isomorphic to $\mathbb{Z}/(q-1)\mathbb{Z}$, for q a power of a prime number.*

DEFINITION 2.24 (Generator)   Let $\mathbb{F}_q$ be a finite field, and $\mathbb{F}_q^{\star}$ be its multiplicative subgroup, where q is a power of a prime number. There exists an element g such that any element of $\mathbb{F}_q^{\star}$ is a power of g. Such an element g is called a **generator** of $\mathbb{F}_q^{\star}$.

Example 2.20

- For a prime number p, $\mathbb{F}_p^{\star} = \{1, 2, \ldots, p-2, p-1\}$. The generators of $\mathbb{F}_p^{\star}$ are the elements of order $p-1$.
- $\mathbb{F}_2^{\star} = \{1\}$ and 1 is a generator of $\mathbb{F}_2^{\star}$.
- $\mathbb{F}_3^{\star} = \{1, 2\}$ and 2 is a generator of $\mathbb{F}_3^{\star}$. The element 1 is not a generator of $\mathbb{F}_3^{\star}$.

The arithmetic of pairings involves the construction of tower extension fields. An extension field of $\mathbb{F}_p$ can be seen as a finite field of characteristic p bigger than $\mathbb{F}_p$ and including $\mathbb{F}_p$. We present a constructive definition of an extension field of degree $k > 1$ of a finite field $\mathbb{F}_p$ based on the use of a quotient ring.

DEFINITION 2.25 [**Extension field**] Let p be a prime number and k a non-zero positive integer. An **extension field** of degree k of $\mathbb{F}_p$, denoted by $\mathbb{F}_{p^k}$, is a finite field of characteristic p given by the quotient $\mathbb{F}_p[X]/(P(X)\mathbb{F}_p[X])$, where $P(X) \in \mathbb{F}_p[X]$ is an irreducible polynomial of degree k. The set $P(X)\mathbb{F}_p[X]$ is the set of polynomials admitting $P(X)$ in their factorization, otherwise $P(X)\mathbb{F}_p[X]$ is the ideal of $\mathbb{F}_p[X]$ generated by $P(X)$ and denoted $(P(X))$.

Remark 2.11

- The extension field $\mathbb{F}_{p^k}$ can be denoted by $\mathbb{F}_p[X]/(P(X)\mathbb{F}_p[X])$ or $\mathbb{F}_p[X]/(P(X))$.
- The extension field $\mathbb{F}_{p^k}$ of degree k of $\mathbb{F}_p$ can be described as the set of polynomials with coefficient in $\mathbb{F}_p$ and of degree strictly less than k.

$$\mathbb{F}_{p^k} = \{R(X) \in \mathbb{F}_p[X], \text{ such that } \deg(R) < k\}.$$

Property 2.2 The cardinal of $\mathbb{F}_{p^k}$ is p^k.

We can construct a basis of $\mathbb{F}_{p^k}$ in order to describe any element of $\mathbb{F}_{p^k}$. The polynomial $P(X)$ used to construct the extension of degree k is irreducible over $\mathbb{F}_p$. But in $\mathbb{F}_{p^k}$, we know that $P(X) = 0$ in $\mathbb{F}_{p^k}$ (see Definition 2.25).

Property 2.3 **Basis of an extension field** Let p be a prime number, k an integer greater than 1. Let $P(X) \in \mathbb{F}_p[X]$ be an irreducible polynomial of degree k. Let $\mathbb{F}_{p^k} = \mathbb{F}_p[X]/(P(X)\mathbb{F}_p[X])$ be the extension field of degree k of $\mathbb{F}_p$ constructed over $P(X)$.

Let γ be a root of $P(X)$ in $\mathbb{F}_{p^k}$. The variable γ can be seen as a class of X in $\mathbb{F}_{p^k}$. A basis $\mathcal{B}_{\mathbb{F}_{p^k}}$ of $\mathbb{F}_{p^k}$ as a vector space is given by the powers of γ strictly smaller than k: $\mathcal{B}_{\mathbb{F}_{p^k}} = \{1, \gamma, \gamma^2, \ldots, \gamma^{k-1}\}$.

Remark 2.12 Thanks to Property 2.3, we can prove that $\text{card}(\mathbb{F}_{p^k}) = p^k$, as presented in Property 2.2.

Example 2.21 The polynomial $P(X) = X^4 - 3$ is irreducible over $\mathbb{F}_{257}$. As $\deg(P(X)) = 4$, we can construct an extension of degree 4 of $\mathbb{F}_{257}$. Let γ be a root of $P(X)$ in $\mathbb{F}_{257^4}$. We can describe the elements of $\mathbb{F}_{257^4}$ as

$$\mathbb{F}_{257^4} \cong \{a_0 + a_1\gamma + a_2\gamma^2 + a_3\gamma^3, \quad a_i \in \mathbb{F}_{257}\}.$$

Example 2.22 Given the irreducible polynomial $P(X) = X^2 + X + 1$, the finite field $\mathbb{F}_{2^2}$, can be defined as:

$$\mathbb{F}_{2^2} = \mathbb{F}_2[X]/(X^2 + X + 1) = \{bz + a \mid a, b \in \mathbb{F}_2\} = \{0, 1, X, X + 1\}.$$

In other words, $\mathbb{F}_{2^2}$ is a finite field with order $2^2 = 4$, whose elements are the set of polynomials in $\mathbb{F}_2[X]$ modulo $X^2 + X + 1$.

The structure of the abelian groups $\mathbb{F}_{2^2}^+$ and $\mathbb{F}_{2^2}^*$ are described in the following Cayley tables:

$\oplus_{X^2+X+1}$	0	1	X	$X+1$
0	0	1	X	$X+1$
1	1	0	$X+1$	X
X	X	$X+1$	0	1
$X+1$	$X+1$	X	1	0

$\odot_{X^2+X+1}$	1	X	$X+1$
1	1	X	$X+1$
X	X	$X+1$	1
$X+1$	$X+1$	1	X

Remark 2.13 We can notice that the finite field $\mathbb{F}_{2^2} = (\{0,1,X,X+1\}, +, \times)$ contains the prime field $\mathbb{F}_2 = (\{0,1\}, +, \times)$.

2.2.6 Arithmetic of Finite Fields

In this section, we briefly present the arithmetic over a finite field $\mathbb{F}_p$. We will also introduce the arithmetic of an extension of a finite field $\mathbb{F}_p$. The arithmetic over extension fields is based on the polynomial arithmetic over $\mathbb{F}_p$. For a more detailed description of this topic, the reader is referred to Chapter 5.

2.2.7 The Arithmetic of $\mathbb{F}_p$

The arithmetic over the finite field $\mathbb{F}_p$ for p a prime number is the modular arithmetic mod p. The addition (respectively, subtraction) in $\mathbb{F}_p$ is usually implemented as the classical addition (respectively, subtraction) possibly followed by a subtraction (respectively addition) of p.

ALGORITHM 2.1 Addition over $\mathbb{F}_p$.

Input : $a \in \mathbb{F}_p$ and $b \in \mathbb{F}_p$
Output: $c = a + b \in \mathbb{F}_p$
1 $c \leftarrow a + b$
2 **if** $c \geq p$ **then**
3 $\quad | \quad c \leftarrow c - p$
4 **end**
5 **return** c

ALGORITHM 2.2 Subtraction over $\mathbb{F}_p$.

Input : $a \in \mathbb{F}_p$ and $b \in \mathbb{F}_p$
Output: $c = a - b \in \mathbb{F}_p$
1 $c \leftarrow a - b$
2 **if** $c < 0$ **then**
3 $\quad | \quad c \leftarrow c + p$
4 **end**
5 **return** c

The most efficient multiplication in $\mathbb{F}_p$ is often performed using Barrett's or Montgomery's algorithms. These two algorithms include the modular reduction in the multiplication.

The choice of the multiplication algorithm is adapted to the bit size of the operands or the targeted device. We present here one of the most used algorithms for the modular multiplication: the Montgomery algorithm. The Montgomery multiplication is performed between two elements of $\mathbb{F}_p$ represented in the Montgomery representation. A more detailed introduction to Montgomery multiplication is presented in Chapter 5.

The inversion in $\mathbb{F}_p$ is an expensive operation. It is avoided as often as possible. When an inversion must be performed, several methods exist. We can cite, for example, the Extended Euclidean algorithm, or the use of Fermat's little theorem.

THEOREM 2.8 *Let p be a prime number and $x \in \mathbb{F}_p^\star$, then $x^{p-2} = x^{-1}$.*

2.2.8 The Arithmetic of $\mathbb{F}_{p^k}$

The arithmetic of $\mathbb{F}_{p^k} = \mathbb{F}_p[X]/(P(X))$ is based on the arithmetic of polynomials modulo $P(X)$, combined with the arithmetic over $\mathbb{F}_p$. In order to have an efficient arithmetic over $\mathbb{F}_{p^k}$, the polynomial $P(X)$ should be chosen sparse on the model of $X^k - \beta$ for a small $\beta \in \mathbb{F}_p$ [15].

DEFINITION 2.26 (k^{th} root of unity) Let $\mathbb{F}$ be a field. An element $x \in \mathbb{F}$ is a kth **root of unity**, if $x^k = 1 \in \mathbb{F}$.

The probability that a random element in $\mathbb{F}_p$ is a kth root in $\mathbb{F}_p$ is highly dependent on k. We can compute this probability when $p \equiv 1 \mod k$.

Property 2.4 Let p be a prime number, k an integer greater than 1. If $p \equiv 1 \mod k$, then a random element $\alpha \in \mathbb{F}_p$ is a kth root with a probability $1/k$.

Proof. Let α be a kth root in $\mathbb{F}_p$. Therefore, there is an element $x \in \mathbb{F}_p$ such that $x^k \equiv \alpha \mod p$. As $p \equiv 1 \mod k$, we have that k divides $(p-1)$. We can then raise the equality $x^k \equiv \alpha \mod p$ to the power $\frac{(p-1)}{k}$:

$$
\begin{aligned}
\alpha &\equiv x^k \mod p, \\
\alpha^{\frac{(p-1)}{k}} &\equiv x^{k\frac{(p-1)}{k}} \mod p, \\
\alpha^{\frac{(p-1)}{k}} &\equiv x^{(p-1)} \mod p, \\
\alpha^{\frac{(p-1)}{k}} &\equiv 1 \mod p.
\end{aligned}
$$

Hence, α is a kth root in $\mathbb{F}_p$ if and only if α belongs to the subgroup of $\mathbb{F}_p$ of order $\frac{(p-1)}{k}$. The cardinality of this subgroup is exactly $\frac{(p-1)}{k}$. □

COROLLARY 2.1 *Let $\beta \in \mathbb{F}_p$ be a random element. If $p \equiv 1 \mod k$, then the polynomial $X^k - \beta$ is irreducible over $\mathbb{F}_p$ with a probability $\frac{p-1}{k}$.*

If $\mathbb{F}_{p^k}$ is constructed using a polynomial $X^k - \beta$, the multiplication in $\mathbb{F}_{p^k}$ of the two elements

$$
U(X) = \sum_{i=0}^{k-1} u_i X^i \quad \text{and } V(X) = \sum_{i=0}^{k-1} v_i X^i
$$

can be performed into two steps. First we compute the polynomial product $W = U \times V$, followed by the polynomial reduction $W(X)$ modulo $P(X)$. The binomial form of $P(X)$ makes the second step quite easy. Let us split into two parts the result $W(X) = U(X) \times V(X)$, $W(X) = \underline{W} + X^k \overline{W}$, where $\deg \underline{W} < k$ and $\deg \overline{W} < k$. The polynomial reduction consists then of the operation $\underline{W} + \beta(\overline{W})$, performed in at most k multiplications in $\mathbb{F}_p$ and k additions in $\mathbb{F}_p$.

The most expensive part of the multiplication in $\mathbb{F}_{p^k}$ is the polynomial multiplication. Several ways to improve on this will be further discussed in Chapter 5, which is devoted to the arithmetic of finite fields.

Example 2.23 Using once again the field extension of Example 2.21, let $a = 3 + \gamma^3$ and $b = 234\gamma + 36\gamma^2$ be two elements of $\mathbb{F}_{257^4}$. In order to compute $c = a \times b \in \mathbb{F}_{257^4}$, we first compute the product:

$$
\begin{aligned}
c &= a \times b, \\
  &= (3 + \gamma^3) \times (234\gamma + 36\gamma^2), \\
  &= 702\gamma + 108\gamma^2 + 234\gamma^4 + 36\gamma^5, \\
  &= 188\gamma + 108\gamma^2 + 234\gamma^4 + 36\gamma^5.
\end{aligned}
$$

Then we perform the polynomial reduction considering that $\gamma^4 = 3$.

$$
\begin{aligned}
c &= 188\gamma + 108\gamma^2 + 234\gamma^4 + 36\gamma^5, \\
  &= 3 \times 234 + (188 + 3 \times 36)\gamma + 108\gamma^2, \\
  &= 702 + 296\gamma + 108\gamma^2, \\
  &= 188 + 39\gamma + 108\gamma^2.
\end{aligned}
$$

2.2.9   Quadratic Residuosity

DEFINITION 2.27 (Norm)   Let p be a prime number and let $n \in \mathbb{N}$. The conjugates of $a \in \mathbb{F}_{p^n}$ are the elements a^{p^i}, where $0 \leq i \leq n - 1$. The **norm** of a, denoted as $|a|$, is given by the product of all the conjugates of a, i.e.,

$$
|a| = \prod_{i=0}^{n-1} a^{p^i}. \tag{2.1}
$$

The problem of computing a field square root of any arbitrary element $a \in \mathbb{F}_p$ consists of finding a second element $b \in \mathbb{F}_p$ such that $b^2 = a$. The following definitions are relevant for computing square roots over finite fields.

DEFINITION 2.28 (Quadratic Residue)   A field element $a \in \mathbb{F}_p$ is called a *quadratic residue*, if there exists a field element x such that:

$$
x^2 = a.
$$

We say that x is the square root of a in the field $\mathbb{F}_p$.

DEFINITION 2.29 (Quadratic Residuosity test (Euler's criterion))   The square root of an element $a \in \mathbb{F}_p^*$ exists if and only if $a^{\frac{p-1}{2}} = 1$. We denote by $\chi_p(a)$ the value of $a^{\frac{p-1}{2}}$. For a non-zero field element a, $\chi_p(a) = \pm 1$. If $\chi_p(a) = 1$, we say that the element a is a quadratic residue in $\mathbb{F}_p$. In $\mathbb{F}_p^*$ there exist exactly $(p - 1)/2$ quadratic residues.

Remark 2.14   The quadratic residuosity test of a field element a, denoted in this book as $\chi_p(a)$, is sometimes called the quadratic character of a.

Therefore, a quadratic residuosity test over the field element $a \in \mathbb{F}_p$ can be performed by computing the exponentiation $a^{\frac{q-1}{2}}$. A computationally cheaper way to perform this test uses the law of quadratic reciprocity, as shown in Algorithm 2.3.

ALGORITHM 2.3   Quadratic residuosity test χ_p in $\mathbb{F}_p$.

Input   : $a \in \mathbb{F}_p^*$ and the field characteristic p
Output: $k = 1$ if a is a quadratic residue in $\mathbb{F}_p$, $k = -1$ otherwise

---

1 tab $\leftarrow \{0, 1, 0, -1, 0, -1, 0, 1\}$
2 $k \leftarrow 1$
3 **while** $a > 0$ **do**
4 $\quad$ $v \leftarrow 0$
5 $\quad$ **while** a *is even* **do**
6 $\quad\quad$ $a \leftarrow a/2$
7 $\quad\quad$ $v \leftarrow v + 1$
8 $\quad\quad$ **if** v *is odd* **then**
9 $\quad\quad\quad$ $k \leftarrow \text{tab}[p \mod 8] \cdot k$
10 $\quad\quad$ **end**
11 $\quad\quad$ $c_1 \leftarrow p \mod 4$
12 $\quad\quad$ $c_2 \leftarrow a \mod 4$
13 $\quad\quad$ **if** $c_1 = c_2$ *and* $c_1 = 3$ **then**
14 $\quad\quad\quad$ $k \leftarrow -k$
15 $\quad\quad$ **end**
16 $\quad\quad$ $t \leftarrow a$
17 $\quad\quad$ $a \leftarrow p \mod t$
18 $\quad\quad$ $p \leftarrow t$
19 $\quad$ **end**
20 **end**
21 **return** k

2.3   Elliptic Curves

Elliptic curves are mathematical tools known from the eighteenth century. They seem to appear first in the work of Fagnano, published on December 23, 1751. They were related to the study of an ellipse perimeter. However, Washington in [20, Chapter 1] traces the discovery of the group addition law back to Diophantus (around 250 A.D.). Washington also points out that the description of the so-called *congruent number problem*, which can be solved with elliptic curves, appeared in Arab manuscripts around 900 A.D. Anyway, in this section, we give a simplified presentation of elliptic curves used nowadays in pairing-based cryptography.

We do not give the proof of theorems, but we give references in which to find them. We begin with the definition of an elliptic curve over finite fields. We are able to calculate the cardinals of those elliptic curves, and we illustrate this with the proper algorithm. Elliptic curves used in pairing-based cryptography must admit an r-torsion group with a reasonable embedding degree. We define those notions and introduce the notion of pairing-friendly elliptic curves. We also present two isogenies useful for the computation of pairings: the Frobenius map and the twist of an elliptic curve. We then discuss the arithmetic of points of an elliptic curve. Finally, we discuss several representations and systems of coordinates for the points of an elliptic curve.

2.3.1   Definition

By definition, an elliptic curve is a smooth curve defined by a polynomial equation of degree three. The most general form of such a curve defined over a field $\mathbb{F}$ is a set of projective points whose coordinates satisfy an equation:

$$f(x, y) = ax^3 + bx^2y + cxy^2 + dy^3 + ex^2 + fxy + gy^2 + hx + iy + j = 0, \qquad (2.2)$$

with coefficients $a, \ldots, j$ in the field $\mathbb{F}$.

In order to ensure that the polynomial $f(x, y)$ is actually of degree three, at least one of the coefficients a, b, c, or d must be non-zero.

It is also required that the polynomial $f(x, y)$ is absolutely irreducible, which means that it is irreducible in the algebraic closure of the field $\mathbb{F}$, in order to ensure that the roots of this polynomial are not roots of a polynomial of lower degree.

In the projective plane $\mathbb{P}_2$, Equation (2.2) leads to an homogeneous equation with indeterminates x, y, and z:

$$ax^3 + bx^2y + cxy^2 + dy^3 + ex^2z + fxysz + gy^2z + hxz^2 + iyz^2 + jz^3 = 0. \qquad (2.3)$$

When the projective plane $\mathbb{P}_2$ is considered as the union of the affine plane together with the projective line at infinity, the roots of Equation (2.3) are exactly the solutions x and y of Equation (2.2) and $z = 1$, plus the point at infinity of the curve that is solution of Equation (2.3) with $z = 0$.

DEFINITION 2.30 (Smooth curve) A curve is said to be **smooth** if it admits a single tangent line on each of its points. This means that, for every point (x, y, z) of the curve, the linear mapping $D_{xyz} : (dx, dy, dz) \mapsto f'_x(x, y, z)dx + f'_y(x, y, z)dy + f'_z(x, y, z)dz$ is not the null mapping, where $f'_x()$ (respectively, $f'_y()$ and $f'_z()$) is the derivate of f with respect to x (respectively, y and z).

DEFINITION 2.31 (Projective Weierstrass equation) An elliptic curve E over a finite field $\mathbb{F}_p$, for p a prime number is the set of points verifying the Weierstrass equation:

$$y^2z + a_1xyz + a_3yz^2 = x^3 + a_2x^2z + a_4xz^2 + a_6z^3, \qquad (2.4)$$

for $a_i \in \mathbb{F}_p$.

Remark 2.15 We note that the indexes of the constants a_1, a_3, a_2, a_4, and a_6 are not chosen at random. They are the complement to 6 of the degree of the monomials, counting a degree two, three, and zero for x, y, and z, respectively.

Depending on the characteristic of the field $\mathbb{F}$, it is possible to simplify the equation given by the Weierstrass form using a change of variables. In characteristic 2 and 3, there are two equations of the short Weierstrass form of an elliptic curve. The two equations are respectively defined for ordinary or supersingular elliptic curve. We give the definitions of ordinary and supersingular elliptic curves in Section 2.3.5.

In characteristic 2, the Weieirstrass equation can be transformed into

$$y^2z + xyz = x^3 + ax^2z + bz^3 \text{ or } y^2z + cyz^2 = x^3 + axz^2 + bz^3.$$

In characteristic 3, the Weierstrass equation can be transformed into

$$y^2z = x^3 + ax^2z + bz^3 \text{ or } y^2z = x^3 + axz^2 + bz^3.$$

In the sequel, we only consider elliptic curves over finite fields of characteristic different from 2 and 3. Indeed, the pairing computation in characteristic 2 or 3 are not considered as secure according to the recent work [12, 2].

DEFINITION 2.32 (Projective definition) Let $\mathbb{F}$ be a field of characteristic different from 2 and from 3. An elliptic curve is the subset of the projective plane $\mathbb{P}_2$ defined by:

$$E = \left\{ (X, Y, Z) \in \mathbb{P}_2(\mathbb{F}) \mid Y^2 Z = X^3 + aXZ^2 + bZ^3 \right\},$$

where $\Delta = 4a^3 + 27b^2$ is non-zero.

The condition $\Delta = 4a^3 + 27b^2 \neq 0$ means that the polynomial $x^3 + ax + b$ has no double root, i.e., that this polynomial and its derivative polynomial are relatively prime.

When Z is non-zero, a projective point (X, Y, Z) admits a representative with $z = 1$ and the equation of the curve becomes:

$$y^2 = x^3 + ax + b.$$

When Z is zero, that is on the line at infinity, the points of the curve satisfy $x^3 = 0$, and so $x = 0$. There exists only one point of the curve on the line at infinity, denoted P_∞ and representing all points whose coordinates are equivalent to $(0, 1, 0)$. This point can be seen as the intersection of the vertical lines.

These remarks lead to the affine definition of elliptic curves. An elliptic curve is then considered as a subset of the affine plane, together with an additional point, called *point at infinity*, which corresponds to the point of the curve on the line at infinity. We denote by 0_E the point at infinity.

DEFINITION 2.33 (Affine definition) Let $\mathbb{F}$ be a field of characteristic different from 2 and from 3. An elliptic curve is a set defined by:

$$E(\mathbb{F}) = \left\{ (x, y) \in \mathbb{F}^2 \mid y^2 = x^3 + ax^2 + b \right\} \cup \left\{ P_\infty \right\},$$

where $\Delta = 4a^3 + 27b^2$ is non-zero and where P_∞ is an additional point called the point at infinity of the curve.

We give the previous definitions considering the affine and projective coordinates for the plane. It is possible to describe an elliptic curve using a different system of coordinates. We can cite, for example, the Jacobian and Chudnosky coordinate systems. The projective and Jacobian coordinates are the most frequently used for practical implementations. Indeed, the operations over the points of an elliptic curve are more efficient using these systems of coordinates. We describe an elliptic curve using the Weierstrass model. Other models of elliptic curves exist, such as the Edwards and Jacobi forms. We do not describe them, as the pairing computations are more efficient in the Weierstrass model. One can refer, for instance, to [7, 21, 8, 9] for the description of the pairing computation in other models of elliptic curves.

Remark 2.16 Let $\mathbb{F}$ be a field and E an elliptic curve. The set of points of $E(\mathbb{F})$ forms a group for the addition. The point at infinity is the neutral element for the addition over an elliptic curve.

2.3.2 Arithmetic over an Elliptic Curve

In the projective plane, an elliptic curve and a line have exactly three points of intersection. Note that in the affine plane, a vertical line and a curve of equation $y^3 + ax + b$ have only two common points. The point at infinity of elliptic curves is the third common point.

This remark allows to define a composition law, denoted as an addition, on the points of an elliptic curve. This law is defined by the following rules:

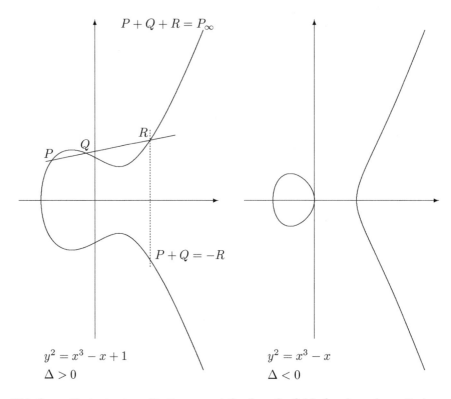

$P + Q + R = P_\infty$

R

P Q

$P + Q = -R$

$y^2 = x^3 - x + 1$
$\Delta > 0$

$y^2 = x^3 - x$
$\Delta < 0$

FIGURE 2.1 This figure illustrates two elliptic curves defined on the field of real numbers. It also displays the addition law. To add points P and Q, draw the line (PQ). It intersects the curve at point R. The sum of P and Q is the symmetric of R relatively to the horizontal axis.

1. The neutral element is the point at infinity.
2. If three points of the curve are aligned on the same line, their sum is the point at infinity (See Figure 2.1).
3. If two points P and Q have the same abscissa, i.e., if $x_P = x_Q$, then the third point of the curve that is on the line (PQ) is the point at infinity. It follows from the second rule above that in this case, the sum of P and Q is the point at infinity. Thus, the points P and Q are opposite. In other words, the opposite of the point $P = (x_P, y_P)$ is the point $-P = (x_P, -y_P)$.
4. If two points P and Q do not have the same abscissa, then the line (PQ) meets the curve on a third point R, which is, according to the second rule, the opposite of the sum $P + Q$. Figure 2.1 illustrates the construction of the sum of P and Q.
5. The double of a point P is obtained similarly by considering the line tangent to the curve passing by the point P.

Affine coordinates

The curve together with this composition law is an Abelian group. The most difficult group axiom to verify is the associativity: For all points P, Q, and R on the curve, one has: $P + (Q + R) = (P + Q) + R$. The addition formula is directly deduced from the above rules. Let P and Q be two non-opposite points of the curve. The slope of the line (PQ) expressed in affine coordinates is:

$$\lambda = \frac{y_Q - y_P}{x_Q - x_P}.$$

The abscissa of the sum S of P and Q is given by:

$$x_S = \lambda^2 - x_P - x_Q.$$

The ordinate of S is:

$$y_S = -\lambda(x_S - x_P) - y_P = -\lambda^3 + 2\lambda x_P + \lambda x_Q - y_P.$$

If $P = Q$, the doubling formula is similar. The only difference is the expression of the slope of the tangent to the curve, which is:

$$\lambda = \frac{3x_P^2 + a}{2y_P}.$$

Projective coordinates

The computation of the slope λ in affine coordinates requires a division. The quotient is computed by the extended Euclidean algorithm, and this is an arithmetic operation considerably more costly than field multiplication. With homogeneous coordinates (X, Y, Z) in the projective plane, it is possible to avoid this division at the price of computing more products. Recall that the projective coordinates (X, Y, Z), with $Z \neq 0$, represent the affine point $(X/Z, Y/Z)$. Conversely, the affine point (x, y) admits homogeneous coordinates $(X, Y, 1)$. As a consequence, we can deduce the equations of the addition and doubling of points over an elliptic curve in projective coordinates from the equations in affine coordinates. Using projective coordinates involves faster computation for adding and doubling over an elliptic curve.

In projective coordinates, the equation of an elliptic curve over $\mathbb{F}_p$, with p a large prime number, is:

$$E(\mathbb{F}_p) = \{(X, Y, Z) \in \mathbb{F}_p^3, \text{ such that } Y^2 Z = X^3 + aXZ^2 + bZ^3.$$

The point at infinity, denoted 0_E, admits the following coordinates: $0_E = (0, 1, 0)$.
The opposite of the point $P = (X_P, Y_P, Z_P)$ is $-P = (X_P, -Y_P, Z_P)$.

Adding

Let $P = (X_P, Y_P, Z_P)$ and $Q = (X_Q, Y_Q, Z_Q)$ be two points of the curve given by homogeneous coordinates.

Let $A = Y_Q Z_P - Y_P Z_Q$, $B = X_Q Z_P - X_P Z_Q$, $C = A^2 Z_P Z_Q - B^3 - 2B^2 X_P Z_Q$.
Then, the three coordinates of the sum S of P and Q are given by:

$$
\begin{aligned}
X_S &= BC, \\
Y_S &= A(B^2 X_P Z_Q - C) - B^3 Y_P Z_Q, \\
Z_S &= B^3 Z_P Z_Q.
\end{aligned}
$$

The computation of the sum of two points requires 12 multiplications and 5 squares in the base field.

Remark 2.17   Mixed addition We denote by mixed addition an addition of two points P and Q, where one of the points is given in projective coordinates and the other in affine coordinates. In the case of a mixed addition, the cost of an addition of a point is 9 multiplications and 2 squares in the base field.

Doubling

Let $P = (X_P, Y_P, Z_P)$, a point of the curve given by homogeneous coordinates.

Let $A = aZ_P^2 + 3X_1^2$, $B = Y_P Z_P$, $C = X_P Y_P B$, and $D = A^2 - 8C$, then the three coordinates of the double S of P are given by:

$$\begin{aligned} X_S &= 2BD, \\ Y_S &= A(4C - D) - 8Y_P^2 B^2, \\ Z_S &= 8B^3. \end{aligned}$$

The computation of the double of a point requires 7 multiplications and 5 squares in the base field.

Remark 2.18 There exist other systems of coordinates over an elliptic curve, for instance, the Jacobian coordinates. The cost of a pairing computation depends on the system of coordinates. In Chapters 3 and 10, the best tradeoff between the cost of a pairing computation and the choice of system of coordinates is presented.

Jacobian coordinates

In Jacobian coordinates, the elliptic curve $E(\mathbb{F}_p)$ is given by

$$E(\mathbb{F}_p) = \{(X, Y, Z) \in \mathbb{F}_p^3, \text{ such that } Y^2 = X^3 + aXZ^4 + bZ^6.$$

The point at infinity is $0_E = (1, 1, 0)$.

The opposite of the point $P = (X_P, Y_P, Z_P)$ is $-P = (X_P, -Y_P, Z_P)$.

Let $P = (X_P, Y_P, Z_P)$ be a point given in Jacobian coordinates. The Jacobian coordinates are homogeneous following the rule: for λ, a non zero integer, then $P \equiv (\lambda^2 X_P, \lambda^3 Y_P, \lambda Z_P)$.

If $Z_P \neq 0$, the point P in Jacobian coordinates is equivalent to the point $(X_P/Z_P^2, Y_P/Z_P^3)$ in affine coordinates.

Adding

Let $P = (X_P, Y_P, Z_P)$ and $Q = (X_Q, Y_Q, Z_Q)$ be two points of the curve given by Jacobian coordinates.

Let $A = X_P Z_Q^2$, $B = X_Q Z_P^2$, $C = Y_P Z_Q^3$, $D = Y_Q Z_P^3$, $E = B - A$, and $F = D - C$.
Then, the three coordinates of the sum S of P and Q are given by:

$$\begin{aligned} X_S &= -E^3 - 2AE^2 + F^2, \\ Y_S &= -CE^3 + F(AE^2 - X_S), \\ Z_S &= EZ_P Z_Q. \end{aligned}$$

The computation of the sum of two points requires 12 multiplications and 4 squares in the base field.

Remark 2.19 **Mixed addition** We denote by mixed addition an addition of two points P and Q where one of the points is given in Jacobian coordinates and the other in affine coordinates. In the case of a mixed addition, the cost of an addition of a point is 8 multiplications and 3 squares in the base field.

Doubling

Let $P = (X_P, Y_P, Z_P)$ be a point of the curve given by Jacobian coordinates.

Let $A = 4X_P Y_P^2$, $B = 3X_P^2 + aZ_P^4$, then the three coordinates of the double S of P are given by:

$$\begin{aligned}
X_S &= -2A + B^2, \\
Y_S &= -8Y_P^4 + B(A - X_S), \\
Z_S &= 2Y_P Z_P.
\end{aligned}$$

The computation of the double of a point requires 4 multiplications and 6 squares in the base field.

Remark 2.20 When $a = -3$, a small optimization of the complexity of a doubling can be performed by computing $B = 3X_P^2 + aZ_P^4$ as $B = 3(X_P - Z_P)(X_P + Z_P)$. In this case, the computation of the double of a point requires 5 multiplications and 4 squares in the base field.

2.3.3 Isogeny over Elliptic Curves

DEFINITION 2.34 (Isogeny) Let E_1 and E_2 be two elliptic curves. Let $y^2 = x^3 + a_1 x + b_1$ be the equation of E_1 and $y^2 = x^3 + a_2 x + b_2$ be the equation of E_2. An isogeny from E_1 to E_2 is a rational mapping from E_1 to E_2 that fixes the point at infinity.

According to this definition, an isogeny is given by two rational fractions in the indeterminates x and y such that for each point of E_1, the obtained values define a point that belongs to E_2.

Example 2.24
1. For all integers n, the multiplication by n is an isogeny from a curve to itself.
2. The so-called Frobenius isogeny is defined on a curve over the finite field $\mathbb{F}_q$ by $(x, y) \mapsto \pi_q(x, y) = (x^q, y^q)$.
3. On a curve E of equation $y^2 = x^3 + b$, the mapping $(x, y) \mapsto (jx, y)$, where j is a primitive cubic root of unity, defines an isogeny of E.
4. On a curve E of equation $y^2 = x^3 + ax$, the mapping $(x, y) \mapsto (-x, iy)$, where i is an element of the field $\mathbb{F}$ such that $i^2 = -1$, defines an isogeny of E.

PROPOSITION 2.2 **Reduced form of an isogeny** *Any isogeny $\varphi(x, y)$ admits a reduced expression:*

$$\varphi(x, y) = \big(r(x), ys(x)\big),$$

where $r(x)$ and $s(x)$ are rational fractions over the field $\mathbb{F}$.

Example 2.25 The reduced form of the Frobenius isogeny is

$$\pi_q(x, y) = \big(x^q, y(x^3 + ax + b)^{(q-1)/2}\big).$$

2.3.4 Cardinality

The cardinality of an elliptic curve over a finite field $\mathbb{F}_q$, for q, for instance ref [6] or [18] of this chapter, a power of prime number p, can, be computed by an algorithm discovered by Schoof in 1985.

DEFINITION 2.35 (Trace) The trace of an elliptic curve over the finite field $\mathbb{F}_q$ is the integer t defined by

$$t = q + 1 - \mathrm{Card}\big(E(\mathbb{F}_q)\big).$$

Knowing the trace of an elliptic curve, the cardinality can be deduced from the trace by: $\text{Card}\big(E(\mathbb{F}_q)\big) = q + 1 - t$.

The Schoof algorithm computes the trace of an elliptic curve over the finite field $\mathbb{F}_q$. It is used to compute the number of points of the curve with coordinates in the field $\mathbb{F}_q$. The Schoof algorithm is presented in Algorithm 2.4, which is based on the following proposition:

ALGORITHM 2.4   Schoof's algorithm.

Input   : E an elliptic curve over a finite field $\mathbb{F}_q$

Output: The trace t of $E(\mathbb{F}_q)$

1 Choose an integer $\ell \geq 3$ and build an ℓ-torsion point $P = (x_P, y_P)$

2 We want to solve the characteristic equation of the Frobenius: $\pi_q^2(P) - [\vartheta]\pi - q(P) - (q \mod l)P = 0_E$. Compute $(x_1, y_1) = \big(x_P^{q^2}, y_P^{q^2}\big) + (q \mod \ell) \cdot (x_P, y_P)$.

3 Compute $\pi_q(P) = (x_P^q, y_P^q)$ and find by exhaustive search an integer ϑ modulo ℓ that satisfies:

$$\vartheta \cdot (x_P^q, y_P^q) = (x_1, y_1).$$

4 Thus the trace t of the curve is congruent to ϑ modulo ℓ.

5 Repeat the previous steps with different values of the integer ℓ, pairwise relatively prime, until their product is greater that $4\sqrt{q}$.

6 The Chinese Remainder Theorem [16] allows us to compute the trace t of the curve and conclude the algorithm.

PROPOSITION 2.3   **Frobenius** *Let E be an elliptic curve over the finite field $\mathbb{F}_q$. The mapping $\pi_q - [1] : P \mapsto \pi_q(P) - P$ is a separable isogeny whose degree is the cardinality of E over $\mathbb{F}_q$.*

THEOREM 2.9 (Hasse boundary)   *The trace t of an elliptic curve over the field $\mathbb{F}_q$ is bounded by:*

$$-2\sqrt{q} \leq t \leq 2\sqrt{q}.$$

Equivalently, the cardinality of the curve is bounded by:

$$q + 1 - 2\sqrt{q} \leq \text{Card}\big(E(\mathbb{F}_q)\big) \leq q + 1 + 2\sqrt{q}.$$

DEFINITION 2.36 (n-torsion group)   Let E be an elliptic curve over an algebraically closed field. For any non-zero integer n, the n-**torsion group** of the elliptic curve E is the kernel of the multiplication by n. It is denoted $E[n]$:

$$E[n] = \ker([n]) = \{P \in E \mid nP = P_\infty\}.$$

Even if the elliptic curve is defined over a finite field, the coordinates of the points of any n-torsion group are considered in the algebraic closure of the field. As any isogeny kernel, the n-torsion group is always finite.

Example 2.26

1. The 1-torsion group is the subgroup reduced to the neutral element P_∞.

2. The 2 torsion group is the set of points of the curve such that $P + P = P_\infty$, that is, points equal to their negate. They are the so-called *special points* of the curve:

$$E[2] = \{P_\infty, P_\alpha, P_\beta, P_\gamma\},$$

where $P_\alpha = (\alpha, 0)$, $P_\beta = (\beta, 0)$, $P_\gamma = (\gamma, 0)$, where α, β, γ are the three distinct roots of the equation $x^3 + ax + b = 0$.

3. The 3-torsion group is the set of the so-called *inflexion* points of the curve, satisfying $3P = P_\infty$.

THEOREM 2.10 (Structure of the n-torsion group) *[19, Corollary 6.4] Let E be an elliptic curve defined over a finite field of characteristic p.*

- *For any non-zero integer n prime with p, the n-torsion group of E is isomorphic to the direct product $\mathbb{Z}/n\mathbb{Z} \times \mathbb{Z}/n\mathbb{Z}$.*
- *If $n = p^e$, for e an integer, then either $E[n] \cong \{O\}$ for all $e = 1, 2, 3, \ldots$, or $E[n] \cong \mathbb{Z}/(p^e)\mathbb{Z}$ for all $e = 1, 2, 3, \ldots$*

COROLLARY 2.2 (Cardinality of the n-torsion group) *For any non-zero integer n, the n-torsion group of an elliptic curve contains n^2 points.*

Remark 2.21 (Base of the n-torsion group) For any non-zero integer n and any elliptic curve, there exist two points S and T of the curve such that any point P of the n-torsion group of the curve can be expressed as

$$P = aS + bT,$$

where a and b are two elements of $\mathbb{Z}/n\mathbb{Z}$.

THEOREM 2.11 (Cassel's theorem) *The group of points of an elliptic curve over a finite field $\mathbb{F}_q$ is either cyclic, or isomorphic to a direct product of two cyclic groups $\mathbb{Z}/n_1\mathbb{Z} \times \mathbb{Z}/n_2\mathbb{Z}$, where n_1 divides n_2.*

Let n be any non-zero integer and $\mathbb{F}_q$ any finite field. The multiplicative subgroup of the n-roots of unity in $\mathbb{F}_q$ is denoted by μ_n. It is by definition:

$$\mu_n = \{x \in \mathbb{F}_q \mid x^n = 1\}.$$

The embeding degree of μ_n relatively to $\mathbb{F}_q$ is the least degree of an extension of $\mathbb{F}_q$ that contains μ_n.

PROPOSITION 2.4 *Let n be an integer that divides the cardinality of an elliptic curve over the finite field $\mathbb{F}_q$ and such that n does not divide $q - 1$. The n-torsion group is included in the set of points of the elliptic curve whose coordinates belong to the extension of degree k of $\mathbb{F}_q$ if and only if n divides $q^k - 1$.*

Example 2.27 If the cardinality of the curve is $q + 1$, which means that the trace of the curve is zero, then n divides $q^2 - 1$. In this case, all the n-torsion points have coordinates in the field $\mathbb{F}_{q^2}$. The embedding degree of μ_n relative to $\mathbb{F}_q$ is 2.

2.3.5 Supersingular and Ordinary Elliptic Curves

Elliptic curves are divided into two families: the supersingular and ordinary elliptic curves [19, Chapter V §3 and 4]. The following definition highlights the difference between ordinary and supersingular elliptic curves.

DEFINITION 2.37 (Supersingular and ordinary elliptic curve)   Let p be a prime number greater than 3, $q = p^n$ for n a positive integer, and let $E(\mathbb{F}_q)$ be an elliptic curve over $\mathbb{F}_q$. The curve E is **supersingular** if the following assertions are verified:

- $\#E(\mathbb{F}_q) \equiv 1 \mod p$, or $\#E(\mathbb{F}_q) = q + 1 \mod p$, or $t = 0$,
- E does not admit a point of order p over $\overline{\mathbb{F}_q}$.

A non-supersingular elliptic curve is said to be **ordinary**.

Example 2.28
- The elliptic curve defined over $\mathbb{F}_{257}$ by the equation $y^2 = x^3 + b$, for $b \in \mathbb{F}_{257}^{\star}$, is a supersingular curve.
- More generally, an elliptic curve defined over $\mathbb{F}_p$ by the equation $y^2 = x^3 + x$ is supersingular if and only if $p \equiv 3 \mod 4$.

Remark 2.22   The supersingular elliptic curves are classified in [4]. The embedding degree of a supersingular elliptic curve is at the most 6.

The security level of pairing-based cryptography will be studied in the following chapters. We can now remark that in order to follow the recommendations of the NIST [1], the supersingular elliptic curves should be discarded.

2.3.6   Twist of an Elliptic Curve

Let E and E' be two elliptic curves. We said that E' is a twist of E if there is an isomorphism between E and E'. In this section, we will only give the necessary tools on twisted elliptic curves for pairing-based cryptography, i.e., the definition and a theorem describing explicitly the equations of possible twisted elliptic curves. A nice theoretical description of the twists is presented in [19], for instance.

DEFINITION 2.38 (Twist of an elliptic curve)   Let E and E' be two elliptic curves defined over $\mathbb{F}_q$, for q, a power of a prime number p. Then, the curve E' is a **twist of degree** d of E if we can define an isomorphism Ψ_d over $\mathbb{F}_{q^d}$ from E' into E and such that d is minimal:

$$\Psi_d : E'(\mathbb{F}_q) \quad \rightarrow \quad E(\mathbb{F}_{q^d}).$$

The possible number of twists for a given elliptic curve is bounded. It highly depends on the group of endomorphisms of the elliptic curve E. Theorem 2.12 gives the classification of the potential twists.

THEOREM 2.12   *Let E be an elliptic curve defined by the short Weierstrass equation $y^2 = x^3 + ax + b$ over an extension $\mathbb{F}_q$ of a finite field $\mathbb{F}_p$, for p a prime number, k a positive integer such that $q = p^k$. According to the value of k, the potential degrees for a twist are $d = 2, 3, 4$ or 6.*

In practice, we will consider that E is defined over $\mathbb{F}_{p^k}$. The curve E is considered in its short Weierstrass form, $y^2 = x^3 + ax + b$.

We will denote by E' the twisted elliptic curve of E defined over $\mathbb{F}_{p^{k/d}}$ a subfield of $\mathbb{F}_{p^k}$, for d a divisor of k. According to our definition, we have that $q = p^{k/d}$. We describe the possible twist using the notation $p^{k/d}$ as it will be the one used for the description of pairings. The equations of the potential twisted elliptic curve are described below. For each potential degree d, we give the corresponding equation of $E'(\mathbb{F}_{p^{k/d}})$, the twisted elliptic curve of E. We explicitly give the morphism $\Psi_d : E' \to E$.

- $d = 2$. Let $\nu \in \mathbb{F}_{p^{k/2}}$ be such that the polynomial $X^2 - \nu$ is irreducible over $\mathbb{F}_{p^{k/2}}$. The equation of E' over $\mathbb{F}_{p^{k/2}}$ is $E' : \nu y^2 = x^3 + ax + b$. The morphism Ψ_2 is given by:

$$\begin{aligned} \Psi_2 : E'(\mathbb{F}_{p^{k/2}}) &\to E(\mathbb{F}_{p^k}) \\ (x, y) &\to (x, y\nu^{1/2}). \end{aligned}$$

- $d = 4$. The elliptic curve E admits a twist of degree 4 if and only if $b = 0$. Let $\nu \in \mathbb{F}_{p^{k/4}}$ be such that the polynomial $X^4 - \nu$ is irreducible over $\mathbb{F}_{p^{k/4}}$. The equation of E' over $\mathbb{F}_{p^{k/4}}$ is $y^2 = x^3 + \frac{a}{\nu}x$. The morphism Ψ_4 is given by:

$$\begin{aligned} \Psi_4 : E'(\mathbb{F}_{p^{k/4}}) &\to E(\mathbb{F}_{p^k}) \\ (x, y) &\to (x\nu^{1/2}, y\nu^{3/4}). \end{aligned}$$

- $d = 3$ (respectively 6), the curve E admits a twist of degree 3 (respectively 6) if and only if $a = 0$. Let $\nu \in \mathbb{F}_{p^{k/d}}$ be such that the polynomial $X^3 - \nu$ (respectively $X^6 - \nu$) is irreducible over $\mathbb{F}_{p^{k/d}}$. The equation of E' is $y^2 = x^3 + \frac{b}{\nu}$. The morphism Ψ_3 is given by:

$$\begin{aligned} \Psi_d : E'(\mathbb{F}_{p^{k/d}}) &\to E(\mathbb{F}_{p^k}) \\ (x, y) &\to (x\nu^{1/3}, y\nu^{1/2}). \end{aligned}$$

We can compute the cardinal of a twisted elliptic curve according to the degree of the twist [14].

PROPOSITION 2.5 ([14])   *Let $E(\mathbb{F}_q)$ be an ordinary elliptic curve with $\mathrm{Card}(E(\mathbb{F}_q)) = q + 1 - t$, admitting a twist E' of degree d. Then the possible group orders of $E'(\mathbb{F}_q)$ are the following:*

$$\begin{aligned} d = 2 \quad & \mathrm{Card}(E(\mathbb{F}_q)) = q + 1 - t \\ d = 3 \quad & \mathrm{Card}(E(\mathbb{F}_q)) = q + 1 - (3f - t)/2 \quad && \text{with } t^2 - 4q = -3f^2 \\ & \mathrm{Card}(E(\mathbb{F}_q)) = q + 1 + (3f + t)/2 \quad && \text{with } t^2 - 4q = -3f^2 \\ d = 4 \quad & \mathrm{Card}(E(\mathbb{F}_q)) = q + 1 + f \quad && \text{with } t^2 - 4q = -f^2 \\ & \mathrm{Card}(E(\mathbb{F}_q)) = q + 1 - f \quad && \text{with } t^2 - 4q = -f^2 \\ d = 6 \quad & \mathrm{Card}(E(\mathbb{F}_q)) = q + 1 + (3f - t)/2 \quad && \text{with } t^2 - 4q = -3f^2 \\ & \mathrm{Card}(E(\mathbb{F}_q)) = q + 1 - (3f + t)/2 \quad && \text{with } t^2 - 4q = -3f^2. \end{aligned}$$

2.4   Conclusion

In this chapter, we present the elementary mathematical notions that will be used in the following chapters. We present definitions, properties and examples of groups, finite fields, and elliptic curves.

References

[1] SP-800-57. *Recommendation for Key Management – Part 1: General.* National Institute of Standards and Technology, U.S. Department of Commerce, July 2012.

[2]   Razvan Barbulescu, Pierrick Gaudry, Antoine Joux, and Emmanuel Thomé. A heuristic quasi-polynomial algorithm for discrete logarithm in finite fields of small characteristic. In P. Q. Nguyen and E. Oswald, editors, *Advances in Cryptology – EUROCRYPT 2014*, volume 8441 of *Lecture Notes in Computer Science*, pp. 1–16. Springer, Heidelberg, 2014.

[3]   Ian F. Blake, Gadiel Seroussi, and Nigel P. Smart. *Elliptic Curves in Cryptography*, volume 265 of *London Mathematical Society Lecture Notes Series*. Cambridge University Press, 1999.

[4]   Ian F. Blake, Gadiel Seroussi, and Nigel P. Smart, editors. *Advances in Elliptic Curve Cryptography*, volume 317 of *London Mathematical Society Lecture Notes Series*. Cambridge University Press, 2004.

[5]   Henri Cohen. *A Course in Computational Algebraic Number Theory*, volume 138 of *Graduate Texts in Mathematics*. Springer-Verlag, 4th printing, 2000.

[6]   Henri Cohen and Gerhard Frey, editors. *Handbook of Elliptic and Hyperelliptic Curve Cryptography*, volume 34 of *Discrete Mathematics and Its Applications*. Chapman & Hall/CRC, 2006.

[7]   M. Prem Laxman Das and Palash Sarkar. Pairing computation on twisted Edwards form elliptic curves. In S. D. Galbraith and K. G. Paterson, editors, *Pairing-Based Cryptography – Pairing 2008*, volume 5209 of *Lecture Notes in Computer Science*, pp. 192–210. Springer, Heidelberg, 2008.

[8]   Sylvain Duquesne and Emmanuel Fouotsa. Tate pairing computation on Jacobi's elliptic curves. In M. Abdalla and T. Lange, editors, *Pairing-Based Cryptography – Pairing 2012*, volume 7708 of *Lecture Notes in Computer Science*, pp. 254–269. Springer, Heidelberg, 2013.

[9]   Sylvain Duquesne, Nadia El Mrabet, and Emmanuel Fouotsa. Efficient computation of pairings on jacobi quartic elliptic curves. *J. Mathematical Cryptology*, 8(4):331–362, 2014.

[10] Steven D. Galbraith. *Mathematics of Public Key Cryptography*. Cambridge University Press, 2012.

[11] N. Jacobson. *Basic Algebra I: Second Edition*. Dover Books on Mathematics. Dover Publications, 2012.

[12] Antoine Joux and Vanessa Vitse. Cover and decomposition index calculus on elliptic curves made practical - application to a previously unreachable curve over $\mathbb{F}_{p^6}$. In D. Pointcheval and T. Johansson, editors, *Advances in Cryptology – EUROCRYPT 2012*, volume 7237 of *Lecture Notes in Computer Science*, pp. 9–26. Springer, Heidelberg, 2012.

[13] Antoine Joux and Vanessa Vitse. Elliptic curve discrete logarithm problem over small degree extension fields - application to the static Diffie-Hellman problem on $E(\mathbb{F}_{q^5})$. *Journal of Cryptology*, 26(1):119–143, 2013.

[14] Marc Joye and Gregory Neven, editors. *Identity-Based Cryptography*, volume 2 of *Cryptology and Information Security Series*. IOS press, 2009.

[15] Neal Koblitz and Alfred Menezes. Pairing-based cryptography at high security levels (invited paper). In N. P. Smart, editor, *Cryptography and Coding*, volume 3796 of *Lecture Notes in Computer Science*, pp. 13–36. Springer, Heidelberg, 2005.

[16] Rudolf Lidl and Harald Niederreiter. *Finite Fields*, volume 20 of *Encyclopedia of Mathematics and Its Applications*. Cambridge University Press, 2nd edition, 1997.

[17] Alfred Menezes. *Elliptic Curve Public Key Cryptosystems*. Kluwer Academic Publishers, 1993.

[18] Victor Shoup. *A Computational Introduction to Number Theory and Algebra*. Cambridge University Press, 2nd edition, 2009.

[19]   Joseph H. Silverman. *The Arithmetic of Elliptic Curves*, volume 106 of *Graduate Texts in Mathematics*. Springer-Verlag, 2nd edition, 2009.

[20]   Lawrence C. Washington. *Elliptic Curves: Number Theory and Cryptography*. Discrete Mathematics and Its Applications. Chapman & Hall/CRC, second edition, 2008.

[21]   Takanori Yasuda, Tsuyoshi Takagi, and Kouichi Sakurai. Application of scalar multiplication of Edwards curves to pairing-based cryptography. In G. Hanaoka and T. Yamauchi, editors, *IWSEC 12: 7th International Workshop on Security, Advances in Information and Computer Security*, volume 7631 of *Lecture Notes in Computer Science*, pp. 19–36. Springer, Heidelberg, 2012.

3

Pairings

Sorina Ionica
Université de Picardie Jules Verne

Damien Robert
*INRIA Bordeaux Sud-Ouest, Université de
Bordeaux*

We recall from Section 1.2 that a pairing is a map $e : \mathbb{G}_1 \times \mathbb{G}_2 \to \mathbb{G}_T$ between finite abelian groups $\mathbb{G}_1$, $\mathbb{G}_2$ and $\mathbb{G}_T$, that satisfies the following conditions:

- e is bilinear, which means that $e(P + Q, R) = e(P, R) \times e(Q, R)$ and $e(P, Q + R) = e(P, Q) \times e(P, R)$;

- e is non-degenerate, which means that for any $P \in \mathbb{G}_1$ there is a $Q \in \mathbb{G}_2$ such that $e(P, Q) \neq 1$, and for any $Q \in \mathbb{G}_2$ there is a $P \in \mathbb{G}_1$ such that $e(P, Q) \neq 1$.

A pairing e is suitable for use in cryptography when furthermore it is easy to compute, but difficult to invert. Inverting a pairing e means given $z \in \mathbb{G}_T$ to find $P \in \mathbb{G}_1$ and $Q \in \mathbb{G}_2$ such that $e(P, Q) = z$.

The most efficient cryptographic pairings currently known come from elliptic curves (or higher-dimensional algebraic varieties). Starting from an elliptic curve E defined over a finite field $\mathbb{F}_q$, we consider the Weil pairing and the Tate pairing associated to it. This allowed cryptographers to construct a map such that:

- $\mathbb{G}_1$ and $\mathbb{G}_2$ are subgroups of the rational points of E defined over an extension $\mathbb{F}_{q^k}$ of $\mathbb{F}_q$;

- $\mathbb{G}_T$ is the group $(\mathbb{F}_{q^k}^*, \times)$ where the group law is given by the field multiplication on $\mathbb{F}_{q^k}$ (or more precisely $\mathbb{G}_T = \mu_r \subset \mathbb{F}_{q^k}^*$ is the subgroup of r-th roots of unity);

- The pairing e can be efficiently computed using Miller's algorithm (see Algorithm 3.2);
- Currently the most efficient way to invert e is to solve the Diffie-Helman problem on $\mathbb{G}_1$, $\mathbb{G}_2$ or $\mathbb{G}_T$.

In this chapter we introduce pairings associated to an elliptic curve E over a finite field $\mathbb{F}_q$ and explain how to compute them efficiently, via an algorithm that evaluates functions on points of the curve. We first explain in Section 3.1 how to represent functions efficiently by looking at their associated divisors, and then give Miller's algorithm, which allows us to evaluate them.

In Section 3.2 we present the general theory of the Weil and Tate pairing, and we review the main recent optimizations for their computation: the Ate, twisted Ate, and optimal pairings, which are preferred in implementations nowadays. Finally, we give concrete formulae to compute them in practice in Section 3.3. Since the group $\mathbb{G}_T$ is a subgroup of the multiplicative group of $\mathbb{F}_{q^k}$, security requirements involve choosing a base field $\mathbb{F}_q$ with large characteristic (see [2] or Chapter 9).

For simplicity, in Sections 3.1 and 3.2, points on the elliptic curve are represented in affine coordinates. Using this representation, formulae for pairing computation are easy to write down. However, note that affine coordinates involve divisions and are not efficient for a practical implementation. We study more efficient representations of points in Section 3.3.

For the cryptographic usage of pairings, only a specific version of Miller's algorithm and the Weil and Tate pairing need to be presented. This is the version we give in Section 3.1 and 3.2, where we omit most proofs. For the sake of completness, we give the general version of pairings along with complete proofs in Section 3.4.

Notation

We recall that an elliptic curve defined over a field with characteristic greater than 5 can always be given in short Weierstrass form, as explained in Chapter 2. In the remainder of this chapter, all elliptic curves are defined over a field K with characteristic greater than 5 and will be given by a short Weierstrass equation. We denote this equation by $y^2 = H(x)$, with $H(x) = x^3 + ax + b$ and $a, b \in K$.

3.1 Functions, Divisors, and Miller's Algorithm

3.1.1 Functions and Divisors on Curves

Pairing computations will rely crucially on evaluating functions on points of elliptic curves. A convenient way to represent functions is by their divisor. We first give a gentle introduction to the theory of divisors by looking at examples of functions on the line before considering elliptic curves.

Let $\mathbb{A}^1$ be the affine line over an algebraically closed field $\overline{K}$. Adding the point at infinity means that we work on the projective line $\mathbb{P}^1 = \mathbb{A}^1 \cup \{\infty\}$. Rational functions $\overline{K}(\mathbb{P}^1)$ on $\mathbb{A}^1$ are simply the rational functions $\overline{K}(t)$. Let $f = P/Q = \dfrac{\prod (t - x_i)^{n_i}}{\prod (t - y_i)^{m_i}} \in \overline{K}(t)$ be such a rational function, where the numerator P and the denominator Q are assumed to be prime with each other. Then the points x_i are *zeroes* of f with multiplicity m_i and the points y_i are *poles* of f with multiplicity n_i. This allows us to define a multiplicity $\mathrm{ord}_x(f)$ for every point $x \in \mathbb{P}^1(\overline{K})$

$$\mathrm{ord}_x(f) = \begin{cases} n & \text{if } x \text{ is a zero of } f \text{ with multiplicity } n, \\ -n & \text{if } x \text{ is a pole of } f \text{ of multiplicity } n, \\ 0 & \text{if } x \text{ is neither a zero nor a pole.} \end{cases}$$

For example, f has no pole in $\mathbb{A}^1$ if and only if it is a polynomial $P(t)$.

Given a rational function $f = P/Q$ as above, we can also define the evaluation of f on the point at infinity ∞. Here is how to compute the evaluation of f at ∞: the change of variables $u = 1/t$ sends ∞ to 0. Define g by $g(u) = f(1/u)$. This gives the relation $f(t) = g(u)$ when $t = 1/u$. We can then define the *order* of f at ∞ as the order of g at 0, and when the order is 0 we can define the *value* of f at ∞ as the value of g at 0. One can then easily check that $\text{ord}_\infty(f) = -\deg f = \deg Q - \deg P$.

We associate a formal sum to a function f:

$$\text{div}(f) = \sum_{x \in \mathbb{P}^1(\overline{K})} \text{ord}_x(f)[x],$$

where we use the notation $[x]$ to represent the point $x \in \mathbb{P}^1(\overline{K})$ in the formal sum. This formal sum is called the divisor of f. Since there is only a finite number of poles or zeroes, it is in fact finite. Moreover, f is characterized by $\text{div}(f)$ up to the multiplication by a constant: If f_1 and f_2 are two rational functions such that $\text{div}(f_1) = \text{div}(f_2)$, then f_1 and f_2 have the same poles and zeroes, so they differ by a multiplicative constant.

More generally, a divisor D is defined to be a formal sum of a finite number of points:

$$D = \sum_{x \in \mathbb{P}^1(\overline{K})} n_i[x_i].$$

To a divisor D one can associate its degree $\deg(D) = \sum n_i$. By the remark above concerning the multiplicity of f at ∞, we get that $\deg \text{div}(f) = 0$. Conversely, given a divisor D of degree 0, it is easy to construct a rational function f such that $\text{div}(f) = D$.

The whole theory extends when we replace the line $\mathbb{P}^1$ by a (geometrically connected smooth) curve C. If $P \in C(\overline{K})$ is a point of C, then there is always a uniformizer t_P, which is a rational function on C with a simple zero at P. Thus if $f \in \overline{K}(C)$ is a rational function on C, then we can always write $f = t_P^m \cdot g$ where g is a function having no poles or zeroes at P. We then define the *multiplicity* $\text{ord}_P(f)$ of f at P to be m. If the multiplicity $\text{ord}_P(f)$ is zero, that is if P is neither a pole nor a zero of f, then one can define the *value* of f at P to be $f(P)$.

In the case of the projective line $\mathbb{P}^1$, a uniformizer at x is $t - x$ and a uniformizer at ∞ is $u = \frac{1}{t}$. Hence this new notion of multiplicity coincides with the one introduced above.

For an elliptic curve E we have the following uniformizers:

- $t_P = x - x_P$, except when $H(x_P) = 0$;
- $t_P = y - y_P$, except when $H'(x_P) = 0$;
- $t_{0_E} = x/y$.

We denote by $\text{Disc}\,P$ the discriminant of a polynomial P, and we recall that the discriminant is non-zero if and only if P does not admit a double root. Since E is an elliptic curve, $\text{Disc}\,H \neq 0$ and we cannot have $H(x_P) = 0$ and $H'(x_P) = 0$ at the same time. Hence there is indeed a uniformizer for every point $P \in E(\overline{K})$.

One can also define a *divisor* on E as a formal finite sum of geometric points $D = \sum n_i[P_i]$ of E, and associate to a rational function $f \in \overline{K}(E)$ a divisor $\text{div}(f) = \sum_{P \in E(\overline{K})} \text{ord}_P(f)[P]$. One can check that $\text{ord}_P(f) = 0$ for all but a finite number of P, so we get a well-defined divisor. The degree $\deg D$ of a divisor $D = \sum n_i[P_i]$ is $\sum n_i$. A divisor D is said to be *principal* when there exists a function f such that $D = \text{div}(f)$. Two divisors D_1 and D_2 are said to be *linearly equivalent* when there exists a function f such that $D_1 = D_2 + \text{div}(f)$. It is easy to check that a divisor D is principal if and only if it is equivalent to the zero divisor, and that two divisors D_1 and D_2 are linearly equivalent if and only if $D_1 - D_2$ is linearly equivalent to the zero divisor.

PROPOSITION 3.1 *Let E be an elliptic curve over an algebraically closed field $\overline{K}$.*

1. *Let $f, g \in \overline{K}(E)$ be two rational functions, then $\mathrm{div}(f) = \mathrm{div}(g)$ if and only g differs from f by a multiplicative constant;*

2. *If $f \in \overline{K}(E)$ is a rational function, then $\mathrm{div}(f)$ is a divisor of degree 0;*

3. *Conversely, if $D = \sum n_i[P_i]$ is a divisor on E of degree 0, then D is the divisor of a function $f \in \overline{K}(E)$ (i.e., D is a principal divisor) if and only if $\sum n_i P_i = 0_E \in E(\overline{K})$ (where the last sum is not formal but comes from the addition on the elliptic curve).*

Proof. See [22, Proposition 3.4]. In fact, for the last item, given a divisor $D = \sum n_i[P_i]$ of degree 0, we give in Section 3.4.2 an explicit algorithm that constructs a rational function f such that $D = [P] - [0_E] + \mathrm{div}(f)$ and $P = \sum n_i P_i \in E(\overline{K})$. If $P = 0_E$ then $D = \mathrm{div}(f)$ is a principal divisor. It remains to show that if $P \neq 0_E$ then the divisor $[P] - [0_E]$ is not principal. But if we had a function f such that $\mathrm{div}(f) = [P] - [0_E]$, then the morphism $E \to \mathbb{P}^1_{\overline{K}} : x \mapsto (1 : f(x))$ associated to f would be birational. (Indeed, since f has one simple zero and one simple pole, one could get every degree of zero divisors as the divisor of a suitable rational function of f. So the function field of $k(E)$ would be $k(f)$.) But this is absurd: E is an elliptic curve so it has genus 1, it cannot have genus 0. □

An elliptic curve E defined over $\mathbb{F}_q$ can also be seen as an elliptic curve $E_{\overline{\mathbb{F}_q}}$ over the algebraic closure $\overline{\mathbb{F}_q}$. We say that a divisor $D = \sum n_i[P_i]$ of $E_{\overline{\mathbb{F}_q}}$ is rational when it is invariant under the action of the Frobenius endomorphism π. If $f \in \mathbb{F}_q(E)$ is a rational function defined over $\mathbb{F}_q$, then $\mathrm{div}(f)$ is rational. Conversely, if $f \in \overline{\mathbb{F}_q}(E)$ has a rational divisor $\mathrm{div}(f)$, then there exists a non-zero constant λ such that $\lambda f \in \mathbb{F}_q(E)$ [22, Chapter II §2].

3.1.2 Miller's Algorithm

Let F be a principal divisor. Then by definition there is a rational function f on E such that $F = \mathrm{div} f$. Then f is uniquely determined up to a constant. If 0_E is neither a pole nor a zero of f, then one can uniquely define f by requiring that $f(0_E) = 1$. More generally, we can define the normalized function associated to a principal divisor as follows: Since $\mathrm{ord}_{0_E}(x/y) = 1$, $(x/y)^{\mathrm{ord}_{0_E}(f)}$ has the same order at 0_E as f. In particular, the function $\left(\frac{f}{(x/y)^{\mathrm{ord}_{0_E}(F)}}\right)$ is defined at 0_E, and we can normalize f uniquely by requiring that the above function has value 1 at 0_E. This gives the following definition.

DEFINITION 3.1 Let F be a principal divisor. We define f_F to be the unique function such that $F = \mathrm{div} f_F$ and $\left(\frac{f_F}{(x/y)^{\mathrm{ord}_{0_E}(F)}}\right)(0_E) = 1$. Such a function is called normalized at 0_E (or simply normalized). If F is rational, then f_F is rational too.

If P and Q are points on E, then $[P] + [Q] - [P+Q] - [0_E]$ is principal. Indeed, it has degree 0 and $P + Q - 2 0_E - (P+Q) + 0_E = 0_E$, so by Proposition 3.1 there exists a function $\mu_{P,Q}$ such that $\mathrm{div}(\mu_{P,Q}) = [P] + [Q] - [P+Q] - [0_E]$.

DEFINITION 3.2 We denote by $\mu_{P,Q}$ the normalized function with principal divisor $[P] + [Q] - [P+Q] - [0_E]$.

If E is given by a short Weierstrass equation, we can construct $\mu_{P,Q}$ explicitly: If $P = -Q$, then $P + Q = 0_E$, and we can choose

$$\mu_{P,Q} = x - x_P. \tag{3.1}$$

Otherwise, let $l_{P,Q}$ be the line going through P and Q (if $P = Q$ then we take $l_{P,Q}$ to be the tangent to the elliptic curve at P). Then by definition of the addition law on E, we have that $\mathrm{div}(l_{P,Q}) = [P] + [Q] + [-P-Q] - 3[0_E]$. Now let $v_{P,Q} = x - x_{P+Q}$ be the vertical line going through $P + Q$ and $-P-Q$. Then $\mathrm{div}(v_{P,Q}) = [P+Q] + [-P-Q] - 2[0_E]$, so that $\mathrm{div}(\frac{l_{P,Q}}{v_{P,Q}}) = [P] + [Q] - [P+Q] - [0_E]$ and one can take $\mu_{P,Q} = \frac{l_{P,Q}}{v_{P,Q}}$.

To compute x_{P+Q}, we know that $-P-Q$ is the third intersection point between the line $l_{P,Q} : y = \alpha x + \beta$ and the elliptic curve $E : y^2 = x^3 + ax + b$. So x_{-P-Q}, x_P, x_Q are all roots of the degree-three equation $x^3 + ax + b - (\alpha x + \beta)^2 = 0$, and we get that $x_{P+Q} = x_{-P-Q} = \alpha^2 - x_P - x_Q$. Putting everything together, we finally obtain

$$\mu_{P,Q} = \frac{y - \alpha(x - x_P) - y_P}{x + (x_P + x_Q) - \alpha^2}, \tag{3.2}$$

with $\alpha = \frac{y_P - y_Q}{x_P - x_Q}$ when $P \neq Q$ and $\alpha = \frac{H'(x_P)}{2y_P}$ when $P = Q$.

One can check that the functions $\mu_{P,Q}$ defined above are normalized (see Section 3.4). Let $R \in E$. The following lemma explains how to evaluate $\mu_{P,Q}$ on R (in the usual cases encountered in cryptographic applications, we refer to Lemma 3.4 for the remaining cases).

LEMMA 3.1 (Evaluating $\mu_{P,Q}$) *Let* $P = (x_P, y_P)$, $Q = (x_Q, y_Q)$, *and* $R = (x_R, y_R)$ *be points on* E.

- *Suppose that* P, Q, *and* $P + Q$ *are all different from* 0_E. *Then* $\mu_{P,Q} = \frac{l_{P,Q}}{v_{P,Q}}$ *where* $l_{P,Q} = y - \alpha x - \beta$ *with* $\alpha = \frac{y_P - y_Q}{x_P - x_Q}$ *when* $P \neq Q$ *and* $\alpha = H'(x_P)$ *when* $P = Q$, $\beta = y_P - \alpha x_P = y_Q - \alpha x_Q$ *and* $v_{P,Q} = x - x_{P+Q}$ *with* $x_{P+Q} = \alpha^2 - x_P - x_Q$. *Assume that* R *is not equal to* P, Q, $P+Q$, $-P-Q$, *or* 0_E; *then we have*

$$\mu_{P,Q}(R) = \frac{y_R - \alpha x_R - \beta}{x_R - x_{P+Q}}. \tag{3.3}$$

  (If $R = -P-Q$ *and* $-P-Q \neq P, Q, P+Q, 0_E$ *then* $\mu_{P,Q}$ *is well defined on* R, *but computing the exact value requires more work; see Lemma 3.4 for the formula.)*
- *If* $P = -Q$ *(but* $P \neq 0_E$*) so that* $P + Q = 0_E$, *then* $\mu_{P,Q} = x - x_P$. *Assume that* R *is different from* 0_E, *then* $\mu_{P,Q}(R) = x_R - x_P$.
- *If* $P = 0_E$ *or* $Q = 0_E$, *then* $\mu_{P,Q} = 1$.

Let $P \neq 0_E$ be a point of r-torsion on E. Then $r[P] - r[0_E]$ is a principal divisor [22, Corollary III.3.5]. As a consequence, we have the following definition.

DEFINITION 3.3 We denote by $f_{r,P}$ the normalized function with principal divisor $r[P] - r[0_E]$.

All pairing computations will involve the following key computation: *Given* $P \neq 0_E$, *a point of* r-torsion on E, *and* $Q \neq P, 0_E$, *a point of the elliptic curve, evaluate* $f_{r,P}(Q)$. To explain how to compute $f_{r,P}$, first we need to extend its definition.

DEFINITION 3.4 Let $\lambda \in \mathbb{N}$ and $P \in E(K)$; we define $f_{\lambda,P} \in K(E)$ to be the function normalized at 0_E such that

$$\mathrm{div}(f_{\lambda,P}) = \lambda[P] - [\lambda P] - (\lambda - 1)[0_E].$$

Note that if $r \in \mathbb{N}$ and $P \in E[r]$, then $f_{r,P}$ is indeed the normalized function with divisor $r[P] - r[0_E]$.

PROPOSITION 3.2   *Let P be as above, and $\lambda, \nu \in \mathbb{N}$. We have*

$$f_{\lambda+\nu,P} = f_{\lambda,P} f_{\nu,P} \mathbf{f}_{\lambda,\nu,P},$$

where $\mathbf{f}_{\lambda,\nu,P} = \mu_{\lambda P,\nu P}$ is the function associated to the divisor $[(\lambda + \nu)P] - [\lambda P] - [\nu P] + [0_E]$ and normalized at 0_E.

Proof. We have seen in Lemma 3.1 that the function $\mu_{\lambda X,\nu X}$ defined in Equations (3.1) and (3.2) is normalized and has for associated divisor $[(\lambda+\nu)X] - [(\lambda)X] - [(\nu)X] + [0_E]$. By definition of $f_{\lambda,X}$, we have that $\operatorname{div}(f_{\lambda+\nu,X}) = (\lambda+\nu)[X] - [(\lambda+\nu)X] - (\lambda+\nu-1)[0_E] = \lambda[X] - [\lambda X] - (\lambda - 1)[0_E] + \nu[X] - [\nu X] - (\nu - 1)[0_E] + [(\lambda+\nu)X] - [\lambda X] - [\nu X] + [0_E] = \operatorname{div}(f_{\lambda,X} f_{\nu,X} \mathbf{f}_{\lambda,\nu,X})$. So $f_{\lambda+\nu,X} = f_{\lambda,X} f_{\nu,X} \mathbf{f}_{\lambda,\nu,X}$, since they have the same associated divisor and are both normalized at 0_E. □

Proposition 3.2 is the main ingredient that we need to compute $f_{r,P}$, using a double-and-add algorithm, whose pseudocode is described in Algorithm 3.2. Here is how this algorithms works: given $P \in E[r]$, we compute rP as we would with a standard double-and-add algorithm. If the current point is $T = \lambda P$, then at each step in the loop we perform a doubling $T \mapsto 2T$, and whenever the current bit of r is a 1, we also do an extra addition $T \mapsto T + P$. The only difference between Miller's algorithm and scalar multiplication is that, at each step in the Miller loop, we also keep track of the function $f_{\lambda,P}$ (corresponding to the principal divisor $\lambda[P] - [T] - (\lambda-1)[0_E]$). During the doubling and addition step we increment this function using Proposition 3.2, until in the end we obtain $f_{r,P}$, which we can evaluate on Q. Note that in practice we do the evaluations directly at each step because representing the full function $f_{r,P}$ would be too expensive.

Remark 3.1

- One should be careful that at the last step, the sum (whether it is a doubling or an addition) gives 0_E, so the corresponding Miller function is simply $x - x_T$.
- There is one drawback in evaluating the intermediate Miller functions $\mu_{\lambda P,\nu P}$ directly on Q: If $Q \notin \{0_E, P\}$, then $f_{r,P}(Q)$ is well defined. But if Q is a zero or pole of $\mu_{\lambda P,\nu P}$, then Algorithm 3.1 fails to give the correct result. A solution to compute $f_{r,P}(Q)$ anyway is to change the addition chain used to try to get other Miller functions $\mu_{\lambda P,\nu P}$ that do not have a pole or zero on Q. Another solution is given in Section 3.4. We note that this situation can happen only when Q is a multiple of P.

Using Lemma 3.1, we get an explicit version of Algorithm 3.1, for an elliptic curve $y^2 = x^3 + ax + b$. For efficiency reasons, we only do one division at the end.

3.2   Pairings on Elliptic Curves

3.2.1   The Weil Pairing

The first pairing on elliptic curves has been defined by Weil. Although it is usually not used in practice for cryptography (rather than the Tate pairing), it is important for historical reasons, and also because the original construction of the Tate pairing uses the Weil pairing.

ALGORITHM 3.1 Miller's algorithm (general version).

Input: $r \in \mathbb{N}$, $I = [\log r]$, $P = (x_P, y_P) \in E[r](K)$, $Q = (x_Q, y_Q) \in E(K)$.

Output: $f_{r,P}(Q)$.

1. Compute the binary decomposition: $r := \sum_{i=0}^{I} b_i 2^i$. Let $T = P, f = 1$.

2. For i in $[I - 1..0]$ compute

  (a) $f = f^2 \mu_{T,T}(Q)$;

  (b) $T = 2T$;

  (c) If $b_i = 1$, then compute

      i. $f = f \mu_{T,P}(Q)$;

      ii. $T = T + P$.

Return f.

THEOREM 3.1 *Let E be an elliptic curve defined over a finite field K, $r \geq 2$ an integer prime to the characteristic of K and P and Q two points of r-torsion on E. Then*

$$e_{W,r} = (-1)^r \frac{f_{r,P}(Q)}{f_{r,Q}(P)} \qquad (3.4)$$

is well defined when $P \neq Q$ and $P, Q \neq 0_E$. One can extend the application to the domain $E[r] \times E[r]$ by requiring that $e_{W,r}(P, 0_E) = e_{W,r}(0_E, P) = e_{W,r}(P, P) = 1$. Furthermore, the application $e_{W,r} : E[r] \times E[r] \to \mu_r$ obtained in this way is a pairing, called the Weil pairing. The pairing $e_{W,r}$ is alternate, which means that $e_{W,r}(P, Q) = e_{W,r}(Q, P)^{-1}$.

Proof. See [22, Section III.8] or Section 3.4.3. □

Note that the Weil pairing is defined over any field K of characteristic prime to r, and takes its values in $\mu_r \subset \overline{K}$. For cryptographic applications, we consider $K = \mathbb{F}_q$, with q a prime number, and we define the *embedding degree* k to be such that $\mathbb{F}_{q^k}$ is the smallest field containing μ_r. In other words, $\mathbb{F}_{q^k} = \mathbb{F}_q(\mu_r)$ or alternatively, k is the smallest integer such that $r \mid q^k - 1$.

Computing the Weil pairing

To compute the Weil pairing in practice we use Algorithm 3.2 twice to compute $f_{r,P}(Q)$ and $f_{r,Q}(P)$. Note that in this case, by Remark 3.1, whenever Miller's algorithm fails because we have an intermediate zero or pole, then Q is a multiple of P so $e_{W,r}(P, Q) = 1$. Indeed, if $Q = \lambda P$ then $e_{W,r}(P, Q) = e_{W,r}(P, P)^\lambda = 1$ because $e_{W,r}(P, P) = 1$ ($e_{W,r}$ is alternate).

3.2.2 The Tate Pairing

The Tate pairing was defined by Tate for number fields in [24, 18] and used by Frey and Rück in the case of finite fields [7]. For simplicity, we assume that $K = \mathbb{F}_q$, with q prime, and that k is the embedding degree corresponding to r (although the construction is valid for any finite field).

ALGORITHM 3.2   Miller's algorithm for affine short Weierstrass coordinates.

Input: $r \in \mathbb{N}$, $I = [\log r]$, $P = (x_P, y_P) \in E[r](K)$, $Q = (x_Q, y_Q) \in E(K)$.

Output: $f_{r,P}(Q)$.

1. Compute the binary decomposition: $r := \sum_{i=0}^{I} b_i 2^i$. Let $T = P$, $f_1 = 1$, $f_2 = 1$.

2. For i in $[I - 1..0]$, compute (except at the last step)

    (a) $\alpha = \frac{3x_T^2 + a}{2y_T}$, the slope of the tangent of E at T;

    (b) $x_{2T} = \alpha^2 - 2x_T$, $y_{2T} = -y_T - \alpha(x_{2T} - x_T)$;

    (c) $f_1 = f_1^2(y_Q - y_T - \alpha(x_Q - x_T))$, $f_2 = f_2^2(x_Q + 2x_T - \alpha^2)$;

    (d) $T = 2T$.

    (e) If $b_i = 1$, then compute

        i. $\alpha = \frac{y_T - y_P}{x_T - x_P}$, the slope of the line going through P and T;

        ii. $x_{T+P} = \alpha^2 - x_T - x_P$, $y_{T+P} = -y_T - \alpha(x_{T+P} - x_T)$;

        iii. $f_1 = f_1(y_Q - y_T - \alpha(x_Q - x_T))$, $f_2 = f_2(x_Q + (x_P + x_T) - \alpha^2)$;

        iv. $T = T + P$.

3. At the last step: $f_1 = f_1(x_Q - x_T)$.

Return
$$\frac{f_1}{f_2}.$$

THEOREM 3.2   *Let E be an elliptic curve, r a prime number dividing $\#E(\mathbb{F}_q)$, $P \in E[r](\mathbb{F}_{q^k})$ a point of r-torsion defined over $\mathbb{F}_{q^k}$, and $Q \in E(\mathbb{F}_{q^k})$ a point of the elliptic curve defined over $\mathbb{F}_{q^k}$, Let R be any point in $E(\mathbb{F}_{q^k})$ such that $\{R, Q + R\} \cap \{P, 0_E\} = \emptyset$. Then*

$$e_{T,r}(P,Q) = \left(\frac{f_{r,P}(Q + R)}{f_{r,P}(R)} \right)^{\frac{q^k - 1}{r}} \tag{3.5}$$

is well defined and does not depend on R.

Furthermore, the application

$$E[r](\mathbb{F}_{q^k}) \times E(\mathbb{F}_{q^k})/rE(\mathbb{F}_{q^k}) \rightarrow \mu_r$$
$$(P,Q) \mapsto e_{T,r}(P,Q)$$

is a pairing, called the Tate pairing.

Proof. See [7]. We give an elementary proof in Section 3.4.4 when all the r-torsion is rational over $\mathbb{F}_{q^k}$. $\square$

When $E(\mathbb{F}_{q^k})$ does not contain a point of r^2-torsion (which is always the case in the cryptographic setting because r is a large prime), then the Tate pairing restricted to the r-torsion is also non-degenerate.

PROPOSITION 3.3   *Assume that $E[r] \subset E(\mathbb{F}_{q^k})$ and that there are no points of r^2-torsion in $E(\mathbb{F}_{q^k})$. Then the inclusion $E[r](\mathbb{F}_{q^k}) \subset E(\mathbb{F}_{q^k})$ induces an isomorphism $E[r] \simeq E(\mathbb{F}_{q^k})/rE(\mathbb{F}_{q^k})$*

so the Tate pairing $e_{T,r}$ is a non-degenerate pairing

$$E[r] \times E[r] \to \mu_r.$$

Proof. Suppose that $P \in E[r](\mathbb{F}_{q^k})$ is equivalent to 0 in $E(\mathbb{F}_{q^k})/rE(\mathbb{F}_{q^k})$. Then by definition there exists a point $P_0 \in E(\mathbb{F}_{q^k})$ such that $P = rP_0$. This means that P_0 is a point of r^2-torsion. By hypothesis there are no non-trivial points of r^2-torsion in $E(\mathbb{F}_{q^k})$, hence we deduce that $E[r] \to E(\mathbb{F}_{q^k})/rE(\mathbb{F}_{q^k})$ is injective. Since both groups have cardinality r^2 (this is shown in the proof of Theorem 3.11), the injection is an isomorphism. □

Computing the Tate pairing

In practice, to compute the Tate pairing, when Q is not a multiple of P one can take $R = 0_E$ so that

$$e_{T,r}(P,Q) = f_{r,P}(Q)^{\frac{q^k-1}{r}}. \tag{3.6}$$

(We can't apply Theorem 3.2 directly with $R = 0_E$, but Theorem 3.11 will show that formula (3.6) is correct). We use Algorithm 3.2 to compute $f_{r,P}(Q)$ and then we do the final exponentiation by a fast exponentiation algorithm. By Remark 3.1 there are no problems during the execution of Miller's algorithm.

Unlike for the Weil pairing, $e_{T,r}(P,P)$ may not be trivial, so if we want to compute $e_{T,r}(P,P)$, or $e_{T,r}(P,Q)$ with Q a multiple of P, then we need to use Equation 3.18 with R a random point in $E(\mathbb{F}_{q^k})$. If we are unlucky and get an intermediate zero or pole, we restart the computation with another random R. An alternative method is to use the general Miller's algorithm described in Section 3.4.2 to compute the Tate pairing.

3.2.3   Using the Weil and the Tate Pairing in Cryptography

For the applications of the Weil and Tate pairing to cryptography, we will always consider an elliptic curve E defined over $\mathbb{F}_q$ and a large prime number r such that $r \mid \#E(\mathbb{F}_q)$. When the embedding degree k is greater than one, then $E[r]$ is defined over $\mathbb{F}_{q^k}$, and we can define two subgroups $\mathbb{G}_1$ and $\mathbb{G}_2$ of interest for pairing computations.

LEMMA 3.2 (The central setting for cryptography)   *Let E be an elliptic curve defined over $\mathbb{F}_q$, r a large prime number such that $r \mid \#E(\mathbb{F}_q)$, and π_q the Frobenius endomorphism. Let k be the embedding degree relative to r, and assume that $k > 1$. Then $E[r] = \mathbb{G}_1 \times \mathbb{G}_2 \subset E(\mathbb{F}_{q^k})$ where*

$$\mathbb{G}_1 = E[r](\mathbb{F}_q) = \{P \in E[r] \mid \pi_q P = P\}, \tag{3.7}$$

$$\mathbb{G}_2 = \{P \in E[r] \mid \pi_q P = [q]P\}. \tag{3.8}$$

$\mathbb{G}_1$ is called the rational subgroup of $E[r]$, while $\mathbb{G}_2$ is called the trace zero subgroup.

Proof. The characteristic polynomial of the Frobenius modulo r is the degree-two polynomial $X^2 - tX + q$ modulo r where t is the trace. Let λ_1 and λ_2 be the two eigenvalues. Since $r \mid \#E(\mathbb{F}_q)$, there is a rational point of r-torsion in $E(\mathbb{F}_q)$ so that $\lambda_1 = 1$. This implies that $\lambda_2 = q$. Furthermore, since $k > 1$, then $q \neq 1 \pmod{r}$. The two eigenvalues are then distinct, so the action of π_q on $E[r]$ is diagonalisable, and we have

$$E[r] = \mathrm{Ker}(\pi_q - \mathrm{Id}) \oplus \mathrm{Ker}(\pi_q - q\,\mathrm{Id}) = \mathbb{G}_1 \oplus \mathbb{G}_2.$$

Furthermore, let ϕ be the endomorphism given by the trace of the Frobenius (i.e., $\phi = 1 + \pi_q + \cdots + \pi_q^{k-1}$). Then ϕ acts on $\mathbb{G}_1$ by multiplication by k (which in the cryptographic setting will be prime to r), and on $\mathbb{G}_2$ the trace acts by multiplication by $\frac{q^k-1}{q-1}$. Since the embedding degree k is greater than 1 by hypothesis, then $r \mid q^k - 1$ and $r \nmid q-1$. Hence $r \mid \frac{q^k-1}{q-1}$. We conclude that the trace restricted to $E[r]$ has $\mathbb{G}_2$ as kernel and $\mathbb{G}_1$ as image [3, 4]. This explains the name trace zero subgroup for $\mathbb{G}_2$.   □

In practice, when using pairing-friendly elliptic curves to compute pairings for cryptographic applications, we will always be in the situation of Lemma 3.2. It will be convenient to restrict the Tate pairing to the subgroups $\mathbb{G}_1$ and $\mathbb{G}_2$ rather than to deal with the full r-torsion. Under some additional hypotheses (which always hold in the cryptographic setting), the Tate pairing restricted to $\mathbb{G}_1 \times \mathbb{G}_2$ or to $\mathbb{G}_2 \times \mathbb{G}_1$ is non-degenerate.

PROPOSITION 3.4 *Assume that we are in the situation of Lemma 3.2. Then the restriction of $e_{W,r}$ to $\mathbb{G}_1 \times \mathbb{G}_2$ or to $\mathbb{G}_2 \times \mathbb{G}_1$ is non-degenerate. If, furthermore there are no points of r^2-torsion in $E(\mathbb{F}_{q^k})$, then the restriction of $e_{T,r}$ to $\mathbb{G}_1 \times \mathbb{G}_2$ or to $\mathbb{G}_2 \times \mathbb{G}_1$ is also non-degenerate. More generally, if $\mathbb{G}_3$ is any cyclic subgroup of $E[r]$ different from $\mathbb{G}_1$ and $\mathbb{G}_2$, then the Weil and Tate pairing restricted to $\mathbb{G}_1 \times \mathbb{G}_3$, $\mathbb{G}_3 \times \mathbb{G}_1$, $\mathbb{G}_2 \times \mathbb{G}_3$, and $\mathbb{G}_3 \times \mathbb{G}_2$ are non-degenerate.*

Proof. Note that the Weil pairing is non-degenerate on $E[r]$, but is trivial on $\mathbb{G}_1 \times \mathbb{G}_1$ and $\mathbb{G}_2 \times \mathbb{G}_2$ (because these groups are cyclic and the Weil pairing is alternate). Then since $E[r] = \mathbb{G}_1 \times \mathbb{G}_2$, the Weil pairing has to be non-degenerate on $\mathbb{G}_1 \times \mathbb{G}_2$ and $\mathbb{G}_2 \times \mathbb{G}_1$. Given $P \in \mathbb{G}_1$ there exists $Q \in \mathbb{G}_2$ such that $e_{W,r}(P,Q) \neq 1$. There exists $T \in \mathbb{G}_1$ such that $Q + T \in \mathbb{G}_3$, and $e_{W,r}(P, Q + T) = e_{W,r}(P,Q) \neq 1$. Hence the Weil pairing on $\mathbb{G}_1 \times \mathbb{G}_3$ is non-degenerate. The same reasoning holds for the other groups. We refer to Section 3.4.4 for the proof for the Tate pairing.   □

In the remainder of this chapter, we will always assume that we are in the setting of Proposition 3.4. Moreover, we focus on the optimization of the computation of the Tate pairing, since it is now preferred to the Weil pairing in cryptographic settings. This choice is explained by the fact that the Miller loop only needs to compute the evaluation of a single $f_{r,P}$ function.

Denominator elimination

The final exponentiation of the Tate pairing kills any element γ which lives in a strict subfield of $\mathbb{F}_{q^k}$. In particular we see that replacing $f_{r,P}$ by $\gamma f_{r,P}$ in Equation (3.5) does not change the result. In the execution of Algorithm 3.2, we can then modify the Miller functions $\mathbf{f}_{\lambda,\nu,P}$ by a factor γ in a strict subfield of $\mathbb{F}_{q^k}$ without affecting the final result.

Suppose that P and Q are in $\mathbb{G}_1$ or $\mathbb{G}_2$ and the embedding degree k is even. Remember that by Lemma 3.1, the Miller function is $\mathbf{f}_{\lambda,\nu,P} = \mu_{\lambda P, \nu P} = \frac{l_{\lambda P, \nu P}}{v_{\lambda P, \nu P}}$. Then by Lemma 3.3 below, $v_{\lambda P, \nu P}(Q) = x_Q - x_{(\lambda+\nu)P}$ lives in a strict subfield of $\mathbb{F}_{q^k}$, so this factor will be killed by the final exponentiation. Hence in this situation we don't need to compute the division by $v_{\lambda P, \nu P}(Q)$ in Miller's algorithm for the Tate pairing; this is called denominator elimination.

LEMMA 3.3 *Let E be an elliptic curve defined over $\mathbb{F}_q$, such that $E[r] \subset E(\mathbb{F}_{q^k})$ with k even. Let $Q \in \mathbb{G}_1$ or $Q \in \mathbb{G}_2$. Then $x_Q \in \mathbb{F}_{q^{k/2}}$.*

Proof. If $Q \in \mathbb{G}_1$ then $Q \in E(\mathbb{F}_q)$, so both x_Q and y_Q are in $\mathbb{F}_q \subset \mathbb{F}_{q^{k/2}}$. Now if $Q \in \mathbb{G}_2$, then by definition of $\mathbb{G}_2$ we know that $\pi_q^{k/2}(Q) = q^{k/2}Q$. By definition of the embedding degree k, $q^k = 1 \mod r$, so $q^{k/2} = \pm 1 \mod r$. But since k is the smallest integer such that $q^k = 1 \mod r$, we then have $q^{k/2} = -1 \mod r$. So $\pi_q^{k/2}(Q) = -Q$, and in particular $\pi_q^{k/2}(x_Q) = x_{-Q} = x_Q$. So x_Q is fixed by $\pi_q^{k/2}$, which means that $x_Q \in \mathbb{F}_{q^{k/2}}$. $\qquad\square$

To sum up, denominator elimination yields Algorithm 3.3 to compute the Tate pairing over $\mathbb{G}_1 \times \mathbb{G}_2$ or $\mathbb{G}_2 \times \mathbb{G}_1$.

ALGORITHM 3.3 Tate's pairing over $\mathbb{G}_1 \times \mathbb{G}_2$ or $\mathbb{G}_2 \times \mathbb{G}_1$.

Input: $r \in \mathbb{N}$ an odd prime dividing $\#E(\mathbb{F}_q)$ s.t. $k > 1$ is the corresponding embedding degree, k is even and there are no points of r^2-torsion in $E(\mathbb{F}_{q^k})$, $P \in \mathbb{G}_1$, $Q \in \mathbb{G}_2$ (or $P \in \mathbb{G}_2$, $Q \in \mathbb{G}_1$).

Output: The reduced Tate pairing $e_{T,r}(P,Q) = f_{r,P}(Q)^{\frac{q^k-1}{r}}$.

1. Compute the binary decomposition: $r := \sum_{i=0}^{I} b_i 2^i$. Let $T = P, f = 1$.

2. For i in $[I-1..0]$ compute (except at the last step)

   (a) $\alpha = \frac{3x_T^2+a}{2y_T}$, the slope of the tangent of E at T.

   (b) $x_{2T} = \alpha^2 - 2x_T$, $y_{2T} = -y_T - \alpha(x_{2T} - x_T)$;

   (c) $f = f^2 l_{T,T}(Q) = f^2(y_Q - y_T - \alpha(x_Q - x_T))$,

   (d) $T = 2T$,

   (e) If $b_i = 1$, then compute

      i. $\alpha = \frac{y_T-y_P}{x_T-x_P}$, the slope of the line going through P and T;

      ii. $x_{T+P} = \alpha^2 - x_T - x_P$, $y_{T+P} = -y_T - \alpha(x_{T+P} - x_T)$;

      iii. $f = f l_{T,P}(Q) = f(y_Q - y_T - \alpha(x_Q - x_T))$

      iv. $T = T + P$,

3. At the last step: $f = f(x_Q - x_T)$.

Return
$$f^{\frac{q^k-1}{r}}.$$

Finding a non-trivial pairing

For all cryptographic applications of pairings, one needs to find two points P and Q on the elliptic curve such that $e(P,Q) \neq 1$. For instance, the original use of the Weil pairing was used in [19] as an attack method by reducing the DLP from elliptic curves to finite fields: the MOV attack (see Chapter 9). For the reduction to work, given $P \in E[r](\mathbb{F}_q)$, one needs to find a point Q such that $e_{W,r}(P,Q) \neq 1$. Then the DLP between (P, nP) over $E(\mathbb{F}_q)$ reduces to a DLP between $(e_{W,r}(P,Q), e_{W,r}(P,Q)^n)$ over a finite field.

When the embedding degree k is greater than 1 as in Lemma 3.2, then taking any $Q \in \mathbb{G}_2 \setminus 0_E$ gives a non-degenerate pairing $e_{W,r}(P,Q)$. The same is true for the Tate pairing by Proposition 3.4. However when the embedding degree k is 1, and $E[r](\mathbb{F}_q) =< P >$ is cyclic, then $e_{W,r}(P,P) = 1$. To get a non-degenerate Weil pairing one needs to find a $Q \in E[r] \setminus E[r](\mathbb{F}_q)$, and such a point lives over an extension of degree r. But if we replace the Weil pairing by the Tate pairing, then in this case $e_{T,r}(P,P) \neq 1$ by Section 3.4.4. This property was the original reason for the use of the Tate pairing in the article [7].

Pairings of type I,II,III

We conclude the discussion in this section by explaining how to instantiate pairings used in cryptographic protocols, following the classification into three types introduced in Section 1.2.2.

- Type III: The Tate pairing restricted to $\mathbb{G}_1 \times \mathbb{G}_2$ (indeed it is non-degenerate by Proposition 3.4).

- Type II: Let $P \in E[r]$ be a point neither in $\mathbb{G}_1$ nor in $\mathbb{G}_2$ and define $\mathbb{G}_3 =< P >$ to be the cyclic subgroup generated by P. Then by Proposition 3.4 the Tate pairing restricted to $\mathbb{G}_1 \times \mathbb{G}_3$ is non-degenerate. Furthermore, since the trace of the Frobenius has image $\mathbb{G}_1$ and kernel $\mathbb{G}_2$, the restriction of the trace to $\mathbb{G}_3$ is an isomorphism between $\mathbb{G}_3$ and $\mathbb{G}_1$, so the the Tate pairing on $\mathbb{G}_1 \times \mathbb{G}_3$ is of Type II.

- Type I: An instantiation of Type I pairings is given by the Tate pairing on $\mathbb{G} \times \mathbb{G}$, where $\mathbb{G} = E[r](\mathbb{F}_q)$, when the embedding degree $k = 1$ and $E[r](\mathbb{F}_q)$ is cyclic as discussed in the paragraph above and in Section 3.4.4.

  Another example is given by supersingular elliptic curves, in the situation of Lemma 3.2. Indeed, for a supersingular elliptic curve E there exists a *distorsion map* $\psi : \mathbb{G}_1 = E[r](\mathbb{F}_q) \to E[r](\mathbb{F}_{q^k})$ such that $\psi(\mathbb{G}_1) \neq \mathbb{G}_1$. In particular, $e_{W,r}(P, \psi(P)) \neq 1$ and $e_{T,r}(P, \psi(P)) \neq 1$, so composing the Weil or Tate pairing with the distorsion map gives a pairing on $\mathbb{G}_1 \times \mathbb{G}_1$. We refer to [26, 8] for more details on the construction of ψ.

3.2.4   Ate and Optimal Ate Pairings

Miller's basic algorithm described in the previous section is an extension of the double-and-add method for finding a point multiple. With the inception of pairing-based protocols in the early 2000s, the cryptographic community put in a lot of effort in simplifying and optimizing this algorithm. The complexity of Miller's algorithm heavily depends on the length of the Miller loop. Major progress in pairing computation was made in 2006, with the introduction of the loop-shortening technique. This construction, called the eta pairing, was first proposed by Barreto et al. on supersingular curves and further simplified and extended to ordinary curves by Hess et al. In this section, we detail this construction and give explicit formulae for its implementation.

By definition, when $Q \in \mathbb{G}_2$, $\pi_q(Q) = qQ$. So one can use the Frobenius endomorphism π_q to speed up the scalar multiplication $Q \mapsto rQ$. Since Miller's algorithm is an extended version of the scalar multiplication, one can try to use this property of the Frobenius to speed up the computation of the Miller function $f_{r,Q}$. The first idea was to replace r by $q^k - 1$ (which is a multiple of r), and use the Frobenius to speed up the computation of $f_{q^k,Q}$. This leads to the following result given by Hess et al. [10].

THEOREM 3.3   *Let E be an elliptic curve defined over $\mathbb{F}_q$ and r a large prime with $r|\#E(\mathbb{F}_q)$. Let $k > 1$ be the embedding degree and let $\mathbb{G}_1 = E[r] \cap Ker(\pi_q - \mathrm{Id})$ and $\mathbb{G}_2 = E[r] \cap Ker(\pi_q - q\,\mathrm{Id})$. Let $\lambda \equiv q \pmod{r}$ and $m = (\lambda^k - 1)/r$. For $Q \in \mathbb{G}_2$ and $P \in \mathbb{G}_1$ we have*

  (i) *$(Q, P) \mapsto (f_{\lambda,Q}(P))^{(q^k-1)/r}$ defines a bilinear map on $\mathbb{G}_2 \times \mathbb{G}_1$.*

  (ii) *Then $e_{T,r}(Q, P)^m = f_{\lambda,Q}(P)^{c(q^k-1)/r}$ where $c = \sum_{i=0}^{k-1} \lambda^{k-1-i} q^i \equiv kq^{k-1} \pmod{r}$, so this map is non-degenerate if $r \nmid m$.*

In particular, let t be the trace of the Frobenius, $T = t - 1$, and $L = (T^k - 1)/r$. Then $T \equiv q \pmod{r}$ so

$$
\begin{aligned}
a_T : \mathbb{G}_2 \times \mathbb{G}_1 &\mapsto \mu_r \\
(Q, P) &\mapsto (f_{T,Q}(P))^{(q^k-1)/r}
\end{aligned}
$$

defines a pairing on $\mathbb{G}_2 \times \mathbb{G}_1$ when $r \nmid L$, which we call the Ate pairing.

By Hasse's Theorem 2.9, the trace of the Frobenius t is such that $|t| \leq 2\sqrt{q}$. If t is suitably small with respect to r, then the Ate pairing can be computed using a Miller loop of shorter size and is thus faster than the Tate pairing. The exact same algorithm as Algorithm 3.3 allows us to compute the Ate pairing by replacing r with T (since denominator elimination holds, too).

Other pairings may be obtained from Theorem 3.3, by setting $\lambda \equiv q^i \pmod{r}$ [27]. Pushing the idea further, one may look at a multiple of cr of r so that we can write $cr = \sum c_i q^i$ with c_i small coefficients. When $Q \in \mathbb{G}_2$, computing the scalar multiplication by cr requires computing the points $c_i Q$, using the Frobenius to compute the $c_i q^i Q$ and then summing everything. The same idea applied to pairings shows that one can then use a suitable combination of Miller functions $f_{c_i, Q}$ to construct a bilinear pairing that is a power m of the Tate pairing. Once again when $r \nmid m$ we get a new pairing.

THEOREM 3.4 *Let* $\lambda = \sum_{i=0}^{\phi(k)-1} c_i q^i$ *such that* $\lambda = mr$, *for some integer* m. *Then* $a_{[c_0,\dots,c_l]} : \mathbb{G}_2 \times \mathbb{G}_1 \to \mu_r$ *defined as*

$$(Q, P) \to \left(\prod_{i=0}^{\phi(k)-1} f_{c_i, Q}^{q^i}(P) \cdot \prod_{i=0}^{\phi(k)-1} \frac{l_{s_{i+1}Q, c_i q^i Q}(P)}{v_{s_i Q}(P)} \right)^{(q^k-1)/r}, \tag{3.9}$$

with $s_i = \sum_{j=i}^{\phi(k)-1} c_j q^j$ *defines a bilinear map. This pairing is non-degenerate if and only if* $mkq^{k-1} \neq ((q^k - 1)/r) \sum_{i=0}^{\phi(k)-1} i c_i q^{i-1} \pmod{r}$, *and we call it the optimal Ate pairing.*

Proof. The optimal Ate pairing was proposed by Vercauteren [25]. See Section 3.4.5 where we follow the lines of his proof. □

3.2.5 Using Twists to Speed up Pairing Computations

The group $\mathbb{G}_1$ is defined over the base field $\mathbb{F}_q$, so it admits an efficient representation. In particular, when computing the Tate pairing over $\mathbb{G}_1 \times \mathbb{G}_2$, the Miller functions are defined over $\mathbb{F}_q$, so most of the operations during the computation are performed in $\mathbb{F}_q$.

We explain here why $\mathbb{G}_2$ also admits an efficient representation: It is isomorphic to a subgroup of order r on a twist defined over a subfield of $\mathbb{F}_{q^k}$. We prove this result here and we will show in the next section that this allows us to do part of the pairing computations in a subfield $\mathbb{F}_{q^e}$, with $e \mid k$, rather than in $\mathbb{F}_{q^k}$.

THEOREM 3.5 *Let* E *be an ordinary elliptic curve over* $\mathbb{F}_q$ *admitting a twist of degree* d. *Assume that* r *is an integer such that* $r \| \#E(\mathbb{F}_q)$ *and let* $k > 2$ *be the embedding degree. Then there is a unique twist* E' *such that* $r \| \#E'(\mathbb{F}_{q^e})$, *where* $e = k/\gcd(k, d)$. *Furthermore, if we denote by* $\mathbb{G}'_2$ *the unique subgroup of order* r *of* $E'(\mathbb{F}_{q^e})$ *and by* $\Psi : E' \to E$ *the twisting isomorphism, the subgroup* $\mathbb{G}_2$ *is given by* $\mathbb{G}_2 = \phi(\mathbb{G}'_2)$ *and verifies the equation*

$$\mathbb{G}_2 = E[r] \cap Ker([\xi_d]\pi_{q^e} - \mathrm{Id}),$$

where $[\xi_d]$ *is an automorphism of order dividing* d.

Proof. Replacing d by $\gcd(k, d)$, we can assume that $d \mid k$ and that $e = k/d$. Take $Q \in \mathbb{G}_2$. By the definition of $\mathbb{G}_2$ we know that $\pi^e(Q) = q^e Q$. But since k is the smallest integer such that $q^k = 1 \pmod{r}$, we have that $q^e = \xi_d \pmod{r}$, where $\xi_d a$ is d-th primitive root of unity in $\mathbb{F}_{q^k}$. Note that we have an isomorphism $[\cdot] : \mu_d \to \mathrm{Aut}(E)$ ([22, Corollary III.10.2]). Points in $\mathbb{G}_2$ are eigenvectors for any endomorphism on the curve, and we denote by $[\xi_d]$ the automorphism such that $[\xi_d]Q = \xi_d^{-1} Q \pmod{r}$.

Let E' be a twist of degree d of E, defined over $\mathbb{F}_{q^e}$, such that $\Psi \circ (\Psi^{-1})^\sigma$ (with $\cdot^\sigma$ the action of the Frobenius on the coefficients of the automorphism) is the automorphism $[\xi_d]$ on E. If we denote by π_{q^e} the Frobenius morphism on E', we observe that $\Psi \circ \pi_{q^e} \circ \Psi^{-1} = \Psi \circ (\Psi^{-1})^\sigma \circ \pi_{q^e}$. Therefore, we have

$$\mathbb{G}_2 \subseteq \mathrm{Ker}([\xi_d]\pi_{q^e} - \mathrm{Id}).$$

Let $\mathbb{G}_2' = \Psi^{-1}(\mathbb{G}_2)$. Then $\Psi \circ \pi_{q^e} \circ \Psi^{-1}(\mathbb{G}_2) = \mathbb{G}_2$. It follows that $\mathbb{G}_2'$ is invariant under π_{q^e}, hence it is defined over $\mathbb{F}_{q^e}$. $\qquad\qquad\square$

Using this result, one can compute the Miller loop for the Ate (or optimal Ate) pairing $a_T(Q,P)$ by working over $\mathbb{G}_2'$ to compute the multiples of $\Psi^{-1}(Q)$, and going back to $\mathbb{G}_2$ only to evaluate the Miller functions on P. Alternatively one can do the full computation on the twist E', as shown in [6].

THEOREM 3.6   *Let E be an elliptic curve defined over $\mathbb{F}_q$. Assume that r is an integer such that $r \| \#E(\mathbb{F}_q)$ and let $k > 2$ be the embedding degree. Let E' be the twist of degree d and $\Psi : E' \to E$ the associated twist isomorphism, as in Theorem 3.5. Consider $Q \in \mathbb{G}_2$, $P \in \mathbb{G}_1$, and let $Q' = \Psi^{-1}(Q)$ and $P' = \Psi^{-1}(P)$. Let $a_T(Q,P)$ be the Ate pairing of Q and P. Then*

$$a_T(Q,P)^{\gcd(d,6)} = a_T(Q',P')^{\gcd(d,6)}$$

where $a_T(Q',P') = f_{T,Q'}(P')^{(q^k-1)/r}$ uses the same parameter loop.

This shows that the pairing on $\mathbb{G}_2 \times \mathbb{G}_1$ may be seen as a $\mathbb{G}_1 \times \mathbb{G}_2$ pairing on a twist defined over $\mathbb{F}_{q^e}$. Indeed, since $\mathbb{G}_2 = E[r] \cap \mathrm{Ker}([\xi_d]\pi_{q^e} - \mathrm{Id})$ by Theorem 3.5, $\mathbb{G}_1 = E[r] \cap \mathrm{Ker}([\xi_d]\pi_{q^e} - q^e \, \mathrm{Id})$, so $\Psi^{-1}(\mathbb{G}_2) = \mathbb{G}_1(E')$ and $\Psi^{-1}(\mathbb{G}_1) = \mathbb{G}_2(E')$, with $\mathbb{G}_1(E')$ and $\mathbb{G}_2(E')$ the subgroups giving the eigenvectors of the Frobenius on E'.

The twist improves the Ate pairing on $\mathbb{G}_2 \times \mathbb{G}_1$ by giving an efficient representation of $\mathbb{G}_2$. Alternatively, it can be used to give a shorter Miller loop for pairings on $\mathbb{G}_1 \times \mathbb{G}_2$ [13].

THEOREM 3.7   *Let $\lambda \equiv q \pmod{r}$ and $m = (\lambda^k - 1)/r$. Assume that E has a twist of degree d and set $n = \gcd(k,d)$, $e = k/n$.*

(i) $(P,Q) \mapsto (f_{\lambda^e,P}(Q))^{(q^k-1)/r}$ defines a bilinear map on $\mathbb{G}_1 \times \mathbb{G}_2$.

(ii) $e_{T,r}(P,Q)^m = f_{\lambda^e,P}(Q)^{c(q^k-1)/r}$ where $c = \sum_{i=0}^{n-1} \lambda^{e(n-1-i)} q^{ei} \equiv nq^{e(n-1)} \pmod{r}$, so this map is non-degenerate if $r \nmid m$.

In particular, if t is the trace of the Frobenius, $T = t - 1$, and $L = (T^k - 1)/r$, then $(P,Q) \mapsto (f_{T^e,P}(Q))^{(q^k-1)/r}$ defines a pairing if $r \nmid L$, which we call the *twisted Ate pairing*.

One can also define a twisted optimal Ate pairing on $\mathbb{G}_1 \times \mathbb{G}_2$. This was given by Hess [12], using a general formula for the pairing function. We present it here in a simplified way, better suited for implementations.

THEOREM 3.8   *Assume that E has a twist of degree d and set $n = \gcd(k,d)$, $e = k/n$. Let $\lambda = \sum_{i=0}^{\phi(k)/e-1} c_i q^{ie}$ such that $\lambda = mr$, for some integer m. Then*

$$a_{[c_0,\dots,c_l]} : \mathbb{G}_1 \times \mathbb{G}_2 \quad \to \quad \mu_r \qquad\qquad\qquad (3.10)$$

$$(P,Q) \quad \to \quad \left(\prod_{i=0}^{\phi(k)/e-1} f_{c_i,P}^{q^{ie}}(Q) \cdot \prod_{i=0}^{\phi(k)/e-1} \frac{l_{s_{i+1}P,c_iq^{ie}P}(Q)}{v_{s_iP}(Q)} \right)^{(q^k-1)/r},$$

where $s_i = \sum_{j=i}^{\phi(k)/e-1} c_j q^{je}$, *defines a bilinear map on* $\mathbb{G}_1 \times \mathbb{G}_2$. *This pairing is non-degenerate if and only if* $mkq^{k-1} \neq ((q^k - 1)/r) \sum_{i=0}^{\phi(k)/e-1} i c_i q^{e(i-1)}$ (mod r).

Proof. See Section 3.4.5. □

3.2.6 The Optimal Ate and Twisted Optimal Ate in Practice

In order for the optimal Ate and twisted optimal Ate pairings to give a short Miller loop, we would like the coefficients c_i to be as small as possible. The idea is to search for the coefficients c_i in Equations 3.9 and 3.10 by computing short vectors in the following lattice:

$$
\begin{pmatrix}
r & 0 & 0 & \cdots & 0 \\
-q & 1 & 0 & \cdots & 0 \\
-q^2 & 0 & 1 & \cdots & 0 \\
\vdots & \vdots & \vdots & & \vdots \\
-q^l & 0 & 0 & \cdots & 1
\end{pmatrix}, \tag{3.11}
$$

where l is either $\phi(k) - 1$ in the optimal Ate pairing case, and $\phi(k)/e - 1$, in the twisted Ate case. The volume of this lattice is r, hence by Minkowski's theorem there is a short vector v in the lattice such that $||v||_\infty \leq r^{1/l+1}$.

Starting from this bound and Theorem 3.4, Vercauteren discusses the existence of pairings that may be computed with a Miller loop of size $(\log r)/\phi(k)$. Note that Theorem 3.4 does not guarantee that the pairing defined in Equation 3.9 can be computed in $(\log r)/\phi(k)$ operations. If the procedure described above produces a short vector with several c_i coefficients different from zero, then computing each $f_{c_i,Q}(P)$ separately costs $O((\log r)/\phi(k))$ operations. Possible optimizations would be to use multi-exponentiation techniques or a parallel version of Miller's algorithm to compute all the $f_{c_i,Q}(P)$ functions at once. However, in the case of parametric families introduced in Chapter 4, the entire computation can be carried with a single basic Miller loop, and the pairing given in Theorem 3.4 is indeed optimal, thanks to the special form of the short vectors we obtain. To explain this idea, we give explicit formulae for this computation in the case of several Brezing-Weng-type constructions of pairing-friendly curves. Since k is small, these formulae can be obtained by computing short vectors for the lattice given by the matrix 3.11, by using an available implementation of the LLL algorithm. We recommend using, for instance, the functions LLL() or BKZ() in Sage [23].

Example 3.1 [25, Vercauteren] We consider the Barreto-Naehrig family of curves that was introduced in [25, Vercauteren]. We are briefly reminded here that these families have embedding degree 12 and are given by the following parametrizations:

$$
\begin{aligned}
r(x) &= 36x^4 + 36x^3 + 18x^2 + 6x + 1, \\
t(x) &= 6x^2 + 1, \\
q(x) &= 36x^4 + 36x^3 + 24x^2 + 6x + 1.
\end{aligned}
$$

By Theorem 3.3, the length of Miller's loop for the Ate pairing is $\frac{\log_2 r}{2}$. We will show that the complexity of the computation of the optimal Ate pairing for this family is $O(\frac{\log_2 r}{4})$. Indeed, in order to apply Theorem 3.4, we compute the following short vector:

$$
[6x + 2, 1, -1, 1].
$$

Note that in this case, 3 out of the 4 coefficients in the short vector are trivial. We conclude that the optimal twisted Ate pairing for this family of curves is given by the simple formula:

$$
(f_{6x+2,Q}(P) \cdot l_{Q_3,-Q_2}(P) l_{-Q_2+Q_3,Q_1}(P) l_{Q_1-Q_2+Q_3,[6x+2]Q}(P))^{\frac{q^{12}-1}{r}},
$$

where $Q_i = Q^{q^i}$, for $i = 1, 2, 3$. Note that the evaluation at Q of vertical lines can actually be ignored because of the final exponentiation. The only costly computation is that of $f_{6x+2,Q}(P)$ and costs $O(\log r/2)$ operations. While the twisted Ate has loop length $\log r$, a search for a short vector giving the optimal twisted Ate pairing gives

$$[6x^2 + 2x, 2x + 1].$$

Hence we need to compute $f_{x,P}(Q)$, $f_{x^2,P}(Q)$ and the complexity of computation is $O(\log r/2)$.

Example 3.2   We consider here the family of curves with $k = 18$ proposed by *Kachisa* et al., whose construction is given in Chapter 4. We briefly recall that this family is parametrized by the following polynomials:

$$
\begin{aligned}
r(x) &= x^6 + 37x^3 + 343, \\
t(x) &= \frac{1}{7}(x^4 + 16x + 7), \\
q(x) &= \frac{1}{21}(x^8 + 5x^7 + 7x^6 + 37x^5 + 188x^4 + 259x^3 + 343x^2 + 1763x + 2401).
\end{aligned}
$$

A similar search for the optimal twisted Ate pairing on curves with embedding degree 18 gives, for example, the short vector

$$[1, x^3 + 18].$$

Hence the complexity of Miller's algorithm is $\frac{\log_2 r}{2}$. The optimal Ate pairing computation for curves with $k = 18$ has complexity $\mathcal{O}(\frac{\log r}{6})$.

Building on these results, Vercauteren [25] introduces the concept of optimal pairing, i.e., no, a pairing that is computed in $\log r/\phi(k)$ Miller iterations. He puts forward the following conjecture:

Optimality conjecture: Any non-degenerate pairing on an elliptic curve without any efficiently computable endomorphisms different from the powers of the Frobenius requires at least $O(\log r/\phi(k))$ basic Miller iterations.

Hess [12] proved the optimality conjecture for all known pairing functions. The pairings given by the formulae in Theorem 3.4 are the fastest known pairings at the time of this writing. On curves endowed with efficiently computable endomorphisms other than the Frobenius (such as automorphisms), it is currently not known how to use the action of these endomorphisms to improve on pairing computation.

Choosing the right pairing

Assume that we are in the situation of Proposition 3.4, and let $P \in \mathbb{G}_1$ and $Q \in \mathbb{G}_2$. Then one may choose among the Tate pairing $e_{r,W}(P, Q)$, the Ate (or Optimal Ate) pairing $a_{r,T}(P, Q)$, the twisted Ate pairing ... Furthermore, when k is even, we can apply denominator elimination for the Tate pairing thanks to the final exponentiation. On the downside, one should remember that the final exponentiation may be expensive too. Indeed, the loop length of the final exponentiation is around $k \log q$ compared to $\log q$ for the Miller step. So the implementation of the final exponentiation step should not be neglected and we will give in Chapter 7 efficient algorithms for its computation.

We conclude that the choice of parameters for applications is a complex matter, with multiple aspects to take into account. Therefore, we devote the whole of Chapter 10 to discussing this problem. In the remainder of this chapter, we give optimized formulae for computing one step of a Miller loop.

3.3 Formulae for Pairing Computation

One of the most efficient ways of computing pairings on an elliptic curve given by a Weierstrass equation is to use Jacobian coordinates [17], [9]. A point $[X, Y, Z]$ in Jacobian coordinates represents the affine point $(X/Z^2, Y/Z^3)$ on the elliptic curve. A point in projective coordinates $[X, Y, Z]$ represents the point $(X/Z, Y/Z)$ on the elliptic curve.

In this section we denote by **s** and **m** the costs of squaring and multiplication in $\mathbb{F}_q$ and by **S** and **M** the costs of these operations in the extension field $\mathbb{F}_{q^k}$, if $k > 1$. We denote by $\mathbf{d}_a$ the cost of the multiplication by a constant a. Sometimes, if q is a sparse prime (such as a generalized Mersenne prime), we may assume that $\mathbf{s/m} = 0.8$. However, when constructing pairing friendly curves, it is difficult to obtain such primes. Hence, we generally have $\mathbf{s/m} \approx 1$.

3.3.1 Curves with Twists of Degree 2

In the remainder of this section, we suppose that the embedding degree is even and that E has a twist of order 2 defined over $\mathbb{F}_{q^{k/2}}$. From Theorem 3.5 and by using the equations of twists given in Subsection 2.3.6, we derive an efficient representation of points in $\mathbb{G}_2$. It follows that the subgroup $\mathbb{G}_2 = \langle Q \rangle \subset E(\mathbb{F}_{q^k})$ can be chosen such that the x-coordinates of all its points lie in $\mathbb{F}_{q^{k/2}}$ and the y-coordinates are products of elements of $\mathbb{F}_{q^{k/2}}$ with $\sqrt{\beta}$, where β is not a square in $\mathbb{F}_{q^{k/2}}$ and $\sqrt{\beta}$ is a fixed square root in $\mathbb{F}_{q^k}$.

For curves with twists of degree 2, the fastest known formulae for Miller's algorithm doubling [14] and addition steps [1] are in Jacobian coordinates. Therefore we represent the point T as $T = [X_1, Y_1, Z_1, W_1]$, where $[X_1, Y_1, Z_1]$ are the Jacobian coordinates of the point T on the Weierstrass curve and $W_1 = Z_1^2$.

The doubling step

We will look at the doubling step in the Miller loop. We represent the point T as $T = (X_1, Y_1, Z_1, W_1)$, where (X_1, Y_1, Z_1) are the Jacobian coordinates of the point T on the Weierstrass curve and $W_1 = Z_1^2$. We compute $2T = (X_3, Y_3, Z_3, W_3)$ as:

$$
\begin{aligned}
X_3 &= (3X_1^2 + aW_1^2)^2 - 8X_1Y_1^2, \\
Y_3 &= (3X_1^2 + aW_1^2)(4X_1Y_1^2 - X_3) - 8Y_1^4, \\
Z_3 &= 2Y_1Z_1, \\
W_3 &= Z_3^2.
\end{aligned}
$$

We write the normalized function $l_{T,T}$ that appears in Algorithm (3.3) as :

$$
l_{T,T}(x_Q, y_Q) = (Z_3W_1y - 2Y_1^2 - (3X_1^2 + aW_1^2)(W_1x - X_1))/(Z_3W_1).
$$

Thanks to elimination in the final exponentiation, the term Z_3W_1 can be ignored. For $k = 2$, we have that $x \in \mathbb{F}_q$ and we can compute the function $l_{T,T}$ as:

$$
l_{T,T}(x, y) = Z_3W_1y - 2Y_1^2 - (3X_1^2 + aW_1^2)(W_1x - X_1).
$$

For $k > 2$, we have that x is in $\mathbb{F}_{q^{k/2}}$ and the computation is slightly different:

$$
l_{T,T}(x, y) = Z_3W_1y - 2Y_1^2 - W_1(3X_1^2 + aW_1^2)x + X_1(3X_1^2 + aW_1^2).
$$

The computations are done in the following order:

$$
\begin{aligned}
A &= W_1^2,\ B = X_1^2,\ C = Y_1^2,\ D = C^2, E = (X_1 + C)^2 - B - D, \\
F &= 3B + aA,\ G = F^2, X_3 = -4E + G,\ Y_3 = -8D + F \cdot (2E - X_3), \\
Z_3 &= (Y_1 + Z_1)^2 - C - W_1, W_3 = Z_3^2,\ H = (Z_3 + W_1)^2 - W_3 - A,\ I = H \cdot y, \\
J &= (F + W_1)^2 - G - A,\ K = J \cdot x,\ L = (F + X_1)^2 - G - B \\
l_{T,T} &= I - 4C - K + L, f = f^2 \cdot l_{T,T}.
\end{aligned}
$$

The operation count gives $10\mathbf{s} + 3\mathbf{m} + 1\mathbf{a} + 1\mathbf{S} + 1\mathbf{M}$ for $k = 2$ and $11\mathbf{s} + (k+1)\mathbf{m} + 1\mathbf{d}_a + 1\mathbf{S} + 1\mathbf{M}$ if $k > 2$.

The mixed addition step

In implementations, it is often possible to choose the point P such that its Z-coordinate is 1, in order to save some operations. The addition of two points $T = [X_1, Y_1, Z_1]$ and $P = [X_2, Y_2, 1]$ is called *mixed addition*.

The result of the addition of $T = [X_1, Y_1, Z_1, W_1]$ and $P = [X_2, Y_2, 1]$ is $T + P = [X_3, Y_3, Z_3, W_3]$ with:

$$
\begin{aligned}
X_3 &= (X_1 + X_2 Z_1^2)(X_1 - X_2 Z_1^2)^2 + (Y_2 Z_1^3 - Y_1)^2, \\
Y_3 &= (Y_2 Z_1^3 - Y_1)(X_1(X_1 - X_2 Z_1^2)^2 - X_3) + Y_1(X_1 - X_2 Z_1^2)^2, \\
Z_3 &= Z_1(X_2 Z_1^2 - X_1), \\
W_3 &= Z_3^2, \\
T_3 &= W_3 x_Q - X_3.
\end{aligned}
$$

The line $l_{T,P}$ is given by the equation:

$$
l_{T,P} = Z_3 y_Q - Y_2 Z_3 - (2Y_2 Z_1^3 - 2Y_1)(x_Q - X_2).
$$

The computations are done in the following order:

$$
\begin{aligned}
&A = Y_2^2, B = X_2 \cdot W_1, D = ((Y_2 + Z_1)^2 - A - W_1) \cdot W_1, H = B - X_1, I = H^2 \\
&E = 4I, J = H \cdot E, L_1 = D - 2Y_1, V = X_1 \cdot E, X_3 = L_1^2 - J - 2V, \\
&Y_3 = (D - Y_1) \cdot (V - X_3) - 2Y_1 \cdot I \\
&Z_3 = (Z_1 + H)^2 - W_1 - I, W_3 = Z_3^2, l_{T,P} = 2Z_3 \cdot y_Q - (Y_2 + Z_3)^2 + A + W_3 - 2L_1 \cdot (x_Q - X_2).
\end{aligned}
$$

The operation count gives $6\mathbf{s} + 6\mathbf{m} + k\mathbf{m} + 1\mathbf{M}$ [1].

3.3.2   Curves with Equation $y^2 = x^3 + ax$

These curves have twists of degree 4 and assume k is divisible by 4. Therefore, by using the equations for twists given in Section 2.3.6 and Theorem 3.5, we derive that a point $Q \in \mathbb{G}_2$ may be written as

$$
(x_Q, y_Q) = (x_Q' \nu^{1/2}, y_Q' \nu^{3/4}),
$$

where $x_{Q'}, y_{Q'}, \nu \in \mathbb{F}_{q^{k/4}}$ and $X^4 - \nu$ is an irreducible polynomial. Moreover, thanks to the simple form of the Weierstrass equation, the doubling and addition formulae for these curves are simpler and faster than in the case of curves allowing only twists of degree 2. The fastest formulae for pairing computation on these curves [6] use Jacobian coordinates. In the doubling step, we compute $2T$ as

$$
\begin{aligned}
X_3 &= (X_1^2 - aZ_1^2)^2, \\
Y_3 &= 2Y_1(X_1^2 - aZ_1^2)((X_1^2 + aZ_1^2)^2 + 4aZ_1^2X_1^2), \\
Z_3 &= 4Y_1^2.
\end{aligned}
$$

The line function is

$$
l_{T,T} = -2(3X_1^2Z_1 + aZ_1^3)x_Q + (4Y_1Z_1)y_Q + 2(X_1^3 - aZ_1^2X_1).
$$

The computation is done using the following sequence of operations:

$$
\begin{aligned}
A &= X_1^2, B = Y_1^2, C = Z_1^2, D = aC, X_3 = (A-D)^2, \\
E &= 2(A+D)^2 - X_3, F = ((A-D+Y_1)^2 - B - X_3), Y_3 = E \cdot F, Z_3 = 4B, \\
G &= -2Z_1(3 \cdot A + D), H = 2((Y_1 + Z_1)^2 - B - C), II = (X_1 + A - D)^2 - X_3 - A, \\
l_{T,T} &= G \cdot x_Q + H \cdot y_Q + II.
\end{aligned}
$$

The total cost is $(2k/d + 2)\mathbf{m} + 8\mathbf{s} + 1\mathbf{d}_a$. In the mixed addition step of $T = (X_1, Y_1, Z_1)$ and $P = (X_2, Y_2, 1)$ the sum is $T + P = (X_3, Y_3, Z_3)$ with

$$
\begin{aligned}
X_3 &= (Y_1 - Y_2Z_1^2)^2 - (X_1 + X_2Z_1)S, \\
Y_3 &= ((Y_1 - Y_2Z_1^2)(X_1S - X_3) - Y_1SU)UZ_1, \\
Z_3 &= (UZ_1)^2,
\end{aligned}
$$

where $S = (X_1 - X_2Z_1)^2Z_1$ and $U = X_1 - X_2Z_1$. This is computed with the following operations:

$$
\begin{aligned}
A &= Z_1^2, E = X_2 \cdot Z_1, G = Y_2 \cdot A, H = X_1 - E, I = 2(Y_1 - G), II = I^2, J = 2Z_1 \cdot H \\
K &= 4J \cdot H, X_3 = 2II - (X_1 + E) \cdot K, Z_3 = J^2 \\
Y_3 &= ((J+I)^2 - Z_3 - II) \cdot (X_1 \cdot K - X_3) - Y_1 \cdot K^2, Z_3 = 2Z_3 \\
l_{T,P} &= I \cdot X_2 - I \cdot x_Q + J \cdot y_Q - J \cdot Y_2.
\end{aligned}
$$

The total cost of the computation is $((2k/d) + 9)\mathbf{m} + 5\mathbf{s}$.

3.3.3 Curves with Equation $y^2 = x^3 + b$

These curves have twists of degree 6 and assume k is divisible by 6. Therefore, by using the equations for twists given in Section 2.3.6 and Theorem 2.12, we derive that a point $Q \in \mathbb{G}_2$ may be written as

$$
(x_Q, y_Q) = (x_{Q'}\nu^{1/3}, y_{Q'}\nu^{1/2}),
$$

where $x_{Q'}, y_{Q'}, \nu \in \mathbb{F}_{q^{k/6}}$ and $X^6 - \nu$ is an irreducible polynomial. The fastest existing formulae on these curves use projective coordinates. Following [6], we compute $2T$ as:

$$
\begin{aligned}
X_3 &= 2X_1Y_1(Y_1^2 - 9bZ_1^2), \\
Y_3 &= Y_1^4 + 18bY_1^2Z_1^2 - 27b^2Z_1^4, \\
Z_3 &= 8Y_1^3Z_1.
\end{aligned}
$$

TABLE 3.1   Cost of one step in Miller's algorithm for even embedding degree.

	Doubling		Mixed addition
	$k = 2$	$k \geq 4$	
$\mathcal{J}$ [14],[1]	$3m + 10s + 1a + 1M + 1S$	$(1+k)m+11s+1a+1M+1S$	$(6+k)m+6s+1M$
$\mathcal{J}, y^2 = x^3 + b$ $d = 2, 6$ [6]	$(2k/d+2)m+7s+1a+1M+1S$	$(2k/d+2)m+7s+1a+1M+1S$	$(2k/d+9)m+2s+1M$
$\mathcal{J}, y^2 = x^3 + ax$ $d = 2, 4$ [6]	$(2k/d+2)m+8s+1a+ 1M+1S$	$(2k/d+2)m+8s+1a+ 1M+1S$	$(2k/d+12)m+4s+1M$

The line equation is

$$l_{T,T} = 3X_1^2 \cdot x_Q - 2Y_1 Z_1 \cdot y_Q + 3bZ_1^2 - Y_1^2.$$

The computation is performed in the following order:

$$
\begin{aligned}
A &= X_1^2, B = Y_1^2, C = Z_1^2, D = 3bC, E = (X_1 + Y_1)^2 - A - B, \\
F &= (Y_1 + Z_1)^2 - B - C, G = 3D, X_3 = E \cdot (B - G), \\
Y_3 &= (B + G)^2 - 12D^2, Z_3 = 4B \cdot F, H = 3A, I = -F, J = D - B. \\
l_{T,T} &= H \cdot x_Q + I \cdot y_Q + J.
\end{aligned}
$$

The total count for the above sequence of operations is $(2k/d)\mathbf{m} + 5\mathbf{s} + 1\mathbf{d}_b$. In the mixed addition step of $T = (X_1, Y_1, Z_1)$ and $P = (X_2, Y_2, 1)$ the sum is $T + P = (X_3, Y_3, Z_3)$ with

$$
\begin{aligned}
X_3 &= (X_1 - Z_1 X_2)(Z_1(Y_1 - Z_1 Y_2)^2 - c(X_1 + Z_1 X_2)(X_S - Z_1 X_2)^2), \\
Y_3 &= (Y_1 - Z_1 Y_2)(c(X_1 + Z_1 X_2)(X_1 Z_2 - Z_1 X_2)^2 - Z_1(Y_1 - Z_1 Y_2)^2) - cY_1(X_1 Z_2 - Z_1 X_2)^3, \\
Z_3 &= cZ_1(X_1 - Z_1 X_2)^3,
\end{aligned}
$$

where $c = 1/b$. The line formula is given by

$$l_{T,P} = (Y_1 - Z_1 Y_2) \cdot (X_2 - x_Q) - (X_1 - Z_1 X_2) \cdot Y_2 + (X_1 - Z_1 X_2) \cdot Z_2 y_Q.$$

The computation is performed using the following sequence of operations:

$$
\begin{aligned}
t_1 &= Z_1 \cdot X_2, t_1 = X_1 - t_1, t_2 = Z_1 \cdot Y_2, t_2 = Y_s - t_2, \{T, P\} = c_1 \cdot t_2 - t_1 \cdot Y_2 + t_1 \cdot y_Q \\
t_3 &= t_1^2, t_3 = c \cdot t_3, X_3 = t_3 \cdot X_1, t_3 = t_1 \cdot t_3, t_4 = t_2^2 \\
t_4 &= t_4 \cdot Z_1, t_4 = t_3 + t_4, t_4 = t_4 - X_3, X_3 = -X_3 + t_4, t_2 = t_2 \cdot X_3, Y_3 = t_3 \cdot Y_1 \\
Y_3 &= t_2 - Y_3, X_3 = t_1 \cdot t_4, Z_3 = Z_1 \cdot t_3,
\end{aligned}
$$

where $c_1 = X_2 - x_Q$. The total cost is $(2k/d + 9)\mathbf{m} + 2\mathbf{s}$. In Table 3.1 we summarize all these results.

3.4   Appendix: The General Form of the Weil and Tate Pairing

The versions of the Tate and Weil pairing we gave required us to evaluate a function on a point. In this section we will give a generalized definition that requires us to evaluate a function on a divisor.

Furthermore, we have seen that during the execution of Miller's algorithm, some intermediate poles and zeroes are introduced. As we pointed out, this is not really a problem in practice, since this situation only happens when computing a pairing between P and Q with Q a multiple of P. As explained in Section 3.2.2, for the Tate pairing we can circumvent the problem by using a random point R.

Another way to circumvent the problem is to define the (extended) evaluation of a function on a point or a divisor even in the case when the supports are non-disjoint. This allows us to generalize Miller's algorithm so that it always works and to give a more general definition of the Weil and Tate pairing. From this more general definition, we can prove their bilinearity and that they are non-degenerate.

3.4.1 Evaluating Functions on a Divisor

If $D = \sum n_i[P_i]$ is a divisor on E, we define the *support* $\operatorname{supp}(D)$ as the set $\{P_i \mid n_i \neq 0\}$. By abuse of langage we define the support of f as the support of div f, so the support of f is simply the union of the zeroes and poles of f.

If the support of f and the support of D are disjoint, then one can define the evaluation of f on $D = \sum n_i P_i$ as

$$f(D) = \prod_i f(P_i)^{n_i}. \tag{3.12}$$

It is easy to check that we have $(fg)(D) = f(D) \cdot g(D)$ and $f(D_1 + D_2) = f(D_1).f(D_2)$.

One can extend this definition even when the supports are non-disjoint by *fixing once and for all* uniformizers t_P for every point $P \in E(K)$. Then one can define the *extended evaluation* of f at P as $(\frac{f}{t_P^{\operatorname{ord}_P(f)}}(P), \operatorname{ord}_P(f))$. We will often simply refer to $\frac{f}{t_P^{\operatorname{ord}_P(f)}}(P)$ as the *value* of f at P and to $\operatorname{ord}_P(f)$ as the *valuation* (or the *order*) of this value. If P is not in the support of f then the extended evaluation of f at P is simply $(f(P), 0)$. One can define a product on the extended values by taking the product of the values and adding the valuations: $(\alpha, n).(\beta, m) = (\alpha\beta, n+m)$. This definition of the product allows us to have the standard property:

$$(fg)(P) = f(P) \cdot g(P).$$

By using Equation (3.12) one can define the extended evaluation of f at a divisor $D = \sum n_i P_i$ as $f(D) = \prod_i f(P_i)^{n_i}$ where this time the product is on extended values. By the definition of $f(D)$ and the product on extended values, we have $(fg)(D) = f(D) \cdot g(D)$ and $f(D_1 + D_2) = f(D_1) \cdot f(D_2)$.

When D and f do not have disjoint supports, one needs to be careful that the extended value $f(D)$ depends on the choice of uniformizers and is not intrinsic to the curve. For example, if P is a point in the support of f with order n, then changing the uniformizer t_P at P by $t'_P = \alpha t_P$ changes the value by α^{-n} (but the order stays the same). So in the following we fix once and for all the following uniformizers for the elliptic curve:

- $t_{0_E} = x/y$;
- $t_P = x - x_P$, except when $H(x_P) = 0$;
- $t_P = y$, when $H(x_P) = 0$ (so $y_P = 0$).

A powerful tool used in computing evaluation of divisors is Weil's reciprocity theorem.

THEOREM 3.9 (Weil's reciprocity theorem) *Let $f, g \in K(E)$. Then*

$$f(\operatorname{div}(g)) = (-1)^{\sum_P \operatorname{ord}_P(f)\operatorname{ord}_P(g)} g(\operatorname{div}(f)).$$

Expressing the above equation in terms of divisors (see Definition 3.1), we get the following reformulation: Let D_1 and D_2 be two degree 0 divisors and define $\epsilon(D_1, D_2) = (-1)^{\sum_P \operatorname{ord}_P(D_1)\operatorname{ord}_P(D_2)}$. If D_1 and D_2 are principal, then

$$f_{D_1}(D_2) = \epsilon(D_1, D_2) f_{D_2}(D_1).$$

Proof. See [21, p. 44–46].   □

3.4.2   Miller's Algorithm for Pairing Computation

Let $f \in k(E)$ be a rational function on E and D a divisor of degree 0. Then $f(D)$ depends only on $\mathrm{div}(f)$, not on f. Indeed, if g has the same divisor as f, there exists $\lambda \in K^*$ such that $g = \lambda f$ so that $g(D) = \lambda^{\deg D} f(D) = f(D)$. One can see the divisor $F = \mathrm{div}\, f$ as an efficient way to encode the rational function f. Recall that we note f_F the normalized function with divisor F.

As we have seen in Section 3.2, all pairing computations involve the following computation: Given $P \neq 0_E$ a point of r-torsion on E, and $Q \neq P, 0_E$ a point of the elliptic curve, evaluate $f_{r,P}(Q)$. We recall that $f_{r,P}$ is the normalized function with divisor $r([P] - [0_E])$.

This computation is a particular case of the following more general framework: *Let $P \neq 0_E$ be a point of r-torsion on E, and $Q \neq 0_E$ a point of the elliptic curve. Let D_P and D_Q be two divisors linearly equivalent to $[P] - [0_E]$ and $[Q] - [0_E]$, respectively. Then evaluate the function f_{rD_P} on the divisor D_Q.*

The evaluation makes sense because $r[P] - r[0_E]$ is a principal divisor by Proposition 3.1, so rD_P is principal too. Taking $D_P = [P] - [0_E]$ and $D_Q = [Q] - [0_E]$, we recover the previous computation since by Definition 3.3, the evaluation of a function associated to $r[P] - r[0_E]$ on $[Q] - [0_E]$ is simply $f_{r,P}(Q)$. One has to take care here that the divisors $r[P] - r[0_E]$ and $[Q] - [0_E]$ do not have disjoint support, so the evaluations above are to be understood as extended evaluations: If $P \neq Q$ then the value $f_{r,P}(Q)$ has valuation $-r$, otherwise the value has valuation $r - r = 0$.

We have seen in Section 3.1 how to use Miller's algorithm to compute $f_{r,P}(Q)$. More generally, given F and D, two degree-zero divisors, we give a general version of Miller's algorithm, which allows us to compute the value $f_F(D)$. The key principle behind this extended Miller's algorithm is to use the functions $\mu_{P,Q}$ introduced in Definition 3.2.

Whenever we have two points P and Q different from 0_E in the support of F, we can decompose F as $F = [P] + [Q] + F'$ and then use the function $\mu_{P,Q}$ to get $F = [P] + [Q] - [P + Q] - [0_E] + [P + Q] + [0_E] + F' = \mathrm{div}(\mu_{P,Q}) + [P + Q] + [0_E] + F' = \mathrm{div}(\mu_{P,Q}) + F_1$, where $F_1 = [P + Q] + [0_E] + F'$. This decomposition of F means that we just need to evaluate $\mu_{P,Q}$ and F_1 on D and then take the product. Since $\mu_{P,Q}$ is an explicit function, evaluating it on D simply means evaluating it on each point in the support of D and then taking the product.

Now to evaluate F_1 on D we proceed as we did for F and decompose F_1 again. Each time we decompose the divisor, we decrease the number of non-zero points in the support (counted with multiplicities). After a finite number of iterations, we find a divisor F_n of degree 0, which has at most one non-zero point in its support (counted with multiplicity). So F_n is of the form $[P] - [0_E]$ and since F is principal, F_n is principal too, and by Proposition 3.1 we have that $P + 0_E = 0_E$, or in other words $P = 0_E$ and $F_n = 0$. Of course, $f_{F_n} = 1$ and $F_n(D) = 1$.

So evaluating F on D decomposes to the evaluation of the functions $\mu_{P,Q}$ appearing in the decomposition of F on the points in the support of D. We give explicit formulae in Lemma 3.4.

LEMMA 3.4 (Evaluating $\mu_{P,Q}$)   *Let $P = (x_P, y_P)$, $Q = (x_Q, y_Q)$, and $R = (x_R, y_R)$ be points on E. Then $\mu_{P,Q} = \frac{l_{P,Q}}{v_{P,Q}}$ where $l_{P,Q} = y - \alpha x - \beta$ with $\alpha = \frac{y_P - y_Q}{x_P - x_Q}$ when $P \neq Q$ and $\alpha = H'(x_P)$ when $P = Q$, $\beta = y_P - \alpha x_P = y_Q - \alpha x_Q$, and $v_{P,Q} = x - x_{P+Q}$ with $x_{P+Q} = \alpha^2 - x_P - x_Q$.*

Assume that P, Q, and $P + Q$ are all different from 0_E. The extended value of $v_{P,Q}(R)$ is given by the following cases (taking into account that $\mathrm{div}(v_{P,Q}) = [P + Q] + [-P - Q] - 2[0_E]$):

- *If R is different from $P+Q$, $-P-Q$ or 0_E, then R is not in the support of $\mathrm{div}(v_{P,Q})$ and we have a value with valuation 0: $v_{P,Q}(R) = x_R - x_{P+Q}$;*

- If $R = 0_E$ then we have a value with valuation -2. By definition, since the uniformizer at 0_E is the function y/x:

$$v_{P,Q}(0_E) = \frac{x - x_{P+Q}}{(y/x)^{-2}}(0_E) = \frac{x^2(x - x_{P+Q})}{y^2}(0_E) = 1$$

because $y^2 = x^3 + ax + b$;

- If $R = P + Q$ or $R = -P - Q$ but $P + Q \neq -P - Q$ (or in other words $P + Q$ is not a point of 2-torsion), then we have a value with valuation 1. The uniformizer is $x - x_R$ because $H(x_R) \neq 0$ since R is not a point of 2-torsion, and the value is

$$v_{P,Q}(R) = \frac{x - x_{P+Q}}{x - x_R}(x_R) = 1$$

because in this case $x_R = x_{P+Q}$;

- If $R = P + Q$ and $P + Q$ is a point of 2-torsion, then this time we have a value with valuation 2. Since $H(x_R) = 0$, the uniformizer is y, so we have

$$v_{P,Q}(R) = \frac{x - x_{P+Q}}{y^2}(x_R) = \frac{1}{H'(x_{P+Q})}.$$

Indeed, if we write $H(x) = (x - x_{P+Q})g(x)$, then since $y^2 = H(x)$ we have $\frac{x - x_{P+Q}}{y^2}(x_R) = \frac{1}{g(x_{P+Q})}$, and we compute $H'(x) = (x - x_{P+Q})g'(x) + g(x)$ so that $H'(x_{P+Q}) = g(x_{P+Q})$.

The extended value of $l_{P,Q}(R)$ is given by the following cases (taking into account that $\operatorname{div}(l_{P,Q}) = [P] + [Q] + [-P - Q] - 3[0_E]$):

- If R is different from P, Q, $-P - Q$, or 0_E, then R is not in the support of $\operatorname{div} l_{P,Q}$ and we have a simple value with valuation 0: $l_{P,Q}(R) = y_R - \alpha x_R - \beta$;

- If $R = 0_E$ then we have a value with valuation -3 and

$$l_{P,Q}(0_E) = \frac{y - \alpha x - \beta}{(x/y)^{-3}}(0_E) = \frac{(y - \alpha x - \beta)x^3}{y^3}(0_E) = 1;$$

- If $R = P$ or $R = Q$ or $R = -P - Q$ but $l_{P,Q}$ is not tangent to E at R, then we have a value with valuation 1. If R is not a point of 2-torsion, then the uniformizer is $t_R = x - x_R$ and the value is

$$l_{P,Q}(R) = \frac{y - \alpha x - \beta}{x - x_R}(R) = \frac{y - y_R - \alpha(x - x_R)}{x - x_R}(R) = \frac{y - y_R}{x - x_R}(R) - \alpha = \frac{H'(x_R)}{2y_R} - \alpha.$$

If R is a point of two torsion, then the uniformizer is $t_R = y$ and the value is

$$l_{P,Q}(R) = \frac{y - \alpha x - \beta}{y}(R) = 1 - \alpha\frac{x - x_R}{y}(R) = 1.$$

- If $R = P$, $R = Q$ or $R = -P - Q$, and $l_{P,Q}$ is tangent to E at R but is not an inflection point, then we have a value of valuation 2. In this case R cannot be a point of 2-torsion, so the uniformizer is $t_R = x - x_R$. To compute the value we must compute the formal series corresponding to y in the completion of $K[E]$ along $x - x_R$ up to order 2: $y = y_R + \alpha(x - x_R) + \alpha_2(x - x_R)^2 + O(x - x_R)^3$. We have $\alpha_2 = \frac{H''(x_R)/2 - \alpha^2}{2y_R}$, so the value is

$$l_{P,Q}(R) = \frac{y - y_R - \alpha(x - x_R)}{(x - x_R)^2}(R) = \alpha_2.$$

- *Finally, when R is an inflection point of H, so that $R = P = Q = -P - Q$ (and in particular is a point of 3-torsion), then we have a value with valuation 3. We compute the formal series corresponding to y in the completion of $K[E]$ along $x - x_R$ up to order 3: $y = y_R + \alpha(x - x_R) + 0(x - x_R)^2 + \alpha_3(x - x_R)^3 + O((x - x_R)^4)$. We have $\alpha_3 = \frac{1}{2y_R}$ and*

$$l_{P,Q}(R) = \frac{y - y_R - \alpha(x - x_R)}{(x - x_R)^3}(R) = \alpha_3.$$

Combining these values we can now compute the extended value of $\mu_{P,Q}(R)$ (taking into account that $\mathrm{div}(\mu_{P,Q}) = [P] + [Q] - [P + Q] - [0_E]$):

- *When R is not equal to P, Q, $P + Q$, $-P - Q$, or 0_E then the valuation is 0 and we have a simple value:*

$$\mu_{P,Q}(R) = \frac{y_R - \alpha x_R - \beta}{x_R - x_{P+Q}}. \tag{3.13}$$

  (If $R = -P - Q$ and R is not in the support of $\mathrm{div}(\mu_{P,Q})$ then the valuation is also 0, but Equation (3.13) is not well defined, so to compute the value we need to look at the particular cases above);

- *When $R = 0_E$ the valuation is -1 and we have*

$$\mu_{P,Q}(0_E) = 1. \tag{3.14}$$

  Since the value is 1 we see that the function $\mu_{P,Q}$ is indeed normalized at 0_E;

- *For all the other cases we refer to the study of the special cases done for $v_{P,Q}$ and $l_{P,Q}$ above.*

Finally, when $P = -Q$ (but $P \neq 0_E$) so that $P + Q = 0_E$, then $\mu_{P,Q} = x - x_P$ and the extended value of $\mu_{P,Q}$ at R is given by the same formulae as the study of $v_{P,Q}(R)$ above.

The second key insight into Miller's algorithm is to speed up the decomposition algorithm above by using a double - and - add algorithm. Indeed, when P is a point on an elliptic curve, the scalar multiplication $P \mapsto r.P$ is computed a lot faster when doing a double-and-add algorithm than when doing a naive decomposition $rP = P + P + \cdots + P$: The complexity is $O(\log r)$ additions rather than $O(r)$. Proposition 3.2 and Algorithm 3.1 outline a similar strategy to evaluate the function f_F where F is the divisor $r[P] - r[0_E]$. More generally, by decomposing a divisor F as $F = F_1 + 2F_2 + 4F_3 + \cdots + 2^n F_n$, one can derive a general double-and-add algorithm for divisor evaluation.

3.4.3   The General Definition of the Weil Pairing

THEOREM 3.10   *Let E be an elliptic curve, r a prime number, and P and Q two points of r-torsion on E. Let D_P be a divisor linearly equivalent to $[P] - [0_E]$ and let D_Q be a divisor linearly equivalent to $[Q] - [0_E]$. Then*

$$e_{W,r}(P,Q) = \epsilon(D_P, D_Q)^r \frac{f_{rD_P}(D_Q)}{f_{rD_Q}(D_P)} \tag{3.15}$$

is well defined, does not depend on the choice of uniformizers nor on the choice of D_P and D_Q ($\epsilon(D_P, D_Q) = \pm 1$ is defined in Theorem 3.9 and has value 1 if D_P and D_Q have disjoint support). Furthermore, the application $E[r] \times E[r] \to \mu_r : (P, Q) \mapsto e_{W,r}(P, Q)$ is a pairing, called the Weil pairing. The pairing $e_{W,r}$ is an alternate pairing, which means that $e_{W,r}(P, Q) = e_{W,r}(Q, P)^{-1}$.

Remark 3.2 We recover Theorem 3.1 by taking $D_P = [P] - [0_E]$ and $D_Q = [Q] - [0_E]$. Indeed, by Proposition 3.2, we get

$$e_{W,r} = (-1)^r \frac{f_{r,P}(Q)}{f_{r,Q}(P)}.$$

Proof. The fact that $e_{W,r}$ is alternate is immediate from Equation (3.15).

We have seen in Section 3.4.2 that the divisor $r[P] - r[0_E]$ is principal. We deduce that if D_P is linearly equivalent to $[P] - [0_E]$ then rD_P is also principal, hence Equation (3.15) is well defined.

Let $D_{P,1}$ and $D_{P,2}$ be two divisors linearly equivalent to $[P] - [0_E]$. Then there exists a rational function $g \in k(E)$ such that $D_{P,1} = D_{P,2} + \operatorname{div} g$. Then

$$\epsilon(D_{P_1}, D_Q)^r \frac{f_{rD_{P,1}}(D_Q)}{f_{rD_Q}(D_{P,1})} = \epsilon(D_{P_1}, D_Q)^r \frac{f_{rD_{P,2}}(D_Q) \cdot g(D_Q)^r}{f_{rD_Q}(D_{P,2}) \cdot f_{rD_Q}(\operatorname{div} g)}. \tag{3.16}$$

But by Weil's reciprocity theorem (Theorem 3.9), we have

$$f_{rD_Q}(\operatorname{div} g) = \epsilon(\operatorname{div} g, rD_Q)g(rD_Q) = \epsilon(\operatorname{div} g, rD_Q)g(D_Q)^r.$$

Since $\epsilon(D_{P_1}, D_Q)^r \epsilon(\operatorname{div} g, rD_Q) = \epsilon(D_{P_1}, D_Q)^r$, Equation (3.16) simplifies to

$$\epsilon(D_{P_1}, D_Q)^r \frac{f_{rD_{P,1}}(D_Q)}{f_{rD_Q}(D_{P,1})} = \epsilon(D_{P_2}, D_Q)^r \frac{f_{rD_{P,2}}(D_Q)}{f_{rD_Q}(D_{P,2})},$$

which shows that $e_{W,r}(P,Q)$ does not depend on the linear equivalence class of D_P. Likewise by (anti-)symmetry, it does not depend on the linear equivalence class of D_Q.

To show that it does not depend on the choice of uniformizers, we can as well take $D_P = [P] - [0_E]$ and $D_Q = [Q] - [0_E]$ (so that $\epsilon(D_P, D_Q) = -1$). Then a function associated to rD_P is the function $f_{rD_P} = f_{r,P}$ defined in Definition 3.4. If R is a point on the elliptic curve, the evaluation $f_{r,P}(R)$ does not depend on the choice of uniformizers, except when R is in the support of $\operatorname{div} f_{r,P}$ (i.e. if $R = P$ or $R = 0_E$).

Going back to the definition of $e_{W,r}(P,Q)$ as

$$e_{W,r}(P,Q) = (-1)^r \frac{f_{r,P}([Q] - [0_E])}{f_{r,Q}([P] - [0_E])} = (-1)^r \frac{f_{r([P]-[0_E])}([Q] - [0_E])}{f_{r([Q]-[0_E])}([P] - [0_E])},$$

we see that the result does not depend on the uniformizers, except possibly when we change the uniformizer for 0_E, and (when $P = Q$) when we change the uniformizer for P. But if we replace the uniformizer x/y for 0_E by $\alpha x/y$, then both the numerator and denominator are multiplied by α^r, hence the result stays the same. Likewise, when $P = Q$ and we change the uniformizer at P (actually from the definition it is obvious that $e_{W,r}(P,P) = 1$, whatever the uniformizer at P).

We are left with showing bilinearity and non-degeneracy. For that it will be convenient to give yet another form of the Weil pairing, which is not convenient for computations but gives easier proofs. If $D = [R]$ is a divisor, we define r^*D as $r^*D = \sum_{S \in E(\overline{K}), rS = R}[S]$. This extends by linearity to define a divisor r^*D for a general divisor D. If D is of degree 0, then r^*D is also of degree 0. Furthermore, if $D = \operatorname{div}(f)$, then $r^*D = \operatorname{div}(f \circ [r])$.

If $D_P = [P] - [0_E]$, then using Proposition 3.1 one can check that r^*D_P is a principal divisor. Let g_P be a function corresponding to r^*D_P. By definition of g_P, if P_0 is a point in E such that $P = rP_0$, then $\operatorname{div} g_P = \sum_{T \in E[r]}[P_0 + T] - [T]$. Now the function $x \mapsto g_P(x+Q)$ has for divisor div $g_P(x+Q) = \sum_{T \in E[r]}[P_0 + T - Q] - [T - Q]$. But since $Q \in E[r]$, then div $g_P(x+Q) = $ div g_P, hence both functions differ by a constant. We claim that this constant is $e_{W,r}(P,Q)$, hence:

$$e_{W,r}(P,Q) = g_P(x+Q)/g_P(x) \tag{3.17}$$

(whenever the right-hand side is well defined).

Fix $D_Q = [Q] - [0_E]$, let Q_0 be such that $Q = rQ_0$, g_Q is a function with divisor $r^* D_Q$ (and normalized at 0_E), and define h_Q to be the function normalized at 0_E with divisor $(r - 1)[Q_0] + [Q_0 - Q] - r[0_E]$ (which exists by Proposition 3.1). Let $H_Q = \prod_{T \in E[r]} h_Q(x+T)$. Then $H_Q = g_Q^r = f_Q \circ r$. Indeed, they all have associated divisor $\sum_{T \in E[r]} r[Q_0 + T] - r[T]$ and are normalized. Now by Theorem 3.9, we have that $h_Q(\text{div } g_P) = (-1)^r g_P(\text{div } h_Q)$, which gives the equation

$$\frac{\prod_{T \in E[r]} h_Q(P_0 + T)}{\prod_{T \in E[r]} h_Q(T)} = (-1)^r g_P^r([Q_0] - [0_E]) \frac{g_P(Q_0 - Q)}{g_P(Q_0)}.$$

Combining with $g_Q^r = H_Q$ we find that

$$g_Q^r([P_0] - [0_E]) = H_Q([P_0] - [0_E]) = (-1)^r g_P^r([Q_0] - [0_E]) \frac{g_P(Q_0 - Q)}{g_P(Q_0)}.$$

Since $g_Q^r = f_Q \circ r$, we have that $f_{r,Q}(D_P) = g_Q^r([P_0] - [0_E])$, and similarly $f_{r,P}(D_Q) = g_P^r([Q_0] - [0_E])$. Putting everything together, we compute

$$e_{W,r}(P,Q) = (-1)^r \frac{f_{rD_P}(D_Q)}{f_{rD_Q}(D_P)} = (-1)^r \frac{f_{r,P}(D_Q)}{f_{r,Q}(D_P)} = (-1)^r \frac{g_P^r([Q_0] - [0_E])}{g_Q^r([P_0] - [0_E])} = \frac{g_P(Q_0)}{g_P(Q_0 - Q)}.$$

which proves Equation (3.17) (with $x = Q_0 - Q$).

Using this reformulation, we compute

$$e_{W,r}(P, Q_1 + Q_2) = \frac{g_P(x + Q_1 + Q_2)}{g_P(x)} = \frac{g_P(x + Q_1 + Q_2)}{g_P(x + Q_2)} \frac{g_P(x + Q_2)}{g_P(x)} = e_{W,r}(P, Q_1) e_{W,r}(P, Q_2)$$

so $e_{W,r}$ is bilinear on the right. Now by (anti-)symmetry, using Equation 3.15, $e_{W,r}$ is also bilinear on the left: $e(P_1 + P_2, Q) = e(P_1, Q)e(P_2, Q)$, so it is indeed bilinear. Furthermore, using bilinearity, $e_{W,r}(P,Q)^r = e_{W,r}(P, 0_E) = 1$, so $e_{W,r}(P,Q)$ is a r-root of unity.

We now show non-degeneracy, following [22, Proposition 8.1]. Once more by symmetry we just need non-degeneracy on the left, that is, given $P \neq 0_E$ we need to show that there exists a Q such that $e_{W,r}(P,Q) \neq 1$. If this were not the case then by Equation 3.17 we would have $g_P(x + Q) = g_P(x)$ for all $Q \in E[r]$. So g_P would be a function invariant by translation by a point of r-torsion; this means that there would exist a rational function g on the curve E such that $g_P = g \circ [r]$ by [22, Theorem 4.10.b]. Then $\text{div}(g_P) = [r]^* \text{div}(g)$, but by definition $\text{div}(g_P) = [r]^* D_P$. So div $g = D_P = [P] - [0_E]$, but D_P is not principal by Proposition 3.1, so this is absurd.   □

3.4.4   The General Definition of the Tate Pairing

THEOREM 3.11 *Let $E/\mathbb{F}_q$ be an elliptic curve, r a prime number dividing $\#E(\mathbb{F}_q)$, $P \in E[r](\mathbb{F}_{q^k})$ a point of r-torsion defined over $\mathbb{F}_{q^k}$, and $Q \in E(\mathbb{F}_{q^k})$ a point of the elliptic curve defined over $\mathbb{F}_{q^k}$. Let D_P be a divisor linearly equivalent to $[P] - [0_E]$ and D_Q be a divisor linearly equivalent to $[Q] - [0_E]$. Then*

$$e_{T,r}(P,Q) = (f_{rD_P}(D_Q))^{\frac{q^k - 1}{r}} \tag{3.18}$$

is well defined, does not depend on the choice of uniformizers or on the choice of D_P and D_Q.

Furthermore the application $E[r](\mathbb{F}_{q^k}) \times E(\mathbb{F}_{q^k})/rE(\mathbb{F}_{q^k}) \to \mu_r : (P,Q) \mapsto e_{T,r}(P,Q)$ is a pairing, called the Tate pairing.

Remark 3.3 There are two versions of the Tate pairing: The first one is to define the pairing as simply $f_{rD_P}(D_Q)$ and see the Tate pairing as a pairing with values in $\mathbb{F}_{q^k}^*/\mathbb{F}_{q^k}^{*,r}$, meaning that we identify two values differing by an r-power. The second one, which we have used in Equation 3.18, is to use the bijection $\mathbb{F}_{q^k}^*/\mathbb{F}_{q^k}^{*,r} \to \mu_r : \gamma \mapsto \gamma^{\frac{q^k-1}{r}}$. Indeed, if $\gamma = \gamma'\alpha^r$, then $(\alpha^r)^{\frac{q^k-1}{r}} = \alpha^{q^k-1} = 1$ so $\gamma^{\frac{q^k-1}{r}} = (\gamma')^{\frac{q^k-1}{r}}$. We call the exponentiation by $\frac{q^k-1}{r}$ the *final exponentiation*, and the value $e_{T,r}(P,Q) = (f_{rD_P}(D_Q))^{\frac{q^k-1}{r}}$ the *reduced Tate pairing*.

There is an important difference to keep in mind between the Weil pairing and the Tate pairing. The Weil pairing is geometric: The value of $e_{W,r}(P,Q)$ does not depend on the field of definition we are working on, whereas the Tate pairing is arithmetic. For instance, if $P \in E[r](\mathbb{F}_{q^k})$ and $Q \in E(\mathbb{F}_{q^k})$, but we look at the Tate pairing over $\mathbb{F}_{q^{rk}}$, then the final exponentiation is to the power of $\frac{q^{rk}-1}{r}$ so that $e_{T,r,\mathbb{F}_{q^{rk}}}(P,Q) = 1$ (the Tate pairing stays non-degenerate over $\mathbb{F}_{q^{rk}}$ but one needs to take Q in $E(\mathbb{F}_{q^{rk}})$ to get a non-trivial pairing with P for the Tate pairing over $\mathbb{F}_{q^{rk}}$).

Proof. We first show that the value does not depend on the linear equivalence class of D_P and D_Q. Unlike the Weil pairing where P and Q played symmetric roles, for the Tate pairing we have to handle the left argument and the right argument separately.

Let $D_{P,2} = D_{P,1} + \operatorname{div}(g)$, where g is a rational function. Let $f_{rD_{P,1}}$ be a function corresponding to the principal divisor $rD_{P,1}$, then a function corresponding to $rD_{P,2}$ is $f_{rD_{P,1}}g^r$. We compute

$$f_{rD_{P,2}}(D_Q)^{\frac{q^k-1}{r}} = f_{rD_{P,1}}(D_Q)^{\frac{q^k-1}{r}} \cdot g(D_Q)^{r\frac{q^k-1}{r}} = f_{rD_{P,1}}(D_Q)^{\frac{q^k-1}{r}}.$$

So we can as well take $D_P = r[P] - r[0_E]$.

Likewise, if $D_{Q,2} = D_{Q,1} + \operatorname{div} h$, then by Theorem 3.9, we have that $f_{rD_P}(\operatorname{div} h) = \epsilon h(\operatorname{div} f_{rD_P}) = \epsilon h(r[P] - r[0_E]) = \epsilon h([P] - [0_E])^r$ where $\epsilon = \epsilon(\operatorname{div} f, \operatorname{div} h)$. Since $\epsilon = \pm 1$, $\epsilon^{\frac{q^k-1}{r}} = 1$ and we compute:

$$f_{rD_P}(D_{Q,2})^{\frac{q^k-1}{r}} = f_{rD_P}(D_{Q,1})^{\frac{q^k-1}{r}} \cdot f_{rD_P}(\operatorname{div} h)^{\frac{q^k-1}{r}} = f_{rD_P}(D_{Q,1})^{\frac{q^k-1}{r}} \cdot h([P] - [0_E])^{r\frac{q^k-1}{r}}$$
$$= f_{rD_P}(D_{Q,1})^{\frac{q^k-1}{r}}.$$

To show that $e_{T,r}$ does not depend on the choice of uniformizers, we can take $D_P = [P] - [0_E]$, $D_Q = [Q] - [0_E]$, and by Proposition 3.2 choose $f_{rD_P} = f_{r,P}$. Since $\operatorname{div}(f_{r,P}) = r[P] - r[0_E]$, changing uniformizers does not affect $f_{r,P}(D_Q)$ except at 0_E and P (when $P = Q$). But if we replace the uniformizer x/y at 0_E by $\gamma x/y$, then the value $f_{r,P}(D_Q)$ is multiplied by γ^r, which is then killed by the final exponentiation. Likewise for the uniformizer at P.

It remains to show that $e_{T,r}$ is a pairing. For simplicity here we assume that $E(\mathbb{F}_{q^k})$ contains all of $E[r]$. For the general case, we refer to [11, 20, 5].

For the bilinearity and the non-degeneracy, as for the Weil pairing it will be more convenient to give an alternative definition of the Tate pairing. Let P and Q be as in the theorem. Let $Q_0 \in E(\overline{\mathbb{F}_q})$ be a point such that $Q = rQ_0$. Let π be the Frobenius endomorphism of $\mathbb{F}_q$, which acts on the points of E. Then π^k is the Frobenius endomorphism of $\mathbb{F}_{q^k}$. Let $Q_1 = \pi^k Q_0 - Q_0$. We compute $rQ_1 = \pi^k rQ_0 - rQ_0 = \pi^k Q - Q = 0_E$ (where we used the fact that scalar multiplication commutes with the Frobenius, and that Q is defined over $\mathbb{F}_{q^k}$, so that $\pi^k Q = Q$). So Q_1 is a point of r-torsion. Furthermore, it does not depend on Q_0: If we replace Q_0 by $Q_0 + T$ where $T \in E[r]$, then we compute $(\pi^k - 1)(Q_0 + T) = Q_1 + (\pi^k - 1)(T) = Q_1$, because $T \in E(\mathbb{F}_{q^k})$. So the application $\frac{\pi^k-1}{r} : E(\mathbb{F}_{q^k}) \to E[r], Q \mapsto Q_1$ is well defined, and it is easy to check that it is an endomorphism of $E_{\mathbb{F}_{q^k}}$. We have

$$e_{T,r}(P,Q) = e_{W,r}\left(P, \frac{\pi^k-1}{r}Q\right). \tag{3.19}$$

Equation (3.19) shows a strong link between the Weil and Tate pairing. To show Equation (3.19), we use Equation (3.17) to get $e_{W,r}(P, \pi^k Q_0 - Q_0) = \frac{g_P(\pi^k Q_0)}{g_P(Q_0)}$. Now since P is defined over $\mathbb{F}_{q^k}$, g_P is in $\mathbb{F}_{q^k}(E)$, so π^k commutes with g_P. We thus get

$$\frac{g_P(\pi^k Q_0)}{g_P(Q_0)} = g_P(Q_0)^{q^k - 1} = (g_P^r(Q_0))^{\frac{q^k-1}{r}} = f_{r,P}(Q)^{\frac{q^k-1}{r}},$$

where in the last equation we have used that $g_P^r = f_{r,P} \circ [r]$. This shows the equivalence between the two definitions of the Tate pairing.

Using Equation (3.19) we see that the Tate pairing is bilinear. For the non-degeneracy, we have to show that $\frac{\pi^k-1}{r} : E(\mathbb{F}_{q^k}) \to E[r]$ is surjective. Indeed, because the Weil pairing is non-degenerate, Equation (3.19) will then show that the Tate pairing is non-degenerate too. The kernel of $\frac{\pi^k-1}{r}$ restricted to $E(\mathbb{F}_{q^k})$ is $rE(\mathbb{F}_{q^k})$, so the image is isomorphic to $E(\mathbb{F}_{q^k})/rE(\mathbb{F}_{q^k})$. Now $E(\mathbb{F}_{q^k})$ is a finite abelian group of the form $\mathbb{Z}/a\mathbb{Z} \oplus \mathbb{Z}/b\mathbb{Z}$ with $a \mid b$, and since $E(\mathbb{F}_{q^k}) \supset E[r]$, we know that $r \mid a$ and $r \mid b$. We deduce that $E(\mathbb{F}_{q^k})/rE(\mathbb{F}_{q^k})$ is isomorphic to $\mathbb{Z}/r\mathbb{Z} \oplus \mathbb{Z}/r\mathbb{Z}$, in particular it has cardinal r^2, so the application is indeed surjective.   □

Taking $D_P = [P] - [0_E]$ and $D_Q = [Q+R] - [R]$ where R is any point in $E(\mathbb{F}_{q^k})$ (this divisor is equivalent to $[Q] - [0_E]$ by Proposition 3.1), we recover the formula from Theorem 3.2:

$$e_{T,r}(P, Q) = \left(\frac{f_{r,P}(Q+R)}{f_{r,P}(R)}\right)^{\frac{q^k-1}{r}}.$$

If we take $R = 0_E$, we find

$$e_{T,r}(P, Q) = f_{r,P}(Q)^{\frac{q^k-1}{r}}.$$

Here Q may be a pole or zero of $f_{r,P}$, so we need to use the general Miller's algorithm to compute the extended evaluation.

Restriction of the Tate pairing to subgroups

We give a proof of Proposition 3.4 that the restriction of the Tate pairing to $\mathbb{G}_1 \times \mathbb{G}_2$ is non-degenerate:

Proof. Recall that since $k > 1$ and the assumptions in Lemma 3.2 hold, $\mathbb{G}_1$ is the subgroup of $E[r]$ of eigenvectors for the eigenvalue 1 and $\mathbb{G}_2$ corresponds to eigenvectors for the eigenvalue $q \neq 1$ mod r. We have already proved in Proposition 3.4 that the restriction of the Weil pairing to $\mathbb{G}_1 \times \mathbb{G}_2$ or to $\mathbb{G}_2 \times \mathbb{G}_1$ is non-degenerate.

Since the endomorphism $\frac{\pi^k-1}{r}$ commutes with the Frobenius π, it stabilizes $\mathbb{G}_1$ and $\mathbb{G}_2$. The alternative definition of the Tate pairing given by Equation (3.19) shows that the Tate pairing restricted to $\mathbb{G}_1 \times \mathbb{G}_2$ or to $\mathbb{G}_2 \times \mathbb{G}_1$ is also non-degenerate.

Likewise, the Tate pairing restricted to $\mathbb{G}_1 \times \mathbb{G}_1$ or to $\mathbb{G}_2 \times \mathbb{G}_2$ is degenerate, because the Weil pairing is degenerate. The same reasoning as in the proof for the Weil pairing shows that the Tate pairing on $\mathbb{G}_1 \times \mathbb{G}_3$ (and the other groups) is non-degenerate.   □

Remark 3.4 By the proof, $e_{T,r}(P, P) = 1$ when $P \in \mathbb{G}_1$ or $P \in \mathbb{G}_2$. However, unlike the Weil pairing, we can have $e_{T,r}(P, P) \neq 1$ when $P \in E[r](\mathbb{F}_{q^k})$ but $P \notin \mathbb{G}_1$ and $P \notin \mathbb{G}_2$. See for instance [16], where the authors study the link between the Tate self pairing and the structure of the isogeny graph.

The case of embedding degree 1

Let E be an elliptic curve defined over $\mathbb{F}_q$ such that $r \mid \#E(\mathbb{F}_q)$. By Lemma 3.2, if the embedding degree k is greater than 1, then $E[r] \subset E(\mathbb{F}_{q^k})$ and we can apply the proof of Theorem 3.11.

If $k = 1$, then $E(\mathbb{F}_q)$ may not contain the full r-torsion, so we can't apply the elementary proof we have given. But even in this case one can still show using Galois cohomology that both Theorem 3.11 and the alternative definition of the Tate pairing given by Equation 3.19 stay true. In this case $\frac{\pi^k - 1}{r}$ is not a well-defined endomorphism, but represents a cocycle in a Galois cohomology class such that Equation 3.19 stays well defined over $\mathbb{F}_q$.

Moreover, when $E(\mathbb{F}_q)$ does not contain points of r^2-torsion, then by a similar argument as in Proposition 3.4, we can show that $e_{T,r} : E[r](\mathbb{F}_q) \times E[r](\mathbb{F}_q) \to \mu_r \subset \mathbb{F}_q^*$ is still a pairing. In particular, when the rational r-torsion is cyclic, if $P \in E[r](\mathbb{F}_q)$ then $e_{T,r}(P, P) \neq 1$.

3.4.5 The Optimal Ate and Twisted Optimal Ate Pairing

In order to prove the formulae for the Ate and twisted Ate, we need the following lemma.

LEMMA 3.5 *Let E be an elliptic curve defined over a finite field $\mathbb{F}_q$*

- *For any point P on the elliptic curve E*

$$f_{ab,P} = f_{a,P}^b \cdot f_{b,aP}. \tag{3.20}$$

- *Let ϕ be an endomorphism of E of degree d, with trivial kernel. Then for any integer λ*

$$f_{\lambda,\phi(P)} = f_{\lambda,P}^d.$$

Proof. The first equation may be proved easily by writing down the divisors for the functions involved. For the second item, see [15]. □

We prove here Theorem 3.4.

Proof. Let $l = \phi(k)$. It is easy to see that:

$$f_{\lambda,Q}(P) = \prod_{i=0}^{l-1} f_{c_i q^i, Q}(P) \prod_{i=0}^{l-1} \frac{l_{s_{i+1}Q, c_i q^i Q}(P)}{v_{s_i Q}(P)}.$$

By Equation 3.20 and Lemma 3.5, we compute $f_{c_i q^i, Q}(P)$ as

$$f_{c_i q^i, Q}(P) = f_{q^i, Q}^{c_i} f_{c_i, q^i Q}(P) = f_{q^i, Q}^{c_i}(P) f_{c_i, Q}^{q^i}(P). \tag{3.21}$$

As a consequence, we obtain that

$$e_{T,r}(Q, P)^m = \prod_{i=0}^{l} \left(f_{q^i, Q}^{c_i}(P) \right)^{(q^k - 1)/r} \cdot a_{[c_0, \dots, c_l]}(Q, P).$$

Since the left-hand side and the factor in brackets are pairings, we conclude that $a_{[c_0, \dots, c_l]}$ is a bilinear map. By Theorem 3.3, we have that the left-hand side is

$$e_{T,r}(Q, P)^m = f_{q,Q}(P)^{mkq^{k-1}((q^k - 1)/r)^{-1}}.$$

The product on the right-hand side right writes as

$$\prod_{i=0}^{l} \left(f_{q^i,Q}^{c_i}(P) \right) = f_{q,Q}(P)^{\sum_{i=0}^{l} ic_i q^{i-1}}.$$

We conclude that if $mkq^{k-1}((q^k-1)/r)^{-1} \not\equiv \sum_{i=0}^{l} ic_i q^{i-1}$, then $a_{[c_0,\ldots,c_l]}$ is a non-degenerate map. This concludes the proof for the optimal Ate pairing. □

For the twisted optimal Ate pairing, the proof of Theorem 3.8 is made similar to the one above by inverting the roles of P and Q. We give it below.

Proof. Note that by Theorem 3.5, we have that

$$\mathbb{G}_2 = \mathrm{Ker}(\xi_d \circ \pi_{q^e} - \mathrm{Id}).$$

It follows easily that

$$\mathbb{G}_1 = \mathrm{Ker}(\xi_d \circ \pi_{q^e} - q^e \, \mathrm{Id}).$$

As a consequence, in (3.21) we compute $f_{c_i q^{ie},P}(Q)$ by applying Lemma 3.5 for the endomorphism $\xi_d \circ \pi_{q^e}$. The rest of the computation follows naturally. □

References

[1] Christophe Arène, Tanja Lange, Michael Naehrig, and Christophe Ritzenthaler. Faster computation of the Tate pairing. *Journal of Number Theory*, 131(5):842–857, 2011.

[2] Razvan Barbulescu, Pierrick Gaudry, Antoine Joux, and Emmanuel Thomé. A heuristic quasi-polynomial algorithm for discrete logarithm in finite fields of small characteristic. In P. Q. Nguyen and E. Oswald, editors, *Advances in Cryptology – EUROCRYPT 2014*, volume 8441 of *Lecture Notes in Computer Science*, pp. 1–16. Springer, Heidelberg, 2014.

[3] Dan Boneh and Matthew K. Franklin. Identity based encryption from the Weil pairing. *SIAM Journal on Computing*, 32(3):586–615, 2003.

[4] Dan Boneh, Ben Lynn, and Hovav Shacham. Short signatures from the Weil pairing. *Journal of Cryptology*, 17(4):297–319, 2004.

[5] Peter Bruin. The Tate pairing for abelian varieties over finite fields. *J. de theorie des nombres de Bordeaux*, 23(2):323–328, 2011.

[6] Craig Costello, Tanja Lange, and Michael Naehrig. Faster pairing computations on curves with high-degree twists. In P. Q. Nguyen and D. Pointcheval, editors, *Public Key Cryptography – PKC 2010*, volume 6056 of *Lecture Notes in Computer Science*, pp. 224–242. Springer, Heidelberg, 2010.

[7] Gerhard Frey and Hans-Georg Rück. A remark concerning m-divisibility and the discrete logarithm in the divisor class group of curves. *Mathematics of Computation*, 62(206):865–874, 1994.

[8] Steven D. Galbraith and Victor Rotger. Easy decision Diffie-Hellman groups. *LMS Journal of Computation and Mathematics*, 7:201–218, 2004.

[9] Robert Granger, Dan Page, and Nigel P. Smart. High security pairing-based cryptography revisited. In F. Hess, S. Pauli, and M. E. Pohst, editors, *Algorithmic Number Theory (ANTS-VII)*, volume 4076 of *Lecture Notes in Computer Science*, pp. 480–494. Springer, 2006.

[10] F. Hess, N.P. Smart, and F. Vercauteren. The eta pairing revisited. Cryptology ePrint Archive, Report 2006/110, 2006. http://eprint.iacr.org/2006/110.

[11] Florian Hess. A note on the Tate pairing of curves over finite fields. *Archiv der Mathematik*, 82(1):28–32, 2004.

[12] Florian Hess. Pairing lattices (invited talk). In S. D. Galbraith and K. G. Paterson, editors, *Pairing-Based Cryptography – Pairing 2008*, volume 5209 of *Lecture Notes in Computer Science*, pp. 18–38. Springer, Heidelberg, 2008.

[13] Florian Hess, Nigel P. Smart, and Frederik Vercauteren. The Eta pairing revisited. *IEEE Transactions on Information Theory*, 52(10):4595–4602, 2006.

[14] Sorina Ionica and Antoine Joux. Another approach to pairing computation in Edwards coordinates. In D. R. Chowdhury, V. Rijmen, and A. Das, editors, *Progress in Cryptology – INDOCRYPT 2008*, volume 5365 of *Lecture Notes in Computer Science*, pp. 400–413. Springer, Heidelberg, 2008.

[15] Sorina Ionica and Antoine Joux. Pairing computation on elliptic curves with efficiently computable endomorphism and small embedding degree. In M. Joye, A. Miyaji, and A. Otsuka, editors, *Pairing-Based Cryptography – Pairing 2010*, volume 6487 of *Lecture Notes in Computer Science*, pp. 435–449. Springer, Heidelberg, 2010.

[16] Sorina Ionica and Antoine Joux. Pairing the volcano. *Mathematics of Computation*, 82(281):581–603, 2013.

[17] Neal Koblitz and Alfred Menezes. Pairing-based cryptography at high security levels (invited paper). In N. P. Smart, editor, *Cryptography and Coding*, volume 3796 of *Lecture Notes in Computer Science*, pp. 13–36. Springer, Heidelberg, 2005.

[18] Stephen Lichtenbaum. Duality theorems for curves over *p*-adic fields. *Inventiones mathematicae*, 7(2):120–136, 1969.

[19] Alfred Menezes, Scott A. Vanstone, and Tatsuaki Okamoto. Reducing elliptic curve logarithms to logarithms in a finite field. In *23rd Annual ACM Symposium on Theory of Computing*, pp. 80–89. ACM Press, 1991.

[20] Edward F Schaefer. A new proof for the non-degeneracy of the Frey-Rück pairing and a connection to isogenies over the base field. *Computational aspects of algebraic curves*, 13:1–12, 2005.

[21] Jean-Pierre Serre. *Groupes algébriques et corps de classes*, volume 7 of *Publications de l'Institut de mathématique de l'Université de Nancago*. Hermann, 2nd edition, 1975.

[22] Joseph H. Silverman. *The Arithmetic of Elliptic Curves*, volume 106 of *Graduate Texts in Mathematics*. Springer-Verlag, 2nd edition, 2009.

[23] William Stein. *SAGE: Software for Algebra and Geometry Experimentation*. http://www.sagemath.org/.

[24] John Tate. WC-groups over *p*-adic fields. Exposé 156, Séminaire Bourbaki, 1957/58.

[25] F. Vercauteren. Optimal pairings. Cryptology ePrint Archive, Report 2008/096, 2008. http://eprint.iacr.org/2008/096.

[26] Eric R. Verheul. Evidence that XTR is more secure than supersingular elliptic curve cryptosystems. *Journal of Cryptology*, 17(4):277–296, 2004.

[27] Chang-An Zhao, Fangguo Zhang, and Jiwu Huang. A note on the Ate pairing. *International Journal of Information Security*, 7(6):379–382, 2008.

4

Pairing-Friendly Elliptic Curves

Safia Haloui
Université Paris 8

Edlyn Teske
Independent Researcher

4.1 Introduction

Pairing-based cryptosystems require elliptic curves that are secure and enable efficient pairing computation. It turns out that, for a random elliptic curve, these two conditions are rarely both satisfied. Therefore, specific construction methods have been developed. In this chapter, we survey the main methods used for generating elliptic curves suitable for pairing-based cryptography.

Let E be an elliptic curve over a finite field $\mathbb{F}_q$. The *embedding degree* k of E with respect to a prime divisor r of $\#E(\mathbb{F}_q)$ coprime to q is defined to be the smallest integer such that r divides $q^k - 1$. If r does not divide $q - 1$, then k is the degree of the smallest extension of $\mathbb{F}_q$ over which the full r-torsion of E is defined, and therefore, over which the Weil or Tate pairings and their variants are defined.

In order to guarantee security, the discrete logarithm problem should be computationally infeasible in cyclic subgroups of $E[r]$ and in $\mathbb{F}_{q^k}^*$. The best-known algorithm for discrete logarithm computation on elliptic curves is the parallelized Pollard rho algorithm, which has a running time $\mathcal{O}(\sqrt{r})$ [37, 48]. On the other hand, the best known algorithms for discrete logarithm computation in finite fields are index calculus attacks, which have a running time subexponential in the field size. Notice that index calculus attacks have recently been improved (see Chapter 9) and are likely to get more efficient in the near future. In particular, it is now recommended to avoid fields of characteristic 2 and 3 (see [1, 20] for the description of concrete attacks in these cases).

TABLE 4.1   Sizes of curve parameters and corresponding embedding degrees to obtain commonly desired levels of security.

Security level in bits	Subgroup size r in bits	Extension field size q^k in bits	Embedding degree k	
			$\rho \approx 1$	$\rho \approx 2$
80	160	960 – 1280	6 – 8	3 – 4
112	224	2200 – 3600	10 – 16	5 – 8
128	256	3000 – 5000	12 – 20	6 – 10
192	384	8000 – 10000	20 – 26	10 – 13
256	512	14000 – 18000	28 – 36	14 – 18

In order to guarantee efficient arithmetic on the elliptic curve, it is also important that r is a large factor of $\#E(\mathbb{F}_q)$. A convenient way to formalize this idea is to consider the quantity

$$\rho = \frac{\log q}{\log r}.$$

When q is large, $\#E(\mathbb{F}_q)$ has roughly the same bit size as q (because of the Hasse-Weil bounds, see Chapter 2), so ρ measures the ratio between the size of $\#E(\mathbb{F}_q)$ and the size of r. If $\#E(\mathbb{F}_q)$ is prime (the "ideal" case), then $\rho \approx 1$. In any case, ρ should be reasonably close to 1. Notice that it is fairly easy to generate curves with $\rho \approx 2$ for any k and r of arbitrary size using the Cocks-Pinch method (see Section 4.4.1).

The value of k is entirely determined by ρ and the choice of the bit sizes of r and q^k, since $\log q^k / \log r = k\rho$. Table 4.1, taken from [18], gives ranges for r and q^k (q prime) and k to match commonly desired levels of security. Notice that in some particular cases, it may be interesting to choose r or q^k with higher size than the sizes proposed by Table 4.1 in order to reach embedding degrees that would allow for curves with particularly efficient pairing computation, or other interesting properties (see Chapter 11).

Based on the above discussion, Freeman, Scott, and Teske [18] gave the following definition:

DEFINITION 4.1   An elliptic curve $E/\mathbb{F}_q$ is *pairing-friendly* if the following two conditions hold:

1. $\#E(\mathbb{F}_q)$ has a prime factor $r \geq \sqrt{q}$,
2. the embedding degree of E with respect to r is less than $\log_2(r)/8$.

The first condition in Definition 4.1 is equivalent to $\rho \leq 2$. The bound $\log_2(r)/8$ in the second condition in Definition 4.1 is chosen to roughly reflect the bounds on k given in Table 4.1.

Pairing-friendly elliptic curves are rare: Balasubramanian and Koblitz [3] proved that for a random elliptic curve E over a random field $\mathbb{F}_q$, the probability that E has embedding degree less than $(\log q)^2$ with respect to some prime $r \approx q$ is vanishingly small, and that the embedding degree should be expected to be around r. Moreover, according to the results of Luca and Shparlinski [29] and Urroz, Luca, and Shparlinski [47] there are very few finite fields on which there exists an elliptic curve having a fixed small embedding degree with respect to some $r \geq \sqrt{q}$. In conclusion, trying to find a pairing-friendly curve by selecting curves randomly and counting their points is hopeless because it is highly unlikely that a random curve has the desired properties.

The first proposed elliptic curves with prescribed embedding degree are supersingular (i.e., have $\gcd(q, t) > 1$) [8, 9, 22]. However, a supersingular elliptic curve always has embedding degree at most 6, and at most 3 if we exclude characteristics 2 and 3 (see Section 4.3 for a discussion about supersingular elliptic curves). In order to achieve higher embedding degrees, we must be able to construct ordinary elliptic curves with a prescribed embedding degree. The only known way to solve this problem is to first find parameters of a curve with the desired properties and then construct a curve with these parameters via the Complex Multiplication method. The CM method is described in Section 4.2.1. Finding the parameters of an ordinary pairing-friendly elliptic curve will be the main topic of this chapter.

4.2 Generating Pairing-Friendly Elliptic Curves

Let $E/\mathbb{F}_q$ be an elliptic curve, r be a prime not dividing q, and k be an integer such that $E[r] \subseteq E(\mathbb{F}_{q^k})$. As explained in Chapter 3, the Weil Pairing $e_{W,r} : E[r] \times E[r] \to \mu_r$ (where μ_r is the group of rth roots of unity in an algebraic closure of $\mathbb{F}_q$) is defined by rational functions. It follows that μ_r is a subgroup of the multiplicative group $\mathbb{F}_{q^k}^*$, or in other words, that r divides $q^k - 1$. Balasubramanian and Koblitz [3] proved that under few assumptions, this last condition is sufficient to ensure that $E[r] \subseteq E(\mathbb{F}_{q^k})$:

PROPOSITION 4.1 ([3]) *Let $E/\mathbb{F}_q$ be an elliptic curve and r be a prime that divides $\#E(\mathbb{F}_q)$ but does not divide $q - 1$. Then $E(\mathbb{F}_{q^k})$ contains r^2 points of r-torsion if and only if $r \mid (q^k - 1)$.*

Proof. We only need to prove sufficiency. If $r \mid (q^k - 1)$, then $r \nmid q$, so that $E[r] \simeq \mathbb{Z}/r\mathbb{Z} \times \mathbb{Z}/r\mathbb{Z}$ (see Chapter 2) and thus $E[r]$ can be seen as a 2-dimensional $\mathbb{F}_r$-vector space. By assumption, $E(\mathbb{F}_q)$ contains a non-trivial r-torsion point P. Let Q be such that (P, Q) is a basis of $E[r]$. In this basis, the matrix of the Frobenius π_q is of the form $\left(\begin{smallmatrix} 1 & a \\ 0 & q \end{smallmatrix} \right)$ (taking into account that its determinant is q). Setting $Q' = Q + bP$, where $b = (q-1)^{-1}a \bmod r$ (b is well-defined since $r \nmid (q-1)$), we get $\pi_q(Q') = qQ + aP + bP = qQ' - (q-1)bP + aP = qQ'$, where the third equality comes from the fact that $(q-1)b = a \bmod r$. Then the matrix of the Frobenius in the basis (P, Q') is $\left(\begin{smallmatrix} 1 & 0 \\ 0 & q \end{smallmatrix} \right)$ and its kth power is the identity (because $q^k = 1 \bmod r$), which means that $E[r] \subseteq E(\mathbb{F}_{q^k})$. $\qquad\square$

DEFINITION 4.2 Let $E/\mathbb{F}_q$ be an elliptic curve and r be a prime divisor of $\#E(\mathbb{F}_q)$ coprime to q. The *embedding degree* k of E with respect to r (or to a subgroup of $E(\mathbb{F}_q)$ of order r) is the smallest integer such that r divides $q^k - 1$.

The discussion above shows that if the embedding degree k is greater than 1, then it is also the degree of the smallest extension on which the full r-torsion is defined.

Remark 4.1 Hitt [21] observed that the Weil and the Tate pairings actually take values in $\mathbb{F}_p(\mu_r)$, where $q = p^m$, which can be a proper subfield of $\mathbb{F}_{q^k}$ if q is not prime. An example of this phenomenon is given in Section 4.3, case $k = 3$ (see also [7]).

Let r be a prime divisor of $\#E(\mathbb{F}_q) = q + 1 - t$ coprime to q. The property "E has embedding degree k with respect to r" only depends on the triple (r, t, q). Conversely, if (r, t, q) is a triple satisfying the conditions defining this property, such that q is a prime or a power of a prime and such that $t \in \left] -2\sqrt{q}, 2\sqrt{q} \right[$ is an integer coprime to q, then there exists an ordinary elliptic curve $E/\mathbb{F}_q$ with $\#E(\mathbb{F}_q) = q + 1 - t$ [14, 49], which will have embedding degree k with respect to r. No general method to get an equation for such a curve exists, but the particular case where

$4q - t^2$ has a small square-free part (less than 10^{12}, taking into account Sutherland's work [45]) can be handled using the *Complex Multiplication (CM) method* due to Atkin and Morain [2].

4.2.1 The Complex Multiplication Method

We give a quick overview of how the CM method works. For simplicity, we assume that q is prime. The classical references are [2, 12] and [27, 43] for the mathematical background.

Let $t \in \left]-2\sqrt{q}, 2\sqrt{q}\right[$ be an integer coprime to q and D the positive square-free integer defined by $4q - t^2 = Dy^2$, $y \in \mathbb{Z}$. The endomorphism ring of an elliptic curve with $q + 1 - t$ points is isomorphic to an order in the imaginary quadratic field $\mathbf{K} = \mathbb{Q}(\sqrt{-D})$ containing the Frobenius $\pi_q = t \pm y\sqrt{-D}$. This property nearly determines its number of points: If $E/\mathbb{F}_q$ has an endomorphism ring isomorphic to an order $\mathcal{O}$ of $\mathbf{K}$ containing π_q, then both the Frobenius of E and π_q have norm q, so they must be equal up to complex conjugation and multiplication by a unit, and therefore, E has a twist with $q + 1 - t$ points.

We are led to consider the problem of finding the equation of a curve having an endomorphism ring isomorphic to an order $\mathcal{O}$ of $\mathbf{K}$ containing π_q, and actually, we can always take $\mathcal{O}$ to be the ring of algebraic integers $\mathcal{O}_\mathbf{K}$ of $\mathbf{K}$.

There are two particular cases where this problem can be directly solved using elementary arguments, namely $D = 1$ or 3. The rings of integers of $\mathbb{Q}(\sqrt{-1})$ and $\mathbb{Q}(\sqrt{-3})$ are the only orders in quadratic imaginary fields containing more than 2 units; they contain 4 and 6 units, respectively. The corresponding curves are easy to identify since the only shapes of Weierstrass equations allowing more than 2 automorphisms are $y^2 = x^3 + cx$, $c \in \mathbb{F}_q^*$ (4 automorphisms) and $y^2 = x^3 + c$, $c \in \mathbb{F}_q^*$ (6 automorphisms).

The problem of finding an elliptic curve E with $\mathrm{End}(E) \simeq \mathcal{O}_\mathbf{K}$ has a natural solution over the field of complex numbers. Indeed, elliptic curves over $\mathbb{C}$ "correspond" to one-dimensional complex tori, so we can take E to be the quotient of $\mathbb{C}$ by $\mathcal{O}_\mathbf{K}$ (or more generally, by any fractional ideal of $\mathcal{O}_\mathbf{K}$). A Weierstrass equation for the corresponding curve can be recovered from its j-invariant $j(E)$. An important result of CM theory is that $j(E)$ is an algebraic integer (actually, more is true: $\mathbf{K}(j(E))$ is the Hilbert class field of $\mathbf{K}$). The minimal polynomial $H_D(x)$ of $j(E)$ (the Hilbert class polynomial of $\mathbf{K}$) can be computed using the theory of modular functions. Finally, it can be proved that the endomorphism ring of E is going to "behave well" when we reduce E modulo a prime [27], Theorem 13.4.12.

ALGORITHM 4.1   The CM method.

Input: q prime, $t \in \left]-2\sqrt{q}, 2\sqrt{q}\right[$ a nonzero integer and D the positive square-free integer defined by $4q - t^2 = Dy^2$, $y \in \mathbb{Z}$.

1   **if** $D = 1$ **then** let $E/\mathbb{F}_q$ be defined by $y^2 = x^3 + x$
2   **if** $D = 3$ **then** let $E/\mathbb{F}_q$ be defined by $y^2 = x^3 + 1$
3   **else**
4 | compute the Hilbert class polynomial $H_D(x)$
5 | compute a root j of $H_D(x)$ in $\mathbb{F}_q$
6 | let $E/\mathbb{F}_q$ be defined by $y^2 = x^3 - 3cx + 2c$, $c = j/(j - 1728)$
7   return the twist of E having $q + 1 - t$ points.

The discussion above is summarized in Algorithm 4.1. The main problem is that step 4 can be done efficiently only if $D < 10^{12}$ [45]. The right twist of E can be found by using the fact that if an elliptic curve $E'/\mathbb{F}_q$ has N points, then for any point $P \in E'(\mathbb{F}_q)$, we have $NP = P_\infty$; another way to proceed is to use the method proposed by Rubin and Silverberg [39].

4.2.2 Conditions on the Parameters

In order to be able to construct pairing-friendly ordinary elliptic curves with prescribed embedding degree k, we need to find triples (r, t, q) such that

1. q is prime or a power of a prime,
2. r is prime,
3. t is coprime to q,
4. r divides $q + 1 - t$,
5. $r \mid (q^k - 1)$ and $r \nmid (q^i - 1)$ for $1 \le i < k$,
6. $4q - t^2 = Dy^2$, for some sufficiently small positive integer D and some integer y.

Conditions 3 and 6 ensure that there exists an ordinary elliptic curve $E/\mathbb{F}_q$ with $\#E(\mathbb{F}_q) = q + 1 - t$, which can be efficiently constructed via Algorithm 4.1 (in particular, condition 6 implies that $t \in \;]{-2\sqrt{q}}, 2\sqrt{q}[$). In some cases, it may be preferable to fix D in advance (for instance, if we want to guarantee that the curve to be constructed has some extra twists). If q is prime, then condition 3 is equivalent to $t \ne 0$; in the case where $t = 0$, it is still possible to efficiently construct an elliptic curve with the desired properties (see Section 4.3). Notice that all the known methods for generating (r, t, q) satisfying these conditions are generally outputting q prime.

It can be interesting to write condition 4 as $q + 1 - t = hr$ for some integer h, called the co-factor. Then, condition 6 becomes

$$4hr - (t - 2)^2 = Dy^2.$$

For any integer $\ell \ge 1$, we denote by $\Phi_\ell(x)$ the ℓth cyclotomic polynomial and $\varphi(\ell)$ its degree (so φ is Euler's totient function). We recall that cyclotomic polynomials have integer coefficients and can be defined recursively by setting $\Phi_1(x) = x - 1$ and using the formula

$$x^\ell - 1 = \prod_{d \mid \ell} \Phi_d(x) \tag{4.1}$$

for $\ell > 1$, so the roots of $\Phi_\ell(x)$ in $\overline{\mathbb{Q}}$ are exactly the primitive ℓth roots of unity. For more details about the theory of cyclotomic polynomials, see Lidl and Niederreiter's book [28].

The next proposition gives a different statement of condition 5 above, which is easier to handle in practice:

PROPOSITION 4.2 *Let k be a positive integer, q be a prime or a power of a prime, and r be a prime not dividing kq. Then the following conditions are equivalent:*

1. *$r \mid (q^k - 1)$ and $r \nmid (q^i - 1)$ for $1 \le i < k$,*
2. *$r \mid \Phi_k(q)$.*

Proof. If $r \mid (q^k - 1)$ and $r \nmid (q^i - 1)$ for $1 \le i < k$, then (4.1) and the fact that r is prime imply that r divides $\Phi_k(q)$. Conversely, if $r \mid \Phi_k(q)$, then $r \mid (q^k - 1)$, and it remains to check that $r \nmid (q^i - 1)$ for $1 \le i < k$. We follow Menezes' proof [31], Lemma 6.2. Let $f(x) = x^k - 1$. Since $r \nmid k$, we have $\gcd(f(x), f'(x)) = 1$ in $\mathbb{F}_r[x]$, so $f(x)$ has only single roots in $\mathbb{F}_r$. Using (4.1) and the fact that q is a root of $\Phi_k(x)$ over $\mathbb{F}_r$, we obtain $\Phi_d(q) \ne 0 \mod r$ for $d \mid k$, $1 \le d < k$. Therefore, $r \nmid (q^d - 1)$, for $d \mid k$, $1 \le d < k$. Finally, $r \nmid (q^i - 1)$ for any i that does not divide k, since in this case we would have $r \mid q^{\gcd(i,k)} - 1$. $\square$

Taking into account that $q = t - 1 \mod r$, Proposition 4.2 tells us that we can replace condition 5 above with the following:

5'. r divides $\Phi_k(t - 1)$.

4.2.3　Classification of Pairing-Friendly Elliptic Curves

A large number of constructions of pairing-friendly elliptic curves with prescribed embedding degree has been proposed. In order to help to navigate through the forest of constructions, Freeman, Scott, and Teske [18] introduced the following classification:

1. Curves not in families:

    (a) supersingular elliptic curves,

    (b) ordinary elliptic curves (Cocks-Pinch curves [13], Dupont-Enge-Morain curves [15]),

2. Families of curves:

    (a) sparse families (MNT curves [33] and their generalizations [19, 42], Freeman's family [17]),

    (b) complete families (cyclotomic families, sporadic families, Scott-Barreto families).

The highest-level distinction they make is between methods that construct individual curves and those that construct families of curves. Supersingular elliptic curves, which are discussed in Section 4.3, do not fall into families. There are also two constructions in the literature that produce ordinary elliptic curves with prescribed embedding degree that are not given in terms of families: the method of Cocks and Pinch [13] and that of Dupont, Enge, and Morain [15]; we discuss them in Section 4.4. The first-mentioned method has the advantage to offer more flexibility for the choice of the size of r and can be modified to construct complete families of curves. The remaining constructions of ordinary elliptic curves with prescribed embedding degree fall into the category of *families of curves*. These methods produce polynomials $r(x), t(x), q(x) \in \mathbb{Q}[x]$ such that the triple $(r(x_0), t(x_0), q(x_0))$ is expected to satisfy conditions from Section 4.2.2 for infinitely many $x_0 \in \mathbb{Z}$; for more details about the notion of families, see Section 4.5. The construction of curves from a family depends on our being able to find integers x, y satisfying an equation of the form $Dy^2 = 4q(x) - t(x)^2$ for some fixed positive integer D. For some constructions, y can be written as a polynomial in x; the corresponding families are referred to as *complete* (see Section 4.5.2 for examples of complete families). The families that are not complete are referred to as *sparse* (see Section 4.5.1 for examples of sparse families).

4.3　Supersingular Elliptic Curves

We start by recalling the well-known Deuring-Waterhouse theorem, which gives the possible traces for a supersingular elliptic curve:

THEOREM 4.1 ([14, 49]) *Let t be an integer. There exists a supersingular elliptic curve over $\mathbb{F}_q$, $q = p^s$, p prime with $(q + 1 - t)$ rational points if and only if one of the following conditions is satisfied:*

1. *$t = 0$, s is odd, or s is even and $p \neq 1 \mod 4$,*
2. *$t = \pm\sqrt{q}$, s is even and $p \neq 1 \mod 3$,*
3. *$t = \pm\sqrt{pq}$, s is odd and $p = 2$ or 3,*
4. *$t = \pm 2\sqrt{q}$, s is even.*

A consequence of this theorem is that supersingular elliptic curves have embedding degree $k \in \{1, 2, 3, 4, 6\}$ in general and $k \leq 3$ if we assume that $p > 3$. Therefore, any supersingular elliptic curve is pairing-friendly in the sense of Definition 4.1 as long as it has a large prime order subgroup. Moreover, contrary to ordinary pairing-friendly elliptic curves with $k \geq 2$, supersingular elliptic curves have the advantage of having distortion maps (i.e., endomorphisms inducing a non-trivial map from $\mathbb{G}_1$ to $\mathbb{G}_2$).

Due to the recent results of Granger, Kleinjung, and Zumbragel [20] and Adj, Menezes, Oliveira, and Rodriguez-Henriquez [1], it is advised to avoid supersingular elliptic curves on fields of characteristic 2 or 3, so we will not consider these cases. We discuss the cases where $k = 2$ or 3 and q is a prime or the square of a prime p. In all the remaining content of this section, we assume that $p > 3$.

Embedding degree $k = 2$. An elliptic curve $E/\mathbb{F}_q$ has embedding degree 2 with respect to an odd prime r if and only if r divides $q + 1$. According to Theorem 4.1, any supersingular elliptic curve over a prime field $\mathbb{F}_q$ has $q + 1$ points, and thus has embedding degree 2 with respect to any odd order subgroup.

Elliptic curves over $\mathbb{F}_q$, where q is prime, having equation $y^2 = x^3 + cx$, $c \in \mathbb{F}_q^*$ or $y^2 = x^3 + c$, $c \in \mathbb{F}_q^*$ are supersingular if and only if $q = 3 \mod 4$ and $q = 2 \mod 3$, respectively [44]. Therefore, if $q \neq 1 \mod 12$, it is easy to get the equation of a supersingular elliptic curve over $\mathbb{F}_q$.

In the case where $q = 1 \mod 12$, following Bröker's method [12], we can use the fact that an elliptic curve over a number field with complex multiplication by $\mathbf{K} = \mathbb{Q}(\sqrt{-D})$ has supersingular reduction modulo a good reduction prime of norm q if and only if $\left(\frac{-D}{q}\right) \neq 1$ (see [27], Theorem 13.4.12). So a supersingular elliptic curve over $E/\mathbb{F}_q$ can be constructed using Algorithm 4.1 with any small prime D such that $\left(\frac{-D}{q}\right) = -1$ and $D = 3 \mod 4$ as an input (the condition $D = 3 \mod 4$ guarantees that $H_D(x)$ has a root in $\mathbb{F}_q$, see [12]).

Therefore, the case $k = 2$ offers lots of flexibility: If we fix some prime r and choose any h such that $q = hr - 1$ is a prime, then we can efficiently construct a supersingular elliptic curve $E/\mathbb{F}_q$ with $hr = q + 1$ points and embedding degree 2 with respect to r. Moreover, Koblitz and Menezes [25] give some explicit determinations of distortion maps in the case where E has equation $y^2 = x^3 + cx$ or $y^2 = x^3 + c$.

Embedding degree $k = 3$. It can be seen from Theorem 4.1 that a supersingular elliptic curve over $\mathbb{F}_q$ has embedding degree 3 with respect to a subgroup of order $r > 3$ if and only if $q = p^s$ with s even and $p = 2 \mod 3$, and $t = \pm\sqrt{q}$. The curves satisfying these conditions are exactly the curves with equation $y^2 = x^3 + c$, where c is a non-cube in $\mathbb{F}_q^*$ [34]. For $q = p^2$, $\#E(\mathbb{F}_q)$ is equal to $\Phi_6(p)$ if $t = p$ and $\Phi_3(p)$ if $t = -p$, so the minimal embedding field is $\mathbb{F}_{p^6} = \mathbb{F}_{q^3}$ in the first case and $\mathbb{F}_{p^3} = \mathbb{F}_{q^{3/2}}$ in the second case.

4.4 General Methods

In this section, we describe the Cocks-Pinch method and the Dupont-Enge-Morain method. These methods can be used to construct curves of arbitrary embedding degree, but they both give curves with $\rho \approx 2$. However, the Cocks-Pinch method can be generalized to produce complete families with $\rho < 2$ (see Section 4.5.2). The Cocks-Pinch method also has the advantage to offer more flexibility in the choice of r.

Let k be a fixed arbitrary integer. Both methods consist in finding some integers r prime, t, y and a discriminant D such that

1. $\Phi_k(t - 1) = 0 \mod r$,
2. $Dy^2 + (t - 2)^2 = 0 \mod r$.

If $q = (t^2 + Dy^2)/4$ is a prime number (or a power of a prime number, but this is unlikely), then we are done since $4(q + 1 - t) = Dy^2 + (t - 2)^2 = 0 \mod r$.

4.4.1   The Cocks-Pinch Method

We fix D. Condition 1 above says that $t - 1$ is a primitive kth root of unity in $\mathbb{F}_r$, in particular the cyclic group $\mathbb{F}_r^*$ has elements of order k, which is equivalent to $k \mid (r-1)$. Condition 2 implies that $-D$ is a square modulo r. If we pick some r satisfying these two conditions, it is possible to compute t and y with the desired properties. The Cocks-Pinch method is summarized in Algorithm 4.2. We see that the best possible ρ-values are achieved when t and y in step 4 are chosen to have an absolute value smaller than r (in general, this yields two possibilities for each parameter). In this case, we can expect that t and y will have the same size as r and that q will have twice the size of r, so $\rho \approx 2$.

ALGORITHM 4.2   The Cocks-Pinch method.

Input: A positive integer k and a square-free positive integer D.

1 Choose a prime r such that $r = 1 \mod k$ and $\left(\frac{-D}{r} \right) = 1$
2 choose a primitive kth root of unity $z \in \mathbb{F}_r$ and let $t' = z + 1$
3 let $y' = (t' - 2)/\sqrt{-D} \in \mathbb{F}_r$
4 take t and y to be representative elements in the classes t' and y'
5 set $q = (t^2 + Dy^2)/4$
6 if q is a prime or a power of a prime, return (r, t, q).

4.4.2   The Dupont-Enge-Morain Method

If we watch conditions 1 and 2 at the beginning of this section as a system of two polynomial equations in t, a natural way to check the existence of solutions is to consider the resultant of the corresponding polynomials. Indeed, two polynomials have a common root in $\overline{\mathbb{F}_r}$ if and only if their resultant is zero modulo r (for more details about resultants, see [26]). If we choose r, y, and D such that this condition on the resultant is satisfied, then the system can be solved by computing a root of the gcd of the two polynomials. This root will be in $\mathbb{F}_r$ if and only if $\Phi_k(x - 1)$ splits in linear factors in $\mathbb{F}_r$, so we shall ask that $k|(r - 1)$ (see Section 4.4.1). The Dupont-Enge-Morain method is summarized in Algorithm 4.3. For the same reason as in Section 4.4.1, we get curves with $\rho \approx 2$.

ALGORITHM 4.3   The Dupont-Enge-Morain method.

Input: A positive integer k.

1 Compute the resultant $R(a) = \text{Res}_x \left(\Phi_k(x - 1), a + (x - 2)^2 \right)$
2 choose some positive integers D (small) and y such that $R(Dy^2)$ has a large prime factor $r = 1 \mod k$
3 compute $g(x) = \gcd \left(\Phi_k(x - 1), Dy^2 + (x - 2)^2 \right)$ in $\mathbb{F}_r[x]$ and choose a root $t' \in \mathbb{F}_r$ of $g(x)$
4 take t to be a representative element in the class t'
5 set $q = (t^2 + Dy^2)/4$
6 if q is a prime or a power of a prime, return (r, t, q) and D.

Freeman, Scott, and Teske [18] proved that the polynomial $R(a) \in \mathbb{Z}[a]$ in Algorithm 4.3 represents primes and that moreover, the condition $r = 1 \mod k$ is automatically satisfied in the case where $r = R(Dy^2)$ is an odd prime. In any case, r has to be roughly of the size of $R(Dy^2)$, since it is only feasible to compute an r of cryptographic size if the remaining factors of $R(Dy^2)$ are small.

Like the Cocks-Pinch method, the Dupont-Enge-Morain method is effective for generating curves with arbitrary embedding degree. However, whereas in the former method we could choose the size of r nearly arbitrarily, in this method r has to be roughly of the size of $R(Dy^2)$. Since the polynomial $R(a)$ has degree $\varphi(k)$, the prime r we find will grow roughly like $a^{\varphi(k)}$. Thus the possible r are more restricted in the Dupont-Enge-Morain method than in the Cocks-Pinch method.

4.5 Families of Curves

Several authors proposed parameters (r, t, q), which are values of some polynomials $r(x), t(x), q(x) \in \mathbb{Q}[x]$ satisfying the equations in Section 4.2.2. This strategy allows us to achieve particularly good ρ-values and provides a better control on the size of the parameters. In order to formalize these ideas, Freeman, Scott, and Teske [18] introduced the notion of families of pairing-friendly curves.

Before stating definitions, notice that $q(x_0)$ has to be a prime number (or a power of a prime, but this is difficult to achieve in practice), and $r(x_0)$ has to be a prime number or a prime number times a small integer. Very little is known about polynomials having an infinity of prime values (it is not even known whether $x^2 + 1$ takes infinitely many prime values), but there are rather precise conjectures on the subject. Of course, a polynomial $f(x) \in \mathbb{Z}[x]$ cannot take an infinity of prime values if it is reducible or has a negative leading coefficient; another case that must be ruled out is when all the values of $f(x)$ have a common divisor greater than 1 (for instance, $f(x) = x^2 + x + 2$ only takes even values). Bouniakowsky's conjecture [10] (stated in the mid-nineteenth century) predicts that if $f(x)$ does not belong to one of the two cases above, then it takes infinitely many prime values (notice that this conjecture is true for degree 1 polynomials: this is Dirichlet's theorem on arithmetic progressions). Schinzel's hypothesis H [40] generalizes this conjecture: It states that if $f_1(x), \ldots, f_\ell(x) \in \mathbb{Z}[x]$ are irreducible polynomials with positive leading coefficient such that all the values of the product $\prod_{i=1}^{\ell} f_i(x)$ do not have a common divisor greater than 1, then $f_1(x), \ldots, f_\ell(x)$ are taking simultaneously prime values infinitely many times. Finally, the Bateman and Horn conjecture [6] extends Schinzel's hypothesis H by predicting the asymptotic distribution of integers where the f_i's are taking prime values. The natural extension of these conjectures to polynomials with rational coefficients motivates the following definition:

DEFINITION 4.3 ([18]) A polynomial $f(x) \in \mathbb{Q}[x]$ *represents primes* if the following conditions are satisfied:

1. $f(x)$ is non-constant, irreducible, and has positive leading coefficient,
2. $f(x_0) \in \mathbb{Z}$ for some $x_0 \in \mathbb{Z}$ (equivalently, for infinitely many $x_0 \in \mathbb{Z}$),
3. $\gcd(\{f(x_0) : x_0, f(x_0) \in \mathbb{Z}\}) = 1$

Notice that if either $f(x_0) = \pm 1$ for some $x_0 \in \mathbb{Z}$ or $f(x)$ takes two distinct prime values, then conditions 2 and 3 of Definition 4.3 are both satisfied.

DEFINITION 4.4 ([18]) A polynomial $f(x) \in \mathbb{Q}[x]$ is *integer-valued* if $f(x_0) \in \mathbb{Z}$ for all $x_0 \in \mathbb{Z}$.

For example, $f(x) = \frac{1}{2}(x^2 + x + 2)$ is integer-valued and represents primes.

DEFINITION 4.5 ([18])   Let $r(x), t(x), q(x) \in \mathbb{Q}[x]$ be nonzero polynomials, k be a positive integer, and D be a square-free positive integer.

1. The triple $(r(x), t(x), q(x))$ *parametrizes a family of elliptic curves with embedding degree k and discriminant D* if the following conditions are satisfied:

   (a) $q(x) = p(x)^d$ for some $d \geq 1$ and $p(x)$ that represents primes,

   (b) $r(x)$ is non-constant, irreducible, integer-valued, and has positive leading coefficient,

   (c) $r(x)$ divides $\Phi_k(t(x) - 1)$,

   (d) $r(x)$ divides $q(x) + 1 - t(x)$,

   (e) the equation $Dy^2 = 4q(x) - t(x)^2$ in (x, y) has infinitely many integer solutions.

2. If these conditions are satisfied, the triple $(r(x), t(x), q(x))$ is referred to as *a family* and an elliptic curve with parameters $(r(x_0), t(x_0), q(x_0))$, $x_0 \in \mathbb{Z}$ is referred to as *belonging to the family $(r(x), t(x), q(x))$*.

3. A family $(r(x), t(x), q(x))$ is *ordinary* if $\gcd(t(x), q(x)) = 1$.

4. A family $(r(x), t(x), q(x))$ is *complete* if there exists $y(x) \in \mathbb{Q}[x]$ such that $Dy(x)^2 = 4q(x) - t(x)^2$; otherwise, the family is *sparse*.

5. The triple $(r(x), t(x), q(x))$ parametrizes a *potential family* of elliptic curves if conditions 1b–1e are satisfied (so $p(x)$ may or may not represent primes).

6. The *ρ-value* of a (potential) family $(r(x), t(x), q(x))$ is

$$\frac{\deg q(x)}{\deg r(x)}.$$

Condition 1e in Definition 4.5 implies that $\deg t(x) \leq (\deg q(x))/2$, so $\deg(q(x) + 1 - t(x)) = \deg q(x)$. Writing condition 1d of Definition 4.5 as $q(x) + 1 - t(x) = h(x)r(x)$, $h(x) \in \mathbb{Q}[x]$, we see that any family $(r(x), t(x), q(x))$ has ρ-value at least 1. Moreover, if a family $(r(x), t(x), q(x))$ contains infinitely many elliptic curves of prime order, then we must have $h(x) = 1$, so $(r(x), t(x), q(x))$ has ρ-value equal to 1. However, notice that the condition $\rho = 1$ is not sufficient to ensure that a family contains elliptic curves of prime order: It can happen that $h(x)$ is equal to an integer $h > 1$.

Remark 4.2   In order to have a good control on the bit-size of the parameters r and q, Freeman, Scott, and Teske [18] recommend using families such that $r(x)$ has small degree (less than 40).

4.5.1   MNT Curves

Miyaji, Nakabayashi, and Takano [33] were the first authors to construct families of ordinary pairing-friendly elliptic curves. Their aim was actually to derive explicit conditions to avoid the MOV attack [32]. They gave a necessary and sufficient condition for an elliptic curve of prime order to have an embedding degree $k = 3$, 4, or 6, by directly solving the equation $\Phi_k(q) = \lambda(q + 1 - t)$; an essential ingredient for the success of this approach is to have $\varphi(k) = 2$. The MNT families are sparse in the sense of Definition 4.5. This method has been generalized to curves whose order is a prime times a small cofactor [19, 42].

THEOREM 4.2 (Miyaji, Nakabayashi, Takano, [33])   *Let q be a prime and $E/\mathbb{F}_q$ be an ordinary elliptic curve such that $\#E(\mathbb{F}_q) = q + 1 - t$ is a prime greater than 3.*

1. E has embedding degree $k = 3$ if and only if $t = -1 \pm 6x$ and $q = 12x^2 - 1$ for some $x \in \mathbb{Z}$,

2. E has embedding degree $k = 4$ if and only if $t = -x$ or $t = x + 1$, and $q = x^2 + x + 1$ for some $x \in \mathbb{Z}$,

3. E has embedding degree $k = 6$ if and only if $t = 1 \pm 2x$ and $q = 4x^2 + 1$ for some $x \in \mathbb{Z}$.

Actually, Miyaji et al. proved the theorem for $q > 64$, but the remaining cases can be demonstrated via a brute-force search [18].

Remark 4.3 Karabina and Teske [24] proved that if r and q are odd primes, then there exists an elliptic curve $E/\mathbb{F}_q$ with embedding degree 6, discriminant D, and $\#E(\mathbb{F}_q) = r$ if and only if there exists an elliptic curve $E'/\mathbb{F}_r$ with embedding degree 4, discriminant D, and $\#E'(\mathbb{F}_r) = q$.

Theorem 4.2 gives triples $(r(x), t(x), q(x))$ that parametrize families of pairing-friendly elliptic curves of discriminant D (for some fixed square-free positive integer D) if and only if the CM equation

$$Dy^2 = 4q(x) - t(x)^2 \tag{4.2}$$

has infinitely many integer solutions. The right-hand side of (4.2) is a quadratic polynomial, and by completing the square, it is possible to transform this equation into a generalized Pell equation

$$X^2 - SDY^2 = M. \tag{4.3}$$

The general strategy to solve (4.3) is to find the minimal positive integer solution (U, V) (i.e., $U > 0$, $V > 0$, and V minimal) to the Pell equation $U^2 - SDV^2 = 1$ by computing the simple continued fraction expansion of $\sqrt{SD}$, and then find a so-called fundamental solution (X_0, Y_0) of (4.3) (see [30, 38]). Such a solution may or may not exist. If it does exist, then for $j \in \mathbb{Z}$, the couple (X_j, Y_j) defined by

$$X_j + Y_j\sqrt{SD} = (U + V\sqrt{SD})^j (X_0 + Y_0\sqrt{SD})$$

yields an infinite sequence of solutions to (4.3).

A drawback of MNT curves is that the consecutive solutions (X_j, Y_j) grow exponentially in size, so very few x-values will give a solution; the MNT families are sparse in the sense of Definition 4.5. In fact, Luca and Shparlinski [29] gave a heuristic argument that for any upper bound $\overline{D}$, there exist only a finite number of MNT curves with discriminant $D \leq \overline{D}$, with no bound on the field size. On the other hand, specific sample curves of cryptographic interest have been found, such as MNT curves of 160-bit, 192-bit, or 256-bit prime order (see, for example, [36] and [41]).

4.5.2 Complete Families

The Brezing and Weng method

The most-used method to produce complete families of pairing-friendly elliptic curves is due to Brezing and Weng [11] (it generalizes the constructions by Barreto, Lynn, and Scott [4]). Their idea is to use the Cocks-Pinch method described in Section 4.4.1 with polynomials as input instead of integers. The Brezing and Weng method is summarized in Algorithm 4.4.

The ρ-value of a family generated by Algorithm 4.4 is

$$\frac{2\max\{\deg t(x), \deg y(x)\}}{\deg r(x)},$$

ALGORITHM 4.4   The Brezing and Weng method.

Input: A positive integer k and a square-free positive integer D.

1 Find an irreducible polynomial $r(x) \in \mathbb{Z}[x]$ with positive leading coefficient such that the number field $\mathbf{K} = \mathbb{Q}[x]/r(x)$ contains $\sqrt{-D}$ and the primitive kth roots of unity

2 choose a primitive kth root of unity $\zeta_k \in \mathbf{K}$

3 let $t(x) \in \mathbb{Q}[x]$ be a polynomial mapping to $\zeta_k + 1$

4 let $y(x) \in \mathbb{Q}[x]$ be a polynomial mapping to $(\zeta_k - 1)/\sqrt{-D}$ (so if $\sqrt{-D} \mapsto s(x)$, then $y(x) = (2 - t(x))s(x)/D \mod r(x)$)

5 set $q(x) = (t(x)^2 + Dy(x)^2)/4$

6 if $q(x)$ represents primes and $y(x_0) \in \mathbb{Z}$ for some $x_0 \in \mathbb{Z}$, return $(r(x), t(x), q(x))$.

---

which is strictly smaller than 2, as long as we choose $t(x)$ and $y(x)$ to have degree strictly less than $r(x)$. In general, it is expected that the degrees of $t(x)$ and $y(x)$ are exactly $\deg r(x) - 1$, so that $\rho \approx 2$. But it turns out that much smaller ρ-values can be achieved for some particular choices of $r(x)$.

There are numerous examples of constructions using this method; see [18] for an extensive survey. We give some examples covering the constructions that are currently considered as good candidates for practical applications.

Examples

An obvious choice is to take $r(x)$ to be a cyclotomic polynomial $\Phi_\ell(x)$, where $k \mid \ell$; Murphy and Fitzpatrick [35] give conditions on ℓ to ensure that the cyclotomic field $\mathbf{K} = \mathbb{Q}[x]/\Phi_\ell(x)$ contains some quadratic subfield and explain how to represent its elements. This gives general constructions that cover a large range of k values (provided by the fact that $q(x) = (t(x)^2 + Dy(x)^2)/4$ represents primes) [18]. Such families are called *cyclotomic families*.

The following example of cyclotomic families appears in [4] and [11] in particular cases, and in [18] in its full generality. It makes use of the fact that if a cyclotomic field contains a cube root of unity, then it also contains $\sqrt{-3}$. This leads to (potential) families in all cases where k is not divisible by 18.

Example 4.1   Let k be a positive integer.

- If $k \equiv 1 \mod 6$, let

$$
\begin{aligned}
r(x) &= \Phi_{6k}(x), \\
t(x) &= -x^{k+1} + x + 1, \\
q(x) &= \frac{1}{3}(x+1)^2(x^{2k} - x^k + 1) - x^{2k+1}.
\end{aligned}
$$

- If $k \equiv 2 \mod 6$, let

$$
\begin{aligned}
r(x) &= \Phi_{3k}(x), \\
t(x) &= x^{k/2+1} - x + 1, \\
q(x) &= \frac{1}{3}(x-1)^2(x^k - x^{k/2} + 1) + x^{k+1}.
\end{aligned}
$$

- If $k \equiv 3 \mod 6$, let

$$
\begin{aligned}
r(x) &= \Phi_{2k}(x), \\
t(x) &= -x^{k/3+1} + x + 1, \\
q(x) &= \frac{1}{3}(x+1)^2(x^{2k/3} - x^{k/3} + 1) - x^{2k/3+1}.
\end{aligned}
$$

- If $k \equiv 4 \mod 6$, let

$$
\begin{aligned}
r(x) &= \Phi_{3k}(x), \\
t(x) &= x^3 + 1, \\
q(x) &= \frac{1}{3}(x^3 - 1)^2(x^k - x^{k/2} + 1) + x^3.
\end{aligned}
$$

- If $k \equiv 5 \mod 6$, let

$$
\begin{aligned}
r(x) &= \Phi_{6k}(x), \\
t(x) &= x^{k+1} + 1, \\
q(x) &= \frac{1}{3}(x^2 - x + 1)(x^{2k} - x^k + 1) + x^{k+1}.
\end{aligned}
$$

- If $k \equiv 0 \mod 6$, let

$$
\begin{aligned}
r(x) &= \Phi_k(x), \\
t(x) &= x + 1, \\
q(x) &= \frac{1}{3}(x - 1)^2(x^{k/3} - x^{k/6} + 1) + x.
\end{aligned}
$$

It has been checked that $q(x)$ is irreducible for $k \leq 1000$ and $18 \nmid k$ [18]. In this case, it is easy to see that $q(x)$ represents primes and thus the triple $(r(x), t(x), q(x))$ parametrizes a complete family with embedding degree k and discriminant 3. Let $\ell = \mathrm{lcm}(6, k)$. Then the ρ-value of any such family is $\rho = (\ell/3 + 6)/\varphi(\ell)$ if $k \equiv 4 \pmod 6$, and $(\ell/3 + 2)/\varphi(\ell)$ otherwise. In particular, we have $\rho \leq 2$ for all $k \leq 1000$ except for $k = 4$ and $\rho < 2$ for all $5 \leq k \leq 1000$ except for $k = 6$ and 10. Algorithm 4.4 is used with $r(x) = \Phi_\ell(x)$, and $\sqrt{-3}$ is represented by $2x^{\ell/6} - 1$ in $\mathbf{K} = \mathbb{Q}[x]/r(x) \cong \mathbb{Q}(\zeta_k, \zeta_6)$.

Using non-cyclotomic polynomials $r(x)$ to define (perhaps trivial) extensions of cyclotomic fields may turn out to be even more efficient in some cases. Such constructions are much less general than the cyclotomic constructions, so they are referred to as *sporadic*.

One possibility is to take $r(x)$ to be an irreducible factor of $\Phi_\ell(u(x))$. If $\Phi_\ell(u(x))$ is irreducible, we gain nothing as we will just be evaluating $r(x)$, $t(x)$, and $q(x)$ at $u(x)$. But if $\Phi_\ell(u(x))$ is reducible, then we may get particularly good results.

The following example was given by Barreto and Naehrig [5] (they originally presented their construction as an MNT-type family). They used the work of Galbraith, McKee, and Valença [19], who gave a list of the quadratic $u(x)$ such that $\Phi_\ell(u(x))$ is reducible, for $\varphi(\ell) = 4$.

Example 4.2 (BN curves)   Let

$$
\begin{aligned}
r(x) &= 36x^4 + 36x^3 + 18x^2 + 6x + 1, \\
t(x) &= 6x^2 + 1, \\
q(x) &= 36x^4 + 36x^3 + 24x^2 + 6x + 1.
\end{aligned}
$$

Then $(r(x), t(x), q(x))$ parametrizes a complete family with embedding degree $k = 12$, discriminant 3 and ρ-value 1. In this case, we have $\Phi_{12}(6x^2) = r(x)r(-x)$ and ζ_{12} is represented by $6x^2$ in $\mathbf{K} = \mathbb{Q}[x]/r(x)$ and $\sqrt{-3} = 2\zeta_{12}^2 - 1$.

Freeman, Scott, and Teske [18] gave several other examples in the same vein by doing a computer search for further factorizations of $\Phi_\ell(u(x))$, including the example below, which has been independently found by Tanaka and Nakamula [46].

Example 4.3　Let

$$
\begin{aligned}
r(x) &= 9x^4 + 12x^3 + 8x^2 + 4x + 1, \\
t(x) &= -9x^3 - 3x^2 - 2x, \\
q(x) &= \frac{1}{4}(81x^6 + 54x^5 + 45x^4 + 12x^3 + 13x^2 + 6x + 1).
\end{aligned}
$$

Then $(r(x), t(x), q(x))$ parametrizes a complete family with embedding degree $k = 8$, discriminant 1, and ρ-value 3/2. In this case, $u(x) = 9x^3 + 3x^2 + 2x + 1$.

Remark 4.4　As noticed by Freeman, Scott, and Teske, Example 4.1 gives a better ρ-value than Example 4.3 for $k = 8$, but Example 4.3 has the advantage of having $D = 1$, which allows the use of quartic twists.

Kachisa, Schaefer, and Scott [23] proposed to take $r(x)$ to be the minimal polynomial of some well-chosen primitive element of a cyclotomic field. In particular, they obtained the following examples:

Example 4.4 (KSS 16 curves)　Let

$$
\begin{aligned}
r(x) &= x^8 + 48x^4 + 625, \\
t(x) &= \frac{1}{35}(2x^5 + 41x + 35), \\
q(x) &= \frac{1}{980}(x^{10} + 2x^9 + 5x^8 + 48x^6 + 152x^5 + 240x^4 + 625x^2 + 2398x + 3125).
\end{aligned}
$$

Then $(r(x), t(x), q(x))$ parametrizes a complete family with embedding degree $k = 16$, discriminant 1, and ρ-value 5/4.

Example 4.5 (KSS 18 curves)　Let

$$
\begin{aligned}
r(x) &= x^6 + 37x^3 + 343, \\
t(x) &= \frac{1}{7}(x^4 + 16x + 7), \\
q(x) &= \frac{1}{21}(x^8 + 5x^7 + 7x^6 + 37x^5 + 188x^4 + 259x^3 + 343x^2 + 1763x + 2401).
\end{aligned}
$$

Then $(r(x), t(x), q(x))$ parametrizes a complete family with embedding degree $k = 18$, discriminant 3, and ρ-value 4/3.

Example 4.6 (KSS 32 curves)　Let

$$
\begin{aligned}
r(x) &= x^{16} + 57120x^8 + 815730721, \\
t(x) &= \frac{1}{3107}(-2x^9 - 56403x + 3107), \\
q(x) &= \frac{1}{2970292}(x^{18} - 6x^{17} + 13x^{16} + 57120x^{10} - 344632x^9 + 742560x^8 + 815730721^2 \\
&\quad - 4948305594x + 10604499373).
\end{aligned}
$$

Then $(r(x), t(x), q(x))$ parametrizes a complete family with embedding degree $k = 32$, discriminant 1, and ρ-value 9/8.

Example 4.7 (KSS 36 curves)   Let

$$
\begin{aligned}
r(x) &= x^{12} + 683x^6 + 117649, \\
t(x) &= \frac{1}{259}(2x^7 + 757x + 259), \\
q(x) &= \frac{1}{28749}(x^{14} - 4x^{13} + 7x^{12} + 683x^8 - 2510x^7 + 4781x^6 + 117649x^2 - 386569x \\
&\quad + 823543).
\end{aligned}
$$

Then $(r(x), t(x), q(x))$ parametrizes a complete family with embedding degree $k = 36$, discriminant 3, and ρ-value $7/6$.

More discriminants

When the Brezing and Weng method is used to construct a family of elliptic curves, the discriminant D needs to be fixed in advance and the first step of Algorithm 4.4 requires us to find some finite extension of a cyclotomic field containing $\sqrt{-D}$. For this reason, the known constructions of complete families offer a rather limited choice for D. For implementation purposes, it is preferable to use curves with $D = 1$ or 3 (see Chapters 6 and 11). However, there are known methods to improve the efficiency of Pollard's rho algorithm on such curves [16]. These methods lead to a decrease in security of only a few bits, but some users may take their existence as a warning that curves with CM discriminant having a particular shape are in some sense special and should be avoided. Motivated by these arguments, Freeman, Scott, and Teske [18] proved the following result, which enables us to get complete families with variable discriminant:

PROPOSITION 4.3 ([18])   *Suppose that $(r(x), t(x), q(x))$ parametrizes a complete potential family of elliptic curves with embedding degree k and discriminant D and let $y(x) \in \mathbb{Q}[x]$ such that $Dy(x)^2 = 4q(x) - t(x)^2$. Suppose that $r(x)$, $t(x)$, and $q(x)$ are even polynomials and $y(x)$ is an odd polynomial, and define $r'(x)$, $t'(x)$, $q'(x)$, and $y'(x)$ by*

$$
r(x) = r'(x^2), \quad t(x) = t'(x^2), \quad q(x) = q'(x^2), \quad y(x) = xy'(x^2).
$$

Let α be an integer such that

1. *αD is square-free,*
2. *$r'(\alpha x^2)$ is irreducible,*
3. *$y'(\alpha x^2)$ takes integer values.*

Then the triple $(r'(\alpha x^2), t'(\alpha x^2), q'(\alpha x^2))$ parametrizes a complete potential family of elliptic curves with embedding degree k, discriminant αD, and the same ρ-value as $(r(x), t(x), q(x))$.

Proof. For any integer $\alpha > 0$ satisfying the conditions of Proposition 4.3, we must verify conditions 1b–1e of Definition 4.5 for the triple $(r'(\alpha x^2), t'(\alpha x^2), q'(\alpha x^2))$. If $r'(\alpha x^2)$ is irreducible, then condition 1b on $r'(\alpha x^2)$ follows from the same condition on $r(x)$. Conditions 1c and 1d are identities on the polynomials $r(x)$, $t(x)$, $q(x)$, so they still hold when we evaluate at $\sqrt{\alpha}x$, and the identity $Dy(x)^2 = 4q(x) - t(x)^2$ evaluated at $\sqrt{\alpha}x$ gives $D\alpha x^2 y'(\alpha x^2)^2 = 4q'(\alpha x^2) - t'(\alpha x^2)^2$, so condition 1e is satisfied if $y'(\alpha x^2)$ takes integer values. Finally, the ρ-value of $(r'(\alpha x^2), t'(\alpha x^2), q'(\alpha x^2))$ is $(2\deg q'(x))/(2\deg r'(x)) = \deg q(x)/\deg r(x)$. $\square$

Freeman, Scott and Teske also give an algorithm for generating variable-discriminant cyclotomic families for any k such that $\gcd(k, 24) \in \{1, 2, 3, 6, 12\}$.

TABLE 4.2 Best ρ-values for families of curves with $k \leq 50$.

k	fixed $D \leq 3$				variable D			
	ρ	D	$\deg r(x)$	Constr.	ρ	D	$\deg r(x)$	Constr.
1	2.000	3	2	4.1	2.000	any	1	(6.17)
2	any$^{\#}$	1,3	–	§4.3	any$^{\#}$	3 mod 4	–	§4.3
3	**1.000**$^{\#}$	3	2	§4.3	**1.000**	some	2	§4.5.1
4	1.500	3	4	(6.9)	**1.000**	some	2	§4.5.1
5	1.500	3	8	4.1	1.750	any odd	8	(6.2)+
6	1.250	1	4	(6.16)	**1.000**	some	2	§4.5.1
7	1.333	3	12	4.1, (6.20)+	1.333	3 mod 4	12	(6.20)+
8	1.250	3	8	4.1	–	–	–	–
9	1.333	3	6	4.1	1.833	any odd	12	(6.2)+
10	1.500	1,3	8	(6.5), (6.24)+	**1.000**	some	4	(§5.3)
11	1.200	3	20	4.1, (6.20)+	1.200	3 mod 4	20	(6.20)+
12	**1.000**	3	4	4.2	1.750	2 mod 8	8	(6.7)+
13	1.167	3	24	4.1	1.250	any odd	24	(6.2)+
14	1.333	3	12	4.1	1.500	any odd	12	(6.3)+
15	1.500	3	8	4.1	1.750	any even	32	(6.7)*+
16	1.250	1	8	4.4	–	–	–	–
17	1.125	3	32	4.1	1.188	any odd	32	(6.2)+
18	1.333	3	6	4.5	1.583	2 mod 4	24	(6.7)+
19	1.111	3	36	4.1	1.111	3 mod 4	36	(6.20)+
20	1.375	3	16	4.1	–	–	–	–
21	1.333	3	12	4.1	1.792	2 mod 4	48	(6.7)+
22	1.300	1	20	(6.3)	1.300	any odd	20	(6.3)+
23	1.091	3	44	4.1, (6.20)+	1.091	3 mod 4	44	(6.20)+
24	1.250	3	8	4.1	–	–	–	–
25	1.300	3	40	4.1	1.350	any odd	40	(6.2)+
26	1.167	3	24	4.1, (6.24)+	1.167	3 mod 4	24	(6.24)+
27	1.111	3	18	4.1	1.472	2 mod 4	72	(6.7)+
28	1.333	1	12	(6.4)	1.917	6 mod 8	24	(6.7)*+
29	1.071	3	56	4.1	1.107	any odd	56	(6.2)+
30	1.500	3	8	4.1	1.813	2 mod 4	32	(6.7)+
31	1.067	3	60	4.1, (6.20)+	1.067	3 mod 4	60	(6.20)+
32	1.063	3	32	4.1	–	–	–	–
33	1.200	3	20	4.1	1.575	2 mod 4	80	(6.7)+
34	1.125	3	32	(6.24)+	1.125	3 mod 4	32	(6.24)+
35	1.500	3	48	4.1, (6.20)+	1.500	3 mod 4	48	(6.20)+
36	1.167	3	12	4.7	1.417	2 mod 8	24	(6.7)+
37	1.056	3	72	4.1	1.083	any odd	72	(6.2)+
38	1.111	3	36	4.1	1.167	any odd	36	(6.3)+
39	1.167	3	24	4.1	1.521	2 mod 4	96	(6.7)+
40	1.375	1	16	(6.15)	–	–	–	–
41	1.050	3	80	4.1	1.075	any odd	80	(6.2)+
42	1.333	3	12	4.1	1.625	2 mod 4	48	(6.7)+
43	1.048	3	84	4.1, (6.20)+	1.048	3 mod 4	84	(6.20)+
44	1.150	3	40	4.1	1.750	6 mod 8	40	(6.7)*+
45	1.333	3	24	4.1	1.729	2 mod 4	96	(6.7)+
46	1.136	1	44	(6.3)	1.136	any odd	44	(6.3)+
47	1.043	3	92	4.1	1.043	3 mod 4	92	(6.20)+
48	1.125	3	16	4.1	–	–	–	–
49	1.190	3	84	4.1	1.214	any odd	84	(6.2)+
50	1.300	3	40	4.1, (6.24)+	1.300	3 mod 4	40	(6.24)+

Note: See pages 4-17 for explanations of the symbols and fonts.

4.6 Curves with a Small ρ-Value: The FST Table

Freeman, Scott, and Teske [18] determined the best-known values of ρ for families of curves with embedding degree $k \leq 50$. We reproduce their table in Table 4.2, which shows two different constructions for each embedding degree k: The first construction listed is the one that yields the smallest ρ-value when the CM discriminant D is 1 or 3. Table 4.2 shows that in a large majority of cases, the optimal ρ-value is achieved by Example 4.1; other constructions do better for some small k, $k \equiv 4 \pmod 6$, and k divisible by 18. The second construction indicates the optimal ρ-values for families with variable CM discriminant, the allowed discriminants D, and the constructions that produce these ρ-values. Note that to date we know of no variable-discriminant construction when $k = 20$ or when k is a multiple of 8; in these cases a family with $D \leq 3$ or a Cocks-Pinch curve must be used.

Freeman, Scott, and Teske checked that all of the families listed in Table 4.2 can be used to produce explicit examples of pairing-friendly elliptic curves, and that for parameters of cryptographic size the ρ-value of a curve is very close to the ρ-value of its family.

All families in the table except for one lead to curves over prime fields, and the minimum embedding field is $\mathbb{F}_{q^k}$ for such curves. The lone exception is the supersingular family with $k = 3$. The minimum embedding field for a curve in this family is either $\mathbb{F}_{q^3}$ or $\mathbb{F}_{q^{3/2}}$; see Section 4.3 for details. If a curve over a prime field is required, Example 4.1 gives a family with ρ-value 2.

Explanation of symbols in Table 4.2

bold Entries in bold in the table indicate that curves of prime order can be constructed with the given embedding degree.

italic Entries in italic indicate that while the ρ-value achieved for the given family may be optimal, the degrees of the polynomials involved are too high to make the construction practical (see Remark 4.2).

\# The ρ-values marked with a \# are achieved by supersingular curves.

\+ A construction marked with a + indicates that the given basic construction is combined with the substitution $x^2 \mapsto \alpha x^2$ (Proposition 4.3) to construct families with the given discriminant.

* For $k = 15$, 28, or 44 and variable D, a variant of the FST-Construction 6.7 is used. We refer to [18] for details.

\- Entries missing from the table for a given embedding degree k indicate that there is no known family of curves of the given type (i.e., small D or variable D) for that particular k. In these cases the Cocks-Pinch method should be used to achieve the desired embedding degree and discriminant, constructing a curve with $\rho \approx 2$.

() The construction numbers in brackets are referring to the corresponding construction in [18].

4.7 Conclusion

The selection of a pairing-friendly elliptic curve for a given application depends on many factors. The most important is the desired security level. Fixing bit sizes for the prime-order subgroup of the elliptic curve and of the extension field determines k (up to ρ), since $\log q^k / \log r = k\rho$. However, notice that it is possible to reach other embedding degrees by choosing r or q of bigger size than the required size. For efficiency reasons, k is usually taken to be a multiple of the order of the automorphism group of E (that is, 4 for $D = 1$, 6 for $D = 3$, and 2 otherwise). Furthermore, taking k of the shape $2^i 3^j$ is a popular choice, since it allows efficient field arithmetic.

A usual recommendation in Elliptic Curve Cryptography is to choose curves that are as general as possible, since any specificity could potentially lead to an attack. On the other hand, some specificities can allow substantial efficiency improvements. For instance, curves with $D = 1$ or 3 have some extra twists that can be used for compression (see Chapter 3).

This question of choosing between genericity and efficiency also occurs in the choice of the method to generate the parameters of a pairing-friendly elliptic curve. The curves obtained from the methods described in Section 4.4 are best fulfilling the genericity criterion, but pairing computations on such curves are rather slow (in particular, these curves have $\rho \approx 2$). The constructions from Section 4.4 are therefore recommended in cases where efficiency is not crucial. The Cocks-Pinch method is usually preferred to the Dupont-Enge-Morain method, since it allows more flexibility in the choice of r.

The parameters of curves from families have a very specific shape, since they are given by polynomials; this property could be used in some (yet unknown) attack. On the other hand, this property can allow some interesting optimizations (see Chapters 6 and 11).

Curves from sparse families (such as MNT curves) and supersingular curves (in characteristic greater than 3) are a good choice for low security levels, since they have a small embedding degree. Notice that generating a curve from a sparse family requires us to solve a Pell equation.

All the complete families given in this chapter have $D = 1$ or 3. Curves in these families have an equation of the shape $y^2 = x^3 + ax$ for $D = 1$ and $y^2 = x^3 + a$ for $D = 3$ (see Section 4.2.1). A result from Freeman, Scott, and Teske (see Section 4.5.2) allows the users considering such curves as suspicious to derive from these families some new families with other discriminants. Table 4.2, taken from [18], gives the best-known values of ρ for families of curves with embedding degree $k \le 50$.

References

[1] Gora Adj, Alfred Menezes, Thomaz Oliveira, and Francisco Rodríguez-Henríquez. Computing discrete logarithms in $\mathbb{F}_{3^{6 \cdot 137}}$ and $\mathbb{F}_{3^{6 \cdot 163}}$ using Magma. In Ç. K. Koç, S. Mesnager, and E. Savas, editors, *Arithmetic of Finite Fields (WAIFI 2014)*, volume 9061 of *Lecture Notes in Computer Science*, pp. 3–22. Springer, 2014.

[2] Arthur O. L. Atkin and François Morain. Elliptic curves and primality proving. *Mathematics of Computation*, 61(203):29–68, 1993.

[3] R. Balasubramanian and Neal Koblitz. The improbability that an elliptic curve has subexponential discrete log problem under the Menezes - Okamoto - Vanstone algorithm. *Journal of Cryptology*, 11(2):141–145, 1998.

[4] Paulo S. L. M. Barreto, Ben Lynn, and Michael Scott. Constructing elliptic curves with prescribed embedding degrees. In S. Cimato, C. Galdi, and G. Persiano, editors, *Security in Communication Networks (SCN 2002)*, volume 2576 of *Lecture Notes in Computer Science*, pp. 257–267. Springer, Heidelberg, 2003.

[5] Paulo S. L. M. Barreto and Michael Naehrig. Pairing-friendly elliptic curves of prime order. In B. Preneel and S. Tavares, editors, *Selected Areas in Cryptography (SAC 2005)*, volume 3897 of *Lecture Notes in Computer Science*, pp. 319–331. Springer, Heidelberg, 2006.

[6] Paul T. Bateman and Roger A. Horn. A heuristic asymptotic formula concerning the distribution of prime numbers. *Mathematics of Computation*, 16(79):363–367, 1962.

[7] Naomi Benger, Manuel Charlemagne, and David Mandell Freeman. On the security of pairing-friendly abelian varieties over non-prime fields. In H. Shacham and B. Waters, editors, *Pairing-Based Cryptography – Pairing 2009*, volume 5671 of *Lecture Notes in Computer Science*, pp. 52–65. Springer, Heidelberg, 2009.

[8] Dan Boneh and Matthew K. Franklin. Identity-based encryption from the Weil pairing. In J. Kilian, editor, *Advances in Cryptology – CRYPTO 2001*, volume 2139 of *Lecture Notes in Computer Science*, pp. 213–229. Springer, Heidelberg, 2001.

[9] Dan Boneh, Ben Lynn, and Hovav Shacham. Short signatures from the Weil pairing. *Journal of Cryptology*, 17(4):297–319, 2004.

[10] Viktor I. Bouniakowsky. Sur les diviseurs numériques invariables des fonctions rationnelles entières. *Mém. Acad. Sc. St-Petersbourg*, VI:305–329, 1857.

[11] Friederike Brezing and Annegret Weng. Elliptic curves suitable for pairing based cryptography. *Designs, Codes and Cryptography*, 37(1):133–141, 2005.

[12] Reinier M. Bröker. *Constructing elliptic curves of prescribed order*. PhD thesis, Leiden University, 2006.

[13] Clifford Cocks and Richard G. E. Pinch. Identity-based cryptosystems based on the Weil pairing. Unpublished manuscript, 2001.

[14] Max Deuring. Die Typen der Multiplikatorenringe elliptischer Funktionenkörper. *Abh. Math. Sem. Hansischen Univ.*, 14:197–272, 1941.

[15] Régis Dupont, Andreas Enge, and François Morain. Building curves with arbitrary small MOV degree over finite prime fields. *Journal of Cryptology*, 18(2):79–89, 2005.

[16] Iwan M. Duursma, Pierrick Gaudry, and François Morain. Speeding up the discrete log computation on curves with automorphisms. In K.-Y. Lam, E. Okamoto, and C. Xing, editors, *Advances in Cryptology – ASIACRYPT '99*, volume 1716 of *Lecture Notes in Computer Science*, pp. 103–121. Springer, Heidelberg, 1999.

[17] David Freeman. Constructing pairing-friendly elliptic curves with embedding degree 10. In F. Hess, S. Pauli, and M. E. Pohst, editors, *Algorithmic Number Theory (ANTS-VII)*, volume 4076 of *Lecture Notes in Computer Science*, pp. 452–465. Springer, 2006.

[18] David Freeman, Michael Scott, and Edlyn Teske. A taxonomy of pairing-friendly elliptic curves. *Journal of Cryptology*, 23(2):224–280, 2010.

[19] Steven D. Galbraith, James F. McKee, and P. C. Valença. Ordinary abelian varieties having small embedding degree. *Finite Fields and Their Applications*, 13(4):800–814, 2007.

[20] Robert Granger, Thorsten Kleinjung, and Jens Zumbrägel. Breaking '128-bit secure' supersingular binary curves - (or how to solve discrete logarithms in $F_{2^{4 \cdot 1223}}$ and $F_{2^{12 \cdot 367}}$). In J. A. Garay and R. Gennaro, editors, *Advances in Cryptology – CRYPTO 2014, Part II*, volume 8617 of *Lecture Notes in Computer Science*, pp. 126–145. Springer, Heidelberg, 2014.

[21] Laura Hitt. On the minimal embedding field. In T. Takagi et al., editors, *Pairing-Based Cryptography – Pairing 2007*, volume 4575 of *Lecture Notes in Computer Science*, pp. 294–301. Springer, Heidelberg, 2007.

[22] Antoine Joux. A one round protocol for tripartite Diffie-Hellman. *Journal of Cryptology*, 17(4):263–276, 2004.

[23] Ezekiel J. Kachisa, Edward F. Schaefer, and Michael Scott. Constructing Brezing-Weng pairing-friendly elliptic curves using elements in the cyclotomic field. In S. D. Galbraith and K. G. Paterson, editors, *Pairing-Based Cryptography – Pairing 2008*, volume 5209 of *Lecture Notes in Computer Science*, pp. 126–135. Springer, Heidelberg, 2008.

[24] Koray Karabina and Edlyn Teske. On prime-order elliptic curves with embedding degrees $k = 3$, 4, and 6. In A. J. van der Poorten and A. Stein, editors, *Algorithmic Number Theory (ANTS-VIII)*, volume 5011 of *Lecture Notes in Computer Science*, pp. 102–117. Springer, 2008.

[25] Neal Koblitz and Alfred Menezes. Pairing-based cryptography at high security levels (invited paper). In N. P. Smart, editor, *Cryptography and Coding*, volume 3796 of *Lecture Notes in Computer Science*, pp. 13–36. Springer, Heidelberg, 2005.

[26] S. Lang. *Algebra*. Graduate Texts in Mathematics. Springer New York, 2005.

[27] Serge Lang. *Elliptic Functions*, volume 112 of *Graduate Texts in Mathematics*. Springer-Verlag, 2nd edition, 1987.

[28] Rudolf Lidl and Harald Niederreiter. *Finite Fields*, volume 20 of *Encyclopedia of Mathematics and Its Applications*. Cambridge University Press, 2nd edition, 1997.

[29] Florian Luca and Igor E. Shparlinski. Elliptic curves with low embedding degree. *Journal of Cryptology*, 19(4):553–562, 2006.

[30] Keith Matthews. The diophantine equation $x^2 - Dy^2 = N, D > 0$. *Expositiones Mathemeticae*, 18(4):323–331, 2000.

[31] Alfred Menezes. An introduction to pairing-based cryptography. In I. Luengo, editor, *Recent Trends in Cryptography*, volume 477 of *Contemporary Mathematics*, pp. 47–65. AMS-RMSE, 2009.

[32] Alfred Menezes, Scott A. Vanstone, and Tatsuaki Okamoto. Reducing elliptic curve logarithms to logarithms in a finite field. In *23rd Annual ACM Symposium on Theory of Computing*, pp. 80–89. ACM Press, 1991.

[33] Atsuko Miyaji, Masaki Nakabayashi, and Shunzo Takano. Characterization of elliptic curve traces under FR-reduction. In D. Won, editor, *Information Security and Cryptology – ICISC 2000*, volume 2015 of *Lecture Notes in Computer Science*, pp. 90–108. Springer, Heidelberg, 2001.

[34] François Morain. Classes d'isomorphismes des courbes elliptiques supersingulières en caractéristique ≥ 3. *Utilitas Mathematica*, 52:241–253, 1997.

[35] Angela Murphy and Noel Fitzpatrick. Elliptic curves for pairing applications. Cryptology ePrint Archive, Report 2005/302, 2005. http://eprint.iacr.org/2005/302.

[36] Dan Page, Nigel P. Smart, and Frederik Vercauteren. A comparison of MNT curves and supersingular curves. *Applicable Algebra in Engineering, Communication and Computing*, 17(5):379–392, 2006.

[37] John M. Pollard. Monte Carlo methods for index computation (mod p). *Mathematics of Computation*, 32(143):918–924, 1978.

[38] John P. Robertson. Solving the generalized Pell equation $x^2 - Dy^2 = N$. Unpublished manuscript, available at http://www.jpr2718.org/pell.pdf, 2004.

[39] Karl Rubin and Alice Silverberg. Choosing the correct elliptic curve in the CM method. *Mathematics of Computation*, 79(269):545–561, 2010.

[40] Andrzej Schinzel and Wacław Sierpiński. Sur certaines hypothèses concernant les nombres premiers. *Acta Arithmetica*, 4(3):185–208, 1958. Erratum 5 (1958), 259.

[41] Michael Scott and Paulo S. L. M. Barreto. Compressed pairings. In M. Franklin, editor, *Advances in Cryptology – CRYPTO 2004*, volume 3152 of *Lecture Notes in Computer Science*, pp. 140–156. Springer, Heidelberg, 2004.

[42] Michael Scott and Paulo S. L. M. Barreto. Generating more MNT elliptic curves. *Designs, Codes and Cryptography*, 38(2):209–217, 2006.

[43] Joseph H. Silverman. *Advanced Topics in the Arithmetic of Elliptic Curves*, volume 151 of *Graduate Texts in Mathematics*. Springer-Verlag, 1994.

[44] Joseph H. Silverman. *The Arithmetic of Elliptic Curves*, volume 106 of *Graduate Texts in Mathematics*. Springer-Verlag, 2nd edition, 2009.

[45] Andrew V. Sutherland. Computing Hilbert class polynomials with the Chinese remainder theorem. *Mathematics of Computation*, 80(273):501–538, 2011.

[46] Satoru Tanaka and Ken Nakamula. Constructing pairing-friendly elliptic curves using factorization of cyclotomic polynomials. In S. D. Galbraith and K. G. Paterson, editors, *Pairing-Based Cryptography – Pairing 2008*, volume 5209 of *Lecture Notes in Computer Science*, pp. 136–145. Springer, Heidelberg, 2008.

[47] Jorge Jiménez Urroz, Florian Luca, and Igor E. Shparlinski. On the number of isogeny classes of pairing-friendly elliptic curves and statistics of MNT curves. *Mathematics of Computation*, 81(278):1093–1110, 2012.

[48] Paul C. van Oorschot and Michael J. Wiener. Parallel collision search with cryptanalytic applications. *Journal of Cryptology*, 12(1):1–28, 1999.

[49] William C. Waterhouse. Abelian varieties over finite fields. *Ann. Sci. École Norm. Sup. (4)*, 2:521–560, 1969.

5

Arithmetic of Finite Fields

Jean Luc Beuchat
ELCA Informatique SA

Luis J. Dominguez Perez
CONACyT / CIMAT-ZAC

Sylvain Duquesne
IRMAR

Nadia El Mrabet
SAS-ENSMSE

Laura Fuentes-Castañeda
Intel

Francisco Rodríguez-Henríquez
CINVESTAV-IPN

A careful implementation of elementary field arithmetic operations is a crucial step to achieving efficient implementations of pairing-based cryptography. In this chapter, we describe the algorithms that perform several basic arithmetic operations over the field $\mathbb{F}_p$, sometimes referred to as base field arithmetic. We also study the main arithmetic operations defined over finite field extensions. In particular, we will carefully study the arithmetic operations over the quadratic and cubic field extensions $\mathbb{F}_{p^2}$ and $\mathbb{F}_{p^3}$, as described in a series of papers such as [13, 14, 22, 28].

The chapter is organized as follows. In Section 5.1 we present elementary definitions and properties of the base field arithmetic. Then, in Section 5.2 we present the construction of the so-called tower field extensions, using quadratic and cubic extensions. We illustrate the optimization of pairings computation based on the cyclotomic subgroup arithmetic in Section 5.3. In Section 5.4, we present the notion of lazy reduction and an alternative representation of finite fields, the residue number system representation. Finally, we provide some Sage code in Section 5.5. The interested reader may download our field arithmetic Sage code from http://sandia.cs.cinvestav.mx/Site/GuidePBCcode.

5.1 Base Field Arithmetic

In this section we will review the algorithms for performing multi-precision addition, subtraction, multiplication, and exponentiation over the field $\mathbb{F}_p$. We give special importance to the multiplication operation, covering the schoolbook integer multiplication, and the Barrett and Montgomery reduction methods. Furthermore, we also discuss the computation of squaring, exponentiation, multiplicative inverse, and square root over the field $\mathbb{F}_p$.

5.1.1   Field Element Representation

Base field arithmetic over a prime field $\mathbb{F}_p$, is the most important layer for achieving high-performance pairing-based implementations. Efficient field arithmetic implementations must consider the processor's word size and the bitsize of the prime p used by the cryptographic application. In contemporary microprocessor architectures, the processor's word size ranges from 8 bits, for the smallest mobile processors, up to 64 bits, for high-end Intel or ARM microprocessors. On the other hand, typical cryptographic applications need to offer at least 80 bits of security level, but most commonly, a 128-bit security level is preferred for commercial applications.* This forces us to choose primes with a considerable size in bits, typically, bigger than 256 bits (see Table 1.1 in Chapter 1 for a summary of the cryptographic key length recommendations for RSA, discrete-logarithm-based, and pairing-based cryptographic schemes).

Let W be the target processor word size, and let $a \in \mathbb{F}_p$ with $|a| \simeq |p|$, where $|a|$ denotes the size in bits of a. Then, a can be represented as an array of n words as,

$$a = (a_{n-1}, a_{n-2},\ \ldots\ , a_0),$$

where $a = a_0 + a_1 2^W + a_2 2^{2 \cdot W} + \ldots + a_{n-1} 2^{(n-1) \cdot W}$, with $n = \lceil (\lfloor \log_2 p \rfloor + 1)/W \rceil$, and $|a_i| \leq W$. We say that a_0 is the least-significant word of a, whereas a_{n-1} corresponds to its most-significant word. As a typical example, consider the case where a 64-bit word size architecture is being targeted and $|a| = 256$. Then, a can be represented as an 4-word array as,

$$a = (a_3, a_2, a_1, a_0).$$

Remark 5.1   Why should we use multi-precision representation? Many mathematical software packages (such as Sage, Magma, Maple, etc.), support the so-called infinite precision computation. This implies that in those packages the arithmetic operands can seemingly be represented as single precision variables. Nevertheless, in order to achieve high performance implementations, which are commonly written in low-level programming languages, such as C or even assembly languages, one cannot but resort to the multi-precision representation of the aritmetic operands. Hence, this will be the scenario to be analyzed in the remainder of this chapter.

5.1.2   Addition and Subtraction

Let $a, b \in \mathbb{F}_p$. The addition in the finite field $\mathbb{F}_p$ is defined as $(a+b) \bmod p$. In order to implement this operation for big numbers, i.e., numbers with a bitsize larger than W bits, which must be represented in a multi-precision array, one performs the addition word by word as shown in Algorithm 5.1. Notice that in Step 3 for $i > 0$, the output carry of the previous addition is taken into account, and hence, the operation $c_i = a_i + b_i + \text{carry}$, is computed. The addition-with-carry operation is easy to code in assembly languages as most microprocessors include an *addc*

---

*The security level of a given cryptographic scheme is defined as the minimum number of computational *steps* that an attacker must perform in order to obtain the scheme's secret key.

ALGORITHM 5.1   Addition in $\mathbb{F}_p$.

Input   : An n-word prime p, and two integers a, $b \in [0, p - 1]$ where
$\qquad a = (a_{n-1}, a_{n-2}, \ldots, a_0)$ and $b = (b_{n-1}, b_{n-2}, \ldots, b_0)$
Output: $c = (a + b) \bmod p$

1 carry $\leftarrow 0$;
2 **for** $i = 0 \rightarrow n - 1$ **do**
3 $\quad \big|$ $(c_i, \text{carry}) \leftarrow \text{Add\_with\_Carry}(a_i, b_i, \text{carry})$;
4 **end**
5 $c \leftarrow (c_{n-1}, \ldots, c_2, c_1, c_0)$;
6 **if** $carry = 1$ or $c > p$ **then**
7 $\quad \big|$ $c \leftarrow c - p$; $\qquad\qquad\qquad\qquad\qquad\qquad$ // Algorithm 5.2
8 **end**
9 **return** c;

ALGORITHM 5.2   Subtraction in $\mathbb{F}_p$.

Input   : An n-word prime p, and two integers a, $b \in [0, p - 1]$ where
$\qquad a = (a_{n-1}, a_{n-2}, \ldots, a_0)$ and $b = (b_{n-1}, b_{n-2}, \ldots, b_0)$
Output: $c = (a - b) \bmod p$
1 $(c_0, \text{borrow}) \leftarrow \text{Subtract}(a_0, b_0)$
2 **for** $i = 1 \rightarrow n - 1$ **do**
3 $\quad \big|$ $(c_i, \text{borrow}) \leftarrow \text{Subtract\_with\_borrow}(a_i, b_i, \text{borrow})$;
4 **end**
5 $c \leftarrow (c_{n-1}, \ldots, c_2, c_1, c_0)$;
6 **if** $borrow$ **then**
7 $\quad \big|$ $c \leftarrow c + p$; $\qquad\qquad\qquad\qquad\qquad\qquad$ // Algorithm 5.1
8 **end**
9 **return** c;

---

native instruction. Conversely, a Java implementation of Algorithm 5.1 would likely require additions and comparisons in order to implement Step 3. Steps 6–8 guarantee that the returned result c is always less than p.

In the case of the subtraction operation, the procedure is essentially the same, except that in this case, the word subtraction operations are performed with borrows, as shown in Algorithm 5.2. Once again, several microprocessors include native support for performing subtraction of two operands with an input borrow. Otherwise, this operation must be emulated using subtractions and comparisons.

5.1.3   Integer Multiplication

Multiplication is the most used and therefore the most crucial operation for the efficient computation of bilinear pairings. It is customary to use it as a unit of metrics in order to estimate the computational cost of cryptographic primitives.

Schoolbook integer multiplication

The integer multiplication $t = a \cdot b$ is performed word by word according to the school-book method, at a computational cost of n^2 and $(n-1)^2$ word multiplications and additions, respectively. In the case when p is a 256-bit prime, and the word size is $W = 64$, the following familiar figure depicts the schoolbook method strategy,

					a_3	a_2	a_1	a_0
$\times$					b_3	b_2	b_1	b_0
					pp_{03}	pp_{02}	pp_{01}	pp_{00}
				pp_{13}	pp_{12}	pp_{11}	pp_{10}	
			pp_{23}	pp_{22}	pp_{21}	pp_{20}		
$+$		pp_{33}	pp_{32}	pp_{31}	pp_{30}			
	t_7	t_6	t_5	t_4	t_3	t_2	t_1	t_0

Notice that the double-word partial product variables pp_{ij}, for $0 \leq i, j < n$, which in the following discussion will be denoted using a (Carry, Sum) pair notation, are produced from the word product $a_j \cdot b_i$. Furthermore, notice that the products t_k, for $0 \leq k \leq 2n - 1$ are single word numbers. Algorithm 5.3 shows the schoolbook procedure for computing the product $a \cdot b$.

ALGORITHM 5.3 The schoolbook multiplication algorithm.

Input : Two integers $a, b \in [0, p - 1]$, where $a = (a_{n-1}, a_{n-2}, \ldots, a_0)$ and
$\quad\quad b = (b_{n-1}, b_{n-2}, \ldots, b_0)$

Output: $t = a \cdot b$

1 Initially $t_i \leftarrow 0$ for all $i = 0, 1, \ldots, 2n - 1$.;
2 **for** $i = 0 \rightarrow n - 1$ **do**
3 $\quad$ $C \leftarrow 0$;
4 $\quad$ **for** $j = 0 \rightarrow n - 1$ **do**
5 $\quad\quad$ $(C, S) \leftarrow t_{i+j} + a_j \cdot b_i + C$;
6 $\quad\quad$ $t_{i+j} \leftarrow S$;
7 $\quad$ **end**
8 $\quad$ $t_{i+j+1} \leftarrow C$;
9 **end**
10 **return** $(t_{2n-1}, t_{2n-2} \cdots t_0)$;

In order to properly implement Step 5 of Algorithm 5.3,

$$(C, S) := t_{i+j} + a_j \cdot b_i + C,$$

where the variables t_{i+j}, a_j, b_i, C, and S each hold a single-word, or a W-bit number. The inner-product operation above requires that we multiply two W-bit numbers and add this product to the previous C variable, which is also a W-bit number, and then add this result to the running partial product word t_{i+j}. From these three operations we obtain a $2W$-bit number, since the maximum possible value is

$$2^W - 1 + (2^W - 1)(2^W - 1) + 2^W - 1 = 2^{2W} - 1.$$

A brief inspection of the steps of this algorithm reveals that the total number of inner-product steps is equal to n^2.

ALGORITHM 5.4   The schoolbook squaring algorithm.

Input : The integer $a \in [0, p-1]$, where $a = (a_{n-1}, a_{n-2}, \ldots, a_0)$
Output: $t = a \cdot a$

1 Initially $t_i \leftarrow 0$ for all $i = 0, 1, \ldots, 2n - 1$.
2 **for** $i = 0 \rightarrow n - 1$ **do**
3 $\quad (C, S) \leftarrow t_{i+j} + a_i \cdot a_i$;
4 $\quad$ **for** $j = i + 1 \rightarrow n - 1$ **do**
5 $\quad\quad (C, S) \leftarrow t_{i+j} + 2a_j \cdot a_i + C$;
6 $\quad\quad t_{i+j} \leftarrow S$;
7 $\quad$ **end**
8 $\quad t_{i+j+1} \leftarrow C$;
9 **end**
10 **return** $(t_{2n-1}, t_{2n-2} \cdots t_0)$;

Remark 5.2   Squaring is easier Notice that integer squaring, i.e., the operation of computing $t = a \cdot a$, is an easier operation than multiplication. Indeed, since $pp_{ij} = a_i \cdot a_j = p_{ji}$, half of the single-precision multiplications can be skipped. The corresponding procedure is shown in Algorithm 5.4. However, we warn the reader that the carry-sum pair produced in Step 5 may be 1 bit longer than a single-precision number that requires W bits. Thus, we need to accommodate this 'extra' bit during the execution of the operations in Steps 5, 6, and 8 of Algorithm 5.4. The resolution of this carry may depend on the way the carry bits are handled by the particular processor architecture. This issue, being rather implementation-dependent, will not be discussed any further in this chapter.

Karatsuba multiplier

The Karatsuba multiplier allows us to reduce the number of word multiplications required for the multi-precision integer product $t = a \cdot b$. Indeed, by exploiting the "divide and conquer" strategy, the Karatsuba multiplier effectively reduces the school-book method quadratic complexity $O(n^2)$ on the number of word multiplications to a superlinear complexity of just $O(n^{log_2 3})$ word multiplications.

Let us assume that $a = (a_1, a_0)$ and $b = (b_1, b_0)$ with $|a| = |b| = 2W$. Then, the product $a \cdot b$ can be obtained as,

$$
\begin{aligned}
a \cdot b = & \ a_0 b_0 + (a_0 b_1 + a_1 b_0) 2^W + a_1 b_1 2^{2W} \\
= & \ a_0 b_0 + ((a_0 + a_1)(b_0 + b_1) - a_0 b_0 - a_1 b_1) 2^W + a_1 b_1 2^{2W},
\end{aligned}
$$

where the product $a \cdot b$ can be computed using just three word multiplications as opposed to four that the traditional schoolbook method would require. This is a performance advantage, since in most platforms, the cost of one-word addition and subtraction are far cheaper than the cost of one-word multiplication. Clearly, this method can be used recursively for operands with larger word size.

However, depending on the particular platform where this method is implemented, there would be a practical limit due to the overhead associated with the way that the addition carries are generated and propagated through the remaining computations. Because of this, the Karatsuba approach is usually performed until a certain threshold, and after that it is combined with the schoolbook method. This strategy is sometimes called a shallow Karatsuba approach. The interested reader is referred to [44, 47, 53] for a careful analysis of how to perform an optimal combination of the Karatsuba and the schoolbook method in software and hardware platforms.[*]

5.1.4 Reduction

Let $a = (a_{n-1}, a_{n-2}, ..., a_0)$ and $b = (b_{n-1}, b_{n-2}, ..., b_0)$. A classical method for obtaining the modular multiplication defined as $c = (a \cdot b) \bmod p$, consists of performing the integer multiplication $t = a \cdot b$ first, and then the reduction step, $c = t \bmod p$. At first glance, in order to obtain the reduction $c = t \bmod p$, it would appear that it is necessary to perform a division by p, which is a relatively expensive operation. Fortunately, there exist better algorithmic approaches to deal with this problem. In this chapter we analyze in detail the Barrett and the Montgomery reductions.

Barrett reduction

Introduced at Crypto 1986 by Paul Barrett [12], the Barrett reduction algorithm finds $t \bmod p$ for a positive integer t and a modulus p, such that $|t| \approx 2|p|$. For cryptographic applications, p is often selected as a prime number. The Barrett reduction becomes advantageous if many reductions are performed using the same modulus p [29]. The method requires us to precompute the per-modulus constant parameter,

$$\mu = \left\lfloor \frac{b^{2k}}{p} \right\rfloor$$

where b is usually selected as a power of two close to the word size processor, and $k = \lfloor log_b \, p \rfloor + 1$.

The Barrett reduction algorithm is based in the following observation: Given $t = Qp + R$, where $0 \le R < p$, the quotient $Q = \left\lfloor \frac{t}{p} \right\rfloor$ can be written as,

$$Q = \left\lfloor \lfloor t/b^{k-1} \rfloor \cdot \left(b^{2k}/p\right) \cdot \left(1/b^{k+1}\right) \right\rfloor = \left\lfloor \lfloor t/b^{k-1} \rfloor \cdot \mu \cdot \left(1/b^{k+1}\right) \right\rfloor.$$

ALGORITHM 5.5   Barrett Reduction as presented in [29].

Input : p, $b \ge 3$, $k = \lfloor log_b \, p \rfloor + 1$, $0 \le t < b^{2k}$, and $\mu = \lfloor b^{2k}/p \rfloor$
Output: $r = t \bmod p$
1 $\hat{q} \leftarrow \left\lfloor \lfloor t/b^{k-1} \rfloor \cdot \mu / b^{k+1} \right\rfloor$;
2 $r \leftarrow (t \bmod b^{k+1}) - (\hat{q} \cdot p \bmod b^{k+1})$;
3 **if** $r < 0$ **then**
4 $\quad | \quad r \leftarrow r + b^{k+1}$;
5 **end**
6 **while** $r \ge p$ **do**
7 $\quad | \quad r \leftarrow r - p$;
8 **end**
9 **return** r;

Notice that once Q has been found, then the remainder R can readily be computed as $R = t - Qp \equiv t \bmod p$. Algorithm 5.5 presents the Barrett reduction, where the remainder R is computed. The correctness of this procedure can be outlined as follows.

---

[*]The Toom-Cook algorithm can be seen as a generalization of the Karatsuba algorithm (see [34, Section 4.3.3.A: Digital methods]).

In step 1, one has that,

$$0 \le \hat{q} = \left\lfloor \frac{\left\lfloor \frac{t}{b^{k-1}} \right\rfloor \cdot \mu}{b^{k+1}} \right\rfloor \le \left\lfloor \frac{z}{p} \right\rfloor = Q,$$

where $\hat{q}$ is an approximation of Q such that $Q \le \hat{q} + 2$. Steps 2–5 guarantee that

$$r = (t - \hat{q} \cdot p) \bmod b^{k+1},$$

and also that r is a positive integer. Furthermore, notice that $0 \le z - Qp < p$, which implies that $0 \le z - \hat{q}p \le z - (Q-2)p < 3p$. Hence, Steps 6–8 require at most 2 subtractions to compute the remainder $R = t \bmod p$.

Remark 5.3 Implementation notes. It can be shown that Algorithm 5.5 requires $(k^2 + 5k + 2)/2$ and $(k^2 + 3k)/2$ word multiplications to compute $\hat{q}$ in Step 1 and $\hat{q} \cdot p$ in Step 2, respectively. Hence, this method can compute $R = t \bmod p$ at a cost of approximately $k^2 + 4k$ word multiplications. This implies that the combination of the schoolbook multiplication plus the Barrett reduction can compute a field multiplication $c = a \cdot b \bmod p$, at a cost of about $2k^2 + 4k$ word multiplications.

Example 5.1 As a toy example, let us consider the case where the processor word size is 4 bits and $p = 2113$. By selecting the parameter b as $b = 2^4$, we represent the prime p in base b as the three 4-bit word number, $p = (8, 4, 1)_{16}$, where $k = |p| = 3$. This fixes the per-modulus constant μ as, $\mu = \lfloor b^{2k}/p \rfloor = (1, F, 0, 3)_{16}$.

Assume now that we want to multiply the operands $a_0, a_1 \in \mathbb{F}_p$, with $a_0 = 1528 = (5, F, 8)_{16}$ and $a_1 = 1657 = (6, 7, 9)_{16}$. This can be accomplished using Algorithm 5.3 to first obtain the integer product $t = a_0 \cdot a_1 = (2, 6, A, 2, 3, 8)_{16}$, followed by the execution of Algorithm 5.5, to complete the computation of the field multiplication as,

$$a_0 \cdot a_1 \bmod p = t \bmod p = 522 = (2, 0, A)_{16}.$$

Indeed, this is the case: The execution of Steps 1 and 2 of Algorithm 5.5 yield $\hat{q} = (4, A, E)_{16}$, and $r = (2, 0, A)_{16}$, respectively. Since we already have that $r < p$, this ends the computation.

The Montgomery multiplier

The Montgomery multiplier [40] is arguably the most popular and efficient method for performing field multiplication over large prime fields. As we will discuss next, this procedure replaces costly divisions by the prime p with divisions by r, where r is chosen as a power of two, $r = 2^k$, and k is a positive integer greater than or equal to the size of the prime p in bits, i.e., $k - 1 < |p| < k$.

The above strategy is accomplished by projecting an integer $a \in \mathbb{F}_p$ to the integer $\tilde{a} \in \mathbb{F}_p$, by means of the bijective mapping, $\tilde{a} = a \cdot r \bmod p$. The integer $\tilde{a}$ is called the p-residue of a, and the integer $\tilde{a}$ is said to be in the Montgomery domain. We define the Montgomery product as,

$$\mathrm{MontPr}(\tilde{a}, \tilde{b}) = \tilde{a} \cdot \tilde{b} \cdot r^{-1} \bmod p.$$

Given the integer a, its p-residue $\tilde{a}$ can be computed using the Montgomery product of a times 1, since, $\mathrm{MontPr}(\tilde{a}, 1) = \tilde{a} \cdot 1 \cdot r^{-1} \bmod p = a \bmod p$.

Algorithm 5.6 shows the procedure to compute the Montgomery product, whose main algorithmic idea is based on the following observation. In Step 1 we compute $t = \tilde{a} \cdot \tilde{b}$. Then, we look for a value q such that $t + q \cdot p$ is a multiple of r, i.e., $t + q \cdot p \equiv 0 \bmod r$. This implies that the required value q is given as, $q \equiv -t \cdot p^{-1} \bmod r$.

Step 3 computes the operation, $u = (t + q \cdot p)/r$. Notice that from the discussion above, we know that u is an integer, and since r has been chosen as a power of two, the division by r

---

ALGORITHM 5.6   Montgomery product.

Input   : An n-word prime p, $r = 2^{n \cdot W}$, a parameter p' as defined in Eq. (5.1), and two p-residues $\tilde{a}, \tilde{b}$.

Output: u = $\mathrm{MontPr}(\tilde{a}, \tilde{b})$

1 $t \leftarrow \tilde{a} \cdot \tilde{b}$;
2 $q \leftarrow t \cdot p' \bmod r$;
3 $u \leftarrow (t + q \cdot p)/r$;
4 **if** $u > p$ **then**
5 $\quad$ **return** $u - p$;
6 **else**
7 $\quad$ **return** u;
8 **end**
9 **return** u;

---

can be implemented as a simple shift operation. Now, since $u \cdot r = (t + q \cdot p) \equiv t \bmod p$, then $u \equiv t \cdot r^{-1} \equiv \tilde{a} \cdot \tilde{b} \cdot r^{-1} \bmod p$, as required. Furthermore, assuming that $0 \le t < p \cdot r$, then since $q < r$, it follows that $u \le 2p$. Then, Steps 3–7 suffice to guarantee that the returned value u will always be less than p.

For efficiency reasons, Algorithm 5.6 requires the pre-computation of a constant p' such that,

$$r \cdot r^{-1} - p \cdot p' = 1, \tag{5.1}$$

which can be easily found by means of the extended Euclidean algorithm (cf. §5.1.6).

As we have seen, the Montgomery product exploits the fact that, owing to the selection of r as a power of two, the division and multiplication by r are fast and efficient operations. However, it is noticed that this algorithm has the associated overhead of performing the pre-computation of the parameter p', as well as the computation of the operands' p-residues, and the mapping of the Montgomery product back to the integers. Hence, this method is mainly used in scenarios where many multiplications must be performed one after the other (cf. §5.1.5).

Example 5.2   Let us consider again the case where the processor word size is 4 bits and $p = 2113 = (8, 4, 1)_{16}$. Since p is a three 4-bit word number, for this case we have that $r = 2^{3 \cdot 4} = 2^{12}$. Then, using the extended Euclidean algorithm, we get $p' = -1985$ since,

$$-r \cdot 1024 + p \cdot 1985 = 1.$$

Using Algorithm 5.6, we want to compute the Montgomery product $\mathrm{MontPr}(\tilde{a}, \tilde{b})$ in the field $\mathbb{F}_p$, with $a = 1528$ and $b = 1657$. We find that the p-residues of the operands a, b, are given as,

$$\tilde{a} \quad = a \cdot r \bmod p = 2095,$$
$$\tilde{b} \quad = b \cdot r \bmod p = 116.$$

The execution of Steps 1 and 2 of Algorithm 5.6 yield $t = 243020$, and $u = 1869$, respectively. Since we already have that $u < p$, this ends the computation. Hence, $\mathrm{MontPr}(\tilde{a}, \tilde{b}) = 1869$.

The Montgomery product shown in Algorithm 5.6 can be used to compute the field multiplication $c = a \cdot b \bmod p$, as shown in Algorithm 5.7.

Example 5.3   We already found in Example 5.2 that for $p = 2113$, $a = 1528$, and $b = 1657$, we get $\tilde{c} = \mathrm{MontPr}(\tilde{a}, \tilde{b}) = 1869$. Using Algorithm 5.7, we find that $\mathrm{MontPr}(\tilde{c}, 1) = 522$. Hence, we conclude that $c = a \cdot b \bmod p = 522$.

ALGORITHM 5.7 Montgomery field multiplication over $\mathbb{F}_p$.

Input : An n-word prime p, $r = 2^{n \cdot W}$, and two integers $a, b \in \mathbb{F}_p$
Output: $c = a \cdot b \bmod p$
1 $\bar{a} \leftarrow a \cdot r \bmod p$;
2 $\bar{b} \leftarrow b \cdot r \bmod p$;
3 $\bar{c} \leftarrow \mathrm{MontPr}(\bar{a}, \bar{b})$;   // Algorithm 5.6
4 $c \leftarrow \mathrm{MontPr}(\bar{c}, 1)$;   // Algorithm 5.6
5 **return** c;

Remark 5.4 A slight improvement in the Montgomery field multiplication Notice
that a more efficient algorithm for obtaining the field multiplication can be devised by observing
the property,

$$\mathrm{MonPro}(\bar{A}, B) = (A \cdot r) \cdot B \cdot r^{-1} = A \cdot B \pmod{n},$$

which modifies Algorithm 5.7 as shown in Algorithm 5.8.

ALGORITHM 5.8 Improved version of the Montgomery field multiplication over $\mathbb{F}_p$.

Input : An n-word prime p, $r = 2^{n * W}$, and two integers $a, b \in \mathbb{F}_p$
Output: $c = a \cdot b \bmod p$
1 $\bar{a} \leftarrow a \cdot r \bmod p$;
2 $c \leftarrow \mathrm{MontPr}(\bar{a}, b)$;   // Algorithm 5.6
3 **return** c;

There exist several variants for efficiently implementing the Montgomery product procedure
shown in Algorithm 5.6. In this chapter we will describe the Montgomery product variants SOS
and CIOS as they were presented in [36].

The SOS method

The Separated Operand Scanning (SOS) method computes the Montgomery multiplication into
two phases. First, the integer product $t = a \cdot b$ is performed via Algorithm 5.3. Then, t is
Montgomery reduced by computing the value u corresponding to Step 3 of Algorithm 5.6, using
the formula $u = (t + m \cdot p)/r$, where $m = t \cdot p' \bmod r$ and $r = 2^{n \cdot W}$. Let us recall that W
represents the processor's word size in bits, and n the number of words required to store the
prime p. The SOS Montgomery reduction procedure is shown in Algorithm 5.9.

The computational cost of the SOS Montgomery product method is of $2n^2 + n$ word multiplications, since n^2 word multiplications are required for obtaining the product $a \cdot b$, and $n^2 + n$
to compute u. Moreover, this method requires $4n^2 + 4n + 2$ word additions [36].

Remark 5.5 Rationale of the SOS Montgomery reduction. Algorithm 5.9 assumes that
after executing Algorithm 5.3, the input parameter t holds the integer product $t = a \cdot b$. Then,
the product $q \cdot p$ is computed and added to t in Steps 1–9. Furthermore, since $q = t \cdot p' \bmod r$,
is computed word by word, one can take advantage of the identity $p'_0 = p' \bmod 2^W$ in order
to use p'_0 instead of p' all the way through.[*] Notice that the function *Add* in Step 8 computes

---

[*]This observation was first pointed out in [24].

ALGORITHM 5.9   Montgomery SOS reduction.

Input : $t = (t_{2n-1}, t_{2n-2}, ..., t_0)$, $p = (p_{n-1}, p_{n-2}, ..., p_0)$ and p_0', the least-significant
            word of p' as defined in Eq. (5.1).

Output: $u = (t + (t \cdot p' \bmod r) \cdot p)/r$

1 **for** $i = 0 \rightarrow n - 1$ **do**
2 $\quad$ $C \leftarrow 0$;
3 $\quad$ $q \leftarrow t_i \cdot p_0' \bmod 2^W$;
4 $\quad$ **for** $j = 0 \rightarrow n - 1$ **do**
5 $\quad\quad$ $(C, S) \leftarrow t_{i+j} + q \cdot p_j + C$;
6 $\quad\quad$ $t_{i+j} \leftarrow S$;
7 $\quad$ **end**
8 $\quad$ $t_{i+n} \leftarrow \text{Add}(t_{i+n}, C)$;
9 **end**
10 **for** $i = 0 \rightarrow n - 1$ **do**
11 $\quad$ $u_i \leftarrow t_{i+n}$;
12 **end**
13 **return** u;

---

the word-level addition of t_{i+n} with C, and it propagates the obtained result until there is no
output carry. Finally, the division by $r = 2^{n\omega}$ is performed by assigning to the variable u the n
most significant words of t (see Steps 10–12). Once that u has been obtained, it must be checked
whether it is greater than p or not. If this is indeed the case, then one extra subtraction by p
must be performed.

**Remark 5.6   SOS Montgomery method combined with the Karatsuba approach is
more efficient.** The combination of the SOS Montgomery method along with the Karatsuba
approach discussed in §5.1.3 can compute one Montgomery product at a computational cost of
about $1.75n^2 + n$ word multiplications.

The CIOS method

The Coarsely Integrated Operand Scanning (CIOS) method integrates the integer multiplication and reduction steps. More concretely, instead of performing the product $a \cdot b$ followed
by the Montgomery reduction, the multiplication and reduction operations are performed in
an interleaved fashion within the main loop. This approach produces correct results, because
$q = t \cdot p' \bmod r$ depends only on the value t_i, which is computed by the i-th iteration of the main
loop of Algorithm 5.10 [36].

Remark 5.7   Implementation notes of the Montgomery CIOS product As a benefit
of the approach of multiplying and then reducing, the auxiliary variable t requires just $n + 2$
words, which has to be compared with the auxiliary variable t of the SOS method that requires
$2n$ words. The computational cost of this procedure is of $2n^2 + n$ and $4n^2 + 4n + 2$ word
multiplications and additions, respectively.

5.1.5   Exponentiation

Let $a \in \mathbb{F}_p^*$ and $e \in \mathbb{Z}^+$. The modular exponentiation in the base field is defined as $a^e \bmod p$.
This operation can be performed using the well-known binary exponentiation algorithm as shown
in Algorithm 5.11, invoking the Montgomery multiplier as its main building block.

ALGORITHM 5.10 Montgomery CIOS product.

Input : $a = (a_{n-1}, a_{n-2}, ..., a_0)$, $b = (b_{n-1}, b_{n-2}, ..., b_0)$, $p = (p_{n-1}, p_{n-2}, ..., p_0)$ and p'_0,
             the least significant word of p' as defined in Eq. (5.1).

Output: $t =$ MontPr(a, b)

1   $t = (t_{n+1}, t_n, ..., t_0) \leftarrow 0$;
2   **for** $i = 0 \rightarrow n - 1$ **do**
3      $C \leftarrow 0$;
4   **for** $j = 0 \rightarrow n - 1$ **do**
5          $(C, S) \leftarrow t_j + a_j \cdot b_i + C$;
6          $t_j \leftarrow S$;
7   **end**
8      $(C, S) \leftarrow t_n + C$, $t_n \leftarrow S$, $t_{n+1} \leftarrow C$;
9      $C \leftarrow 0$;
10     $q \leftarrow t_0 \cdot p'_0 \bmod 2^\omega$;
11     $(C, S) \leftarrow t_0 + q \cdot p_0$;
12   **for** $j = 1 \rightarrow n - 1$ **do**
13        $(C, S) \leftarrow t_j + q \cdot p_j + C$;
14        $t_{j-1} \leftarrow S$;
15   **end**
16     $(C, S) \leftarrow t_n + C$, $t_{n-1} \leftarrow S$, $t_n \leftarrow t_{n+1} + C$;
17 **end**
18 **if** $t > p$ **then**
19   **return** $t - p$;
20 **else**
21   **return** t;
22 **end**

5.1.6   Multiplicative Inverse

For $a, b \in \mathbb{F}_p^*$, we say that b is the multiplicative inverse of a, denoted as $b = a^{-1} \in \mathbb{F}_p^*$, iff $a \cdot b \equiv 1 \bmod p$. From the definitions of finite fields, $b = a^{-1}$ always exists and it is unique (see Chapter 2).

The two most popular algorithms for finding the multiplicative inverse are the extended Euclidean algorithm, and the computation of the inverse via Fermat's little theorem. The latter is computed based on the identity, $a^{-1} = a^{p-2} \bmod p$, as it was briefly discussed in Chapter 2 (see Theorem 2.8 and its discussion). Since this technique tends to be expensive it will not be discussed any further in this Chapter.[*]

Multiplicative inverse via the extended Euclidean algorithm

Given two integers a and b, we say that the greatest common divisor of a and b is the largest integer $d = gcd(a, b)$ that divides both a and b. Based on the property $gcd(a, b) = gcd(b, a \bmod$

---

[*]But the interested reader may want to check out [51, Appendix A], where multiplicative inverses were computed via Fermat's little theorem in a real implementation.

ALGORITHM 5.11   Montgomery exponentiation.

Input : A n-word prime p, $r = 2^{n \cdot W}$, $a \in \mathbb{F}_p$, $e = (e_\ell, ..., e_0)_2$
Output: $A = a^e \bmod p$

1 $\bar{a} \leftarrow a \cdot r \bmod p$;
2 $\bar{A} \leftarrow r \bmod p$;
3 **for** $i = \ell \to 0$ **do**
4 $\quad$ $\bar{A} \leftarrow \text{MontPr}(\bar{A}, \bar{A})$;
5 $\quad$ **if** $e_i = 1$ **then**
6 $\quad\quad$ $\bar{A} \leftarrow \text{MontPr}(\bar{A}, \bar{a})$;
7 $\quad$ **end**
8 **end**
9 **return** $A \leftarrow \text{MontPr}(\bar{A}, 1)$;

---

b), the ancient Extended Euclidean Algorithhm (EEA)[†] is able to find the unique integers s, t that satisfy Bezout's formula,

$$a \cdot s + b \cdot t = d,$$

where $d = gcd(a, b)$.

The literature is replete with different versions of the extended Euclidean algorithm. The interested reader is referred to [29, Chapter 2], [34], for a comprehensive survey of the most efficient variants of this algorithm and their related computational complexity. In this Chapter we restrict ourselves to presenting the binary algorithm for inversion in $\mathbb{F}_p$ as reported in [29]. The corresponding procedure is shown in Algorithm 5.12.

Remark 5.8   Computational complexity of the binary algorithm for inversion. One important advantage of the procedure shown in Algorithm 5.12 is that it completely avoids costly divisions by trading all of them with divisions by two, which can be trivially implemented as right shift operations. Furthermore, it is known that the number of iterations performed by Algorithm 5.12 will always lie between ℓ and 2ℓ, where, $\ell = |p|$.

Remark 5.9   Side-channel security of the binary algorithm for inversion. As will be studied in Chapter 12, we will often face adversaries trying to exploit side-channels attacks. One standard countermeasure to thwart these attacks is to implement algorithms that enjoy a constant-time nature in their execution, meaning that the algorithm execution time does not depend on any secret parameter values. Unfortunately, Algorithm 5.12 does not enjoy this property, as the total number of iterations in its execution is highly correlated with the operand a. Fortunately, most side-channel attacks can effectively be thwarted by using blinding techniques (see Remark 8.3 in Chapter 8). For example, one can randomly multiply by a blind factor ω and invoke Algorithm 5.12 to compute the inverse of $a \cdot \omega$. Then, we simply multiply the result so obtained by ω to get the desired inverse a^{-1}.

[†]This algorithm was presented in Euclid's *Elements*, published 300 B.C. Nevertheless, some scholars believe that it might have been previously known by Aristotle and Eudoxus, some 100 years earlier than Euclid's time. Knuth pointed out that it can be considered the oldest non-trivial algorithm that has survived to modern era [34].

ALGORITHM 5.12   Binary algorithm for inversion over $\mathbb{F}_p$.

Input : A prime p and $a \in \mathbb{F}_p$

Output: $a^{-1} \bmod p$

1 $u \leftarrow a$, $v \leftarrow p$;
2 $x_1 \leftarrow 1$, $x_2 \leftarrow 0$;
3 **while** $u \neq 1$ *AND* $v \neq 1$ **do**
4 $\quad$ **while** u *is even* **do**
5 $\quad\quad$ $u \leftarrow u/2$;
6 $\quad\quad$ **if** x_1 *is even* **then**
7 $\quad\quad\quad$ $x_1 \leftarrow x_1/2$;
8 $\quad\quad$ **else**
9 $\quad\quad\quad$ $x_1 \leftarrow (x_1 + p)/2$;
10 $\quad\quad$ **end**
11 $\quad$ **end**
12 $\quad$ **while** v *is even* **do**
13 $\quad\quad$ $v \leftarrow v/2$;
14 $\quad\quad$ **if** x_2 *is even* **then**
15 $\quad\quad\quad$ $x_2 \leftarrow x_2/2$;
16 $\quad\quad$ **else**
17 $\quad\quad\quad$ $x_2 \leftarrow (x_2 + p)/2$;
18 $\quad\quad$ **end**
19 $\quad$ **end**
20 $\quad$ **if** $u \geq v$ **then**
21 $\quad\quad$ $u \leftarrow u - v, x_1 \leftarrow x_1 - x_2$;
22 $\quad$ **else**
23 $\quad\quad$ $v \leftarrow v - u, x_2 \leftarrow x_2 - x_1$;
24 $\quad$ **end**
25 **end**
26 **if** $u == 1$ **then**
27 $\quad$ **return** $x_1 \bmod p$;
28 **else**
29 $\quad$ **return** $x_2 \bmod p$;
30 **end**

Montgomery multiplicative inverse

As we have seen in the previous section, in the Montgomery arithmetic one selects the value $r \geq 2^\ell$, where $\ell = |p|$ is the size in bits of the prime p. Let $\tilde{a} = a \cdot r \bmod p$, the Montgomery multiplicative inverse $\tilde{a}$ is defined as $\tilde{a}^{-1} = a^{-1} \cdot r \bmod p$. The Montgomery multiplicative inverse [29] is computed into two steps. The first one consists of finding a partial multiplicative inverse $a^{-1}2^k$ with $k \in [\ell, 2\ell]$. The second step consists of finding the Montgomery inverse multiplicative as defined above by repeated divisions by two until the required value is obtained.

5.1.7   Square Root

Taking square roots in the base field is an important building block for computing the hashing into elliptic curves, as we will review in detail in Chapter 8. This hashing operation, which was first introduced in the landmark paper [16], is a crucial building block for the vast majority of pairing-based protocols.

The problem of computing the square root of an element $a \in \mathbb{F}_p$ consists of finding an element $x \in \mathbb{F}_p$, such that $x^2 = a \bmod p$. According to Fermat's little theorem, it follows that $a^{p-1} = 1 \bmod p$. Hence, $a^{\frac{p-1}{2}} = \pm 1 \bmod p$. Let $g \in \mathbb{F}_p$ be a generator. Then, for $a = g^i$ with $i \in \mathbb{Z}^+$, one has that $a^{\frac{p-1}{2}} = (g^i)^{\frac{p-1}{2}} \pm 1 \bmod p$. For all i even it holds that $(g^{\frac{i}{2}})^{p-1} = 1 \bmod p$, where $g^{\frac{i}{2}}$ is a root of a, otherwise $(g^i)^{\frac{p-1}{2}} = -1 \bmod p$, which implies that the element a has no square root over $\mathbb{F}_p$. We denote by $\chi_p(a)$ the value of $a^{\frac{p-1}{2}}$. If $\chi_p(a) = 1$, we say that the element a is a quadratic residue (QR) in $\mathbb{F}_p$. It is known that in $\mathbb{F}_p^*$ there exist exactly $(p-1)/2$ quadratic residues.

There exist several algorithms for computing the square root of a field element depending on the form of the prime p that defines the field. One can consider two general cases, when $p \equiv 3 \bmod 4$, and when $p \equiv 1 \bmod 4$. The computation of the square root in the former case can be performed with a simple exponentiation, as $x = a^{\frac{p+1}{4}}$. This can easily be verified from Fermat's little theorem. Since $a^p \equiv a \bmod p$, we have that,

$$x^2 = a^{\frac{p+1}{2}} = (a^{p+1})^{\frac{1}{2}} = (a^p a)^{\frac{1}{2}} = (a^2)^{\frac{1}{2}} = a.$$

Therefore $x^2 = a$, as required. Note that the above derivation only holds iff $4 | p + 1$. Algorithm 5.13 due to Shanks [48], computes the square root of an arbitrary field element $a \in \mathbb{F}_p$, with $p \bmod 4 = 3$. Notice that the quadratic residuosity test (See §2.2.9) has been integrated into Algorithm 5.13. If a is a quadratic residue, the procedure returns its square root and false otherwise.

ALGORITHM 5.13   Shanks algorithm.

Input   : $a \in \mathbb{F}_p^*$
Output: false or $x \in \mathbb{F}_p^*$ such that $x^2 = a$
1 $a_1 \leftarrow a^{\frac{p-3}{4}}$
2 $a_0 \leftarrow a_1^2 a$
3 **if** $a_0 = -1$ **then**
4 $\quad$ **return** false:
5 **end**
6 $x \leftarrow a_1 a$
7 **return** x

In the second case, when $p \equiv 1 \bmod 4$, no simple and general algorithm is known. However, fast algorithms for computing a square root in $\mathbb{F}_p$ when $p \equiv 5 \bmod 8$ or $p \equiv 9 \bmod 16$ have been reported [1]. For the case when $p \equiv 5 \bmod 8$, Atkin developed in 1992 an efficient and deterministic square root algorithm that is able to find the square root of a QR using only one field exponentiation plus a few multiplications in $\mathbb{F}_p$ [4]. A modification of the Atkin's algorithm was presented by Müller in [41], which allows one to compute square roots in $\mathbb{F}_p$ when $p \equiv 9 \bmod 16$, at a price of two exponentiations. For the case when $p \equiv 1 \bmod 16$, no specialized algorithm is known. Hence, for this class of finite fields one must use a general algorithm, such as the Tonelli-Shanks algorithm [50] or a modified version of the Cipolla-Lehmer algorithm [20] as presented by Müller in [41].

Algorithm 5.14 presents a variant of the Tonelli-Shanks procedure where the quadratic test of an arbitrary field element $a \in \mathbb{F}_p$ has been incorporated into the algorithm. It is noticed that the computational complexity of Algorithm 5.14 varies depending on whether the input is or is not a quadratic residue in $\mathbb{F}_p$.

Remark 5.10   Computational complexity of the Tonelli-Shanks algorithm. By taking into account the average contribution of QR and QNR inputs, and using the complexity analysis

---

ALGORITHM 5.14 Tonelli-Shanks algorithm.

Input : $a \in \mathbb{F}_p^*$
Output: false or $x \in \mathbb{F}_p^*$ such that $x^2 = a$

1 **PRECOMPUTATION**
2 Obtain (s, t) such that $p - 1 = 2^s t$ where t is odd;
3 $c_0 \leftarrow 1$;
4 **while** $c_0 = 1$ **do**
5 $\quad$ Select $c \in \mathbb{F}_p^*$ randomly;
6 $\quad$ $z \leftarrow c^t$;
7 $\quad$ $c_0 = c^{2^{s-1}}$;
8 **end**
9 **MAIN COMPUTATION**
10 $\omega \leftarrow a^{\frac{t-1}{2}}$;
11 $a_0 \leftarrow (\omega^2 a)^{2^{s-1}}$;
12 **if** $a_0 = -1$ **then**
13 $\quad$ **return** false;
14 **end**
15 $v \leftarrow s, x \leftarrow a\omega, b \leftarrow x\omega$;
16 **while** $b \neq 1$ **do**
17 $\quad$ Find the least integer $m \geq 0$ such that $b^{2^m} = 1$;
18 $\quad$ $\omega \leftarrow z^{2^{v-m-1}}, z \leftarrow \omega^2, b \leftarrow bz, x \leftarrow x\omega, v \leftarrow m$;
19 **end**
20 **return** x;

---

given in [39, 1] for the classical Tonelli-Shanks algorithm, it is not difficult to see that the average computational cost of Algorithm 5.14 is given as

$$\frac{1}{2}\Big[\lfloor\log_2(p)\rfloor + 4\Big]M_p + \Big[\lfloor\log_2(p)\rfloor + \frac{1}{8}\big(s^2 + 3s - 16\big) + \frac{1}{2^s}\Big]S_p, \tag{5.2}$$

where M_p and S_p stand for field multiplication and squaring over $\mathbb{F}_p$, respectively.

5.2 Tower Fields

ain arithmetic operations must be computed in extension fields of the form $\mathbb{F}_{p^k}$, for moderate values of k. Hence, a major design aspect is to represent the field so that its arithmetic can be performed at the highest efficiency. With the aim of producing a more efficient arithmetic over field extensions $\mathbb{F}_{p^n}$, several authors [5, 11], have proposed the idea of expressing a prime field extension $\mathbb{F}_{p^n} = \mathbb{F}_p[z]/f(z)$ as $\mathbb{F}_q$, where $q = p^m$ and $m|n$, such that

$$\begin{aligned}
\mathbb{F}_{p^n} &= \mathbb{F}_q[v]/h(v) &&, \text{where } h(v) \in \mathbb{F}_q[v] \text{ is a polynomial of degree } \tfrac{n}{m} \\
\mathbb{F}_q &= \mathbb{F}_p[u]/g(u) &&, \text{where } g(u) \in \mathbb{F}_p[u] \text{ is a polynomial of degree } m.
\end{aligned}$$

This approach is known as *tower fields*. Tower fields have been universally used in software and hardware implementations of bilinear pairings [3, 13, 15, 22, 25, 31, 35, 42, 54]. See also [21, Chapter 11] for a historical description of this strategy.

It is noted that the extension n and the prime p are the main factors that determine the optimal structure of a tower field [13]. Concretely, given the positive parameters a and b, we say that a finite field $\mathbb{F}_{p^n}$ is *pairing friendly* [35], if $n = 2^a 3^b$, such that $\mathbb{F}_{p^n}$ can be expressed through the quadratic extensions of a and the cubic extensions b of the base field [13].

Moreover, for all $n = 2^a 3^b$, if $4 \nmid n$, then the tower field can be built by means of irreducible binomials. This is a great advantage from the implementation point of view, as the field arithmetic becomes much more efficient when using low hamming weight irreducible polynomials. On the other hand, if $n \equiv 0 \mod 4$, one requires the condition $p^n \equiv 1 \mod 4$ in order to use this same representation [37, Theorem 3.75].

Example 5.4   Let $p = 97$. Then the finite field $\mathbb{F}_{p^6}$ can be expressed as a cubic extension of the quadratic extension field $\mathbb{F}_{p^2}$ as

$$
\begin{aligned}
\mathbb{F}_{p^6} &= \mathbb{F}_{p^2}[v]/(v^3 - u), \\
\mathbb{F}_{p^2} &= \mathbb{F}_p[u]/(u^2 + 5),
\end{aligned}
$$

where $-u$ and 5 must not be a square or a cube over $\mathbb{F}_{p^2}$ and $\mathbb{F}_p$, respectively.

In the rest of this section, we describe the field arithmetic associated to tower fields over quadratic and cubic extensions of a pairing-friendly finite field $\mathbb{F}_q$, where $q = p^m$ for $m > 0$. We use the notation $\oplus$, $\ominus$, $\otimes$, and $\oslash$ for denoting the addition, subtraction, multiplication, and division over the field $\mathbb{F}_q$, respectively.

5.2.1   Field Arithmetic over Quadratic Extensions

The quadratic extension over a field $\mathbb{F}_q$ is represented as

$$
\mathbb{F}_{q^2} = \mathbb{F}_q[u]/(u^2 - \beta), \tag{5.3}
$$

where $u^2 - \beta$ is an irreducible binomial over $\mathbb{F}_q$, which is always guaranteed when $\beta \in \mathbb{F}_q$ is not a square.

Addition

Given two elements $a, b \in \mathbb{F}_{q^2}$, the operation $a + b$, can be computed as shown in Algorithm 5.15, at a cost of 2 additions in the base field $\mathbb{F}_q$.

ALGORITHM 5.15   Field addition in the quadratic extension $\mathbb{F}_{q^2}$.

Input : $a = (a_0 + a_1 u), \quad b = (b_0 + b_1 u) \in \mathbb{F}_{q^2}$
Output: $c = a + b \in \mathbb{F}_{p^n}$

1 $c_0 \leftarrow a_0 \oplus b_0$;
2 $c_1 \leftarrow a_1 \oplus b_1$;
3 **return** $c = c_0 + c_1 u$;

Multiplication

The field multiplication of two field elements $a, b \in \mathbb{F}_{q^2}$, is defined as

$$
(a_0 + a_1 u) \cdot (b_0 + b_1 u) = (a_0 b_0 + a_1 b_1 \beta) + (a_0 b_1 + a_1 b_0) u.
$$

This operation can be efficiently implemented using the Karatsuba approach, where

$$(a_0 b_1 + a_1 b_0) = (a_0 + a_1) \cdot (b_0 + b_1) - a_0 b_0 - a_1 b_1.$$

ALGORITHM 5.16   Field multiplication in the quadratic extension $\mathbb{F}_{q^2}$.

Input : $a = (a_0 + a_1 u)$, $b = (b_0 + b_1 u) \in \mathbb{F}_{q^2}$
Output: $c = a \cdot b \in \mathbb{F}_{q^2}$

1 $v_0 \leftarrow a_0 \otimes b_0$;
2 $v_1 \leftarrow a_1 \otimes b_1$;
3 $c_0 \leftarrow v_0 \oplus \beta v_1$;
4 $c_1 \leftarrow (a_0 \oplus a_1) \otimes (b_0 \oplus b_1) \ominus v_0 \ominus v_1$;
5 **return** $c = c_0 + c_1 u$;

Algorithm 5.16 requires a total of 3 multiplications and 5 additions in the base field $\mathbb{F}_q$, as well as one m_β computation, where m_β denotes the cost of performing the product of an arbitrary element $a_0 \in \mathbb{F}_q$ by the constant β of the irreducible binomial used to generate the quadratic extension (see Equation (5.3)).

Squaring

In the particular case when $\beta = 1$, i.e., $u^2 = -1$, the operation a^2, with $a \in \mathbb{F}_{q^2}$, can be computed using the so-called squaring complex method that yields the following identity [30, Chapter 12]:

$$(a_0 + a_1 u)^2 = (a_0 + a_1) \cdot (a_0 - a_1) + 2 a_0 a_1 u. \tag{5.4}$$

Algorithm 5.17 is a generalization of Equation (5.4), which computes $(a_0 + a_1 u)^2$, for $\beta \neq 1$, at a cost of $2 m_\beta$, 2 multiplications, and 5 base field additions, $\mathbb{F}_q$.

ALGORITHM 5.17   Field squaring in the quadratic field $\mathbb{F}_{q^2}$.

Input : $a = (a_0 + a_1 u) \in \mathbb{F}_{q^2}$
Output: $c = a^2 \in \mathbb{F}_{q^2}$

1 $v_0 \leftarrow a_0 \ominus a_1$;
2 $v_3 \leftarrow a_0 \ominus \beta a_1$;
3 $v_2 \leftarrow a_0 \otimes a_1$;
4 $v_0 \leftarrow (v_0 \otimes v_3) \oplus v_2$;
5 $c_1 \leftarrow v_2 \oplus v_2$;
6 $c_0 \leftarrow v_0 \oplus \beta v_2$;
7 **return** $c = c_0 + c_1 u$;

Square root

This task can be achieved using the complex method proposed by Scott in [46] and described in detail in [1]. The corresponding procedure is presented in Algorithm 5.18.

ALGORITHM 5.18   Complex method for computing square roots over a quadratic extension field $\mathbb{F}_{q^2}$.

Input   : Irreducible binomial $f(u) = u^2 - \beta$ such that $\mathbb{F}_{q^2} \cong \mathbb{F}_q[u]/(u^2 - \beta)$, $\beta \in \mathbb{F}_q$, with $q = p^n$, $a = a_0 + a_1 u \in \mathbb{F}_{q^2}^*$

Output: If it exists $x = x_0 + x_1 u \in \mathbb{F}_{q^2}$ such that, $x^2 = a$, false otherwise

1 **if** $a_1 = 0$ **then**
2 | **return** $\sqrt{a_0}$;   (in $\mathbb{F}_q$)
3 **end**
4 $\lambda \leftarrow a_0^2 - \beta \cdot a_1^2$;
5 **if** $\chi_q(\lambda) = -1$ **then**
6 | **return** false;
7 **end**
8 $\lambda \leftarrow \sqrt{\lambda}$;   (in $\mathbb{F}_q$)
9 $\delta \leftarrow \frac{a_0 + \lambda}{2}$;
10 $\gamma \leftarrow \chi_q(\delta)$;
11 **if** $\gamma = -1$ **then**
12 | $\delta \leftarrow \frac{a_0 - \lambda}{2}$;
13 **end**
14 $x_0 \leftarrow \sqrt{\delta}$;   (in $\mathbb{F}_q$)
15 $x_1 \leftarrow \frac{a_1}{2x_0}$;
16 $x \leftarrow x_0 + x_1 u$;
17 **return** x;

Taking advantage of the fact that the field characteristic p for several popular pairing-friendly elliptic curves (including the Barreto-Naehrig curves) is usually selected so that $p \equiv 3 \mod 4$ [42, 2], then it is possible to perform the square root computation of the steps 2 and 8 in Algorithm 5.18, by performing the powering $\sqrt{\lambda} = \lambda^{(p+1)/4}$ with $\lambda \in \mathbb{F}_p$. This approach has the added advantage of keeping a constant-time behavior of Algorithm 5.18, which is an important feature to avoid side-channel attacks, as was mentioned in Remark 5.9.

Field inversion

The multiplicative inverse of a field element in the multiplicative group of the field extension $\mathbb{F}_{q^2}$ can be computed using the following identity:

$$(a_0 + a_1 u)^{-1} = \frac{a_0 - a_1 u}{(a_0 - a_1 u) \cdot (a_0 + a_1 u)} = \frac{a_0 - a_1 u}{a_0^2 - a_1^2 \beta}.$$

The above computation costs one multiplication by the constant β, 2 multiplications, 2 squarings, 2 additions, and 1 multiplicative inversion over the base field $\mathbb{F}_q$.

5.2.2   Field Arithmetic over Cubic Extensions

The cubic extension of a finite field $\mathbb{F}_q$ is represented by the polynomials in $\mathbb{F}_q[w]$ modulo the irreducible binomial, $w^3 - \alpha \in \mathbb{F}_q[w]$, i.e.,

$$\mathbb{F}_{q^3} = \mathbb{F}_q[w]/(w^3 - \alpha), \tag{5.5}$$

ALGORITHM 5.19 Field inversion in the quadratic extension $\mathbb{F}_{q^2}$.

Input : $a = (a_0 + a_1 u) \in \mathbb{F}_{q^2}$
Output: $c = a^{-1} \in \mathbb{F}_{q^2}$

1 $v_0 \leftarrow a_0^2$;
2 $v_1 \leftarrow a_1^2$;
3 $v_0 \leftarrow v_0 \ominus \beta v_1$;
4 $v_1 \leftarrow v_0^{-1}$;
5 $c_0 \leftarrow a_0 \otimes v_1$;
6 $c_1 \leftarrow -a_1 \otimes v_1$;
7 **return** $c = c_0 + c_1 u$;

---

where α is not a cube in $\mathbb{F}_q$.

Addition

Field addition of two elements $a, b \in \mathbb{F}_{q^3}$ can be computed using Algorithm 5.20, at a cost of 3 additions in the base field $\mathbb{F}_q$.

ALGORITHM 5.20 Field addition in the cubic extension $\mathbb{F}_{q^3}$.

Input : $a = (a_0 + a_1 w + a_2 w^2)$, $b = (b_0 + b_1 w + b_2 w^2) \in \mathbb{F}_{q^3}$
Output: $c = a + b \in \mathbb{F}_{p^n}$

1 $c_0 \leftarrow a_0 \oplus b_0$;
2 $c_1 \leftarrow a_1 \oplus b_1$;
3 $c_2 \leftarrow a_2 \oplus b_2$;
4 **return** $c = c_0 + c_1 w + c_2 w^2$;

Multiplication

The product of two field elements in the cubic extension of $\mathbb{F}_q$, can be efficiently computed using once again the Karatsuba approach, at a cost of 2 multiplications by α, 6 multiplications, and 15 additions in the base field $\mathbb{F}_q$.

ALGORITHM 5.21 Field multiplication in the cubic extension $\mathbb{F}_{q^3}$.

Input : $a = (a_0 + a_1 w + a_2 w^2)$, $b = (b_0 + b_1 w + b_2 w^2) \in \mathbb{F}_{q^3}$
Output: $c = a \cdot b \in \mathbb{F}_{q^3}$

1 $v_0 \leftarrow a_0 \otimes b_0$;
2 $v_1 \leftarrow a_1 \otimes b_1$;
3 $v_2 \leftarrow a_2 \otimes b_2$;
4 $c_0 \leftarrow ((a_1 \oplus a_2) \otimes (b_1 \oplus b_2) \ominus v_1 \ominus v_2)\alpha \oplus v_0$;
5 $c_1 \leftarrow (a_0 \oplus a_1) \otimes (b_0 + b_1) \ominus v_0 \ominus v_1 \oplus \alpha v_2$;
6 $c_2 \leftarrow (a_0 \oplus a_2) \otimes (b_0 \oplus b_2) \ominus v_0 \ominus v_2 \oplus v_1$;
7 **return** $c = c_0 + c_1 w + c_2 w^2$;

Squaring

Chung and Hasan presented in [19] an optimal formula for computing field squarings. Algorithm 5.22 computes $a^2 \in \mathbb{F}_{q^3}$ at a cost of 2 multiplications by α, 2 multiplications, 3 squarings, and 10 additions in the base field $\mathbb{F}_q$.

ALGORITHM 5.22   Field squaring in the cubic extension $\mathbb{F}_{q^3}$.

    Input : $a = (a_0 + a_1 w + a_2 w^2) \in \mathbb{F}_{q^3}$
    Output: $c = a^2 \in \mathbb{F}_{q^3}$

1 $v_4 \leftarrow 2(a_0 \otimes a_1)$;
2 $v_5 \leftarrow a_2^2$;
3 $c_1 \leftarrow (\alpha v_5 \oplus + v_4)$;
4 $v_2 \leftarrow v_4 \ominus v_5$;
5 $v_3 \leftarrow a_0^2$;
6 $v_4 \leftarrow a_0 \ominus a_1 \oplus a_2$;
7 $v_5 \leftarrow 2(a_1 \otimes a_2)$;
8 $v_4 \leftarrow v_4^2$;
9 $c_0 \leftarrow \alpha v_5 \oplus v_3$;
10 $c_2 \leftarrow (v_2 \oplus v_4 \oplus v_5 \ominus v_3)$;
11 **return** $c = c_0 + c_1 w + c_2 w^2$;

Inversion

Algorithm 5.23 is based on the method described by Scott in [46]. In this approach we begin by precomputing the temporary values,

$$A = a_0^2 - \alpha a_1 a_2, \quad B = \alpha a_2^2 - a_0 a_1, \quad C = a_1^2 - a_0 a_2, \quad F = \alpha a_1 C + a_0 A + \alpha a_2 B,$$

then the operation

$$(a_0 + a_1 w + a_2 w^2)^{-1} = (A + Bw + Cw^2)/F,$$

is computed at a cost of 4 multiplications by the constant α, 9 multiplications, 3 squarings, 5 additions, and one inverse in the base field $\mathbb{F}_q$.

5.2.3   Cost Summary

Table 5.1 shows the cost of the basic field arithmetic operations in the quadratic and cubic extensions of the base field $\mathbb{F}_q$, where 'M', 'S', 'A', 'I', stand for multiplication, squaring, addition and inversion in the base field $\mathbb{F}_q$.

TABLE 5.1   Summary of the field arithmetic cost for the cubic and quadratic extensions of a base field $\mathbb{F}_q$.

Operation	Cost in $\mathbb{F}_{q^2}$	Cost in $\mathbb{F}_{q^3}$
Addition	2A	2A
Multiplication	$3M+5A+m_\beta$	$6M + 15A + 2m_\alpha$
Squaring	$2M+5A+2m_\beta$	$2M+3S+10A+2m_\alpha$
Inversion	$2M+2S+2A+I+m_\beta$	$9M+3S+5A+4m_\alpha$

ALGORITHM 5.23   Field inversion in the cubic extension $\mathbb{F}_{q^3}$.

Input   : $a = (a_0 + a_1 w + a_2 w^2) \in \mathbb{F}_{q^3}$
Output: $c = a^{-1} \in \mathbb{F}_{q^3}$

1 $v_0 \leftarrow a_0^2$;
2 $v_1 \leftarrow a_1^2$;
3 $v_2 \leftarrow a_2^2$;
4 $v_3 \leftarrow a_0 \otimes a_1$;
5 $v_4 \leftarrow a_0 \otimes a_2$;
6 $v_5 \leftarrow a_1 \otimes a_2$;
7 $A \leftarrow v_0 \ominus \alpha v_5$;
8 $B \leftarrow \alpha v_2 \ominus v_3$;
9 $C \leftarrow v_1 \ominus v_4$;
10 $v_6 \leftarrow a_0 \otimes A$;
11 $v_6 \leftarrow v_6 \oplus (\alpha a_2 \otimes B)$;
12 $v_6 \leftarrow v_6 \oplus (\alpha a_1 \otimes C)$;
13 $F \leftarrow 1/v_6$;
14 $c_0 \leftarrow A \otimes F$;
15 $c_1 \leftarrow B \otimes F$;
16 $c_2 \leftarrow C \otimes F$;
17 **return** $c = c_0 + c_1 w + c_2 w^2$;

5.3   Cyclotomic Groups

The final result of any pairing is naturally inside a cyclotomic subgroup. Hence, the exponentiation techniques to be studied in this section are especially tailored for the efficient exponentiation in the group G_T, as it was defined in Chapter 3. It is worth mentioning that due to the action of the easy part of the final exponentiation, which will be studied in Chapter 7, the field elements become *unitary*. This has the important consequence that inversions can be computed almost for free.

5.3.1   Basic Definitions

DEFINITION 5.1   (Roots of unity) Let $n \in \mathbb{N}$. A solution z in any field $\mathbb{F}$ to the equation

$$z^n - 1 = 0$$

is called a root of unity in $\mathbb{F}$.

Example 5.5   For all $n \in \mathbb{N}$, the only n-th roots of unity in $\mathbb{R}$, the field of the reals, are ± 1.

Example 5.6   Let $n = 4$ and $\mathbb{F}_p$ be a finite field of characteristic $p = 7$. The set of the 4-th roots of unity in the field $\mathbb{F}_7$ is $\{1, 6\}$, since:

$$1^4 \equiv 1 \mod 7,$$

$$6^4 \equiv 1 \mod 7.$$

DEFINITION 5.2 (Splitting field) Let $\mathbb{F}$ be a field and $\mathbb{K}$ be an extension field of $\mathbb{F}$. For any polynomial $f(z) \in \mathbb{F}[z]$ of degree $n \geq 0$, $\mathbb{K}$ is called a splitting field for $f(z)$ over $\mathbb{F}$ iff there are elements $a \in \mathbb{F}$ and $a_0, a_1, \ldots a_n \in \mathbb{K}$ such that

$$f(z) = a(z - a_0) \cdot (z - a_1) \cdots (z - a_n),$$

and $\mathbb{K}$ is the smallest extension field that contains both $\mathbb{F}$ and also $\{a_0, a_1, \ldots, a_n\}$.

Example 5.7 Let $f(z) = z^4 - 1 \in \mathbb{R}[z]$. Since

$$z^4 - 1 = (z - 1) \cdot (z - i) \cdot (z + 1) \cdot (z + i)$$

and $\mathbb{C}$ is the smallest extension of $\mathbb{R}$ that contains $\{1, -1, i, -i\}$, then $\mathbb{C}$ is a splitting field for $z^4 - 1 = 0$ over $\mathbb{R}$.

Example 5.8 Let $f(z) = z^4 - 1 \in \mathbb{Q}[z]$. A splitting field for $f(z)$ over $\mathbb{Q}$ is $Q(i) = \{r + si | r, s \in \mathbb{Q}\}$.

Example 5.9 Let $f(z) = z^4 - 1 \in \mathbb{F}_7[z]$. A splitting field for $f(z)$ over $\mathbb{F}_7$ is $\mathbb{F}_7[x]/x^2 + 1$, since

$$z^4 - 1 = (z - 6) \cdot (z - 1) \cdot (z - 6x) \cdot (z + 6x),$$

and $\{1, 6, -x, -6x\} \in \mathbb{F}_7[x]/x^2 + 1$.

Note that a splitting field for $f(z)$ over $\mathbb{F}$ depends not only on the polynomial but on the field $\mathbb{F}$ as well.

The splitting field $\mathbb{K}$ over $\mathbb{F}$ for a polynomial $z^n - 1$ is called the **field of n-th roots of unity over** $\mathbb{F}$. Let $\mathbb{K}^\times$ be the group, under multiplication, of all nonzero elements in the field $\mathbb{K}$. The n-th roots of unity in $\mathbb{K}$ constitute a subgroup of $\mathbb{K}^\times$ with finite order n, denoted by,

$$\mathbb{K}^{(n)} = \{\zeta \in \mathbb{K} : \zeta^n = 1\}.$$

Since it exists at least one generator $\zeta_j \in \mathbb{K}^{(n)}$ such that

$$\mathbb{K}^{(n)} = (\zeta_j, \zeta_j^2, \zeta_j^3, \ldots, \zeta_j^n).$$

this group is known to be cyclic.

In particular, the number of generators in the field $\mathbb{K}^{(n)}$ is given by $\varphi(n)$, where $\varphi(\cdot)$ denotes the Euler's totient function. Any generator ζ_j of $\mathbb{K}^{(n)}$ is called a **primitive n-th root of unity in** $\mathbb{K}$.

Example 5.10 The set of the primitive 4-th roots of the unity in $\mathbb{C}$ is $\{i, -i\}$, i.e., the group $\mathbb{C}^{(4)}$ is generated by these two elements, as shown next:

$$\mathbb{C}^{(4)} = \{i^1 = i, \ i^2 = -1, \ i^3 = -i, \ i^4 = 1\}$$
$$\mathbb{C}^{(4)} = \{(-i)^1 = -i, \ (-i)^2 = -1, \ (-i)^3 = i, \ (-i)^4 = 1\}.$$

DEFINITION 5.3 (Cyclotomic polynomial) Let $n \in \mathbb{N}$ and $\mathbb{K}$ be the field of n-th roots of unity over $\mathbb{F}$. The n-th cyclotomic polynomial $\Phi_n(z)$ in $\mathbb{K}$ is an irreducible polynomial of degree $\varphi(n)$ and coefficients in $\{1, -1\}$, whose roots are the primitive n-th roots of unity in $\mathbb{K}$:

$$\Phi_n(z) = \prod_{j=0}^{\varphi(n)} (z - \zeta_j).$$

Example 5.11 The 4-th cyclotomic polynomial in $\mathbb{C}$ is given by:

$$\Phi_4(z) = z^2 + 1 = (z - i)(z + i).$$

Note that for the field of 4-th roots of unity $\mathbb{F}_7[x]/x^2 + 1$ over $\mathbb{F}_7$, it is also true that

$$\Phi_4(z) = z^2 + 1 = (z - 6x)(z + 6x).$$

Since the set of primitive n-th roots of unity is a subset of the roots of $z^n - 1$, it follows that $\Phi_n(z)$ is a divisor of $z^n - 1$. By definition, let $n, j \in \mathbb{Z}^+$ such that $j|n$ and $j <= n$,

$$z^n - 1 = \prod_j \Phi_j(z). \tag{5.6}$$

Example 5.12 Let $n = 12$.

$$\{j \in \mathbb{N} : j|12 \text{ and } j \leq 12\} = \{1, 2, 3, 4, 6, 12\}, \quad \text{therefore,}$$
$$z^{12} - 1 = \Phi_1(z) \cdot \Phi_2(z) \cdot \Phi_3(z) \cdot \Phi_4(z) \cdot \Phi_6(z) \cdot \Phi_{12}(z)$$

where

$$\Phi_1(z) = z - 1,$$
$$\Phi_2(z) = z + 1,$$
$$\Phi_3(z) = z^2 + z + 1,$$
$$\Phi_4(z) = z^2 + 1,$$
$$\Phi_6(z) = z^2 - z + 1,$$
$$\Phi_{12}(z) = z^4 - z^2 + 1.$$

DEFINITION 5.4 (Cyclotomic group) Given a prime number p, let $\mathbb{F}_{p^n}$ be an extension field of $\mathbb{F}_p$ and let $\mathbb{F}_{p^n}^\times$ be the group, under multiplication, of all non-zero elements in the field $\mathbb{F}_{p^n}$. The n-th cyclotomic group $\mathbb{G}_{\Phi_n(p)}$ is a subgroup of $\mathbb{F}_{p^n}^\times$ defined by:

$$\mathbb{G}_{\Phi_n(p)} = \{\alpha \in \mathbb{F}_{p^n} : \alpha^{\Phi_n(p)} = 1\}$$

In the implementation of asymmetric parings, the properties of the n-th cyclotomic subgroups have been extensively exploited to speed up the computation of the last step, known as *the final exponentiation* step, which is carefully studied in Chapter 7. Because of that, there has been a widespread interest in improving the computational efficiency of the arithmetic in the group $\mathbb{G}_{\Phi_n(p)}$. The following section describes two of the most interesting approaches that optimized the squaring in the cyclotomic group, namely, the improvement presented by Granger et al. in [26] and the algorithm proposed by Karabina in [32].

5.3.2   Squaring in Cyclotomic Subgroups

According to the definition of the n-th cyclotomic group, it follows that $\mathbb{G}_{\Phi_n(p)} \subset \mathbb{F}_{p^n}^\times$, where $\mathbb{F}_{p^n}^\times$ is the multiplicative group comprised by all non-zero elements in the field $\mathbb{F}_{p^n}$. In this section, we are interested in extension fields $\mathbb{F}_{p^n}$ with $6|n$ such that $\mathbb{F}_{p^n}$ can be represented by the tower field,

$$\mathbb{F}_{p^n=(q^2)^3} = \mathbb{F}_{q^2}[v]/v^3 - \gamma$$
$$\mathbb{F}_{q^2} = \mathbb{F}_q[w]/w^2 - \xi$$

where $q = p^{2^{a-1}3^{b-1}}$.

Therefore, any element $\alpha \in \mathbb{F}_{p^n}$ is written as:

$$\alpha = (a_0 + a_1 w) + (b_0 + b_1 w)v + (c_0 + c_1 w)v^2 = (a + bv + cv^2),$$

where a_i, b_i, $c_i \in \mathbb{F}_q$. With this representation, the computation of α^2 can be done in a conventional way at a cost of 3 squarings and 3 multiplications in $\mathbb{F}_{q^2}$,

$$\alpha^2 = (a + bv + cv^2)^2$$
$$= (a^2 + 2bc\gamma) + (2ab + c^2\gamma)v + (2ac + b^2)z^2.$$

Granger-Scott squaring-friendly fields

Granger and Scott introduced in [26] the concept of Squaring-Friendly-Fields (SFF), which are extension fields of the form $\mathbb{F}_{q^6}$ for which $q = p^i \equiv 1 \bmod 6$ is a prime power. Let $\alpha \in \mathbb{G}_{\Phi_n(p)} \subset \mathbb{F}_{p^n}^\times$, such that $\mathbb{F}_{p^n}$ is a SFF and $6|n$. Since

$$\alpha^{\Phi_n(p)=p^{n/3}-p^{n/6}+1} = 1,$$

Granger et al. demonstrated that the following algebraic relations hold,

$$bc = a^2 - \bar{a}/\gamma,$$
$$ab = c^2\gamma - \bar{b},$$
$$ac = b^2 - \bar{c},$$

where $\bar{a}$, $\bar{b}$ and $\bar{c}$ are the conjugates of a, b, and c, respectively. In this way, the effort of computing

$$\alpha^2 = (3a^2 - 2\bar{a}) + (3c^2\gamma + 2\bar{b})v + (3b^2 - 2\bar{c})$$

is reduced to 3 squares in $\mathbb{F}_{q^2}$, i.e., around 6 multiplications in $\mathbb{F}_q$.

Karabina's compressed squaring formulae

Several proposals have strived to compute the arithmetic in the cyclotomic subgroup using a compressed representation of the elements. Although the arithmetic becomes more efficient, the main challenge of these methods lies in the effort that the decompression of the elements requires.

The Karabina approach of [32] introduced new compress formulas that improve the squaring method proposed by Granger et al. for exponentiation algorithms where the exponent has a low Hamming weight. In addition, the decompression in Karabina's formulas has a very low cost, which makes the implementation of this method in the Final Exponentiation more convenient.

Without going into details of how the formulas were defined, let $\alpha \in \mathbb{G}_{\Phi_n(p)} \subset \mathbb{F}_{p^n}^{\times}$ such that $\mathbb{F}_{p^n}$ is a SSF and $6|n$. Using the tower field described in the last section, the computation of

$$\alpha^2 = ((a_0 + a_1 w) + (b_0 + b_1 w)v + (c_0 + c_1 w)v^2)^2$$

can be done in three steps:

1. Compress the element α using the following formula:

$$\mathcal{C}(\alpha) = (b_0 + b_1 w)v + (c_0 + c_1 w)v^2.$$

2. Compute $\mathcal{C}(\alpha)^2 = (B_0 + B_1 w)v + (C_0 + C_1 w)v^2$, where

$$B_0 = 2b_0 + 3((c_0 + c_1)^2 - c_0^2 - c_1^2), \qquad B_1 = 3(c_0^2 + c_1^2\xi) - 2b_1,$$

$$C_0 = 3(b_0^2 + b_1^2\xi) - 2c_0, \qquad C_1 = 2c_1 + 3(b_0 + b_1)^2 - b_0^2 - b_1^2).$$

Note that the cost of computing $\mathcal{C}(\alpha)^2$ is around 6 squares in $\mathbb{F}_q$.

3. Recover the full representation of the element by computing

$$\alpha^2 = \mathcal{D}(\mathcal{C}(\alpha^2)) = (A_0 + A_1 w) + (B_0 + B_1 w)z + (C_0 + C_1 w)z^2, \text{ where}$$

$$\left\{ \begin{array}{ll} A_1 = \frac{C_1^2\xi + 3C_0^2 - 2B_1}{4B_0}, \quad A_0 = (2A_1^2 + B_0 C_1 - 3B_1 C_0)\xi + 1 & \text{if } B_0 \neq 0 \\[2mm] A_1 = \frac{2C_0 C_1}{B_1}, \qquad A_0 = (2A_1^2 - 3B_1 C_0)\xi + 1 & \text{if } B_0 = 0 \end{array} \right\}.$$

The decompression procedure as described above has a cost of around 3 squares, 3 multiplications, and 1 inversion in $\mathbb{F}_q$.

5.3.3 Exponentiation in the Cyclotomic Group

Let $\mathbb{G} = (\mathbb{G}, \cdot, 1)$ be a cyclic finite group described multiplicatively, and let g be a generator of $\mathbb{G}$, the exponentiation g^x, where x is an integer number, and can be computed by representing the exponent x as a binary number, $x = \sum_{i=0}^{\ell} x_i 2^i$, such that,

$$g^x = g^{\sum_{i=0}^{\ell} x_i 2^i} = \prod_{i=0}^{\ell} g^{x_i 2^i} = \prod_{i=0}^{\ell} [g^{2^i}]^{x_i} = \prod_{x_i=1} g^{2^i}.$$

This computation requires $\ell = \log_2(x)$ and $w_H(x)$ squaring and multiplication operations, respectively, where $w_H(x)$ denotes the Hamming weight of x. In the particular case where $g \in \mathbb{G}_{\Phi_n(p)}$, the Karabina method presented in [32] can be applied for the efficient computation of g^x, by using the following procedure:

1. Obtain the signed representation of the exponent $x = \sum_{i=0}^{\ell} 2^i x_i$, such that $x_i = \{0, 1, -1\}$.
2. Compute $\mathcal{C}(g^{2^i})$ for $0 \leq i \leq \ell$ and store the values $h_j = \mathcal{C}(g^{2^i})$ whenever the bit $x_i \neq 0$.
3. Decompress the stored values h_j, i.e., compute, $\mathcal{D}(\mathcal{C}(g^{\pm 2^i}))$ if $x_i \neq 0$.
4. Finally, compute g^x as,
$g^x = \prod g^{\pm 2^i}$ if $x_i \neq 0$.

5.4   Lazy Reduction and RNS Representation

The arithmetic operations we are performing in pairing-based cryptography nowadays involve large characteristic, at least 256 bits. When the computation can be parallelized, the Residue Number System (RNS) representation can be a nice alternative to the classic representation of finite field presented in the previous Section. The RNS representation breaks a large integer into a set of smaller integers. The arithmetic operations over large integers are decomposed into several smaller calculation that can be performed in parallel. Several works study the efficiency of the multiplication in RNS representation [43, 33, 49, 10]. The complexity of a multiplication in the RNS representation is $O(n)$, where n is the number of words in a given basis of an integer. It implies that a multiplication in this representation is cheaper when compared with the Montgomery algorithm.

Unfortunately, the Montgomery reduction in RNS representation is more expensive than an ordinary Montgomery reduction. This drawback is balanced by reducing the number of reductions using the lazy reduction technic. Lazy reduction consists of performing several arithmetic operations before performing a reduction. It is adapted for expressions like $AB \pm CD \pm EF$ in $\mathbb{F}_p$. Some implementations of pairings using lazy reduction were proposed by Scott [46] and generalized by Aranha et al. in [3]. In [23, 18], the authors combined the lazy reduction with the use of the RNS representation in order to perform a pairing implementation.

5.4.1   The Residue Number Systems (RNS)

The Residue Number Systems are a corollary of the Chinese Remainder Theorem (CRT). Assume that we want to construct the arithmetic over $\mathbb{F}_p$ using the RNS representation. Let m_i, for $i = 1, n$ be coprime numbers such that $M = \prod_{i=1}^{n} m_i > p$. A number $a \in \mathbb{F}_p$ is represented in RNS representation by its residues $(a_1, a_2, \ldots, a_n)$ such that $a_i = a \mod m_i$. The a_i are called the RNS-digits of a in the base $\mathcal{B}_M$. The main advantage of this representation is that arithmetical operations $(+, -, \times, /)$ over large integers a and b in $\mathbb{F}_p$ are transformed into operations on the small residue values. Each operation over the residues can be performed in parallel, as they are independent. A good introduction to RNS representation can be found in [34]. In particular, the multiplication over $\mathbb{F}_p$ is reduced to n multiplications of independent RNS-digits. The choice of the m_i has a direct influence on the complexity of the multiplication. A classical optimization is to choose the m_i such that $m_i = 2^h - c_i$, where c_i is small and sparse, i.e., $c_i < 2^{h/2}$. The reduction modulo m_i is then obtained with few shifts and additions [8, 10, 17]. As a consequence, an RNS digit-product can be considered to be equivalent to a 1.1 word-product (word = h-bits). Of course, this estimation is highly dependent on the platform of implementation.

In practice, the number of moduli in an RNS base and the number of words in a normal binary system to represent field size p are equivalent in the complexity analysis, and we use n to denote both of those parameters.

RNS Montgomery reduction

We now focus on the multiplication modulo p using the Montgomery algorithm presented in [6, 7].

The main difference between the Mongotmery multiplication and the multiplication in the RNS representation is the number of bases. The Montgomery multiplication is applied in a classical radix B number system, the value B^n occurs in the reduction, division, and Montgomery factor. In RNS, this value is replaced by M. However, an auxiliary RNS basis is needed to handle the inverse of M. Since M^{-1} does not exist in base M, a new base, $\mathcal{B}_{M'} = \{m'_1, m'_2, \ldots, m'_s\}$, where M is co-prime with M', is introduced to perform the division (i.e., $(T + QN)/R$). Note that all the moduli from both $\mathcal{B}_M$ and $\mathcal{B}_{M'}$ are pairwise coprime as M and M' are coprime. The overhead is two base extensions. Algorithm 5.24 presents the computation of a Montgomery representation in the RNS arithmetic. Hence, some operations as the initial product must be performed on the two bases, which cost $2n$ words-products.

Let a and b be two numbers given in RNS representation; the RNS multiplication evaluates $r = abM^{-1} \mod p$ in RNS. As in the classical Montgomery algorithm given in Algorithm 5.6, this problem can be overcome by using Montgomery representation where $a' = a \times M \mod p$, which is stable for Montgomery product and addition. Of course, the conversion is done only once at the beginning by performing Montgomery product with a and $(M^2 \mod p)$ as operands, and once at the end of the complete cryptographic computing with 1 as second operand. Hence, this transformation will be neglected in the following. Moreover, as the RNS is not redundant, this representation is well suited for cryptography without any conversion [9].

Algorithm 5.24 presents the RNS Montgomery reduction (r can be considered the result of an RNS product on the two bases), where all the operations considered are in RNS. We explicit with respect to on which basis the operations are performed.

ALGORITHM 5.24 RNS Montgomery Reduction of $r \mod p$.

Input : Two RNS bases $\mathcal{B}_M = (m_1, \ldots, m_n)$, and $\mathcal{B}_{M'} = (m'_{n+1}, \ldots, m'_{2n})$, such that $M = \prod_{i=1}^{n} m_i < M' = \prod_{i=1}^{n} m_{n+i}$ and $\gcd(M, M') = 1$. A prime number p such that $4p < M$ and $\gcd(p, M) = 1$. The prime p is represented in basis $\mathcal{B}_{M'}$ and $-p^{-1}$ is precomputed in basis $\mathcal{B}_M$ A positive integer r represented in RNS in both bases, with $r < Mp$.

Output: A positive integer $\rho \equiv rM^{-1} \pmod{p}$ represented in RNS in both bases, with $\rho < 2p$.

1 $q \leftarrow (r) \times (-p^{-1})$ in $\mathcal{B}_M$; $[q$ in $\mathcal{B}_M] \longrightarrow [q$ in $\mathcal{B}_{M'}]$ [*First base extension*];

2 $\rho \leftarrow (r + q \times p) \times M^{-1}$ in $\mathcal{B}_{M'}$;

3 $[\rho$ in $\mathcal{B}_M] \longleftarrow [\rho$ in $\mathcal{B}_{M'}]$ [*Second base extension*];

The base extension can be performed using the Kawamura method [7], which is described in Algorithm 5.25.

The set of instructions 1 and 3 of Algorithm 5.24 are RNS additions or multiplications, which are performed independently for each element of the basis, so they are very efficient (linear). The set of instructions 2 and 4 represent RNS base extensions that are quadratic and then costly. To reduce this cost, we can use two different full RNS extensions as shown in [6, 7].

We can show that the overall complexity of Algorithm 5.24 is $\frac{7}{5}n^2 + \frac{8}{5}n$ RNS digit-products [8].

If we operate with an architecture of n basic word-arithmetic cells, the RNS arithmetic can be easily performed in a parallel manner due to the independence of the RNS digit operations. A parallel evaluation of the multiplication (in both bases) requires only 2 steps, whereas Algorithm 5.24 can be done in $\frac{12}{5}n + \frac{3}{5}$ steps [8].

Advantages of the RNS

The number of operations needed for the reduction in RNS representation is higher when compared with the number of operations in classical representation. Indeed, for the classical Montgomery reduction, we perform $n^2 + n$ word-products. However, the RNS representation presents important advantages:

- Assuming that for ECC size the multiplication needs n^2 word-products, the RNS approach is asymptotically quite interesting for a modular multiplication that represents $2n^2 + n$ word-products in classical systems and $\left(\frac{7}{5}n^2 + \frac{18}{5}n\right) \times 1.1$ in RNS.

- As shown in [27], the RNS arithmetic is easy to implement, particularly in hardware. It provides a reduced cost for multiplication and addition and a competitive modular

---

ALGORITHM 5.25   Kawamura's Base extension [7].

Input   : Two RNS bases $\mathcal{B}_M = (m_1, \ldots, m_n)$, and $\mathcal{B}_{M'} = (m'_{n+1}, \ldots, m'_{2n})$, such that $M = \prod_{i=1}^n m_i < M' = \prod_{i=1}^n m_{n+i}$ and $\gcd(M, M') = 1$. The representation Q_M of an integer in the base M

Output: The representation $Q_{M'}$ of Q in the base M'

1   **for** $i = 1$ *to* n **do**
2     $\xi_i = |x_i m_i / M|_{m_i}$
3   **end**
4   $\delta = errinit$
5   **for** $i = 1$ *to* n **do**
6     $(Q_{M'})_i = 0$
7   **end**
8   **for** $i = 1$ *to* n **do**
9     $\delta = \delta + eval(\xi_i, m_i)$
10   **for** $j = 1$ *to* n' **do**
11       $(Q_{M'})_j = ((Q_{M'})_j + \xi_i M / m_i) \mod m'_j$ **if** $\delta \geq 1$ **then**
12       $(Q_{M'})_j = ((Q_{M'})_j - M) \mod m'_j$
13   **end**
14   **end**
15     $\delta = \delta - \lfloor \delta \rfloor$
16   **end**

---

reduction. Furthermore, due to the independence of the modular operations, computations can be performed in a random way and the architecture can be parallelized.

- A RNS based architecture is flexible: With a given structure of n modular digit operators, it is possible to handle any values of p such that $4p < M$. Hence, the same architecture can be used for different levels of security and several base fields for each of these levels.

- There is a large gap between the cost of the reduction and the cost of the multiplication ($\frac{7}{5}n^2$ vs. $2n$), which is much smaller in classical systems ($n^2 + n$ vs. n^2). We can take a great advantage of this gap by accumulating multiplications before reduction. This method is called lazy reduction.

5.4.2   Lazy Reduction

The Lazy reduction technique is used in order to optimize arithmetical implementations in general [52, 38, 45], and for pairing-based cryptography [46, 3].

The method consists of delaying the modular reduction step after computing several products, which must be summed. It is well suited for expressions like $AB \pm CD \pm EF$ in $\mathbb{F}_p$, where p is an n-word prime number. Lazy reduction performs only one reduction for patterns like $AB \pm CD \pm EF$; hence, it trades expensive reductions with multi-precision additions. In an RNS context, as a multiplication takes only $2n$ word operations, the lazy reduction dramatically decreases the complexity.

Example 5.13   Assume we want to evaluate the expression $AB + CD \in \mathbb{F}_p$. A classical implementation involves two modular multiplications and then requires $4n^2 + 2n$ word-products, using the digit-serial Montgomery modular multiplication [36]. In a lazy reduction implementation, we first compute the two multiplications and add them before a unique reduction step. Thus it requires only $3n^2 + n$ word-products.

The combined use of lazy reduction and RNS arithmetic is very interesting, as it overcomes the gap of complexity between the multiplication and the reduction step. Indeed, while the classical computation of $AB + CD$ requires $\frac{14}{5}n^2 + \frac{36}{5}n$ RNS digit-products, the use of lazy reduction requires only $\frac{7}{5}n^2 + \frac{28}{5}n$ RNS digit-products. Hence, lazy reduction is particularly well adapted to RNS arithmetic. This has already been used in [8] for elliptic curve cryptography and by [23, 18] for pairing-based cryptography.

Remark 5.11 Obviously, using the lazy reduction technique implies that the reduction algorithm can take larger integers as input. In the Example 5.13, the reduction algorithm must handle numbers less than $2p^2$ instead of less than p^2. As a consequence, M must be larger than p.

There are several means to ensure that $M > p$:

- In cryptography, the size of p is chosen as an exact multiple of the word size of the architecture. In this case, we need an additional word to handle M. This solution can be costly if n is small, as, for example, on 64-bit architecture.

- We can choose the size of p to be a little bit smaller than an exact multiple of the word size of the architecture. The drawback is that we obtain security levels that are not standard. But the consequences are not catastrophic. For example, a 254-bit prime p is used in [3, 14], ensuring 127 bits of security, which is one bit then less the standard recommendation.

- Another method is to use larger words, like 36-bit words on FPGA. It induces sufficiently extra bits to handle M for cryptographic applications [27].

5.4.3   Fast Arithmetic in $\mathbb{F}_{p^d}$ Combining Lazy Reduction and RNS

Efficient arithmetic in finite extensions of prime fields is usually done with sparse polynomials with small coefficients, so that the cost of the reduction modulo this polynomial is given by some additions, as it is described in Sections 2.2.8 and 5.2. This means that if the irreducible polynomial defining $\mathbb{F}_{p^d}$ is well chosen, the cost of the reduction step in $\mathbb{F}_{p^d}$ arithmetic is negligible compared to a multiplication in $\mathbb{F}_p[X]$. In pairing-based cryptography, such cheap reduction always holds. Hence we will focus only on multiplication in $\mathbb{F}_p[X]$.

We use the results from Section 5.2 to make our comparison in the following examples.

Example 5.14   Example for a degree-2 extension. Let p be a prime number such that $p \equiv 3 \mod 4$. Then -1 is not a square in $\mathbb{F}_p$. The degree-2 extension can be constructed as $\mathbb{F}_{p^2} = \mathbb{F}_p[u]/(u^2 + 1)$. We want to compute the product of $P = a_0 + a_1 u$ and $Q = b_0 + b_1 u$. Using schoolbook multiplication, we have

$$PQ = a_0 b_0 - a_1 b_1 + (a_0 b_1 + a_1 b_0)u.$$

In this case, the lazy reduction is interesting since the $ab + cd$ pattern occurs. This multiplication in $\mathbb{F}_{p^2}$ involves 4 multiplications in $\mathbb{F}_p$ but only 2 modular reductions. One could remark that as elements in $\mathbb{F}_{p^2}$ have 2 independent components, it is not possible to have less than 2 reductions in $\mathbb{F}_p$ in the general case. Using the Karatsuba method, we perform 3 multiplications in $\mathbb{F}_p$ and also 2 modular reductions, thanks to the formula

$$PQ = a_0 b_0 - a_1 b_1 + ((a_0 + a_1)(b_0 + b_1) - a_0 b_0 - a_1 b_1)u.$$

This formula makes the use of the RNS representation interesting. Indeed, the expensive step of the RNS, namely the reduction step, is used linearly when $\mathbb{F}_{p^d}$ arithmetic is performed, whereas the cheaper step, namely the multiplication step, is used quadratically or sub-quadratically in d, the degree of the extension.

More precisely, we have Property 5.1 from [23].

PROPOSITION 5.1   *Let p be a prime number that can be represented by n words in radix representation and an RNS-digit in RNS representation. Let $\mathbb{F}_{p^d}$ be a finite extension of $\mathbb{F}_p$ defined by a sparse polynomial with small coefficients. We assume that the multiplication in $\mathbb{F}_p[X]$ requires d^λ multiplications in $\mathbb{F}_p$, with $1 < \lambda \leq 2$, and that we use lazy reduction in $\mathbb{F}_p$. A multiplication in $\mathbb{F}_{p^d}$ then requires*

- *$(d^\lambda + d)n^2 + dn$ word multiplications in radix representation,*
- *$1.1 \times \left(\frac{7d}{5}n^2 + \frac{10d^\lambda + 8d}{5}n \right)$ word multiplications if RNS is used.*

Most of the gain is due to the accumulation of many products before reducing, and not only 2 as in [8]. Of course, both the classical and the RNS reduction algorithms must be adapted. Indeed, input data can have a large size compared to p because of this accumulation process. More precisely, input data have maximal size $d'p^2$ where d' has the same size as d (it is not equal to d only because of the polynomial reduction step). Then it is sufficient to choose the RNS basis such that $M > d'p$. Moreover, if we want to use the output of the reduction algorithm (which is in $[0, 2p[$) as an input without a final comparison and subtraction, each product becomes less than $4p$ so that we have to choose $M > 4d'p$. This is not restrictive in practice as long as d is not too large, as explained in [23].

For values of d and n greater than or equal to 6, the gain is spectacular. For instance, if $n = d = 6$ and $\lambda = 1.5$ (which is a mean between Karatsuba and Toom–Cook complexities), a multiplication in $\mathbb{F}_{p^6}$ requires 781 word multiplications in radix representation, while it requires only 590 in RNS. Of course this is just a rough estimation to give an idea of the expected gain. Each particular situation must be studied in detail. We illustrate this method with an example from [23] over a degree-6 extension.

Example 5.15   Example of degree-6 extension in 192 bits

In this example, we give an explicit example of degree-6 extension of a 192-bit prime field. This example comes from [23] and is linked to an MNT curve suitable for pairing-based cryptography; see Chapter 4. Let $\mathbb{F}_p$ be defined by the prime number

$$p = 4691249309589066676602717919800805068538803592363589996389.$$

In this case, $\mathbb{F}_{p^6}$ can be defined by a quadratic extension of a cubic extension thanks to the polynomials $u^3 - 2$ and $w^2 - \alpha$ where α is a cubic root of 2.

$$\mathbb{F}_{p^3} = \mathbb{F}_p[u]/(u^3 - 2) = \mathbb{F}_p[\alpha] \text{ and}$$

$$\mathbb{F}_{p^6} = \mathbb{F}_{p^3}[w]/(w^2 - \alpha) = \mathbb{F}_{p^3}[\beta].$$

As we want to use lazy reduction, the arithmetic of this extension must be completely unrolled. Hence let

$$A = a_0 + a_1\alpha + a_2\alpha^2 + \left(a_3 + a_4\alpha + a_5\alpha^2\right)\beta \text{ and}$$
$$B = b_0 + b_1\alpha + b_2\alpha^2 + \left(b_3 + b_4\alpha + b_5\alpha^2\right)\beta$$

be two elements of $\mathbb{F}_{p^6}$. Using Karatsuba on the quadratic extension leads to

$$AB = \left(a_0 + a_1\alpha + a_2\alpha^2\right)\left(b_0 + b_1\alpha + b_2\alpha^2\right) + \alpha\left(a_3 + a_4\alpha + a_5\alpha^2\right)\left(b_3 + b_4\alpha + b_5\alpha^2\right) +$$
$$\left[\left(a_0 + a_3 + (a_1 + a_4)\alpha + (a_2 + a_5)\alpha^2\right)\left(b_0 + b_3 + (b_1 + b_4)\alpha + (b_2 + b_5)\alpha^2\right)\right.$$
$$\left. - \left(a_0 + a_1\alpha + a_2\alpha^2\right)\left(b_0 + b_1\alpha + b_2\alpha^2\right) - \left(a_3 + a_4\alpha + a_5\alpha^2\right)\left(b_3 + b_4\alpha + b_5\alpha^2\right)\right]\beta\,.$$

Using Karatsuba again to compute each of these 3 products leads to

$$AB = a_0 b_0 + 2\left(a_4 b_4 + (a_1 + a_2)(b_1 + b_2) - a_1 b_1 + (a_3 + a_5)(b_3 + b_5) - a_3 b_3 - a_5 b_5\right)$$
$$+ [a_3 b_3 + (a_0 + a_1)(b_0 + b_1) - a_0 b_0 - a_1 b_1 + 2(a_2 b_2 + (a_4 + a_5)(b_4 + b_5) - a_4 b_4 - a_5 b_5)]\alpha$$
$$+ [a_1 b_1 + 2a_5 b_5 + (a_0 + a_2)(b_0 + b_2) - a_0 b_0 - a_2 b_2 + (a_3 + a_4)(b_3 + b_4) - a_3 b_3 - a_4 b_4]\alpha^2$$
$$+ [(a_0 + a_3)(b_0 + b_3) - a_0 b_0 - a_3 b_3 + 2((a_1 + a_2 + a_4 + a_5)(b_1 + b_2 + b_4 + b_5) - (a_1 + a_4)(b_1 + b_4)$$
$$- (a_2 + a_5)(b_2 + b_5) - (a_1 + a_2)(b_1 + b_2) + a_1 b_1 + a_2 b_2 - (a_4 + a_5)(b_4 + b_5) + a_4 b_4 + a_5 b_5)]\beta$$
$$+ [(a_0 + a_1 + a_3 + a_4)(b_0 + b_1 + b_3 + b_4) - (a_0 + a_3)(b_0 + b_3) - (a_1 + a_4)(b_1 + b_4) - (a_0 + a_1)(b_0 + b_1)$$
$$+ a_0 b_0 + a_1 b_1 - (a_3 + a_4)(b_3 + b_4) + a_3 b_3 + a_4 b_4 + 2((a_2 + a_5)(b_2 + b_5) - a_2 b_2 - a_5 b_5)]\alpha\beta$$
$$+ [(a_1 + a_4)(b_1 + b_4) - a_1 b_1 - a_4 b_4 + (a_0 + a_2 + a_3 + a_5)(b_0 + b_2 + b_3 + b_5) - (a_0 + a_3)(b_0 + b_3)$$
$$- (a_2 + a_5)(b_2 + b_5) - (a_0 + a_2)(b_0 + b_2) + a_0 b_0 + a_2 b_2 - (a_3 + a_5)(b_3 + b_5) + a_3 b_3 + a_5 b_5]\alpha^2\beta\,.$$

It is easy to verify that this formula requires 18 multiplications in $\mathbb{F}_p$. Of course it also requires many additions but this is due to the Karatsuba method, not to lazy reduction. As we use the lazy reduction technique, it requires only 6 reductions thanks to the accumulation of all the operations in each component. However, this accumulation implies that the input of the reduction step can be very large. More precisely, thanks to the existence of the schoolbook method for computing AB, we can easily prove that if the components of A and B (i.e., the a_i and the b_i) are between 0 and $2p$ (which is the case when Algorithm 5.6 or 5.24 is used for reduction) then each component of AB is between 0 and $44p^2$. This means that B^n in Montgomery representation and M in RNS representation must be greater than $44p$ to perform lazy reduction in this degree-6 field.

5.5 SAGE Appendix

In this chapter we have detailed the finite field arithmetic that is required for implementing bilinear pairings. The methods described in this chapter can be seen as high-level algorithms that apply completely independent of the underlying hardware architecture.

A low-level implementation will have to deal with the fact that a computer word is not enough to represent arbitrary finite field elements. Hence, a low-level implementation would need to represent these rather large integers into several processor words to perform multi-precision arithmetic. This process is dependent on the target hardware or software platform. A low-level implementation is discussed at length in Chapter 11, where the specific details of a pairing function implementation as well as general comments on selected ready-to-use pairing libraries are given.

In this section, a finite field arithmetic SAGE code is presented. This code will hopefully be useful to the reader for better understanding the field arithmetic algorithms discussed in this chapter.

5.5.1   Arithmetic in $\mathbb{F}_p$

In this Section we present the SAGE code for the algorithms performing multi-precision addition, subtraction, multiplication, exponentiation, inversion, and square root over the field $\mathbb{F}_p$.

Listing 1 File `fp.sage`. Addition in $\mathbb{F}_p$. Algorithm 5.1.

```
1   #Input  : a, b \in F_p
2   #Output: a+b \in F_p
3   def Fp_addC(a,b):
4       c = a + b
5
6       if c >= p:
7           c = c - p
8
9       return c
10
11
12
13  #Input  : a, b \in Z
14  #Output: a+b
15  def Fp_addNR(a,b):
16      c = a + b
17
18      return c
```

Listing 2 File `fp.sage`. Substraction in $\mathbb{F}_p$. Algorithm 5.2.

```
1   #Input  : a, b \in F_p
2   #Output: a-b \in F_p
3   def Fp_subC(a,b):
4       c = a - b
5       if a < b:
6           c = c + p
7
8       return c
```

Listing 3 File `params.sage`. Barrett reduction parameters for Algorithm 5.5, p-prime for $\mathbb{F}_p$.

```
1   # Barrett reduction
2   B_W = 64
3   B_b = 2^B_W
4   B_k = Integer(math.ceil(math.log(p,B_b)+1))
5   B_mu = B_b^(2*B_k) // p
6   B_mask = 2^(B_W*(B_k+1))-1
7   B_expo = B_b^(B_k + 1)
```

Listing 4 File `fp.sage`. Barrett reduction in $\mathbb{F}_p$ for Algorithm 5.5, see parameters in Listing 3.

```
1   #rs = (z & B_mask) - ((qh * p) & B_mask)
2   #Input : a \in Z
3   #Output: a \in F_p
4   def Fp_BarrettRed(z):
5       qh = ((z >> (B_W * (B_k - 1))) * B_mu) >> (B_W * (B_k + 1))
6       t0 = Fp_mod(Fp_mulNR(qh, p), B_expo)
7       rs = Fp_subNR(Fp_mod(z, B_expo), t0)
8
9       if rs < 0:
10          rs = Fp_addNR(rs, B_expo)
11      while (rs >= p):
12          rs = Fp_subNR(rs, p)
13
14
15      return rs
```

Listing 5 File `params.sage`. Montgomery reduction parameters for Algorithm 5.6, p-prime for $\mathbb{F}_p$.

```
1   #Montgomery space
2   M_r = 1 << 256
3   M_pp = -(Fp_inv(p, M_r))
4   M_rp = Fp_inv(M_r, p)
```

Listing 6 File `fp.sage`. Montgomery product for Algorithm 5.6, see parameters in Listing 5.

```
1   #Input: a,b \in Mont(F_p)
2   #Output: a.b \in Mont(F_p)
3   def MontPr(a, b):
4       t = Fp_mulNR(a,b)
5       q = Fp_mod(Fp_mulNR(t, M_pp), M_r)
6       u = Fp_addNR(t, Fp_mulNR(q, p)) >> 256
7       if u >= p:
8           return Fp_subNR(u, p)
9       else:
10          return u
11
12      return u
```

Listing 7 File `fp.sage`. Montgomery multiplication in $\mathbb{F}_p$ for Algorithm 5.7, see parameters in Listing 5.

```
1   #Input: a,b \in F_p
2   #Output: a.b \in F_p
3   def Fp_mulM(a,b):
4       ap = Fp_mod(a<<256, p)
5       bp = Fp_mod(b<<256, p)
6       cp = MontPr(ap, bp)
7       c  = MontPr(cp, Fp_one())
8
9       return c
```

Listing 8 File `fp.sage`. Improved Montgomery multiplication in $\mathbb{F}_p$ for Algorithm 5.8, see parameters in Listing 5.

```
1   #Input: a,b \in F_p
2   #Output: a.b \in F_p
3   def Fp_mulMI(a,b):
4       ap = Fp_mod(a<<256, p)
5       c = MontPr(ap, b)
6
7       return c
```

Listing 9 File `fp.sage`. Montgomery exponentiation in $\mathbb{F}_p$ for Algorithm 5.11, see parameters in Listing 5.

```
1   #Input: a,e \in F_p
2   #Output: a^e \in F_p
3   def Fp_expM(a,e):
4       ap = Fp_mod(a<<256, p)
5       aa = Fp_mod(M_r, p)
6
7       lbin = e.nbits()
8       bin_e= e.bits()
9       for i in range(lbin-1,-1,-1):
10          aa = MontPr(aa, aa)
11          if bin_e[i]==1:
12              aa = MontPr(aa, ap)
13
14      return MontPr(aa, Fp_one())
```

Listing 10 File `fp.sage`. Inversion in $\mathbb{F}_p$ for Algorithm 5.12.

```
1  #Input  : a \in F_p
2  #Output: a^-1 \in F_p
3  def Fp_invBin(a, q=None):
4      if q==None:
5          n = p
6      else:
7          n = q
8
9      u  = a
10     v  = n
11     x1 = 1
12     x2 = 0
13     while (u != 1) and (v != 1):
14         while (u&1) == 0:
15             u = Fp_divBy2(u)
16             if (x1&1)==0:
17                 x1 = Fp_divBy2(x1)
18             else:
19                 x1 = Fp_divBy2(Fp_addNR(x1, n))
20         while (v&1)==0:
21             v = Fp_divBy2(v)
22             if (x2&1)==0:
23                 x2 = Fp_divBy2(x2)
24             else:
25                 x2 = Fp_divBy2(Fp_addNR(x2, n))
26         if u >= v:
27             u  = Fp_subNR(u, v)
28             x1 = Fp_subNR(x1, x2)
29         else:
30             v  = Fp_subNR(v, u)
31             x2 = Fp_subNR(x2, x1)
32
33     if u == 1:
34         return Fp_BarretRed(x1)
35     else:
36         return Fp_BarretRed(x2)
```

Listing 11 File `fp.sage`. Square root in $\mathbb{F}_p$ for $p \equiv 3 \bmod 4$, Algorithm 5.13.

```
1  #Input  : a \in F_p, for p=3%4
2  #Output: a^(1/2) \in F_p
3  def Fp_SQRTshanks(f):
4      global pm3o4
5      g = Fp_exp(f,pm3o4)
6      a = Fp_mulC(Fp_square(g), f)
7      g = Fp_mulC(g, f)
8      if a == (p-1):
9          return -1
10
11     return g
```

5.5.2 Arithmetic in $\mathbb{F}_{p^2}$

In this section we present the SAGE code for the algorithms performing multi-precision addition, multiplication, squaring, square root, and inversion over the field $\mathbb{F}_{p^2}$.

Listing 12 File `fp2.sage`. Addition in $\mathbb{F}_{p^2}$. Algorithm 5.15.

```
1  #Input  : a, b \in F_{p^2} -> a = (a0 + a1u), b = (b0 + b1u)
2  #Output: c = a + b \in F_{p^2}
3  def Fp2_addC(a,b):
4      c0 = Fp_addC(a[0], b[0])
5      c1 = Fp_addC(a[1], b[1])
6
7      return [c0, c1]
```

Listing 13 File `fp2.sage`. Multiplication in $\mathbb{F}_{p^2}$. Algorithm 5.16.

```
1  #Input  : a, b \in F_{p^2} -> a = (a0 + a1u), b = (b0 + b1u)
2  #Output: a.b \in F_{p^2}
3  def Fp2_mul(a, b):
4      v0 = Fp_mulC(a[0], b[0])
5      v1 = Fp_mulC(a[1], b[1])
6      c0 = beta_mul(v1)
7      c0 = Fp_addC(v0, c0)
8      c1 = Fp_addC(a[0], a[1])
9      t0 = Fp_addC(b[0], b[1])
10     c1 = Fp_mulC(c1, t0)
11     c1 = Fp_subC(c1, v0)
12     c1 = Fp_subC(c1, v1)
13
14     return [c0, c1]
```

Listing 14 File `fp2.sage`. Squaring in $\mathbb{F}_{p^2}$. Algorithm 5.17.

```
1  #Input  : a \in F_{p^2} -> a = (a0 + a1u)
2  #Output: a^2 \in F_{p^2}
3  def Fp2_square(a):
4      v0 = Fp_subC(a[0], a[1])
5      v3 = beta_mul(a[1])
6      v3 = Fp_subC(a[0], v3)
7      v2 = Fp_mulC(a[0], a[1])
8      v0 = Fp_mulC(v0, v3)
9      v0 = Fp_addC(v0, v2)
10     v1 = Fp_addC(v2, v2)
11     v3 = beta_mul(v2)
12     v0 = Fp_addC(v0, v3)
13
14     return [v0, v1]
```

Listing 15 File `fp2.sage`. Square root in $\mathbb{F}_{p^2}$, with $\beta = -1$, $p \equiv 3 \bmod 4$. Algorithm 5.18.

```
1   #Input  : a \in F_{p^2} -> a = (a0 + a1u)
2   #Output: a^(1/2) \in F_{p^2}
3   def Fp2_SQRT(v):
4       u = Fp2_copy(v)
5       if u[1] == Fp_zero():
6           u[0] = Fp_SQRTshanks(u[0])
7           return u
8
9       t0 = Fp_square(u[0])
10      t1 = Fp_square(u[1])
11      t0 = Fp_subC(t0, beta_mul(t1))
12
13      L = Legendre(t0, p)
14      if L == -1:
15          return [-1, 0]
16      t0 = Fp_SQRTshanks(t0)
17
18      t1 = Fp_addC(u[0], t0)
19      t1 = Fp_mulC(t1, Half)
20      L = Legendre(t1, p)
21      if L == -1:
22          t1 = Fp_subC(u[0], t0)
23          t1 = Fp_mulC(t1, Half)
24
25      u[0] = Fp_SQRTshanks(t1)
26      t1   = Fp_addC(u[0], u[0])
27      t1   = Fp_inv(t1)
28      u[1] = Fp_mulC(u[1], t1)
29
30      return u
```

Listing 16 File `fp2.sage`. Inversion in $\mathbb{F}_{p^2}$ for Algorithm 5.19.

```
1   #Input  : a \in F_{p^2} -> a = (a0 + a1u)
2   #Output: a^-1 \in F_{p^2}
3   def Fp2_inv(a):
4       t0 = Fp_square(a[0])
5       t1 = Fp_square(a[1])
6       t1 = beta_mul(t1)
7       t0 = Fp_subC(t0, t1)
8       t1 = Fp_inv(t0)
9       c0 = Fp_mulC(a[0], t1)
10      c1 = Fp_mulC(Fp_negC(a[1]), t1)
11
12      return [c0, c1]
```

5.5.3 Arithmetic in $\mathbb{F}_{p^3}$

In this section we present the SAGE code for the algorithms performing multi-precision addition, multiplication, squaring, and inversion over the field $\mathbb{F}_{p^3}$.

Listing 17 File `fp3.sage`. Addition in $\mathbb{F}_{p^3}$. Algorithm 5.20.

```
1  #Input  : a, b \in F_{p^3} -> a = (a0 + a1w + a2w^2),
2  # b = (b0 + b1w + b2w^2)
3  #Output: c = a + b \in F_{p^3}
4  def Fp3_addC(a,b):
5          c0 = Fp_addC(a[0], b[0])
6          c1 = Fp_addC(a[1], b[1])
7          c2 = Fp_addC(a[2], b[2])
```

Listing 18 File `fp3.sage`. Multiplication in $\mathbb{F}_{p^3}$. Algorithm 5.21.

```
1   #Input  : a, b \in F_{p^3} -> a = (a0 + a1w + a2w^2),
2   # b = (b0 + b1w + b2w^2)
3   #Output: c = a.b \in F_{p^3}
4   def Fp3_mul(a, b):
5       v0 = Fp_mulC(a[0], b[0])
6       v1 = Fp_mulC(a[1], b[1])
7       v2 = Fp_mulC(a[2], b[2])
8       c0 = Fp_mulC(Fp_addC(a[1], a[2]), Fp_addC(b[1], b[2]))
9       c0 = Fp_subC(Fp_subC(c0, v1), v2)
10      c0 = Fp_addC(beta3_mul(c0), v0)
11      c1 = Fp_mulC(Fp_addC(a[0], a[1]), Fp_addC(b[0], b[1]))
12      c1 = Fp_subC(Fp_subC(c1, v0), v1)
13      c1 = Fp_addC(c1, beta3_mul(v2))
14      c2 = Fp_mulC(Fp_addC(a[0], a[2]), Fp_addC(b[0], b[2]))
15      c2 = Fp_subC(Fp_subC(c2, v0), v2)
16      c2 = Fp_addC(c2, v1)
17
18      return [c0, c1, c2]
```

Listing 19 File `fp3.sage`. Squaring in $\mathbb{F}_{p^3}$. Algorithm 5.22.

```
1   #Input  : a \in F_{p^3} -> a = (a0 + a1w + a2w^2)
2   #Output: a^2 \in F_{p^3}
3   def Fp3_square(a):
4       v4 = Fp_mulC(a[0], a[1])
5       v4 = Fp_addC(v4, v4)
6       v5 = Fp_square(a[2])
7       c1 = Fp_addC(beta3_mul(v5), v4)
8       v2 = Fp_subC(v4, v5)
9       v3 = Fp_square(a[0])
10      v4 = Fp_addC(Fp_subC(a[0], a[1]), a[2])
11      v5 = Fp_mulC(a[1], a[2])
12      v5 = Fp_addC(v5, v5)
13      v4 = Fp_square(v4)
14      c0 = Fp_addC(beta3_mul(v5), v3)
15      c2 = Fp_subC(Fp_addC(Fp_addC(v2, v4), v5),v3)
16
17      return [c0, c1, c2]
```

Listing 20 File `fp3.sage`. Inversion in $\mathbb{F}_{p^3}$ for Algorithm 5.23.

```
1   #Input  : a\in F_{p^3} -> a = (a0 + a1w + a2w^2)
2   #Output: a^-1 \in F_{p^3}
3   def Fp3_inv(a):
4       v0 = Fp_square(a[0])
5       v1 = Fp_square(a[1])
6       v2 = Fp_square(a[2])
7       v3 = Fp_mulC(a[0], a[1])
8       v4 = Fp_mulC(a[0], a[2])
9       v5 = Fp_mulC(a[1], a[2])
10      A  = Fp_subC(v0, beta3_mul(v5))
11      B  = Fp_subC(beta3_mul(v2), v3)
12      C  = Fp_subC(v1, v4)
13      v6 = Fp_mulC(a[0], A)
14      v6 = Fp_addC(v6, Fp_mulC(beta3_mul(a[2]), B))
15      v6 = Fp_addC(v6, Fp_mulC(beta3_mul(a[1]), C))
16      F  = Fp_invBin(v6)
17      c0 = Fp_mulC(A, F)
18      c1 = Fp_mulC(B, F)
19      c2 = Fp_mulC(C, F)
20
21      return [c0, c1, c2]
```

5.5.4   Arithmetic in Cyclotomic Subgroups

In this section we present the SAGE code for the special Granger–Scott, and Karabina's squaring in $\mathbb{G}_{Phi_6}$.

Listing 21 File `ep.sage`. Granger-Scott squaring in $\mathbb{G}_{\Phi_6}$.

```
1    # Granger-Scott special squaring
2    def Fp12_sqru(a):
3            z0 = Fp2_copy(a[0][0])
4            z4 = Fp2_copy(a[0][1])
5            z3 = Fp2_copy(a[0][2])
6            z2 = Fp2_copy(a[1][0])
7            z1 = Fp2_copy(a[1][1])
8            z5 = Fp2_copy(a[1][2])
9
10           [t0, t1] = Fp4_square([z0, z1])
11
12           #For A
13           z0 = Fp2_subC(t0, z0)
14           z0 = Fp2_addC(z0, z0)
15           z0 = Fp2_addC(z0, t0)
16
17           z1 = Fp2_addC(t1, z1)
18           z1 = Fp2_addC(z1, z1)
19           z1 = Fp2_addC(z1, t1)
20
21           [t0, t1] = Fp4_square([z2, z3])
22           [t2, t3] = Fp4_square([z4, z5])
23
24           #For C
25           z4 = Fp2_subC(t0, z4)
26           z4 = Fp2_addC(z4, z4)
27           z4 = Fp2_addC(z4, t0)
28
29           z5 = Fp2_addC(t1, z5)
30           z5 = Fp2_addC(z5, z5)
31           z5 = Fp2_addC(z5, t1)
32
33           #For B
34           t0 = chi_mul(t3)
35           z2 = Fp2_addC(t0, z2)
36           z2 = Fp2_addC(z2, z2)
37           z2 = Fp2_addC(z2, t0)
38
39           z3 = Fp2_subC(t2, z3)
40           z3 = Fp2_addC(z3, z3)
41           z3 = Fp2_addC(z3, t2)
42
43
44           return [[z0, z4, z3], [z2, z1, z5]]
```

Listing 22 File `ep.sage`. Karabina's compressed squaring in $\mathbb{G}_{\Phi_6}$.

```
1   # Karabina's compressed squaring
2   # Input: g \in \G_{\varpsi(6)}, g = [0, 0, g2, g3, g4, g5]
3   # Output: g^2 = [0, 0, g2, g3, g4, g5]^2
4   def Fp12_sqrKc(g):
5           T0 = Fp2_square(g[4])
6           T1 = Fp2_square(g[5])
7   #7
8           T2 = chi_mul(T1)
9   #8
10          T2 = Fp2_addC(T2, T0)
11  #9
12          t2 = Fp2_copy(T2)
13  #1
14          t0 = Fp2_addC(g[4], g[5])
15          T2 = Fp2_square(t0)
16  #2
17          T0 = Fp2_addC(T0, T1)
18          T2 = Fp2_subC(T2, T0)
19  #3
20          t0 = Fp2_copy(T2)
21          t1 = Fp2_addC(g[2], g[3])
22          T3 = Fp2_square(t1)
23          T2 = Fp2_square(g[2])
24  #4
25          t1 = chi_mul(t0)
26  #5
27          g[2] = Fp2_addC(g[2], t1)
28          g[2] = Fp2_addC(g[2], g[2])
29  #6
30          g[2] = Fp2_addC(g[2], t1)
31          t1 = Fp2_subC(t2, g[3])
32          t1 = Fp2_addC(t1, t1)
33  #11
34          T1 = Fp2_square(g[3])
35  #10
36          g[3] = Fp2_addC(t1, t2)
37  #12
38          T0 = chi_mul(T1)
39  #13
40          T0 = Fp2_addC(T0, T2)
41  #14
42          t0 = Fp2_copy(T0)
43          g[4] = Fp2_subC(t0, g[4])
44          g[4] = Fp2_addC(g[4], g[4])
45  #15
46          g[4] = Fp2_addC(g[4], t0)
47  #16
48          T2 = Fp2_addC(T2, T1)
49          T3 = Fp2_subC(T3, T2)
50  #17
51          t0 = Fp2_copy(T3)
52          g[5] = Fp2_addC(g[5], t0)
53          g[5] = Fp2_addC(g[5], g[5])
54  #18
55          g[5] = Fp2_addC(g[5], t0)
56
57          f = [Fp2_zero(), Fp2_zero(), g[2], g[3], g[4], g[5]]
58
59          return f
```

Listing 23 File `ep.sage`. Karabina's compressed squaring in $\mathbb{G}_{\Phi_6}$, reordering functions.

```
1   # Karabina's compressed squaring reordering
2   # Input: [a0 + a1.w] + [b0 + b1.w]v + [c0 + c1.w]v^2
3   # Output: [g0, g1, g2, g3, g4, g5]
4   def fptogs(fp):
5           gs     = [Fp2_zero() for i in range(0,6)]
6           gs[0] = Fp2_copy(fp[0][0])
7           gs[1] = Fp2_copy(fp[1][1])
8           gs[2] = Fp2_copy(fp[1][0])
9           gs[3] = Fp2_copy(fp[0][2])
10          gs[4] = Fp2_copy(fp[0][1])
11          gs[5] = Fp2_copy(fp[1][2])
12
13          return gs
14
15
16
17  # Karabina's compressed squaring format change
18  # Input: [g0, g1, g2, g3, g4, g5]
19  # Output: [g0 + g1.w] + [g2 + g3.w]v + [g4 + g5.w]v^2
20  def gstofp(gs):
21          f = Fp12_zero()
22          f[0][0] = Fp2_copy(gs[0])
23          f[0][1] = Fp2_copy(gs[1])
24          f[0][2] = Fp2_copy(gs[2])
25          f[1][0] = Fp2_copy(gs[3])
26          f[1][1] = Fp2_copy(gs[4])
27          f[1][2] = Fp2_copy(gs[5])
28
29          return f
30
31
32
33  # Karabina's compressed squaring reordering function
34  # Input: [g0, g1, g2, g3, g4, g5]
35  # Output: [a0 + a1.w] + [b0 + b1.w]v + [c0 + c1.w]v^2
36  def reorderfpK(fp):
37          f = Fp12_zero()
38          f[0][0] = Fp2_copy(fp[0][0])
39          f[0][1] = Fp2_copy(fp[1][1])
40          f[0][2] = Fp2_copy(fp[1][0])
41          f[1][0] = Fp2_copy(fp[0][2])
42          f[1][1] = Fp2_copy(fp[0][1])
43          f[1][2] = Fp2_copy(fp[1][2])
44
45          return f
```

Listing 24 File `ep.sage`. Karabina's compressed squaring in $\mathbb{G}_{\Phi_6}$, recover g_1 component.

```
1   # Karabina's Sqr recover g_1 from g2,g3,g4,g5
2   # Input:  [0, 0, g2, g3, g4, g5]
3   # Output: [g1_num, g1_den]
4   def Fp12_sqrKrecg1(fp):
5           t = [Fp2_zero(), Fp2_zero()]
6           if fp[0][2] != Fp2_zero():
7                   C12   = chi_mul(Fp2_square(fp[1][2]))
8                   C02   = Fp2_square(fp[1][1])
9                   t[0]  = Fp2_copy(C02)
10                  C02   = Fp2_addC(C02, C02)
11                  C02   = Fp2_addC(C02, t[0])
12                  t[0]  = Fp2_addC(C12, C02)
13                  t[0]  = Fp2_subC(t[0], fp[1][0])
14                  t[0]  = Fp2_subC(t[0], fp[1][0])
15                  t[1]  = Fp2_addC(fp[0][2], fp[0][2])
16                  t[1]  = Fp2_addC(t[1], t[1])
17          else:
18                  t[0]  = Fp2_mul(fp[1][1], fp[1][2])
19                  t[0]  = Fp2_addC(t[0], t[0])
20                  t[1]  = Fp2_copy(fp[1][0])
21
22          return t
23
24
25
26  # Karabina's Sqr inversion of g1 denominator
27  #Input:  t  = [[g1_num, g1_den]_1, [g1_num, g1_den]_2, ...]
28  #Output: t0 = [[g1]_1, [g_1]_2, ...]
29  def Fp12_sqrKinvg1(t):
30          n  = len(t)
31          t0 = [Fp2_zero() for i in range(0, n)]
32
33          f  = t[0][1]
34          for i in range(1, n):
35                  f = Fp2_mul(f, t[i][1])
36          f = Fp2_inv(f)
37
38          if n==1:
39                  t0[0] = f
40                  return t0
41
42          if n==2:
43                  t0[0] = Fp2_mul(t[0][0], Fp2_mul(f, t[1][1]))
44                  t0[1] = Fp2_mul(t[1][0], Fp2_mul(f, t[0][1]))
45                  return t0
46
47          #for n>=3 use Simultaneous Montgomery Inversion
48          d = [Fp2_one() for i in range(0, 2*n)]
49          d[1]   = t[0][1]
50          d[n+1] = t[n-1][1]
51          for i in range(2,n):
52                  d[i]   = Fp2_mul(d[i-1], t[i-1][1])
53                  d[n+i] = Fp2_mul(d[n+i-1], t[n-i][1])
54
55          for i in range(1, n-1):
56                  d[i] = Fp2_mul(d[i], d[2*n-i-1])
57
58          d[0] = d[2*n-1]
59          for i in range(0,n):
60                  d[i]  = Fp2_mul(d[i], f)
61                  t0[i] = Fp2_mul(t[i][0], d[i])
62
63          return t0
```

Listing 25 File `ep.sage`. Karabina's compressed squaring in $\mathbb{G}_{\Phi_6}$, recover g_0 component.

```
1   # Karabina's Sqr recover g_0 from g1,g2,g3,g4,g5
2   # Input:   [0, g1, g2, g3, g4, g5]
3   # Output:  [g0, g1, g2, g3, g4, g5]
4   def Fp12_sqrKrecg0(fp):
5           t0 = Fp2_square(fp[0][1])
6           t1 = Fp2_mul(fp[1][0], fp[1][1])
7           t0 = Fp2_subC(t0, t1)
8           t0 = Fp2_addC(t0, t0)
9           t0 = Fp2_subC(t0, t1)
10          t1 = Fp2_mul(fp[0][2], fp[1][2])
11          t0 = Fp2_addC(t0, t1)
12          g  = chi_mul(t0)
13          g  = Fp2_addC(g, Fp2_one())
14
15          return g
```

Listing 26 File `ep.sage`. Karabina's compressed squaring in $\mathbb{G}_{\Phi_6}$, decompression function.

```
1   def Fp12_sqrKd(gs, poles):
2           n = len(gs)
3           t = [[Fp2_one() for i in range(0,2)] for j in range(0, n)]
4
5           fp = [gstofp(gs[i]) for i in range(0, n)]
6           #recover g1
7           for i in range(0, n):
8                   t[i] = Fp12_sqrKrecg1(fp[i])
9           t = Fp12_sqrKinvg1(t)
10
11          #recover g0
12          for i in range(0,n):
13                  fp[i][0][1] = Fp2_copy(t[i])
14                  t[i] = Fp12_sqrKrecg0(fp[i])
15                  fp[i][0][0] = Fp2_copy(t[i])
16                  fp[i] = reorderfpK(fp[i])
17
18          #multiply
19          if poles[0] > 0:
20                  f = Fp12_copy(fp[0])
21          else:
22                  f = Fp12_conj(fp[0])
23          for i in range(1, n):
24                  if poles[i] > 0:
25                          f = Fp12_mul(f, Fp12_copy(fp[i]))
26                  else:
27                          f = Fp12_mul(f, Fp12_conj(fp[i]))
28
29          return f
```

Listing 27 File `ep.sage`. Exponentiation in $\mathbb{G}_{\Phi_6}$ using Karabina's compressed squaring.

```
1  #Input: f \in \G_{\varpsi_6}, z \in N w/very low-Hammming weight,
2  # or the parameter of the curve
3  #Output: f^z
4  # g0 = EasyExpo(rand12())
5  # Fp12_conj(Fp12_exp(g0,abs(z))) == Fp12_expK(g0)
6  def Fp12_expK(f, z0=None):
7          global z
8          if z0!=None:
9                  z = z0
10
11         #get exponent (binary) poles
12         zbits  = z.bits()
13         znbits = z.nbits()
14         poles = []
15         for i in range(0, znbits):
16                 if zbits[i] != 0:
17                         poles.append(i*zbits[i])
18
19         #special exponent case: 2^0
20         if poles[0] == 0:
21                 poles.remove(0)
22
23         n = len(poles)
24         g  = [Fp2_zero() for i in range(0,6)]
25         g  = fptogs(f)
26
27         #Karabina's compressed squaring
28         gs = []
29         a  = [Fp2_copy(g[i]) for i in range(0,6)]
30         pi = 0
31         for i in range(0, n):
32                 for j in range(0, abs(poles[i])-pi):
33                         a = Fp12_sqrKc(a)
34                 gs.append([Fp2_copy(a[j]) for j in range(0,6)])
35                 pi = abs(poles[i])
36         g = Fp12_sqrKd(gs, poles)
37
38         #special exponent case: 2^0
39         if zbits[0] > 0:
40                 g = Fp12_mul(g, Fp12_copy(f))
41         elif zbits[0] < 0:
42                 g = Fp12_mul(g, Fp12_conj(f))
43
44         return g
```

References

[1] Gora Adj and Francisco Rodríguez-Henríquez. Square root computation over even extension fields. *IEEE Transactions on Computers*, 63(11):2829–2841, 2014.

[2] Diego F. Aranha, Paulo S. L. M. Barreto, Patrick Longa, and Jefferson E. Ricardini. The realm of the pairings. In T. Lange, K. Lauter, and P. Lisonek, editors, *Selected Areas in Cryptography – SAC 2013*, volume 8282 of *Lecture Notes in Computer Science*, pp. 3–25. Springer, Heidelberg, 2014.

[3]   Diego F. Aranha, Koray Karabina, Patrick Longa, Catherine H. Gebotys, and Julio López. Faster explicit formulas for computing pairings over ordinary curves. In K. G. Paterson, editor, *Advances in Cryptology – EUROCRYPT 2011*, volume 6632 of *Lecture Notes in Computer Science*, pp. 48–68. Springer, Heidelberg, 2011.

[4]   A.O.L. Atkin. Probabilistic primality testing, summary by F. Morain. Research Report 1779, INRIA, 1992.

[5]   Daniel V. Bailey and Christof Paar. Optimal extension fields for fast arithmetic in public-key algorithms. In H. Krawczyk, editor, *Advances in Cryptology – CRYPTO '98*, volume 1462 of *Lecture Notes in Computer Science*, pp. 472–485. Springer, Heidelberg, 1998.

[6]   Jean-Claude Bajard, Laurent-Stéphane Didier, and Peter Kornerup. An RNS montgomery modular multiplication algorithm. *IEEE Trans. Computers*, 47(7):766–776, 1998.

[7]   Jean-Claude Bajard, Laurent-Stéphane Didier, and Peter Kornerup. Modular multiplication and base extensions in residue number systems. In *IEEE Symposium on Computer Arithmetic*, pp. 59–65. IEEE Computer Society, 2001.

[8]   Jean-Claude Bajard, Sylvain Duquesne, and Milos D. Ercegovac. Combining leak-resistant arithmetic for elliptic curves defined over f_p and RNS representation. *IACR Cryptology ePrint Archive*, 2010:311, 2010.

[9]   Jean-Claude Bajard and Laurent Imbert. A full RNS implementation of RSA. *IEEE Trans. Computers*, 53(6):769–774, 2004.

[10] Jean-Claude Bajard, Marcelo E. Kaihara, and Thomas Plantard. Selected RNS bases for modular multiplication. In *IEEE Symposium on Computer Arithmetic*, pp. 25–32. IEEE Computer Society, 2009.

[11] Selçuk Baktir and Berk Sunar. Optimal tower fields. *IEEE Transactions on Computers*, 53(10):1231–1243, 2004.

[12] Paul Barrett. Implementing the Rivest Shamir and Adleman public key encryption algorithm on a standard digital signal processor. In A. M. Odlyzko, editor, *Advances in Cryptology – CRYPTO '86*, volume 263 of *Lecture Notes in Computer Science*, pp. 311–323. Springer, Heidelberg, 1987.

[13] Naomi Benger and Michael Scott. Constructing tower extensions of finite fields for implementation of pairing-based cryptography. In M. A. Hasan and T. Helleseth, editors, *Arithmetic of Finite Fields (WAIFI 2010)*, volume 6087 of *Lecture Notes in Computer Science*, pp. 180–195. Springer, 2010.

[14] Jean-Luc Beuchat, Jorge E. González-Díaz, Shigeo Mitsunari, Eiji Okamoto, Francisco Rodríguez-Henríquez, and Tadanori Teruya. High-speed software implementation of the optimal Ate pairing over Barreto-Naehrig curves. In M. Joye, A. Miyaji, and A. Otsuka, editors, *Pairing-Based Cryptography – Pairing 2010*, volume 6487 of *Lecture Notes in Computer Science*, pp. 21–39. Springer, Heidelberg, 2010.

[15] Jean-Luc Beuchat, Emmanuel López-Trejo, Luis Martínez-Ramos, Shigeo Mitsunari, and Francisco Rodríguez-Henríquez. Multi-core implementation of the Tate pairing over supersingular elliptic curves. In J. A. Garay, A. Miyaji, and A. Otsuka, editors, *Cryptology and Network Security (CANS 2009)*, volume 5888 of *Lecture Notes in Computer Science*, pp. 413–432. Springer, Heidelberg, 2009.

[16] Dan Boneh, Ben Lynn, and Hovav Shacham. Short signatures from the Weil pairing. In C. Boyd, editor, *Advances in Cryptology – ASIACRYPT 2001*, volume 2248 of *Lecture Notes in Computer Science*, pp. 514–532. Springer, Heidelberg, 2001.

[17] R. P. Brent and H. T. Kung. The area-time complexity of binary multiplication. *J. ACM*, 28(3):521–534, 1981.

[18] Ray C. C. Cheung, Sylvain Duquesne, Junfeng Fan, Nicolas Guillermin, Ingrid Verbauwhede, and Gavin Xiaoxu Yao. FPGA implementation of pairings using residue number system and lazy reduction. In *CHES*, volume 6917 of *Lecture Notes in Computer Science*, pp. 421–441. Springer, 2011.

[19] Jaewook Chung and M. Anwar Hasan. Asymmetric squaring formulæ. In *18th IEEE Symposium on Computer Arithmetic (ARITH 2007)*, pp. 113–122. IEEE Computer Society, 2007.

[20] M. Cipolla. Un metodo per la risoluzione della congruenza di secondo grado. *Rend. Accad. Sci. Fis. Mat. Napoli*, vol. 9:154–163, 1903.

[21] Henri Cohen and Gerhard Frey, editors. *Handbook of Elliptic and Hyperelliptic Curve Cryptography*, volume 34 of *Discrete Mathematics and Its Applications*. Chapman & Hall/CRC, 2006.

[22] Augusto Jun Devegili, Michael Scott, and Ricardo Dahab. Implementing cryptographic pairings over Barreto-Naehrig curves (invited talk). In T. Takagi et al., editors, *Pairing-Based Cryptography – Pairing 2007*, volume 4575 of *Lecture Notes in Computer Science*, pp. 197–207. Springer, Heidelberg, 2007.

[23] Sylvain Duquesne. RNS arithmetic in $\mathbb{F}_{p^k}$ and application to fast pairing computation. *Journal of Mathematical Cryptology*, 5(1):51–88, 2011.

[24] Stephen R. Dussé and Burton S. Kaliski, Jr. A cryptographic library for the Motorola DSP56000. In I. Damgard, editor, *Advances in Cryptology – EUROCRYPT'90*, volume 473 of *Lecture Notes in Computer Science*, pp. 230–244. Springer, Heidelberg, 1991.

[25] Junfeng Fan, Frederik Vercauteren, and Ingrid Verbauwhede. Faster arithmetic for cryptographic pairings on Barreto-Naehrig curves. In C. Clavier and K. Gaj, editors, *Cryptographic Hardware and Embedded Systems – CHES 2009*, volume 5747 of *Lecture Notes in Computer Science*, pp. 240–253. Springer, Heidelberg, 2009.

[26] Robert Granger and Michael Scott. Faster squaring in the cyclotomic subgroup of sixth degree extensions. In P. Q. Nguyen and D. Pointcheval, editors, *Public Key Cryptography – PKC 2010*, volume 6056 of *Lecture Notes in Computer Science*, pp. 209–223. Springer, Heidelberg, 2010.

[27] Nicolas Guillermin. A high speed coprocessor for elliptic curve scalar multiplications over \mathbb$F$p\mathbb{F}_p. In *CHES*, volume 6225 of *Lecture Notes in Computer Science*, pp. 48–64. Springer, 2010.

[28] D. Hankerson, A. Menezes, and M. Scott. Software implementation of pairings. In M. Joye and G. Neven, editors, *Identity-based Cryptography*, Cryptology and Information Security Series, chapter 12, pp. 188–206. IOS Press, 2009.

[29] D. Hankerson, A. Menezes, and S. Vanstone. *Guide to Elliptic Curve Cryptography*. Springer-Verlag New York, Inc., Secaucus, NJ, USA, 2003.

[30] Marc Joye and Gregory Neven, editors. *Identity-Based Cryptography*, volume 2 of *Cryptology and Information Security Series*. IOS press, 2009.

[31] David Kammler, Diandian Zhang, Peter Schwabe, Hanno Scharwächter, Markus Langenberg, Dominik Auras, Gerd Ascheid, and Rudolf Mathar. Designing an ASIP for cryptographic pairings over Barreto-Naehrig curves. In C. Clavier and K. Gaj, editors, *Cryptographic Hardware and Embedded Systems – CHES 2009*, volume 5747 of *Lecture Notes in Computer Science*, pp. 254–271. Springer, Heidelberg, 2009.

[32] Koray Karabina. Squaring in cyclotomic subgroups. *Mathematics of Computation*, 82(281):555–579, 2013.

[33] Shin-ichi Kawamura, Masanobu Koike, Fumihiko Sano, and Atsushi Shimbo. Cox-rower architecture for fast parallel montgomery multiplication. In *EUROCRYPT*, volume 1807 of *Lecture Notes in Computer Science*, pp. 523–538. Springer, 2000.

[34] Donald E. Knuth. *The Art of Computer Programming, Third Edition.* Addison-Wesley, 1997.

[35] Neal Koblitz and Alfred Menezes. Pairing-based cryptography at high security levels (invited paper). In N. P. Smart, editor, *Cryptography and Coding*, volume 3796 of *Lecture Notes in Computer Science*, pp. 13–36. Springer, Heidelberg, 2005.

[36] Çetin Kaya Koç, Tolga Acar, and Burton S. Kaliski Jr. Analyzing and comparing montgomery multiplication algorithms. *IEEE Micro*, 16(3):26–33, 1996.

[37] Rudolf Lidl and Harald Niederreiter. *Finite Fields*, volume 20 of *Encyclopedia of Mathematics and Its Applications.* Cambridge University Press, 2nd edition, 1997.

[38] Chae Hoon Lim and Hyo Sun Hwang. Fast implementation of elliptic curve arithmetic in $GF(p^n)$. In H. Imai and Y. Zheng, editors, *PKC 2000: 3rd International Workshop on Theory and Practice in Public Key Cryptography*, volume 1751 of *Lecture Notes in Computer Science*, pp. 405–421. Springer, Heidelberg, 2000.

[39] S. Lindhurst. An analysis of Shanks's algorithm for computing square roots in finite fields. *CRM Proc. and Lecture Notes*, Vol. 19:231–242, 1999.

[40] Peter L. Montgomery. Modular multiplication without trial division. *Mathematics of Computation*, 44(170):519–521, 1985.

[41] S. Müller. On the computation of square roots in finite fields. *J. Design, Codes and Cryptography*, vol. 31:301–312, 2004.

[42] Geovandro C. C. F. Pereira, Marcos A. Simplício, Jr., Michael Naehrig, and Paulo S. L. M. Barreto. A family of implementation-friendly BN elliptic curves. *Journal of Systems and Software*, 84(8):1319–1326, 2011.

[43] Karl C. Posch and Reinhard Posch. Modulo reduction in residue number systems. *IEEE Trans. Parallel Distrib. Syst.*, 6(5):449–454, 1995.

[44] Francisco Rodríguez-Henríquez and Çetin K. Koç. On fully parallel Karatsuba multipliers for $GF(2^m)$. In A. Tria and D. Choi, editors, *International Conference on Computer Science and Technology (CST 2003)*, pp. 405–410. ACTA Press, 2003.

[45] Michael Scott. Computing the Tate pairing. In A. Menezes, editor, *Topics in Cryptology – CT-RSA 2005*, volume 3376 of *Lecture Notes in Computer Science*, pp. 293–304. Springer, Heidelberg, 2005.

[46] Michael Scott. Implementing cryptographic pairings. In T. Takagi et al., editors, *Pairing-Based Cryptography – Pairing 2007*, volume 4575 of *Lecture Notes in Computer Science*, pp. 177–196. Springer, Heidelberg, 2007.

[47] Mike Scott. Missing a trick: Karatsuba revisited. Cryptology ePrint Archive, Report 2015/1247, 2015. http://eprint.iacr.org/.

[48] Daniel Shanks. Five number-theoretic algorithms. In R. S. D. Thomas and H. C. Williams, editors, *Proceedings of the Second Manitoba Conference on Numerical Mathematics*, pp. 51–70. Utilitas Mathematica, 1972.

[49] Robert Szerwinski and Tim Güneysu. Exploiting the power of gpus for asymmetric cryptography. In *CHES*, volume 5154 of *Lecture Notes in Computer Science*, pp. 79–99. Springer, 2008.

[50] A. Tonelli. Bemerkung uber die auflosung quadratischer congruenzen. *Götinger Nachrichten*, pp. 344–346, 1891.

[51] Thomas Unterluggauer and Erich Wenger. Efficient pairings and ECC for embedded systems. In L. Batina and M. Robshaw, editors, *Cryptographic Hardware and Embedded Systems – CHES 2014*, volume 8731 of *Lecture Notes in Computer Science*, pp. 298–315. Springer, Heidelberg, 2014.

[52] Damian Weber and Thomas F. Denny. The solution of McCurley's discrete log challenge. In H. Krawczyk, editor, *Advances in Cryptology – CRYPTO '98*, volume 1462 of *Lecture Notes in Computer Science*, pp. 458–471. Springer, Heidelberg, 1998.

[53] André Weimerskirch and Christof Paar. Generalizations of the karatsuba algorithm for efficient implementations. Cryptology ePrint Archive, Report 2006/224, 2006. http://eprint.iacr.org/2006/224.

[54] Gavin Xiaoxu Yao, Junfeng Fan, Ray C. C. Cheung, and Ingrid Verbauwhede. Faster pairing coprocessor architecture. In M. Abdalla and T. Lange, editors, *Pairing-Based Cryptography – Pairing 2012*, volume 7708 of *Lecture Notes in Computer Science*, pp. 160–176. Springer, Heidelberg, 2013.

6

Scalar Multiplication and Exponentiation in Pairing Groups

Joppe Bos
NXP Semiconductors

Craig Costello
Microsoft Research

Michael Naehrig
Microsoft Research

6.1   Introduction

Algorithms to compute cryptographic pairings involve computations on elements in all three pairing groups, $\mathbb{G}_1$, $\mathbb{G}_2$, and $\mathbb{G}_T$. However, protocols often compute only a single pairing operation but need to compute many operations in any or all of the groups $\mathbb{G}_1$, $\mathbb{G}_2$, and $\mathbb{G}_T$ [14, 31, 52]. In this chapter, we discuss ways to enhance the performance of group operations that are not the pairing computation.

This chapter is an extension of the work done by the authors in [16] and contains excerpts from this paper. Like [16], it uses specific pairing-friendly curve families that target the 128-, 192-, and 256-bit security levels as case studies (see Chapter 4 for more information related to pairing-friendly curves). These families are popular choices and have common properties that make pairing implementations particularly efficient. Specifically, we discuss scalar multiplications in

$\mathbb{G}_1$ and $\mathbb{G}_2$, and group exponentiations in $\mathbb{G}_T$ for BN curves [6], which have embedding degree $k = 12$, KSS curves [42] ($k = 18$), and BLS curves [5] (for both $k = 12$ and $k = 24$). We note that most of the discussion that focuses on these families can be easily translated to curves from other families, and that the main algorithm for multi-scalar multiplications using endomorphisms (see Algorithm 6.2) has been presented in a general form so that it can be used as a basis for any endomorphism-accelerated scalar multiplication (or group exponentiation in its multiplicative form).

These curve families allow us to use Gallant-Lambert-Vanstone (GLV) [28] and Galbraith-Lin-Scott (GLS) [26] decompositions of dimensions 2, 4, 6, and 8 to speed up scalar multiplications and exponentiations in all three pairing groups. Extending the work in [16], this chapter contains optimal lattice bases for GLV and GLS decompositions, as well as a discussion on trace-based methods for compressed exponentiations in the group $\mathbb{G}_T$.

Using non-Weierstrass models for elliptic curve group operations can give rise to significant speedups (cf. [17, 50, 10, 11, 36]) and aid in realizing certain types of side-channel countermeasures (see Chapter 12 for more details). Such alternative models have not found the same success within pairing computations, since Miller's algorithm [49] not only requires group operations, but also relies on the computation of functions with divisors corresponding to these group operations. These functions are somewhat inherent in the Weierstrass group law, which is why Weierstrass curves remain faster for the pairings themselves [18]. Nevertheless, this does not mean that alternative curve models cannot be used to give speedups in the standalone group operations in pairing-based protocols. Here, we revisit the findings of [16] about which curve models are applicable in the most popular pairing scenarios and which of them achieve speedups when employed in place of the Weierstrass model.

This chapter contains descriptions and pseudo-code of algorithms for scalar multiplications and exponentiations in the three pairing groups. Most of them are presented in a constant-time fashion, showing how to implement the algorithm such that its execution time is independent of secret input data, such as the scalar or exponent.

6.2 Preliminaries

A cryptographic pairing $e : \mathbb{G}_1 \times \mathbb{G}_2 \to \mathbb{G}_T$ is a bilinear map that relates the three groups $\mathbb{G}_1$, $\mathbb{G}_2$, and $\mathbb{G}_T$, each of prime order r. In this chapter we define these groups as follows. For distinct primes p and r, let k be the smallest positive integer such that $r \mid p^k - 1$. Assume that $k > 1$. For an elliptic curve $E/\mathbb{F}_p$ such that $r \mid \#E(\mathbb{F}_p)$, we can choose $\mathbb{G}_1 = E(\mathbb{F}_p)[r]$ to be the order-r subgroup of $E(\mathbb{F}_p)$. We have $E[r] \subset E(\mathbb{F}_{p^k})$, and $\mathbb{G}_2$ can be taken as the (order-r) subgroup of $E(\mathbb{F}_{p^k})$ of p-eigenvectors of the p-power Frobenius endomorphism on E. Let $\mathbb{G}_T$ be the group of r-th roots of unity in $\mathbb{F}_{p^k}^*$. The *embedding degree* k is very large (i.e., $k \approx r$) for general curves, but must be kept small (i.e., $k < 50$) if computations in $\mathbb{F}_{p^k}$ are to be feasible in practice —this means that so-called *pairing-friendly* curves must be constructed in a special way. In Section 6.2.1 we recall the best-known techniques for constructing such curves with embedding degrees that target the 128-, 192-, and 256-bit security levels.

6.2.1 Parameterized Families of Pairing-Friendly Curves with Sextic Twists

The most suitable pairing-friendly curves for our purposes come from parameterized families, such that the parameters to find a suitable curve $E(\mathbb{F}_p)$ can be written as univariate polynomials. For the four families we consider, we give below the polynomials $p(x)$, $r(x)$, and $t(x)$, where $t(x)$ is such that $N(x) = p(x) + 1 - t(x)$ is the cardinality of the desired curve, which has $r(x)$ as a factor. All of the curves found from these constructions have j-invariant zero, which means they can be written in Weierstrass form as $y^2 = x^3 + b$. Instances of these pairing-friendly families can be found by searching through integer values x of an appropriate size until we find $x = x_0$ such that $p = p(x_0)$ and $r = r(x_0)$ are simultaneously prime, at which point we can simply test different values for b until the curve $E : y^2 = x^3 + b$ has an N-torsion point.

To target the 128-bit security level, we use the BN family [6] ($k = 12$) for which

$$p(x) = 36x^4 + 36x^3 + 24x^2 + 6x + 1, \quad t(x) = 6x^2 + 1, r(x) = p(x) + 1 - t(x). \tag{6.1}$$

At the 192-bit security level, we consider BLS curves [5] with $k = 12$ for which

$$p(x) = (x - 1)^2(x^4 - x^2 + 1)/3 + x, \quad t(x) = x + 1, \quad r(x) = x^4 - x^2 + 1, \tag{6.2}$$

where $x \equiv 1 \pmod 3$, and KSS curves [42] with $k = 18$, which are given by

$$p(x) = (x^8 + 5x^7 + 7x^6 + 37x^5 + 188x^4 + 259x^3 + 343x^2 + 1763x + 2401)/21,$$
$$t(x) = (x^4 + 16x + 7)/7, \quad r(x) = (x^6 + 37x^3 + 343)/7^3, \tag{6.3}$$

with $x \equiv 14 \pmod{42}$. At the 256-bit security level, we use curves from the BLS family [5] with embedding degree $k = 24$, which have the parametrization

$$p(x) = (x - 1)^2(x^8 - x^4 + 1)/3 + x, \quad t(x) = x + 1, \quad r(x) = x^8 - x^4 + 1, \tag{6.4}$$

with $x \equiv 1 \pmod 3$.

For the above families, which all have $k = 2^i 3^j$, the best practice to construct the full extension field $\mathbb{F}_{p^k}$ is to use a tower of (intermediate) quadratic and cubic extensions [45, 7], see also Chapter 5 of this book. Since any curve from these families has j-invariant $j = 0$ and $6 \mid k$, we can always use a *sextic twist* $E'(\mathbb{F}_{p^{k/6}})$ to represent elements of $\mathbb{G}_2 \subset E(\mathbb{F}_{p^k})[r]$ as elements of an isomorphic group $\mathbb{G}_2' = E'(\mathbb{F}_{p^{k/6}})[r]$. The isomorphism $\Psi : E \to E'$ that maps points on the curve to the twist is called the twisting isomorphism. Its inverse Ψ^{-1} is the so-called untwisting isomorphism. This shows that group operations in $\mathbb{G}_2$ can be performed on points with coordinates in an extension field with degree one sixth the size, which is the best we can do for elliptic curves [56, Proposition X.5.4]. For further details, see Chapter 2 of this book.

In all cases considered in this work, the most preferable sextic extension from $\mathbb{F}_{p^{k/6}}$ to $\mathbb{F}_{p^k}$ is constructed by choosing $\xi \in \mathbb{F}_{p^{k/6}}$ such that $\mathbb{F}_{p^{k/6}} = \mathbb{F}_p(\xi)$ and the polynomial $x^6 - \xi$ is irreducible in $\mathbb{F}_{p^{k/6}}[x]$, and then taking $z \in \mathbb{F}_{p^k}$ as a root of $x^6 - \xi$ to construct $\mathbb{F}_{p^k} = \mathbb{F}_{p^{k/6}}(z)$. We describe the individual towers in the four cases as follows: The BN and BLS cases with $k = 12$ preferably take $p \equiv 3 \pmod 4$, so that $\mathbb{F}_{p^2}$ can be constructed as $\mathbb{F}_{p^2} = \mathbb{F}_p[u]/(u^2 + 1)$. For $k = 18$ KSS curves, we prefer that 2 is not a cube in $\mathbb{F}_p$, so that $\mathbb{F}_{p^3}$ can be constructed as $\mathbb{F}_{p^3} = \mathbb{F}_p[u]/(u^3 - 2)$, before taking $\xi = u$ to extend to $\mathbb{F}_{p^{18}}$. For $k = 24$ BLS curves, we again prefer to construct $\mathbb{F}_{p^2}$ as $\mathbb{F}_{p^2} = \mathbb{F}_p[u]/(u^2 + 1)$, on top of which we take $\mathbb{F}_{p^4} = \mathbb{F}_{p^2}[v]/(v^2 - (u + 1))$ (it is shown that $v^2 - u$ cannot be irreducible [19, Proposition 1]), and use $\xi = v$ for the sextic extension. All of these constructions agree with the towers used in the software implementation "speed-record" literature [2, 53, 19, 1].

6.2.2 The GLV and GLS Algorithms

The GLV [28] and GLS [26] methods both use an efficient endomorphism to speed up elliptic curve scalar multiplications. Such a map provides a shortcut to a certain multiple of a given curve point at almost no computational cost. In the GLV setting, this multiple can be used for a two-dimensional scalar decomposition. A given scalar (or exponent) is decomposed into two scalars of roughly half the size of the original one. This means that the problem of computing a scalar multiple is transformed into computing two more efficient scalar multiples. To actually make the overall computation more efficient, one uses a double-scalar multiplication that shares a significant amount of computation between the two easier scalar multiplications. Depending on the nature of the endomorphism, higher-dimensional decompositions are possible. This means that the scalar is decomposed into more than two smaller scalars and the scalar multiplication is reduced to several short scalar multiplications. The way to then actually decrease the computational cost is to use a multi-scalar multiplication.

The GLV method relies on endomorphisms that arise from E having *complex multiplication* (CM) by an order of small discriminant, i.e., endomorphisms that are specific to the special shape of the curve E and that are unrelated to the Frobenius endomorphism. On the other hand, the GLS method works over extension fields where the p-power Frobenius becomes nontrivial, so it does not rely on E having a special shape. However, if E is both defined over an extension field and has a small CM discriminant, then the two can be combined [26, § 3] to give higher-dimensional decompositions, which can further enhance performance.

Since in this chapter we have $E/\mathbb{F}_p : y^2 = x^3 + b$ and $p \equiv 1 \pmod 3$ (the latter being imposed by the parameterizations of p coming from the pairing-friendly curve families), we can use the GLV endomorphism $\phi : (x, y) \mapsto (\zeta x, y)$ in $\mathbb{G}_1$ where $\zeta^3 = 1$ and $\zeta \in \mathbb{F}_p \setminus \{1\}$. In this case ϕ satisfies $\phi^2 + \phi + 1 = 0$ in the endomorphism ring $\text{End}(E)$ of E, so on $\mathbb{G}_1$ it corresponds to scalar multiplication by λ_ϕ, where $\lambda_\phi^2 + \lambda_\phi + 1 \equiv 0 \pmod r$, meaning we get a 2-dimensional decomposition in $\mathbb{G}_1$. Since $\mathbb{G}_2'$ is always defined over an extension field herein, we can combine the GLV endomorphism above with the Frobenius map to get higher-dimensional GLS decompositions. The standard way to do this in the pairing context [27] is to use the untwisting isomorphism Ψ^{-1} to move points from $\mathbb{G}_2'$ to $\mathbb{G}_2$, where the p-power Frobenius π_p can be applied (since E is defined over $\mathbb{F}_p$, while E' is not), before using the twisting isomorphism Ψ to move this result back to $\mathbb{G}_2'$. We define ψ as $\psi = \Psi \circ \pi_p \circ \Psi^{-1}$, which (even though Ψ and Ψ^{-1} are defined over $\mathbb{F}_{p^k}$) can be explicitly described over $\mathbb{F}_{p^{k/6}}$. The GLS endomorphism ψ satisfies $\Phi_k(\psi) = 0$ in $\text{End}(E')$ [27, Lemma 1], where $\Phi_k(\cdot)$ is the k-th cyclotomic polynomial, so it corresponds to scalar multiplication by λ_ψ, where $\Phi_k(\lambda_\psi) \equiv 0 \pmod r$. For the curves with $k = 12$, we thus obtain a 4-dimensional decomposition in $\mathbb{G}_2' \subset E'(\mathbb{F}_{p^2})$; for $k = 18$ curves, we get a 6-dimensional decomposition in $\mathbb{G}_2' \subset E'(\mathbb{F}_{p^3})$; and for $k = 24$ curves, we get an 8-dimensional decomposition in $\mathbb{G}_2' \subset E'(\mathbb{F}_{p^4})$.

To compute the scalar multiple $[m]P$ of P, an n-dimensional decomposition starts by computing the $n - 1$ additional points

$$\psi^i(P) = [\lambda^i]P,$$

for $1 \leq i \leq n - 1$. The simplest way to compute these points is via repeated application of ψ, although optimized explicit formulas for higher powers of ψ can sometimes be slightly more efficient than those for ψ itself. Next, one seeks to find a *short* multi-scalar $(a_1, a_2, \ldots, a_n)$ (we show how to do this optimally in the next subsection) such that

$$[a_1]P + [a_2]\psi(P) + \cdots + [a_n]\psi^{n-1}(P) = [m]P, \tag{6.5}$$

so that $[m]P$ can be computed by the much smaller multi-scalar multiplication.

The typical way to do this is to start by making all of the a_i positive: We simultaneously negate any $(a_i, \psi^{i-1}(P))$ pair for which $a_i < 0$; this can be done in a timing-attack resistant way using bitmasks (using bitmasks to select values in constant time is an often-used approach, see for instance [44]). We then precompute all possible sums $P + \sum_{i=1}^{n-1} [b_i]\psi^i(P)$, for the 2^{n-1} combinations of $b_i \in \{0, 1\}$, and store them in a lookup table. When simultaneously processing the j-th bits of the n mini-scalars, this allows us to update the running value with only one point addition, before performing a single point doubling. In each case, however, this standard approach requires individual attention for further optimization (see Section 6.4).

We aim to create constant-time programs: implementations that have an execution time independent of any secret material (e.g., the scalar). See for additional information about this concept Chapter 11.6, and how to load elements in a secure manner from the look-up tables. This means that we always execute exactly the same amount of point additions and duplications independent of the input. In order to achieve this in the setting of multi-scalar multiplications or multi-exponentiations, we use the recoding techniques from [24, 23]. This recoding technique not only guarantees that the program performs a constant number of point operations, but that the recoding itself is done in constant time as well. Furthermore, an advantage of this method is that the lookup table size is reduced by a factor of two, since we only store lookup elements for which the multiple of the first point P is odd. Besides reducing the memory, this reduces the time to create the lookup table.

Fixed-base scenarios

We note that this chapter focuses on the *dynamic* scenario where precomputations are unable to be exploited, e.g., when the group element P (that is to be multiplied) is not known in advance. In scenarios where P is a fixed system parameter or long-term public key, endomorphisms are essentially redundant for GLV- and GLS-style multi-scalar multiplications, since the fixed point multiples $[\lambda]P$ that are quickly computed on-the-fly as $\psi(P)$ can now be computed offline using a regular scalar multiplication without endomorphisms. If storage permits large tables of precomputed values, then optimized algorithms for this fixed-base scenario perform much faster than those exploiting endomorphisms without any precomputation. For more discussion of the fixed-base case, we refer to the original paper by Lim and Lee [47] and the more recent works that also consider constant-time implementations [33, 24, 32, 23].

6.3 Lattice Bases and Optimal Scalar Decompositions

In this section we show, for a given endomorphism-eigenvalue pair (ψ, λ) in an order r subgroup generated by P, how to *decompose* an integer scalar $m \in \mathbb{Z}$ into a corresponding multi-scalar $(a_1, a_2, \ldots, a_n)$ that satisfies Equation (6.5). The textbook way to find such a multi-scalar is via the *Babai rounding* technique [3], which we recall by following the exposition in [58, §1]. We start by defining the *lattice of decompositions of 0* (also known as the "GLV lattice" [25, p. 229]) as

$$\mathcal{L} = \langle (z_1, z_2, \ldots, z_n) \in \mathbb{Z}^n \mid z_1 + z_2\lambda + \cdots + z_n\lambda^{n-1} \equiv 0 \pmod{r} \rangle,$$

so that the set of decompositions for $m \in \mathbb{Z}/r\mathbb{Z}$ is the lattice coset $(m, 0, \ldots, 0) + \mathcal{L}$. For a given basis $(\mathbf{b}_1, \mathbf{b}_2, \ldots, \mathbf{b}_n)$ of $\mathcal{L}$, and on input of any $m \in \mathbb{Z}$, we compute $(\tilde{\alpha}_1, \tilde{\alpha}_2, \ldots, \tilde{\alpha}_n) \in \mathbb{Q}^n$ as the unique solution to $(m, 0, \ldots, 0) = \sum_{i=1}^{n} \tilde{\alpha}_i \mathbf{b}_i$, and subsequently compute the multi-scalar

$$(a_1, a_2, \ldots, a_n) = (m, 0, \ldots, 0) - \sum_{i=1}^{n} \lfloor \tilde{\alpha}_i \rceil \cdot \mathbf{b}_i. \tag{6.6}$$

It follows that $(a_1, a_2, \ldots, a_n) - (m, 0, \ldots, 0) \in \mathcal{L}$, so $m \equiv a_1 + a_2\lambda + \cdots + a_n\lambda^{n-1} \pmod{r}$. Since $-1/2 \le x - \lfloor x \rceil \le 1/2$, we have that $\|(a_1, a_2, \ldots, a_n)\|_\infty \le \frac{1}{2}\|\sum_{i=1}^{n} \mathbf{b}_i\|_\infty$. Since the lattice $\mathcal{L}$ is dependent on the curve parameters (and is therefore fixed), the task is then to find a basis of the lattice that facilitates short scalar decompositions.

In [20] it is shown how to generate bases that are optimal with respect to Babai rounding, i.e., bases that minimize the (maximum possible) infinity norm $\|(a_1, a_2, \ldots, a_n)\|_\infty$ across all integer inputs $m \in \mathbb{Z}$. We follow the same recipe to derive parameterized bases for the four popular families considered in this chapter (see §6.2.1). In each family, $\mathcal{L}_\phi$ and $\mathcal{L}_\psi$ are used to denote the lattice of decompositions of 0 corresponding to the endomorphisms ϕ on $\mathbb{G}_1$ and ψ on $\mathbb{G}_2$, respectively.

The first step, then, is to write down parameterized bases that generate $\mathcal{L}_\phi$ and $\mathcal{L}_\psi$. Since $\mathcal{L}_\phi = \langle (r, 0), (-\lambda_\phi, 1) \rangle$ and $\mathcal{L}_\psi = \langle (r, 0, \ldots, 0), (-\lambda_\psi, 1, 0, \ldots, 0), \ldots (-\lambda_\psi^{n-1}, 0, \ldots, 0, 1) \rangle$, writing down parameterized bases amounts to finding parameterizations for $\lambda_\phi = \lambda_\phi(x)$ and $\lambda_\psi = \lambda_\psi(x)$ that coincide with the parameterizations in §6.2.1. From here we can find "reduced" bases (in the sense of minimizing the infinity norm of the multi-scalars resulting from Babai rounding [20]). In what follows we do this for each of the four families under consideration, in order to derive explicit formulas for the multi-scalar $(a_1, a_2, \ldots, a_n)$ in terms of the input scalar $m \in \mathbb{Z}$. In all of these families we find $\lambda_\phi(x)$ and $\lambda_\psi(x)$ through solving $\Phi_3(\lambda_\phi(x)) \equiv 0$ and $\Phi_k(\lambda_\psi(x)) \equiv 0$ in $\mathbb{Z}[x]/\langle r(x)\rangle$, where in the latter case we can instantly write $\lambda_\psi(x) := t(x) - 1$ due to the embedding degree condition, which guarantees that $r \mid \Phi_k(t-1)$.

We note that, in $\mathbb{G}_2$, this technique was originally proposed by Galbraith and Scott, and that short parameterized bases for $\lambda_\psi(x)$ in the BN and BLS families can already be found in [27].

6.3.1   Decompositions for the $k = 12$ BN Family

For the BN family we have $\lambda_\phi = -(36x^3 + 18x^2 + 6x + 2)$ and $\lambda_\psi = 6x^2$. The lattice $\mathcal{L}_\phi$ for 2-dimensional decompositions in $\mathbb{G}_1$ is generated by the optimal basis $\mathcal{L}_\phi = \langle \mathbf{b}_1, \mathbf{b}_2\rangle$, where

$$\mathbf{b}_1 = (2x+1, 6x^2 + 4x + 1) \quad \text{and} \quad \mathbf{b}_2 = (6x^2 + 2x, -2x - 1).$$

Babai rounding decomposes a scalar $m \in \mathbb{Z}$ into $(a_1, a_2) = (m, 0) - \alpha_1 \cdot \mathbf{b}_1 - \alpha_2 \cdot \mathbf{b}_2$, where $\alpha_i = \lfloor \tilde{\alpha}_i \rceil$ for $\tilde{\alpha}_i = \hat{\alpha}_i \cdot m/r$, and

$$\hat{\alpha}_1 = 2x + 1 \quad \text{and} \quad \hat{\alpha}_2 = 6x^2 + 4x + 1.$$

It follows that, for any $m \in \mathbb{Z}$, we have $\|(a_1, a_2)\|_\infty \leq |6x^2 + 6x + 2|$.

The lattice $\mathcal{L}_\psi$ for 4-dimensional decompositions in $\mathbb{G}_2$ is generated by the optimal basis $\mathcal{L}_\psi = \langle \mathbf{b}_1, \mathbf{b}_2, \mathbf{b}_3, \mathbf{b}_4\rangle$, where

$$\mathbf{b}_1 = (-x, 2x, 2x+1, -x), \qquad \mathbf{b}_2 = (-2x-1, x, x+1, x),$$
$$\mathbf{b}_3 = (2x+1, 0, 2x, 1), \qquad \mathbf{b}_4 = (-1, 2x+1, 1, 2x).$$

Under this basis, scalars $m \in \mathbb{Z}$ decompose into $(a_1, a_2, a_3, a_4) = (m, 0, 0, 0) - \sum_{i=1}^{4} \alpha_i \mathbf{b}_i$, where $\alpha_i = \lfloor \tilde{\alpha}_i \rceil$ for $\tilde{\alpha}_i = \hat{\alpha}_i \cdot m/r$, and

$$\hat{\alpha}_1 = 2x + 1, \quad \hat{\alpha}_2 = -(12x^3 + 6x^2 + 2x + 1), \quad \hat{\alpha}_3 = 2x(3x^2 + 3x + 1), \quad \text{and} \quad \hat{\alpha}_4 = 6x^2 - 1.$$

For any $m \in \mathbb{Z}$, it follows that $\|(a_1, a_2, a_3, a_4)\|_\infty \leq |5x + 3|$.

6.3.2   Decompositions for the $k = 12$ BLS Family

For the BLS family with $k = 12$ we have $\lambda_\phi = x^2 - 1$ and $\lambda_\psi = x$. The lattice $\mathcal{L}_\phi$ for 2-dimensional decompositions in $\mathbb{G}_1$ is generated by the optimal basis $\mathcal{L}_\phi = \langle \mathbf{b}_1, \mathbf{b}_2\rangle$, where

$$\mathbf{b}_1 = (x^2 - 1, -1) \quad \text{and} \quad \mathbf{b}_2 = (1, x^2).$$

Babai rounding decomposes a scalar $m \in \mathbb{Z}$ into $(a_1, a_2) = (m, 0) - \alpha_1 \cdot \mathbf{b}_1 - \alpha_2 \cdot \mathbf{b}_2$, where $\alpha_i = \lfloor \tilde{\alpha}_i \rceil$ for $\tilde{\alpha}_i = \hat{\alpha}_i \cdot m/r$, and

$$\hat{\alpha}_1 = x^2 \quad \text{and} \quad \hat{\alpha}_2 = 1.$$

It follows that $\|(a_1, a_2)\|_\infty \leq |x^2 + 1|$ for all $m \in \mathbb{Z}$.

The lattice $\mathcal{L}_\psi$ for 4-dimensional decompositions in $\mathbb{G}_2$ is generated by the optimal basis $\mathcal{L}_\psi = \langle \mathbf{b}_1, \mathbf{b}_2, \mathbf{b}_3, \mathbf{b}_4\rangle$, where

$$\mathbf{b}_1 = (x, 1, 0, 0), \quad \mathbf{b}_2 = (0, x, 1, 0), \quad \mathbf{b}_3 = (0, 0, x, 1), \quad \mathbf{b}_4 = (1, 0, -1, -x).$$

Under this basis, scalars $m \in \mathbb{Z}$ decompose into $(a_1, a_2, a_3, a_4) = (m, 0, 0, 0) - \sum_{i=1}^{4} \alpha_i \mathbf{b}_i$, where $\alpha_i = \lfloor \tilde{\alpha}_i \rceil$ for $\tilde{\alpha}_i = \hat{\alpha}_i \cdot m/r$, and

$$\hat{\alpha}_1 = x(x^2 + 1), \quad \hat{\alpha}_2 = -(x^2 + 1), \quad \hat{\alpha}_3 = x, \quad \text{and} \quad \hat{\alpha}_4 = -1.$$

It follows that, for any $m \in \mathbb{Z}$, we have $\|(a_1, a_2, a_3, a_4)\|_\infty \leq |x + 2|$.

6.3.3 Decompositions for the $k = 18$ KSS Family

For the KSS family with $k = 18$ we have $\lambda_\phi = x^3 + 18$ and $\lambda_\psi = (x^4 + 16x)/7$. The lattice $\mathcal{L}_\phi$ for 2-dimensional decompositions in $\mathbb{G}_1$ is generated by the optimal basis $\mathcal{L}_\phi = \langle \mathbf{b}_1, \mathbf{b}_2 \rangle$, where

$$\mathbf{b}_1 = \left(-\frac{x^3}{7^3}, \frac{18x^3 + 343}{7^3} \right) \quad \text{and} \quad \mathbf{b}_2 = \left(\frac{19x^3 + 343}{7^3}, \frac{x^3}{7^3} \right).$$

Under this basis, Babai rounding decomposes a scalar $m \in \mathbb{Z}$ into $(a_1, a_2) = (m, 0) - \alpha_1 \cdot \mathbf{b}_1 - \alpha_2 \cdot \mathbf{b}_2$, where $\alpha_i = \lfloor \tilde{\alpha}_i \rceil$ for $\tilde{\alpha}_i = \hat{\alpha}_i \cdot m/r$, and

$$\hat{\alpha}_1 = -x^3 \quad \text{and} \quad \hat{\alpha}_2 = 18x^3 + 343.$$

For any $m \in \mathbb{Z}$, it follows that $\|(a_1, a_2)\|_\infty \leq |(20x^3 + 343)/7^3|$.

The lattice $\mathcal{L}_\psi$ for 6-dimensional decompositions in $\mathbb{G}_2$ is generated by the optimal basis $\mathcal{L}_\psi = \langle \mathbf{b}_1, \ldots, \mathbf{b}_6 \rangle$, where

$$\mathbf{b}_1 = \left(0, 0, \frac{2x}{7}, 1, 0, \frac{x}{7} \right), \qquad \mathbf{b}_2 = \left(0, \frac{2x}{7}, 1, 0, \frac{x}{7}, 0 \right), \qquad \mathbf{b}_3 = \left(\frac{2x}{7}, 1, 0, \frac{x}{7}, 0, 0 \right),$$

$$\mathbf{b}_4 = \left(-\frac{x}{7}, 0, 0, \frac{3x}{7}, 1, 0 \right), \qquad \mathbf{b}_5 = \left(0, -\frac{x}{7}, 0, 0, \frac{3x}{7}, 1 \right), \qquad \mathbf{b}_6 = \left(-1, 0, -\frac{x}{7}, 1, 0, \frac{3x}{7} \right).$$

Under this basis, scalars $m \in \mathbb{Z}$ decompose into $(a_1, \ldots, a_6) = (m, 0, 0, 0, 0, 0) - \sum_{i=1}^{6} \alpha_i \mathbf{b}_i$, where $\alpha_i = \lfloor \tilde{\alpha}_i \rceil$ for $\tilde{\alpha}_i = \hat{\alpha}_i \cdot m/r$, and

$$\hat{\alpha}_1 = 19x^3 + 343, \qquad \hat{\alpha}_2 = -x(8x^3 + 147), \qquad \hat{\alpha}_3 = x^2(3x^3 + 56),$$
$$\hat{\alpha}_4 = -(x^3 + 21)x^2, \qquad \hat{\alpha}_5 = (5x^3 + 98), \qquad \hat{\alpha}_6 = -(18x^3 + 343).$$

For any $m \in \mathbb{Z}$, we have $\|(a_1, a_2, a_3, a_4, a_5, a_6)\|_\infty \leq |4x/7 + 2|$.

6.3.4 Decompositions for the $k = 24$ BLS Family

For the BLS family with $k = 24$ we have $\lambda_\phi = x^4 - 1$ and $\lambda_\psi = x$. The lattice $\mathcal{L}_\phi$ for 2-dimensional decompositions in $\mathbb{G}_1$ is generated by the optimal basis $\mathcal{L}_\phi = \langle \mathbf{b}_1, \mathbf{b}_2 \rangle$, where

$$\mathbf{b}_1 = (x^4 - 1, -1) \quad \text{and} \quad \mathbf{b}_2 = (1, x^4).$$

Under this basis, Babai rounding decomposes a scalar $m \in \mathbb{Z}$ into $(a_1, a_2) = (m, 0) - \alpha_1 \cdot \mathbf{b}_1 - \alpha_2 \cdot \mathbf{b}_2$, where $\alpha_i = \lfloor \tilde{\alpha}_i \rceil$ for $\tilde{\alpha}_i = \hat{\alpha}_i \cdot m/r$, and

$$\hat{\alpha}_1 = x^4 \quad \text{and} \quad \hat{\alpha}_2 = 1.$$

For any $m \in \mathbb{Z}$, it follows that $\|(a_1, a_2)\|_\infty \leq |x^4 + 1|$.

The lattice $\mathcal{L}_\psi$ for 8-dimensional decompositions in $\mathbb{G}_2$ is generated by the optimal basis $\mathcal{L}_\psi = \langle \mathbf{b}_1, \ldots, \mathbf{b}_8 \rangle$, where

$$\mathbf{b}_1 = (x, -1, 0, 0, 0, 0, 0, 0), \qquad \mathbf{b}_2 = (0, x, -1, 0, 0, 0, 0, 0), \qquad \mathbf{b}_3 = (0, 0, x, -1, 0, 0, 0, 0),$$
$$\mathbf{b}_4 = (0, 0, 0, x, -1, 0, 0, 0), \qquad \mathbf{b}_5 = (0, 0, 0, 0, x, -1, 0, 0), \qquad \mathbf{b}_6 = (0, 0, 0, 0, 0, x, -1, 0),$$
$$\mathbf{b}_7 = (0, 0, 0, 0, 0, 0, x, -1), \qquad \mathbf{b}_8 = (1, 0, 0, 0, -1, 0, 0, x).$$

Under this basis, scalars $m \in \mathbb{Z}$ decompose into $(a_1, \ldots, a_8) = (m, 0, 0, 0, 0, 0, 0, 0) - \sum_{i=1}^{8} \alpha_i \mathbf{b}_i$, where $\alpha_i = \lfloor \tilde{\alpha}_i \rceil$ for $\tilde{\alpha}_i = \hat{\alpha}_i \cdot m/r$, and

$$\hat{\alpha}_1 = x^3(x^4 + 1), \qquad \hat{\alpha}_2 = -x^2(x^4 + 1), \qquad \hat{\alpha}_3 = x(x^4 + 1), \qquad \hat{\alpha}_4 = -(x^4 + 1),$$
$$\hat{\alpha}_5 = x^3, \qquad \hat{\alpha}_6 = -x^2, \qquad \hat{\alpha}_7 = x, \qquad \hat{\alpha}_8 = -1.$$

For any $m \in \mathbb{Z}$, we have $\|(a_1, a_2, a_3, a_4, a_5, a_6, a_7, a_8)\|_\infty \leq |x + 2|$.

6.3.5   Handling Round-Off Errors

In all of the above cases, the bounds given on the length of the multi-scalars assume that the roundings $\alpha_i = \lfloor \hat{\alpha}_i \cdot m/r \rceil$ are computed perfectly. In practice however, it is difficult to efficiently compute all such roundings exactly, particularly in a constant-time routine (cf. [20]). Thus, implementations of scalar decompositions usually use a fast method of approximating the round-offs, which (for our purposes) means that the derived bounds on the multi-scalars may no longer apply.

To fully account for this, we present the following solution. The roundings $\alpha_i = \lfloor \hat{\alpha}_i \cdot m/r \rceil$ can instead be approximated by

$$\tilde{\alpha}_i = \left\lfloor \ell_i \cdot \frac{m}{\nu} \right\rceil, \quad \text{with} \quad \ell_i = \left\lfloor \frac{\hat{\alpha}_i \cdot \nu}{r} \right\rceil,$$

so that, for a fixed constant ν, the ℓ_i are precomputed integer constants that are independent of the scalar m. In [20, Lemma 1] it is shown that, so long as $\nu > r$, the approximation satisfies $\tilde{\alpha}_i = \alpha_i - \epsilon$ for $\epsilon \in \{0, 1\}$. A good choice of ν is a power of 2 that is greater than r, which allows the runtime *division* of m by ν to be efficiently computable as a shift. For example, the implementations in [15, 20] choose $\nu = (2^w)^v$, where w is the word-size of the target architecture and v is minimal such that $\nu > r$.

To give guaranteed upper bounds on the multi-scalars subject to this approximation, we assume that there exists an integer scalar m where the approximation is *off-by-one* for all of the roundings, i.e., $\tilde{\alpha}_i = \alpha_i - 1$ for all $1 \leq i \leq n$. It then follows from Equation (6.6) that we can simply double each of the bounds derived in §§6.3.1–6.3.4. Put another way, using the above method to approximate the Babai roundoffs adds one bit to the maximum possible size of our multi-scalars, or one iteration to the main loop of the multi-scalar multiplication or multi-exponentiation.

6.4   Multi-Scalar Multiplication in $\mathbb{G}_1$ and $\mathbb{G}_2$

We now turn to describing the full algorithm for elliptic curve scalar multiplications in $\mathbb{G}_1$ and $\mathbb{G}_2$ using endomorphisms—see Algorithm 6.2. Although much of this chapter focuses on specific families of pairing-friendly curves, Algorithm 6.2 is presented in as much generality as is needed to perform endomorphism-accelerated group exponentiations in any suitable cryptographic subgroup (i.e., scalar multiplications in the elliptic curve groups $\mathbb{G}_1$ and $\mathbb{G}_2$, or exponentiations in the multiplicative group $\mathbb{G}_T$ otherwise). As such, it assumes as input an element P belonging to a cyclic group $\mathbb{G}$ of order r, equipped with an endomorphism ψ such that $\psi|_\mathbb{G} = [\lambda]_\mathbb{G}$ for $\lambda \in \mathbb{Z}$.

We start by presenting Algorithm 6.1, which is a simple multi-scalar recoding algorithm that is called as a subroutine of Algorithm 6.2. Let μ denote the maximum bitlength of the decomposed multi-scalars, i.e., $\mu = \lceil \log_2 \|(a_1, \ldots, a_n)\|_\infty \rceil$. We denote with $a_i[j]$ the $(j + 1)^{\text{th}}$ least significant (signed) bit of the ith scalar. The main purpose of recoding the multi-scalars is to facilitate a constant-time implementation. In particular, Algorithm 6.1 essentially rewrites a multi-scalar $(a_1, \ldots, a_n)$ derived in the previous section, where the $a_i[j]$ are bit-values ($a_i[j] \in$

$\{0, 1\}$ for $1 \leq j < \mu$), as the multi-scalar $(b_1, \ldots, b_n)$, where the $b_i \in \{-1, 0, 1\}^{\mu+1}$ are signed bit sequences of length $\mu + 1$ such that

$$\text{(i)} \quad a_i = \sum_{j=0}^{\mu} b_i[j] 2^j, \qquad \text{for } 1 \leq i \leq n,$$

$$\text{(ii)} \quad b_1[j] \in \{-1, 1\}, \qquad \text{for } 0 \leq j \leq \mu, \quad \text{and}$$

$$\text{(iii)} \quad b_i[j] \in \{0, b_1[j]\}, \qquad \text{for } 0 \leq j \leq \mu \text{ and } 2 \leq i \leq n.$$

Note that the scalars b_i are one signed bit longer ($\mu + 1$ bits) compared to the maximal bit-size of a_i. Property (i) above ensures that the multi-scalar multiplication gives the correct result since the scalars represent the same values; Property (ii) ensures that every addition in Step 2 of Algorithm 6.2 performs an addition with a non-zero element of $\mathbb{G}$ in the lookup table; and, together with Property (iii), is what allows the lookup table to be of size 2^{n-1} (in contrast to a naive multi-scalar multiplication, which requires a lookup table of size 2^n). This scalar recoding algorithm was adapted to the setting of endomorphism-accelerated multi-scalar multiplications by Faz-Hernandez, Longa, and Sanchez [23], based on the algorithm of Feng, Zhu, Xu, and Li [24].

For ease of exposition we have presented Algorithm 6.2 assuming that the decomposition $m \mapsto (a_1, \ldots, a_n)$ gives rise to an odd a_1; this is because of the nature of the recoding in Algorithm 6.1, which demands an odd a_1 as input. We present two ways to achieve this in practice. Firstly, we can find a short vector $\mathbf{v}$ in the lattice $\mathcal{L}$ of decompositions of 0 that has an odd first element; e.g., any of the short basis vectors $\mathbf{b}_i$ (where $\mathbf{b}_i[1]$ is odd) from Section 6.3. At the end of the decomposition producing $(a_1, \ldots, a_n)$, we then compute $(a_1, \ldots, a_n) + \mathbf{v}$ and select (using bitmasks) whichever of the two multi-scalars has an odd first element. This was the method used in [20]; it does not necessitate any changes to Algorithm 6.2 but, depending on the size of $\mathbf{v}$, may require new analysis on the upper bounds of the multi-scalars (i.e., it may add one bit to the main loop). The second option is to select, from $(a_1, \ldots, a_n)$ and $(a_1 + 1, \ldots, a_n)$, the multi-scalar whose first component is odd and, if necessary, subtract the input point P after the main loop (of course, in a constant-time routine this subtraction will always be performed). This is the method presented in [23].

In Steps 2 and 10 of Algorithm 6.2, where point additions are required, an optimized implementation will take advantage of different styles of point additions. Namely, if we assume that the input point P is in affine space, then (in the context of pairing-specific endomorphisms at least) it is often the case that $\psi^i(P)$ will also be in affine space. Thus, several of the additions used to build the lookup table in Step 2 will be "affine" additions, which take two affine points as input and output a projective point. The other additions will have one or both of the input points as a projective point and output a projective point, which are often called "mixed" and "full" additions, respectively. Given that affine additions are usually cheaper than mixed additions, and that full additions are the most costly, an optimized route to building the lookup table is a straightforward exercise for a given dimension (see [16] for the fine-grained details). A related optimization that is often applied to elliptic curve scalar multiplications is to ensure that each of the additions in the main loop are mixed additions; this is done by normalizing every point in the lookup table before entering the main loop. This can be computed efficiently using Montgomery's simultaneous inversion technique [50], which allows us to compute n independent modular inversions using $3(n-1)$ modular multiplications and a single inversion. In our experiments of 2-, 4-, 6−, and 8-dimensional decompositions, it always proved to be advantageous to perform this conversion (after Step 2 in Algorithm 6.2).

In an implementation dedicated to a particular cryptographic group or decomposition of dimension n, it is likely that dedicated changes to Algorithm 6.2 will optimize performance further. In particular, Algorithm 6.2 uses repeated *double-and-always-add* iterations, but depending on n, this may be suboptimal. In particular, our experiments in [16] showed that a

---

ALGORITHM 6.1 n-dimensional multi-scalar recoding.

Input: n non-negative integers $a_j = \sum_{i=0}^{\mu-1} a_j[i] \cdot 2^i$ where $1 \leq j \leq n$, $a_j[i] \in \{0,1\}$ for $0 \leq i < \mu$, $a_j[\mu] = 0$, a_1 is odd, and $\mu = \lceil \log_2 \|(a_1, \ldots, a_n)\|_\infty \rceil$.

Output: n non-negative integers $b_j = \sum_{i=0}^{\mu} b_j[i] \cdot 2^i$, where $1 \leq j \leq n$, $b_1[i] \in \{-1,1\}$, $b_k[i] \in \{0, b_1[i]\}$ for $0 \leq i \leq \mu$ and $2 \leq k \leq n$ such that $b_j = a_j$.

1   $b_1[\mu] \leftarrow 1$
2   **for** $i = 0$ **to** $\mu - 1$ **do**
3      $b_1[i] \leftarrow 2a_1[i+1] - 1$
4   **for** $j = 2$ **to** n **do**
5          $b_j[i] \leftarrow b_1[i] \cdot a_j[0]$
6          $a_j \leftarrow \lfloor a_j/2 \rfloor - \lfloor b_j[i]/2 \rfloor$
7   **end**
8   **end**
9   **for** $j = 2$ **to** n **do**
10      $b_j[\mu] \leftarrow a_j[0]$
11   **end**
12   **return** $(b_j[\mu], \ldots, b_j[0])$ *for* $1 \leq j \leq n$

ALGORITHM 6.2   Scalar multiplication using a degree-n $(n > 1)$ endomorphism ψ.

Input: Point P and integer scalar $m \in [0, r)$.

Output: $[m]P$.

     /* Compute endomorphisms: */

1   Compute $\psi^i(P)$ for $1 \leq i \leq n - 1$.

     /* Precompute lookup table: */

2   $T[u] \leftarrow P + \sum_{i=1}^{n-1} [u_i]\psi^i(P)$, for all $0 \leq u < 2^{n-1}$ where $u = \sum_{i=1}^{n-1} u_i \cdot 2^{i-1}$ and $u_i \in \{0, 1\}$.

     /* Scalar decomposition: */

3   Decompose m into the multi-scalar $(a_1, \ldots, a_n)$—see Section 6.3.

     /* Make subscalars positive: */

4   For any $a_i < 0$, simultaneously take $(a_i, \psi^{i-1}(P)) = (-a_i, -\psi^{i-1}(P))$.

     /* Scalar recoding: */

5   Recode $(a_1, \ldots, a_n)$ into $(b_1, \ldots, b_n)$ via Algorithm 6.1.
6   Let $d_i = \sum_{j=2}^{n} |b_j[i]| \cdot 2^{j-2}$, $0 \leq i \leq \mu$.

     /* Main loop: */

7   $Q \leftarrow b_1[\mu] \cdot T[d_\mu]$
8   **for** $i = \mu - 1$ **to** 0 **do**
9      $Q \leftarrow [2]Q$
10      $Q \leftarrow Q + b_1[i] \cdot T[d_i]$
11   **end**
12   **return** Q

---

window size of $w = 1$ (which corresponds to repeated double-and-always-add iterations) was preferable for 4-and 6-dimensional decompositions, but for the 2-dimensional decompositions in $\mathbb{G}_1$, larger windows of size $w = 2$ or $w = 3$ gave rise to faster scalar multiplications. On the other hand, the lookup table computed in Step 2 of Algorithm 6.2 is of size 2^{n-1}, so the $n = 8$ decompositions in $\mathbb{G}_2$ for the $k = 24$ BLS family would give rise to a single lookup table of size 128. Rather than computing and storing such a large table, works optimizing 8-dimensional decompositions [15, 16] have instead computed two lookup tables and processed the main loop via a repeated sequence of *double-add-add* operations; here the two *add* operations correspond to additions between the running value and an element from the first and second lookup tables, respectively. The lookup tables both contain 8 elements: the first contains $P + \sum_{i=1}^{3} [u_i]\psi^i(P)$ and the second* contains $\psi^4(P) + \sum_{i=1}^{3} [u_i]\psi^{i+4}(P)$, where (in both cases) $u = (u_3, u_2, u_1)_2$ for $0 \leq u \leq 7$. Thus, when the endomorphism ψ is less expensive than a point addition, each of the 8 elements in the second table can be obtained via application of ψ to the corresponding element in the first table—see [15, 16]. In both the $n = 2$ and $n = 8$ cases, the precomputation and main loop execution in Algorithm 6.2 must be updated according to the above discussion—see [16].

6.4.1 Three Non-Weierstrass Elliptic Curve Models

Unlike the general Weierstrass model, which covers all isomorphism classes of elliptic curves over a particular field, the non-Weierstrass elliptic curves usually only cover a subset of all such classes. Whether or not an elliptic curve E falls into the classes covered by a particular model is commonly determined by the existence of a rational point with a certain order on E. In the most popular scenarios for ECC, these orders are either 2, 3, or 4. In this section we consider the fastest model that is applicable in the pairing context in each of these cases. Since we focus on the fastest curve model for a particular setting we exclude the analysis of other (slower) curve models such as, for instance, Huff curves [41].

- $\mathcal{W}$ - **Weierstrass**: All curves in this paper have j-invariant zero and Weierstrass form $y^2 = x^3 + b$. The fastest formulas on such curves use Jacobian coordinates [10].

- $\mathcal{J}$ - **Extended Jacobi quartic**: If an elliptic curve has a point of order 2, then it can be written in (extended) Jacobi quartic form as $\mathcal{J}: y^2 = dx^4 + ax^2 + 1$ [12, §3]—these curves were first considered for cryptographic use in [12, §3]. The fastest formulas work on the corresponding projective curve given by $\mathcal{J}: Y^2Z^2 = dX^4 + aX^2Z^2 + Z^4$ and use the 4 extended coordinates $(X : Y : Z : T)$ to represent a point, where $x = X/Z$, $y = Y/Z$ and $T = X^2/Z$ [39].

- $\mathcal{H}$ - **Generalized Hessian**: If an elliptic curve (over a finite field) has a point of order 3, then it can be written in generalized Hessian form as $\mathcal{H}: x^3 + y^3 + c = dxy$ [22, Theorem 2]. The authors of [40, 57] studied Hessian curves [35] of the form $x^3 + y^3 + 1 = dxy$ for use in cryptography, and this was later generalized to include the parameter c [9, 22]. The fastest formulas for ADD/MIX/AFF are from [9] while the fastest DBL formulas are from [37]—they work on the homogeneous projective curve given by $\mathcal{H}: X^3 + Y^3 + cZ^3 = dXYZ$, where $x = X/Z$, $y = Y/Z$. We note that the j-invariant zero version of $\mathcal{H}$ has $d = 0$ (see Section 6.4.3), so in Table 6.1 we give updated costs that include this speedup.

- $\mathcal{E}$ - **Twisted Edwards**: If an elliptic curve has a point of order 4, then it can be

---

*The signed recoding adopted from [23] ensures that the coefficient of P (resp. $\psi^4(P)$) is always in $\{-1, 1\}$, and that the *sign* of the other three coefficients (if they are non-zero) is the same as that of P (resp. $\psi^4(P)$). This is why only 8 elements (rather than 16) are needed in each lookup table.

TABLE 6.1   The costs of necessary operations for computing scalar multiplications on four models of elliptic curves.

model / coords	requires	DBL cost	ADD cost	MIX cost	AFF cost
$\mathcal{W}$ / Jac.	–	$\mathbf{7}_{2,5,0,14}$	$\mathbf{16}_{11,5,0,13}$	$\mathbf{11}_{7,4,0,14}$	$\mathbf{6}_{4,2,0,12}$
$\mathcal{J}$ / ext.	pt. of order 2	$\mathbf{9}_{1,7,1,12}$	$\mathbf{13}_{7,3,3,19}$	$\mathbf{12}_{6,3,3,18}$	$\mathbf{11}_{5,3,3,18}$
$\mathcal{H}$ / proj.	pt. of order 3	$\mathbf{7}_{6,1,0,11}$	$\mathbf{12}_{12,0,0,3}$	$\mathbf{10}_{10,0,0,3}$	$\mathbf{8}_{8,0,0,3}$
$\mathcal{E}$ / ext.	pt. of order 4, or $4 \mid \#E$ and $\#K \equiv 1 \pmod 4$	$\mathbf{9}_{4,4,1,7}$	$\mathbf{10}_{9,0,1,7}$	$\mathbf{9}_{8,1,0,7}$	$\mathbf{8}_{7,0,1,7}$

Note: Costs are reported as $\mathbf{T_{M,S,d,a}}$ where $\mathbf{M}$ is the cost of a field multiplication, $\mathbf{S}$ is the cost of a field squaring, $\mathbf{d}$ is the cost of multiplication by a curve constant, $\mathbf{a}$ is the cost of a field addition (we have counted multiplications by 2 as additions), and $\mathbf{T}$ is the total number of multiplications, squarings, and multiplications by curve constants.

written in twisted Edwards form as $\mathcal{E}: ax^2 + y^2 = 1 + dx^2 y^2$ [8, Theorem 3.3]. However, if the field of definition, K, has $\#K \equiv 1 \pmod 4$, then $4 \mid \#E$ is enough to write E in twisted Edwards form [8, §3] (i.e. we do not necessarily need a point of order 4). Twisted Edwards curves [8] are a generalization of the so-called Edwards curves model [21], which were introduced to cryptography in [11], and the most efficient formulas to compute the group law are from [38].

For each model, the cost of the required group operations are summarized in Table 6.1. The total number of field multiplications are reported in bold for each group operation—this includes multiplications, squarings, and multiplications by constants. We note that in the context of plain ECC these models have been studied with small curve constants; in pairing-based cryptography, however, we must put up with whatever constants we get under the transformation to the non-Weierstrass model. The only exception we found in this work is for the $k = 12$ BLS curves, where $\mathbb{G}_1$ can be transformed to a Jacobi quartic curve with $a = -1/2$, which gives a worthwhile speedup [38].

6.4.2   Applicability of Alternative Curve Models for $k \in \{12, 18, 24\}$

Propositions 1–4 in [16] prove the existence or non-existence of points of orders 2, 3, and 4 in the groups $E(\mathbb{F}_p)$ and $E'(\mathbb{F}_{p^{k/6}})$ for the pairing-friendly families considered here. These statements are deduced from properties of the group orders $\#E(\mathbb{F}_p)$ and $\#E'(\mathbb{F}_{p^{k/6}})$ and the curve equation. This section recalls some of the details and Table 6.2 summarizes the alternative curve models that are available for $\mathbb{G}_1$ and $\mathbb{G}_2$ for these curve families.

One can study $\#E(\mathbb{F}_p)$ directly from the polynomial parameterizations in Section 6.2.1, while for $\#E'(\mathbb{F}_{p^e})$ (where $e = k/6$) one argues as follows. With the explicit recursion in [13, Corollary VI.2] one can determine the trace of Frobenius t_e of E over $\mathbb{F}_{p^e}$ and the index f_e. The trace t_e determines the number of $\mathbb{F}_{p^e}$-rational points on E. These parameters are related by the CM equation $4p^e = t_e^2 + 3f_e^2$ (since all curve families considered here have CM discriminant $D = -3$). This allows us to compute the order of the correct sextic twist, which by [34, Prop. 2] is one of $n'_{e,1} = p^e + 1 - (3f_e + t_e)/2$ or $n'_{e,2} = p^e + 1 - (-3f_e + t_e)/2$. For $k = 12$ and $k = 24$ BLS curves, it is assumed that $p \equiv 3 \pmod 4$ so that $\mathbb{F}_{p^2}$ can be constructed (optimally) as $\mathbb{F}_{p^2} = \mathbb{F}_p[u]/(u^2 + 1)$. Finally, since $p \equiv 3 \pmod 4$, $E(\mathbb{F}_p)$ must contain a point of order 4 if E is to be written in twisted Edwards form; however, since E' is defined over $\mathbb{F}_{p^e}$, if e is even then $4 \mid E'$ is enough to write E' in twisted Edwards form (see Section 6.4.1). Using the group orders deduced that way and the curve equation, [16] concludes on the following (non-)existence statements.

If $E/\mathbb{F}_p$ is a BN curve with sextic twist $E'/\mathbb{F}_{p^2}$, the groups $E(\mathbb{F}_p)$ and $E'(\mathbb{F}_{p^2})$ do not contain points of order 2, 3, or 4. This only leaves the Weierstrass model $\mathcal{W}$ for the scalar multiplications

TABLE 6.2 Optional curve models for $\mathbb{G}_1$ and $\mathbb{G}_2$ in popular pairing-friendly curve families.

family-k	$\mathbb{G}_1$		$\mathbb{G}_2$	
	algorithm	models avail.	algorithm	models avail.
BN-12	2-GLV	$\mathcal{W}$	4-GLS	$\mathcal{W}$
BLS-12	2-GLV	$\mathcal{H}, \mathcal{J}, \mathcal{W}$	4-GLS	$\mathcal{W}$
KSS-18	2-GLV	$\mathcal{W}$	6-GLS	$\mathcal{H}, \mathcal{W}$
BLS-24	2-GLV	$\mathcal{H}, \mathcal{J}, \mathcal{W}$	8-GLS	$\mathcal{E}, \mathcal{J}, \mathcal{W}$

in $\mathbb{G}_1$ and $\mathbb{G}_2$. Next, if $p \equiv 3 \pmod 4$ and $E/\mathbb{F}_p$ is a BLS curve with $k = 12$ and sextic twist $E'/\mathbb{F}_{p^2}$, the group $E(\mathbb{F}_p)$ contains a point of order 3 and can contain a point of order 2, but not 4, while the group $E'(\mathbb{F}_{p^2})$ does not contain a point of order 2, 3, or 4. Thus, one can only use $\mathcal{W}$ in $\mathbb{G}_2$, but the group $\mathbb{G}_1$ can be represented using the Hessian model $\mathcal{H}$, the Jacobi quartic model $\mathcal{J}$, or $\mathcal{W}$. Furthermore, if $E/\mathbb{F}_p$ is a KSS curve with $k = 18$ and sextic twist $E'/\mathbb{F}_{p^3}$, the group $E(\mathbb{F}_p)$ does not contain a point of order 2, 3, or 4, while the group $E'(\mathbb{F}_{p^3})$ contains a point of order 3 but does not contain a point of order 2 or 4. This leaves the models $\mathcal{W}$ for $\mathbb{G}_1$ and $\mathcal{H}, \mathcal{W}$ for $\mathbb{G}_2$. Lastly, if $p \equiv 3 \pmod 4$ and $E/\mathbb{F}_p$ is a BLS curve with $k = 24$ and sextic twist $E'/\mathbb{F}_{p^4}$, the group $E(\mathbb{F}_p)$ can contain points of order 2 or 3 (although not simultaneously), but not 4, while the group $E'(\mathbb{F}_{p^4})$ can contain a point of order 2, but does not contain a point of order 3 or 4, leaving all possible models for $\mathbb{G}_1$ and $\mathcal{J}, \mathcal{W}$ and the twisted Edwards model $\mathcal{E}$ for $\mathbb{G}_2$.

6.4.3 Translating Endomorphisms to the Non-Weierstrass Models

This section investigates whether the GLV and GLS endomorphisms from Section 6.2.2 translate to the Jacobi quartic and Hessian models. Whether the endomorphisms translate desirably depends on how efficiently they can be computed on the non-Weierstrass model. It is not imperative that the endomorphisms do translate desirably, but it can aid efficiency: If the endomorphisms are not efficient on the alternative model, then our scalar multiplication routine also incurs the cost of passing points back and forth between the two models—this cost is small but could be non-negligible for high-dimensional decompositions. On the other hand, if the endomorphisms are efficient on the non-Weierstrass model, then the groups $\mathbb{G}_1$ and/or $\mathbb{G}_2$ can be defined so that all scalar multiplications take place directly on this model, and the computation of the pairing can be modified to include an initial conversion back to Weierstrass form.

We essentially show that the only scenario in which the endomorphisms are efficiently computable on the alternative model is the case of the GLV endomorphism ϕ on Hessian curves.

Endomorphisms on the Hessian model

We modify the maps given in [22, §2.2] to the special case of j-invariant zero curves, where we have $d = 0$ on the Hessian model. Assume that $(0: \pm \alpha: 1)$ are points of order 3 on $\mathcal{W}: Y^2 Z = X^3 + \alpha^2 Z^3$, which is birationally equivalent to $\mathcal{H}: U^3 + V^3 + 2\alpha Z^3 = 0$. We define the constants $h_0 = \zeta - 1$, $h_1 = \zeta + 2$, $h_2 = -2(2\zeta + 1)\alpha$, where $\zeta^3 = 1$ and $\zeta \neq 1$. The map $\tau: \mathcal{W} \to \mathcal{H}$, $(X: Y: Z) \mapsto (U: V: W)$ is given as

$$U \leftarrow h_0 \cdot (Y + \alpha Z) + h_2 \cdot Z, \qquad V \leftarrow -U - 3(Y + \alpha Z), \qquad W \leftarrow 3X, \qquad (6.7)$$

where $\tau(0: \pm \alpha: 1) = \mathcal{O} \in \mathcal{H}$. The inverse map $\tau^{-1}: \mathcal{H} \to \mathcal{W}$, $(U: V: W) \mapsto (X: Y: Z)$ is

$$X \leftarrow h_2 \cdot W, \qquad Z \leftarrow h_0 \cdot V + h_1 \cdot U, \qquad Y \leftarrow -h_2 \cdot (U + V) - \alpha \cdot Z. \qquad (6.8)$$

It follows that the GLV endomorphism $\phi_{\mathcal{W}} \in \text{End}(\mathcal{W})$ translates into $\phi_{\mathcal{H}} \in \text{End}(\mathcal{H})$, where $\phi_{\mathcal{W}}: (X: Y: Z) \mapsto (\zeta X: Y: Z)$ becomes $\phi_{\mathcal{H}}: (U: V: W) \mapsto (U: V: \zeta W)$. However, we note that when computing $\phi_{\mathcal{H}}$ on an affine point, it can be advantageous to compute $\phi_{\mathcal{H}}$ as $\phi_{\mathcal{H}} : (u \colon v \colon 1) \mapsto (\zeta^2 u \colon \zeta^2 v \colon 1)$, where ζ^2 is the (precomputed) other cube root of unity, which produces an affine result.

For GLS on Hessian curves, there is no obvious or simple way to perform the analogous untwisting or twisting isomorphisms directly between $\mathcal{H}'(\mathbb{F}_{p^{k/6}})$ and $\mathcal{H}(\mathbb{F}_{p^k})$, which suggests that we must pass back and forth to the Weierstrass curve/s to determine the explicit formulas for the GLS endomorphism on $\mathcal{H}'$. The composition of these maps $\psi_{\mathcal{H}'} = \tau \circ \Psi_{\mathcal{W}} \circ \pi_p \circ \Psi_{\mathcal{W}}^{-1} \circ \tau^{-1}$ does not appear to simplify to a form anywhere near as efficient as the GLS endomorphism is on the Weierstrass curve. Consequently, our GLS routine will start with a Weierstrass point in $\mathcal{W}'(\mathbb{F}_{p^{k/6}})$, where we compute $d-1$ applications of $\psi \in \text{End}(\mathcal{W}')$, before using (6.7) to convert the d points to $\mathcal{H}'(\mathbb{F}_{p^{k/6}})$, where the remainder of the routine takes place (save the final conversion back to $\mathcal{W}'$). Note that since we are converting affine Weierstrass points to $\mathcal{H}'$ via (6.7), this only incurs two multiplications each time. However, the results are now projective points on $\mathcal{H}'$, meaning that the more expensive full addition formulas must be used to generate the remainder of the lookup table.

Endomorphisms on the Jacobi quartic model

Unlike the Hessian model where the GLV endomorphism was efficient, for the Jacobi quartic model it appears that neither the GLV nor GLS endomorphisms translate to be of a similar efficiency as they are on the Weierstrass model. Thus, in all cases where Jacobi quartic curves are a possibility, we start and finish on $\mathcal{W}$, and only map to $\mathcal{J}$ after computing all applications of ϕ or ψ on the Weierstrass model. We adapt maps given in [12] to our special case as follows. Let $(-\theta \colon 0 \colon 1)$ be a point of order 2 on $\mathcal{W} : Y^2 Z = X^3 + \theta^3 Z^3$ and let $a = 3\theta/4$ and $d = -3\theta^2/16$. The curve $\mathcal{W}$ is birationally equivalent to the (extended) Jacobi quartic curve $\mathcal{J} : V^2 W^2 = dU^4 + 2aU^2 W^2 + W^4$, with the map $\tau : \mathcal{W} \to \mathcal{J}$, $\tau : (X \colon Y \colon Z) \mapsto (U \colon V \colon W)$ given as

$$U \leftarrow 2YZ, \qquad W \leftarrow X^2 - X\theta Z + \theta^2 Z^2, \qquad V \leftarrow 6XZ\theta + W - 4aZ(\theta Z + X), \qquad (6.9)$$

where $\tau((-\theta \colon 0 \colon 1)) = (0 \colon -1 \colon 1) \in \mathcal{J}$. The inverse map $\tau^{-1} : \mathcal{J} \to \mathcal{W}$, $\tau^{-1} : (U \colon V \colon W) \mapsto (X \colon Y \colon Z)$ is given by

$$X \leftarrow (2V+2)U + 2aU^3 - \theta U^3, \qquad Y \leftarrow (4V+4) + 4aU^2, \qquad Z \leftarrow U^3, \qquad (6.10)$$

where $\tau^{-1}((0 \colon -1 \colon 1)) = (-\theta \colon 0 \colon 1) \in \mathcal{W}$ and the neutral point on $\mathcal{J}$ is $\mathcal{O}_{\mathcal{J}} = (0 \colon 1 \colon 1)$.

Endomorphisms on the twisted Edwards model

Similarly to the Jacobi-quartic model, endomorphisms on $\mathcal{E}$ are not nearly as efficiently computable as they are on $\mathcal{W}$, so we only pass across to $\mathcal{E}$ after the endomorphisms are applied on $\mathcal{W}$. Here we give the back-and-forth maps that are specific to our case(s) of interest. Namely, since we are unable to use twisted Edwards curves over the ground field (see Table 6.2), let $\mathcal{W}/\mathbb{F}_{p^e} : Y^2 Z = X^3 + b'Z^3$ for $p \equiv 3 \pmod 4$ and e being even. Since we have a point of order 2 on $\mathcal{W}$, i.e., $(\alpha \colon 0 \colon 1)$ with $\alpha = \sqrt[3]{-b'} \in \mathbb{F}_{p^e}$, then take $s = 1/(\alpha\sqrt{3}) \in \mathbb{F}_{p^e}$. The twisted Edwards curve $\mathcal{E} : aU^2 W^2 + V^2 W^2 = W^4 + dU^2 V^2$ with $a = (3\alpha s+2)/s$ and $d = (3\alpha s - 2)/s$ is isomorphic to $\mathcal{W}$, with the map $\tau : \mathcal{W} \to \mathcal{E}$, $(X \colon Y \colon Z) \mapsto (U \colon V \colon W)$ given as

$$U \leftarrow s(X - \alpha Z)(sX - s\alpha Z + Z), \quad V \leftarrow sY(sX - s\alpha Z - Z), \quad W \leftarrow sY(sX - s\alpha Z + Z),$$

with inverse map $\tau : \mathcal{E} \to \mathcal{W}$, $(U \colon V \colon W) \mapsto (X \colon Y \colon Z)$ given as

$$X \leftarrow -U(-W - V - \alpha s(W - V)), \qquad Y \leftarrow (W+V)W, \qquad Z \leftarrow sU(W - V).$$

6.4.4 Curve Choices for Pairings at the 128-, 192-, and 256-Bit Security Levels

Next, we propose specific curves from the four parameterized families that can be used in practice for implementing pairings, and the scalar multiplications as described in this chapter at the three standard levels of 128, 192, and 256 bits of security. These curves differ from the original examples discussed in [16]. The curves given here are those proposed in [4], and have the additional benefits of being subgroup secure in the sense of the definition introduced in [4], and thus have an additional security benefit.

For the specific parameterized families considered in this chapter, ensuring subgroup security sacrifices slightly the efficiency of the pairing computation compared to previously chosen curves in the literature due to the slightly larger Hamming weight for the defining parameter, but otherwise the curves agree in terms of the field sizes and towering options (as analyzed in [4]). For the BN and both BLS families, subgroup security enforces the cofactor $\#E'(\mathbb{F}_p^{k/d})/r$ for $\mathbb{G}_2$ in $E'(\mathbb{F}_p^{k/d})$ to be prime. Thus, there are no small cofactors in the group order $\#E'(\mathbb{F}_p^{k/d})$ and hence no points of small order, which in turn means that alternate curve models cannot be used for scalar multiplications in $\mathbb{G}_2$ and we are restricted to the Weierstrass model. For the KSS family, there is a cofactor of 3 enforced by the polynomial parameterization, and therefore the KSS curve presented below allows for a Hessian curve model for $\mathbb{G}_2$.

It seems that the additional benefits of subgroup security come at the price of restricting the possible improvements for the scalar multiplications through alternate curve models. However, this is actually not the case. The results of the implementation experiments conducted for [16] and presented there have shown that Jacobi quartic curves were unable to outperform the Weierstrass or Hessian curves in any of the scenarios considered there. The reason is that the small number of operations saved in a Jacobi quartic group addition are not enough to outweigh the slower Jacobi quartic doublings (see Table 6.1), and because of the extra computation incurred by the need to pass back and forth between $\mathcal{J}$ and $\mathcal{W}$ to compute the endomorphisms (see Section 6.4.3). On the other hand, while employing the Hessian form also requires us to pass back and forth to compute the endomorphisms, the Hessian group law operations are significantly faster than Weierstrass operations across the board, so Hessian curves reigned supreme whenever they were able to be employed. Table 6.3 summarizes the optimal curve model choices in each scenario based on the experiments from [16]. It shows that the use of the Hessian curve model in $\mathbb{G}_2$ is only advantageous for the KSS family.

In Table 6.3 we also summarize the bounds on the lengths of the multi-scalars computed via Babai rounding with the bases given in Section 6.3. We note that these are the theoretical bounds that assume that the Babai roundoff is computed exactly; if the Babai rounding is computed using the fast approximation discussed in §6.3.5, then the exact bound becomes one bit larger, i.e., $|z|$ becomes $|2z|$. We also present the optimal window size that was found experimentally for the example curves presented below. We note that these window sizes may change for curves within the same family but with different parameter sizes.

TABLE 6.3 Optimal scenarios for group exponentiations.

sec. level	family-k	\multicolumn{4}{c}{exp. in $\mathbb{G}_1$}		\multicolumn{4}{c}{exp. in $\mathbb{G}_2$}
		n	$\|a_i\|_\infty$	w	curve	n	$\|a_i\|_\infty$	w	curve
128-bit	BN-12	2	$|6x^2 + 6x + 2|$	2	Weierstrass	4	$|5x + 3|$	1	Weierstrass
192-bit	BLS-12	2	$|x^2 + 1|$	3	Hessian	4	$|x + 2|$	1	Weierstrass
	KSS-18	2	$|(20x^3 + 373)/7|$	3	Weierstrass	6	$|4x/7 + 2|$	1	Hessian
256-bit	BLS-24	2	$|x^4 + 1|$	3	Hessian	8	$|x + 2|$	1	Weierstrass

Note: For both GLV on $\mathbb{G}_1$ and GLS on $\mathbb{G}_2$ in all four families, we give the decomposition dimension n, the upper bounds on the multi-scalars $\|a_i\|_\infty$, the optimal window size w for the example curves, and the optimal curve model.

The $k = 12$ BN curve

The value $x_0 = 2^{62} + 2^{59} + 2^{55} + 2^{15} + 2^{10} - 1$ in Equation (6.1) results in 254-bit primes $p = p(x_0)$ and $n = n(x_0)$. The curve $E/\mathbb{F}_p : y^2 = x^3 + 5$ over $\mathbb{F}_p$ has $n = \#E(\mathbb{F}_p)$, the sextic twist can be represented by $E'/\mathbb{F}_{p^2} : y^2 = x^3 + 5(u + 1)$, where $\mathbb{F}_{p^2} = \mathbb{F}_p[u]/(u^2 + 1)$.

The $k = 12$ BLS curve

Setting $x_0 = -2^{106} - 2^{92} - 2^{60} - 2^{34} + 2^{12} - 2^9$ in Equation (6.2) gives a 635-bit prime $p = p(x_0)$ and a 425-bit prime $r = r(x_0)$. Let $\mathbb{F}_{p^2} = \mathbb{F}_p[u]/(u^2 + 1)$ and let $\xi = u + 1$. The Weierstrass forms corresponding to $\mathbb{G}_1$ and $\mathbb{G}_2$ are $\mathcal{W}/\mathbb{F}_p : y^2 = x^3 - 2$ and $\mathcal{W}'/\mathbb{F}_{p^2} : y^2 = x^3 - 2/\xi$. Only $\mathbb{G}_1$ has options for alternative models (see Table 6.2): let $\alpha \in \mathbb{F}_p$ with $\alpha^2 = -2$, then the Hessian curve $\mathcal{H}/\mathbb{F}_p : x^3 + y^3 + 2\alpha = 0$ is isomorphic to $\mathcal{W}$ over $\mathbb{F}_p$. The example curve originally proposed in [16] also had an isomorphic Jacobi quartic model. The new subgroup-secure example does not have a Jacobi quartic model over $\mathbb{F}_p$ because the cofactor $\#E'(\mathbb{F}_{p^{k/d}})/r$ for $\mathbb{G}_2$ is prime.

The $k = 18$ KSS curve

Setting $x_0 = 2^{64} + 2^{47} + 2^{43} + 2^{37} + 2^{26} + 2^{25} + 2^{19} - 2^{13} - 2^7$ in Equation (6.3) gives a 508-bit prime $p = p(x_0)$ and a 376-bit prime $r = r(x_0)$. Let $\mathbb{F}_{p^3} = \mathbb{F}_p[u]/(u^3 - 2)$. The Weierstrass forms for $\mathbb{G}_1$ and $\mathbb{G}_2$ are $\mathcal{W}/\mathbb{F}_p : y^2 = x^3 + 2$ and $\mathcal{W}'/\mathbb{F}_{p^3} : y^2 = x^3 + 2/u$. Let $\alpha \in \mathbb{F}_p$ with $\alpha^2 = 2/u$, then the Hessian curve $\mathcal{H}'/\mathbb{F}_{p^3} : x^3 + y^3 + 2\alpha = 0$ is isomorphic to $\mathcal{W}'$ over $\mathbb{F}_{p^3}$.

The $k = 24$ BLS curve

Setting $x_0 = -(2^{63} - 2^{47} - 2^{31} - 2^{26} - 2^{24} + 2^8 - 2^5 + 1)$ in Equation (6.4) gives a 629-bit prime $p = p(x_0)$ and a 504-bit prime $r = r(x_0)$. Let $\mathbb{F}_{p^2} = \mathbb{F}_p[u]/(u^2 + 1)$ and $\mathbb{F}_{p^4} = \mathbb{F}_{p^2}[v]/(v^2 - (u + 1))$. The Weierstrass forms corresponding to $\mathbb{G}_1$ and $\mathbb{G}_2$ are $\mathcal{W}/\mathbb{F}_p : y^2 = x^3 + 1$ and $\mathcal{W}'/\mathbb{F}_{p^4} : y^2 = x^3 + 1/v$. This gives us the option of a Hessian model in $\mathbb{G}_1$: the curve $\mathcal{H}/\mathbb{F}_p : x^3 + y^3 + 2 = 0$ is isomorphic to $\mathcal{W}$ over $\mathbb{F}_p$. Since this curve is subgroup secure, the cofactor of $\mathbb{G}_2$ in $E'(\mathbb{F}_{p^{k/d}})$ is prime, the curve does not have an alternate curve model over $\mathbb{F}_{p^{k/d}}$, and we are left with the Weierstrass model. Note that this is different than for the example curve presented in [16].

6.5   Exponentiations in $\mathbb{G}_T$

The group $\mathbb{G}_T$ is the group of r-th roots of unity in the multiplicative group of the k-th degree extension $\mathbb{F}_{p^k}$. Since $r \mid \Phi_k(p)$ due to the embedding degree condition, it is contained in the cyclotomic subgroup of $\mathbb{F}_{p^k}^*$ of order $\Phi_k(p)$. This fact can be used for various efficiency improvements and compression techniques. For example, there exist squaring algorithms [30, 43] that are much faster than the general squaring in $\mathbb{F}_{p^k}^*$. Furthermore, trace-based or norm-based compression methods exist (see [55, 29, 51]). All these techniques make use of the fact that the computation of the p-power Frobenius endomorphism and any of its powers in $\mathbb{F}_{p^k}$ is very cheap.

6.5.1   Multi-Exponentiation via the Frobenius Endomorphism

One way to carry out exponentiations in $\mathbb{G}_T$ is to use the fact that for the sake of an endomorphism-aided arithmetic, its structure is analogous to that of $\mathbb{G}_2$. Galbraith and Scott [27] remark that exponentiations in $\mathbb{G}_T \subset \mathbb{F}_{p^k}$ can be implemented with the same $\varphi(k)$-dimensional decomposition that is used for GLS in $\mathbb{G}_2$. This means that the same techniques for multi-exponentiation can be applied directly, simply translated to the multiplicative setting of $\mathbb{G}_T$.

Algorithm 6.2 carries over directly to $\mathbb{G}_T$. The endomorphism ψ on the curve group is replaced by the p-power Frobenius endomorphism π_p, which provides a $\varphi(k)$-dimensional scalar decomposition with a relation analogous to the minimal polynomial for ψ in the case of $\mathbb{G}_2$. For example, while the GLS map ψ on curves with $k = 12$ gives $\psi^4(Q') - \psi^2(Q') + Q' = \mathcal{O}$ for all $Q' \in \mathbb{G}_2'$, in $\mathbb{G}_T$, the p-power Frobenius map π_p gives $f \cdot \pi_p^4(f)/\pi_p^2(f) = 1$ for all $f \in \mathbb{G}_T$. The scalar decompositions for each family are the same as those for $\mathbb{G}_2$ given in Section 6.3. The scalar recoding technique also carries across analogously, since inversions of $\mathbb{G}_T$-elements (which are conjugations over $\mathbb{F}_{p^{k/2}}$) are almost for free [27, §7], just as in the elliptic curve groups. It is again done using Algorithm 6.1. For precomputing the lookup table and the group operations in the main loop, one simply replaces the elliptic curve operations with the corresponding multiplicative operations in $\mathbb{G}_T$, using special algorithms for the squarings such as the cyclotomic squarings from [30]. This method is a way of obtaining an implementation protected against timing attacks in a straightforward manner, just as in the case for the group $\mathbb{G}_2$.

6.5.2 Compressed Single Exponentiation via the Trace Map

It was pointed out to us by Michael Scott that compressing pairing values using the finite field trace as described in [55] and applying the implicit exponentiation method for the XTR cryptosystem [46] as presented in [59] might be significantly faster than a multi-exponentiation via the GLS scalar decomposition. Scott provides a few further details in [54, Section 8.1]. The group $\mathbb{G}_T$ is a relatively small subgroup (namely of order r) in a quite large multiplicative group $\mathbb{F}_{p^k}^*$, and the standard representation of elements as $\mathbb{F}_{q^k}$-elements is quite wasteful with respect to the small number of group elements. Therefore, it is useful to compress group elements to a smaller representation. In addition to the smaller storage requirements, an exponentiation approach like XTR is more efficient than general arithmetic using the full element representation.

The idea is to use the XTR techniques in the relative field extension $\mathbb{F}_{p^k}$ of degree 6 over $\mathbb{F}_{p^{k/6}}$ (see for example [48]). This can be done for all the examples of pairing-friendly curve families considered in this chapter because they all satisfy $6 \mid k$. One works with the relative trace map

$$\text{Tr} : \mathbb{F}_{p^k} \to \mathbb{F}_{p^{k/3}}, \ a \mapsto a + a^{p^{k/3}} + a^{p^{2k/3}},$$

to compress an element $a \in \mathbb{G}_T$ to $c = \text{Tr}(a)$, which only requires one third of the memory that is required for storing a as a full $\mathbb{F}_{p^k}$ element. The trace evaluates to the same value for all three conjugates of a, namely $c = \text{Tr}(a) = \text{Tr}(a^{p^{k/3}}) = \text{Tr}(a^{p^{2k/3}})$ and therefore, one works with equivalence classes of three elements from this point on. Note that this might require special attention for some protocols.

The XTR algorithm now provides an algorithm to implicitly exponentiate an element in $\mathbb{G}_T$. Namely, given $c = \text{Tr}(a)$ and a scalar $m \in \mathbb{Z}$, one can efficiently compute $c_m = \text{Tr}(a^m)$. To see how this works, we briefly revisit Sections 2.3 and 2.4 of [59]. The former collects several facts that can be used to compute with trace values c_u, $u \in \mathbb{Z}$.

For example, $c_{-u} = c_u^{p^{k/3}}$, which means that implicit negations are almost for free. The main facts for XTR single exponentiation are formulas for computing $c_{2u-1}, c_{2u},$ and c_{2u+1} given the values $c_{u-1}, c_u, c_{u+1},$ and c_1. They are

$$\begin{cases} c_{2u} = c_u^2 - 2c_u^{p^{k/3}} \\ c_{2u-1} = c_{u-1}c_u - c_1^{p^{k/3}}c_u^{p^{k/3}} + c_{u+1}^{p^{k/3}} \\ c_{2u+1} = c_{u+1}c_u - c_1 c_u^{p^{k/3}} + c_{u-1}^{p^{k/3}} \end{cases} .$$

Working with triples $S_u = (c_{u-1}, c_u, c_{u+1})$, one can compute c_m via computing the triple S_{2v+1} where $v = \lfloor \frac{m-1}{2} \rfloor$ and then returning c_{2v+1} or c_{2v+2} depending on whether m was odd or even. The single exponentiation in [59, Section 2.4] is a simple left-to-right binary ladder that starts with the triple $S_1 = (3, c_1, c_2 = c_1^2 - 2c_1^{p^{k/3}})$ and updates to $S_{2u-1} = (c_{2u-2}, c_{2u-1}, c_{2u})$ or $S_{2u+1} = (c_{2u}, c_{2u+1}, c_{2u+2})$ using the above formulas, depending on whether the current bit in

v is 0 or 1. The regular structure of the updates makes it easy to implement this algorithm in constant time, provided that memory reads and writes are implemented in constant time.

Algorithm 6.3 describes the single exponentiation in a constant-time manner. The algorithm assumes that it is asserted that the scalar m is chosen or modified such that the bit length l of v is constant for all allowed choices of m. This might be achieved, for example, by allowing only scalars with certain bits set, by recoding the scalar, or by adding a suitable small multiple of the group order r.

ALGORITHM 6.3 XTR single exponentiation in $\mathbb{G}_T$.

Input: Trace value $c_1 = \mathrm{Tr}(a) \in \mathbb{F}_{p^{k/3}}$ and integer scalar $m \in [0, r)$.
Output: $c_m = \mathrm{Tr}(a^m) \in \mathbb{F}_{p^{k/3}}$.

1 $v \leftarrow \lfloor \frac{m-1}{2} \rfloor$

2 Write $v = \sum_{i=0}^{l-1} v_i 2^i$, $v_i \in \{0, 1\}$

3 $\hat{c}_1 \leftarrow c_1^{p^{k/3}}$

4 $S \leftarrow (3, c_1, c_1^2 - 2\hat{c}_1)$
   /*   */

5 Main loop: **for** $i = l - 1$ **to** 0 **do**

6 $\quad d_1 \leftarrow (1 - v_i)\hat{c}_1 + v_i c_1$

7 $\quad t_0 \leftarrow S[2v_i]$

8 $\quad t_1 \leftarrow S[1]$

9 $\quad t_2 \leftarrow S[2(v_i + 1) \mod 4]$

10 $\quad s_0 \leftarrow t_0^2 - 2t_0^{p^{k/3}}$

11 $\quad s_1 \leftarrow t_0 t_1 - d_1 t_1^{p^{k/3}} + t_2^{p^{k/3}}$

12 $\quad s_2 \leftarrow t_1^2 - 2t_1^{p^{k/3}}$

13 **end**

14 **return** $S[2 - (m \mod 2)]$

6.5.3   Compressed Double Exponentiation via the Trace Map

Section 3.1 in [59] describes a double exponentiation algorithm to compute $c_{a_1 l + a_2 k}$ for $a_1, a_2, k, l \in \mathbb{Z}$ given the precomputed values c_k and c_l and exponents a_1 and a_2. It uses the same formulas as in the previous section to update the trace values, working with quadruples $(c_u, c_v, c_{u-v}, c_{u-2v})$. This algorithm can be used via a two-dimensional scalar decomposition using the $p^{k/3}$-power Frobenius endomorphism $\pi_{p^{k/3}}$.

For the four pairing-friendly families considered in this chapter, $\pi_{p^{k/3}}$ satisfies the equation $\pi_{p^{k/3}}^2 - \pi_{p^{k/3}} + 1 = 0$. Recall that the endomorphism used in the GLV method for the group $\mathbb{G}_1$ satisfies $\phi^2 + \phi + 1 = 0$. Therefore, one can use the two-dimensional GLV decomposition algorithm for the group $\mathbb{G}_1$ (see Section 6.3) to obtain the mini-scalars (a_1, a_2) and then negate the scalar a_2 that is the coefficient of the endomorphism $\pi_{p^{k/3}}$. Now, one uses the XTR double exponentiation algorithm with the parameters $l = 1$ and $k = p^{k/3} \mod r$. One can precompute the required trace values as follows:

$$c_1 = \mathrm{Tr}(a), \qquad c_{p^{k/3}} = \mathrm{Tr}(a^{p^{k/3}}), \qquad c_{p^{k/3}-1} = \mathrm{Tr}(a^{p^{k/3}}/a), \qquad c_{p^{k/3}-2} = \mathrm{Tr}(a^{p^{k/3}}/a^2).$$

Note that the inversions in the cyclotomic subgroup are conjugations over $\mathbb{F}_{p^{k/2}}$ and are therefore almost for free. All other operations required for the precomputation are cheap Frobenius operations and additions to compute the traces.

Due to its dependencies on intermediate values, it does not seem straightforward to make this algorithm run in constant time, independent of its input data. However, it can be a possibly more efficient alternative for exponentiations with public exponents.

References

[1] Diego F. Aranha, Laura Fuentes-Castañeda, Edward Knapp, Alfred Menezes, and Francisco Rodríguez-Henríquez. Implementing pairings at the 192-bit security level. In M. Abdalla and T. Lange, editors, *Pairing-Based Cryptography – Pairing 2012*, volume 7708 of *Lecture Notes in Computer Science*, pp. 177–195. Springer, Heidelberg, 2013.

[2] Diego F. Aranha, Koray Karabina, Patrick Longa, Catherine H. Gebotys, and Julio López. Faster explicit formulas for computing pairings over ordinary curves. In K. G. Paterson, editor, *Advances in Cryptology – EUROCRYPT 2011*, volume 6632 of *Lecture Notes in Computer Science*, pp. 48–68. Springer, Heidelberg, 2011.

[3] László Babai. On Lovász' lattice reduction and the nearest lattice point problem. *Combinatorica*, 6(1):1–13, 1986.

[4] Paulo S. L. M. Barreto, Craig Costello, Rafael Misoczki, Michael Naehrig, Geovandro C. C. F. Pereira, and Gustavo Zanon. Subgroup security in pairing-based cryptography. In K. E. Lauter and F. Rodríguez-Henríquez, editors, *Progress in Cryptology – LATINCRYPT 2015*, volume 9230 of *Lecture Notes in Computer Science*, pp. 245–265. Springer, Heidelberg, 2015.

[5] Paulo S. L. M. Barreto, Ben Lynn, and Michael Scott. Constructing elliptic curves with prescribed embedding degrees. In S. Cimato, C. Galdi, and G. Persiano, editors, *Security in Communication Networks (SCN 2002)*, volume 2576 of *Lecture Notes in Computer Science*, pp. 257–267. Springer, Heidelberg, 2003.

[6] Paulo S. L. M. Barreto and Michael Naehrig. Pairing-friendly elliptic curves of prime order. In B. Preneel and S. Tavares, editors, *Selected Areas in Cryptography (SAC 2005)*, volume 3897 of *Lecture Notes in Computer Science*, pp. 319–331. Springer, Heidelberg, 2006.

[7] Naomi Benger and Michael Scott. Constructing tower extensions of finite fields for implementation of pairing-based cryptography. In M. A. Hasan and T. Helleseth, editors, *Arithmetic of Finite Fields (WAIFI 2010)*, volume 6087 of *Lecture Notes in Computer Science*, pp. 180–195. Springer, 2010.

[8] Daniel J. Bernstein, Peter Birkner, Marc Joye, Tanja Lange, and Christiane Peters. Twisted Edwards curves. In S. Vaudenay, editor, *Progress in Cryptology – AFRICACRYPT 2008*, volume 5023 of *Lecture Notes in Computer Science*, pp. 389–405. Springer, Heidelberg, 2008.

[9] Daniel J. Bernstein, Chitchanok Chuengsatiansup, David Kohel, and Tanja Lange. Twisted hessian curves. In K. E. Lauter and F. Rodríguez-Henríquez, editors, *Progress in Cryptology – LATINCRYPT 2015*, volume 9230 of *Lecture Notes in Computer Science*, pp. 269–294. Springer, Heidelberg, 2015.

[10] Daniel J. Bernstein and Tanja Lange. Explicit-formulas database. http://www.hyperelliptic.org/EFD.

[11] Daniel J. Bernstein and Tanja Lange. Faster addition and doubling on elliptic curves. In K. Kurosawa, editor, *Advances in Cryptology – ASIACRYPT 2007*, volume 4833 of *Lecture Notes in Computer Science*, pp. 29–50. Springer, Heidelberg, 2007.

[12] Olivier Billet and Marc Joye. The Jacobi model of an elliptic curve and side-channel analysis. In M. Fossorier, T. Høholdt, and A. Poli, editors, *Applied Algebra, Algebraic Algorithms and Error-Correcting Codes (AAECC 2003)*, volume 2643 of *Lecture Notes in Computer Science*, pp. 34–42. Springer, 2003.

[13] Ian F. Blake, Gadiel Seroussi, and Nigel P. Smart. *Elliptic Curves in Cryptography*, volume 265 of *London Mathematical Society Lecture Notes Series*. Cambridge University Press, 1999.

[14] Dan Boneh, Xavier Boyen, and Eu-Jin Goh. Hierarchical identity based encryption with constant size ciphertext. In R. Cramer, editor, *Advances in Cryptology – EUROCRYPT 2005*, volume 3494 of *Lecture Notes in Computer Science*, pp. 440–456. Springer, Heidelberg, 2005.

[15] Joppe W. Bos, Craig Costello, Hüseyin Hisil, and Kristin Lauter. High-performance scalar multiplication using 8-dimensional GLV/GLS decomposition. In G. Bertoni and J.-S. Coron, editors, *Cryptographic Hardware and Embedded Systems – CHES 2013*, volume 8086 of *Lecture Notes in Computer Science*, pp. 331–348. Springer, Heidelberg, 2013.

[16] Joppe W. Bos, Craig Costello, and Michael Naehrig. Exponentiating in pairing groups. In T. Lange, K. Lauter, and P. Lisonek, editors, *Selected Areas in Cryptography – SAC 2013*, volume 8282 of *Lecture Notes in Computer Science*, pp. 438–455. Springer, Heidelberg, 2014.

[17] David V. Chudnovsky and Gregory V. Chudnovsky. Sequences of numbers generated by addition in formal groups and new primality and factorization tests. *Advances in Applied Mathematics*, 7(4):385–434, 1986.

[18] Craig Costello, Tanja Lange, and Michael Naehrig. Faster pairing computations on curves with high-degree twists. In P. Q. Nguyen and D. Pointcheval, editors, *Public Key Cryptography – PKC 2010*, volume 6056 of *Lecture Notes in Computer Science*, pp. 224–242. Springer, Heidelberg, 2010.

[19] Craig Costello, Kristin Lauter, and Michael Naehrig. Attractive subfamilies of BLS curves for implementing high-security pairings. In D. J. Bernstein and S. Chatterjee, editors, *Progress in Cryptology – INDOCRYPT 2011*, volume 7107 of *Lecture Notes in Computer Science*, pp. 320–342. Springer, Heidelberg, 2011.

[20] Craig Costello and Patrick Longa. FourQ: Four-dimensional decompositions on a Q-curve over the Mersenne prime. In T. Iwata and J. H. Cheon, editors, *Advances in Cryptology – ASIACRYPT 2015, Part I*, volume 9452 of *Lecture Notes in Computer Science*, pp. 214–235. Springer, Heidelberg, 2015.

[21] Harold. M. Edwards. A normal form for elliptic curves. *Bulletin of the American Mathematical Society*, 44(3):393–422, 2007.

[22] Reza Rezaeian Farashahi and Marc Joye. Efficient arithmetic on hessian curves. In P. Q. Nguyen and D. Pointcheval, editors, *Public Key Cryptography – PKC 2010*, volume 6056 of *Lecture Notes in Computer Science*, pp. 243–260. Springer, Heidelberg, 2010.

[23] Armando Faz-Hernández, Patrick Longa, and Ana H. Sánchez. Efficient and secure algorithms for GLV-based scalar multiplication and their implementation on GLV-GLS curves. *Journal of Cryptographic Engineering*, 5(1):31–52, 2015.

[24] Min Feng, Bin B. Zhu, Maozhi Xu, and Shipeng Li. Efficient comb elliptic curve multiplication methods resistant to power analysis. Cryptology ePrint Archive, Report 2005/222, 2005. http://eprint.iacr.org/2005/222.

[25] Steven D. Galbraith. *Mathematics of Public Key Cryptography*. Cambridge University Press, 2012.

[26] Steven D. Galbraith, Xibin Lin, and Michael Scott. Endomorphisms for faster elliptic curve cryptography on a large class of curves. *Journal of Cryptology*, 24(3):446–469, 2011.

[27] Steven D. Galbraith and Michael Scott. Exponentiation in pairing-friendly groups using homomorphisms. In S. D. Galbraith and K. G. Paterson, editors, *Pairing-Based Cryptography – Pairing 2008*, volume 5209 of *Lecture Notes in Computer Science*, pp. 211–224. Springer, Heidelberg, 2008.

[28] Robert P. Gallant, Robert J. Lambert, and Scott A. Vanstone. Faster point multiplication on elliptic curves with efficient endomorphisms. In J. Kilian, editor, *Advances in Cryptology – CRYPTO 2001*, volume 2139 of *Lecture Notes in Computer Science*, pp. 190–200. Springer, Heidelberg, 2001.

[29] Robert Granger, Dan Page, and Martijn Stam. On small characteristic algebraic tori in pairing-based cryptography. *LMS Journal of Computation and Mathematics*, 9:64–85, 2006.

[30] Robert Granger and Michael Scott. Faster squaring in the cyclotomic subgroup of sixth degree extensions. In P. Q. Nguyen and D. Pointcheval, editors, *Public Key Cryptography – PKC 2010*, volume 6056 of *Lecture Notes in Computer Science*, pp. 209–223. Springer, Heidelberg, 2010.

[31] Jens Groth. Short pairing-based non-interactive zero-knowledge arguments. In M. Abe, editor, *Advances in Cryptology – ASIACRYPT 2010*, volume 6477 of *Lecture Notes in Computer Science*, pp. 321–340. Springer, Heidelberg, 2010.

[32] Mike Hamburg. Fast and compact elliptic-curve cryptography. Cryptology ePrint Archive, Report 2012/309, 2012. http://eprint.iacr.org/2012/309.

[33] Mustapha Hedabou, Pierre Pinel, and Lucien Bénéteau. Countermeasures for preventing comb method against SCA attacks. In R. H. Deng et al., editors, *Information Security Practice and Experience (ISPEC 2005)*, volume 3439 of *Lecture Notes in Computer Science*, pp. 85–96. Springer, 2005.

[34] Florian Hess, Nigel P. Smart, and Frederik Vercauteren. The Eta pairing revisited. *IEEE Transactions on Information Theory*, 52(10):4595–4602, 2006.

[35] Otto Hesse. Über die Elimination der Variabeln aus drei algebraischen Gleichungen vom zweiten Grade mit zwei Variabeln. *Journal für die reine und angewandte Mathematik*, 10:68–96, 1844.

[36] Hüseyin Hişil. *Elliptic curves, group law, and efficient computation*. PhD thesis, Queensland University of Technology, 2010.

[37] Hüseyin Hisil, Gary Carter, and Ed Dawson. New formulae for efficient elliptic curve arithmetic. In K. Srinathan, C. P. Rangan, and M. Yung, editors, *Progress in Cryptology – INDOCRYPT 2007*, volume 4859 of *Lecture Notes in Computer Science*, pp. 138–151. Springer, Heidelberg, 2007.

[38] Hüseyin Hisil, Kenneth Koon-Ho Wong, Gary Carter, and Ed Dawson. Twisted Edwards curves revisited. In J. Pieprzyk, editor, *Advances in Cryptology – ASIACRYPT 2008*, volume 5350 of *Lecture Notes in Computer Science*, pp. 326–343. Springer, Heidelberg, 2008.

[39] Hüseyin Hisil, Kenneth Koon-Ho Wong, Gary Carter, and Ed Dawson. Jacobi quartic curves revisited. In C. Boyd and J. M. G. Nieto, editors, *Information Security and Privacy (ACISP 2009)*, volume 5594 of *Lecture Notes in Computer Science*, pp. 452–468. Springer, Heidelberg, 2009.

[40] Marc Joye and Jean-Jacques Quisquater. Hessian elliptic curves and side-channel attacks. In Ç. K. Koç, D. Naccache, and C. Paar, editors, *Cryptographic Hardware and Embedded Systems – CHES 2001*, volume 2162 of *Lecture Notes in Computer Science*, pp. 402–410. Springer, Heidelberg, 2001.

[41] Marc Joye, Mehdi Tibouchi, and Damien Vergnaud. Huff's model for elliptic curves. In G. Hanrot, F. Morain, and E. Thomé, editors, *Algorithmic Number Theory (ANTS-IX)*, volume 6197 of *Lecture Notes in Computer Science*, pp. 234–250. Springer, 2010.

[42] Ezekiel J. Kachisa, Edward F. Schaefer, and Michael Scott. Constructing Brezing-Weng pairing-friendly elliptic curves using elements in the cyclotomic field. In S. D. Galbraith and K. G. Paterson, editors, *Pairing-Based Cryptography – Pairing 2008*, volume 5209 of *Lecture Notes in Computer Science*, pp. 126–135. Springer, Heidelberg, 2008.

[43] Koray Karabina. Squaring in cyclotomic subgroups. *Mathematics of Computation*, 82(281):555–579, 2013.

[44] Emilia Käsper. Fast elliptic curve cryptography in OpenSSL. In G. Danezis, S. Dietrich, and K. Sako, editors, *Financial Cryptography and Data Security (FC 2011 Workshops, RLCPS and WECSR 2011)*, volume 7126 of *Lecture Notes in Computer Science*, pp. 27–39. Springer, Heidelberg, 2012.

[45] Neal Koblitz and Alfred Menezes. Pairing-based cryptography at high security levels (invited paper). In N. P. Smart, editor, *Cryptography and Coding*, volume 3796 of *Lecture Notes in Computer Science*, pp. 13–36. Springer, Heidelberg, 2005.

[46] Arjen K. Lenstra and Eric R. Verheul. The XTR public key system. In M. Bellare, editor, *Advances in Cryptology – CRYPTO 2000*, volume 1880 of *Lecture Notes in Computer Science*, pp. 1–19. Springer, Heidelberg, 2000.

[47] Chae Hoon Lim and Pil Joong Lee. More flexible exponentiation with precomputation. In Y. Desmedt, editor, *Advances in Cryptology – CRYPTO '94*, volume 839 of *Lecture Notes in Computer Science*, pp. 95–107. Springer, Heidelberg, 1994.

[48] Seongan Lim, Seungjoo Kim, Ikkwon Yie, Jaemoon Kim, and Hongsub Lee. XTR extended to $GF(p^{6m})$. In S. Vaudenay and A. M. Youssef, editors, *Selected Areas in Cryptography (SAC 2001)*, volume 2259 of *Lecture Notes in Computer Science*, pp. 301–312. Springer, Heidelberg, 2001.

[49] Victor S. Miller. The Weil pairing, and its efficient calculation. *Journal of Cryptology*, 17(4):235–261, 2004.

[50] Peter L. Montgomery. Speeding the Pollard and elliptic curve methods of factorization. *Mathematics of Computation*, 48(177):243–264, 1987.

[51] Michael Naehrig, Paulo S. L. M. Barreto, and Peter Schwabe. On compressible pairings and their computation. In S. Vaudenay, editor, *Progress in Cryptology – AFRICACRYPT 2008*, volume 5023 of *Lecture Notes in Computer Science*, pp. 371–388. Springer, Heidelberg, 2008.

[52] Bryan Parno, Jon Howell, Craig Gentry, and Mariana Raykova. Pinocchio: Nearly practical verifiable computation. In *2013 IEEE Symposium on Security and Privacy*, pp. 238–252. IEEE Computer Society Press, 2013.

[53] Geovandro C. C. F. Pereira, Marcos A. Simplício, Jr., Michael Naehrig, and Paulo S. L. M. Barreto. A family of implementation-friendly BN elliptic curves. *Journal of Systems and Software*, 84(8):1319–1326, 2011.

[54] Michael Scott. Unbalancing pairing-based key exchange protocols. Cryptology ePrint Archive, Report 2013/688, 2013. http://eprint.iacr.org/2013/688.

[55] Michael Scott and Paulo S. L. M. Barreto. Compressed pairings. In M. Franklin, editor, *Advances in Cryptology – CRYPTO 2004*, volume 3152 of *Lecture Notes in Computer Science*, pp. 140–156. Springer, Heidelberg, 2004.

[56] Joseph H. Silverman. *The Arithmetic of Elliptic Curves*, volume 106 of *Graduate Texts in Mathematics*. Springer-Verlag, 2nd edition, 2009.

[57] Nigel P. Smart. The hessian form of an elliptic curve. In Ç. K. Koç, D. Naccache, and C. Paar, editors, *Cryptographic Hardware and Embedded Systems – CHES 2001*, volume 2162 of *Lecture Notes in Computer Science*, pp. 118–125. Springer, Heidelberg, 2001.

[58] Benjamin Smith. Easy scalar decompositions for efficient scalar multiplication on elliptic curves and genus 2 Jacobians. In S. Ballet, M. Perret, and A. Zaytsev, editors, *Algorithmic Arithmetic, Geometry, and Coding Theory*, volume 637 of *Contemporary Mathematics*, pp. 127–145. AMS, 2015.

[59] Martijn Stam and Arjen K. Lenstra. Speeding up XTR. In C. Boyd, editor, *Advances in Cryptology – ASIACRYPT 2001*, volume 2248 of *Lecture Notes in Computer Science*, pp. 125–143. Springer, Heidelberg, 2001.

7

Final Exponentiation

Jean-Luc Beuchat
ELCA Informatique SA

Luis J. Dominguez Perez
CONACyT / CIMAT-ZAC

Laura Fuentes-Castaneda
Intel México

Francisco Rodriguez-Henriquez
CINVESTAV-IPN

Let us define k, the embedding degree of the elliptic curve $E(\mathbb{F}_p)$, as the smallest integer such that $r|p^k - 1$, where r is a large prime divisor of $\#E(\mathbb{F}_p)$. Then, as was already presented in Chapter 1 and studied in more detail in Chapter 3, an asymmetric bilinear pairing is defined as the mapping $\hat{e} : \mathbb{G}_2 \times \mathbb{G}_1 \to \mathbb{G}_T$, where the groups $\mathbb{G}_1$ and $\mathbb{G}_2$ are given as, $\mathbb{G}_1 = E(\mathbb{F}_p)[r]$, $\mathbb{G}_2 = \tilde{E}(\mathbb{F}_{p^{k/d}}[r])$, and $\mathbb{G}_T$ is the subgroup of $\mathbb{F}_{p^k}^*$ where each one of its elements is r-th roots of the unity.

By far the most popular cryptographic instantiation of a bilinear pairing is the Tate pairing or variants of it. The standard procedure to compute the Tate pairing and its variants is divided into two major steps. First, the Miller function value $f = f_{r,P}(Q) \in \mathbb{F}_{p^k}^*$ is calculated. This gives an output value f that belongs to the quotient group $\mathbb{F}_{p^k}^*/(\mathbb{F}_{p^k}^*)^r$. Second, in order to obtain a unique representative in this quotient group, the value f is raised to the power $e = (p^k - 1)/r$. This last step is known as the *final exponentiation* step.

A naive way to compute the final exponentiation consists of representing the exponent e mentioned above in base two, and then invoking the binary exponentiation procedure shown in Algorithm 7.1 to compute the power f^e.

Nevertheless, in most of the pairing-friendly elliptic curves studied in this book, the size in bits of the field characteristic p and the order of the pairing groups r, are about the same, which implies that the size in bits of the exponent e is very close to $\ell = (k - 1)|p|$ bits, where $|p| = \lceil \log_2(p) \rceil$ represents the bit length of the prime p. Since the computational complexity of Algorithm 7.1 is given as approximately $\ell - 1$ and $\frac{\ell}{2}$ squarings and multiplications over the

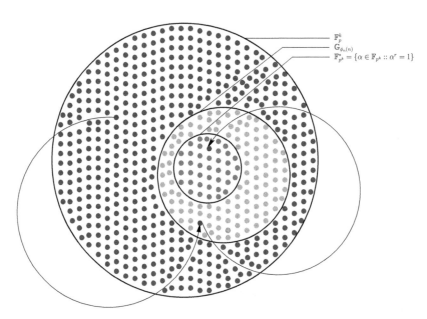

$$\mathbb{F}_p^k$$
$$\mathbb{G}_{\phi_n(n)}$$
$$\mathbb{F}_{p^k}^* = \{\alpha \in \mathbb{F}_{p^k} :: \alpha^r = 1\}$$

FIGURE 7.1 Final exponentiation of an arbitrary element in $\mathbb{F}_{p^k}$ to the group of the r-th roots of unity, μ_r.

ALGORITHM 7.1 Binary method for exponentiation.

Input : $f, x = (x_{n-1}, x_{n-2}, \ldots, x_1, x_0)$	**2 for** $(i = n - 2)$ ***down to*** 0 **do**
Output: f^x	3 $\quad$ $g \leftarrow g^2$;
	4 $\quad$ **if** $x_i = 0$ **then**
1 $g \leftarrow f$;	5 $\quad\quad$ $g = g \cdot f$;
	6 $\quad$ **end**
	7 end
	8 return g;

---

field $\mathbb{F}_{p^k}^*$, this would be a costly computation, considerably more expensive than the cost of calculating the Miller function value f.

In this chapter we will review several techniques that will allow us to greatly reduce the computational cost of the final exponentiation step. The main idea is illustrated in Figure 7.1, where an arbitrary field element $f \in \mathbb{F}_{p^k}^*$ is first mapped to an element f' that belongs to the cyclotomic subgroup $\mathbb{G}_{\Phi_k(p)}$ (see its definition in Section 5.3 of Chapter 5). Then, f' is raised to the power $\Phi_k(p)/r$ to obtain a field element that belongs to μ_r, the subgroup of the r-th roots of the unity. This finishes the final exponentiation step, as the resulting element belongs to the pairing group $\mathbb{G}_T$.

The chapter is organized as follows. In §7.1, a general strategy for computing the final exponentiation is outlined. The important cases of the Barreto–Naehrig and the Kachisa-Schaefer-Scott pairing-friendly elliptic curves are carefully studied in §7.2 and §7.3. Then, in §7.4 the final exponentiation computation for other families of pairing-friendly curves, relevant for implementing pairings at higher security levels, are presented. In §7.5, we compare the computational costs of the different approaches studied in this chapter, and make some concluding remarks.

Finally, Appendix 7.6 presents our SAGE implementation of some of the algorithms discussed in this chapter.

The interested reader can download our complete SAGE library for the final exponentiation step, covering several families of pairing-friendly curves, from:
http://sandia.cs.cinvestav.mx/Site/GuidePBCcode.

7.1 General Approach for Computing the Final Exponentiation

Let $E/\mathbb{F}_p$, be a pairing-friendly elliptic curve such that r is a large prime divisor of $\#E(\mathbb{F}_p)$. The computation of the final exponentiation can be highly optimized by taking advantage of the definition of the embedding degree k. Indeed, the fact that k is the smallest integer such that $r|p^k - 1$ implies that

$$p^k \equiv 1 \bmod r \text{ and } p^i \not\equiv 1 \bmod r, \ \forall i \in \mathbb{Z}^+ \mid i < k.$$

In other words, p is a primitive k^{th} root of the unity modulo r. Let $\Phi_k(z)$ be the k-th cyclotomic polynomial (see §5.3 of Chapter 5). Since the roots of $\Phi_k(z)$ are the kth primitive roots of the unity, it holds that $\Phi_k(p) \equiv 0 \bmod r$, i.e. $r|\Phi_k(p)$.

Moreover, since Φ_k divides the polynomial $z^k - 1$, and since the set of the primitive k^{th} roots of the unity in $\Phi_k(p)$ is a subset of the roots of the polynomial $p^k - 1$, it follows that $\Phi_k(p)|p^k - 1$. As pointed out in [18], this observation allows us to break the exponent $e = (p^k - 1)/r$ into two parts, as

$$e = (p^k - 1)/r = [(p^k - 1)/\Phi_k(p)] \cdot [\Phi_k(p)/r]. \tag{7.1}$$

Computing the map $f \mapsto f^{(p^k-1)/\Phi_k(p)}$ is called the easy part of the final exponentiation since it is relatively inexpensive. On the other hand, raising to the power $d = \Phi_k(p)/r$ is considered the hard part of the final exponentiation.

The main early efforts for improving the efficiency of the final exponentiation computation were presented in [2, 18, 10], whereas the current state-of-the-art procedures for computing the hard part of the final exponentiation can be found in [6, 8, 9, 15, 16, 17, 21, 22]. The remainder of this section is devoted to presenting a general approach for the efficient computation of the easy and hard part of the final exponentiation.

7.1.1 Easy Part of the Final Exponentiation

Although the first part of the final exponentiation is considered inexpensive, its computation carries out important consequences to the rest of the computation.

From Definition 5.3 of Chapter 5, the exponent $(p^k - 1)/\Phi_k(p)$ can be represented as

$$(p^k - 1)/\Phi_k(p) = \prod_{j|k, j<k} \Phi_j(p), \tag{7.2}$$

where $j \in \mathbb{Z}^+$. Let f be a non-zero field element of the extension field $\mathbb{F}_{p^k}$. Then, since for a small j, $\Phi_j(p)$ is a polynomial evaluated in p with coefficients in $\{-1, 0, 1\}$, the cost of computing $f^{\prod \Phi_j(p)}$ is of only some few multiplications, inversions, and very cheap p-th powerings in $\mathbb{F}_{p^k}$, which are equivalent to applications of the Frobenius operator.

Moreover, whenever the embedding degree k is an even number, the exponent $(p^k - 1)/\Phi_k(p)$ can be broken into two parts, as

$$\frac{p^k - 1}{\Phi_k(p)} = (p^{k/2} - 1) \cdot \left[\frac{p^{k/2} + 1}{\Phi_k(p)}\right]. \tag{7.3}$$

Koblitz and Menezes introduced in [18] the concept of Pairing-Friendly Fields (PFFs). A PFF is an extension field $\mathbb{F}_{p^k}$, with $p \equiv 1 \bmod 12$ and $k = 2^i 3^j$ with $i \geq 1, j \geq 0$. In the case that $6|k$, the k-th cyclotomic polynomial Φ_k has the following structure [11],

$$\Phi_{2^i 3^j}(z) = z^{2 \cdot 2^{i-1} 3^{j-1}} - z^{2^{i-1} 3^{j-1}} + 1,$$

which implies,

$$\Phi_k(p) = p^{k/3} - p^{k/6} + 1. \tag{7.4}$$

Substituting the value of $\Phi_k(p)$ from Eq. 7.4 into Eq. 7.2, and after performing some algebra, one obtains,

$$(p^k - 1)/\Phi_k(p) = (p^{k/2} - 1) \cdot (p^{k/6} + 1). \tag{7.5}$$

The exponentiation $g = f^{(p^{k/2}-1)(p^{k/6}+1)} \in \mathbb{F}_{p^k}$, is known as the easy part of the final exponentiation.

Remark 7.1   The multiplicative inverse of the field element $g = f^{(p^k-1)/\Phi_k(p)}$ can be computed by a simple conjugation. Recall that for an even embedding degree k, we have that $p^k - 1 = (p^{k/2} - 1) \cdot (p^{k/2} + 1)$. Let us define $\tilde{g} = f^{p^{k/2}-1}$. Then, using the fact that the order of $f \in \mathbb{F}_{p^k}$ does not divide $p^{k/2} - 1$, one has that

$$(\tilde{g})^{p^{k/2}+1} = (f)^{(p^{k/2}-1) \cdot (p^{k/2}+1)} = 1.$$

From the above equation one concludes that $(\tilde{g})^{p^{k/2}} = 1/\tilde{g}$. Notice that by choosing a suitable quadratic irreducible polynomial with a primitive root i, the field $\mathbb{F}_{p^k}$ can be seen as a quadratic extension of $\mathbb{F}_{p^{k/2}}$. This field towering permits us to represent the element $\tilde{g}$ as $\tilde{g} = \tilde{g}_0 + i \cdot \tilde{g}_1$, where $\tilde{g}_0, \tilde{g}_1 \in \mathbb{F}_{p^{k/2}}$. By taking advantage of the Frobenius operator properties, the operation $(\tilde{g})^{p^{k/2}}$ can be performed by a simple conjugation of the element $\tilde{g}$, as

$$(\tilde{g})^{p^{k/2}} = \tilde{g}_0 - i \cdot \tilde{g}_1 = 1/\tilde{g}.$$

Remark 7.2   The field element $g = f^{(p^k-1)/\Phi_k(p)}$ is a member of the k-th cyclotomic group $G_{\Phi_k(p)}$. This follows directly from the definition of the cyclotomic group,

$$g^{\Phi_k(p)} = (f^{(p^k-1)/\Phi_k(p)})^{\Phi_k(p)} = 1.$$

Remark 7.3   The field element $g = f^{(p^k-1)/\Phi_k(p)}$ is unitary. A direct inspection of the norm definition as stated in Definition 2.27 of Chapter 2 along with the properties of the Frobenius operator shows that the norm of the element $g \in \mathbb{F}_{p^k}$ with respect to $\mathbb{F}_p$ is given as $|g| = 1$

7.1.2   Hard Part of the Final Exponentiation

As we have seen in Chapter 4, the integer curve parameters p, t, and r of a family of pairing-friendly elliptic curves are represented as polynomials in the ring $\mathbb{Q}[x]$, where $x \in \mathbb{Z}$. In order to instantiate a particular pairing-friendly elliptic curve, one needs to identify an integer value x_0, such that both polynomials $p(x)$ and $r(x)$, yield prime numbers when evaluated at x_0 (see §4.5 of Chapter 4).

Let $\varphi(\cdot)$ denote the Euler Totient function and let $\deg p(x)$ denote the degree of the polynomial $p(x)$. Taking advantage of this polynomial representation of the parameters of the curve, and using the fact that a p-th powering is considerably cheaper than a multiplication in $\mathbb{F}_{p^k}$, the exponent $d = \Phi_k(p)(x)/r(x)$, corresponding to the hard part of the final exponentiation, is usually represented as a polynomial in base $p(x)$ of degree $\varphi(k) - 1$ as

$$d(x) = \sum_{i=0}^{\varphi(k)-1} \lambda_i(x)p^i(x), \tag{7.6}$$

where

$$\lambda_i(x) = \sum_{j=0}^{\deg p(x)-1} \lambda_{ij}x^j. \tag{7.7}$$

Using the polynomial representation of the exponent d as given in Equation (7.6), the powering f^d can be computed efficiently by once again exploiting the cheap cost of the Frobenius operator combined with vectorial addition chain methods as briefly described next (see § 7.2 for a full example).

In a precomputation phase, the exponentiations $f^x, \ldots, f^{x^{\deg p(x)-1}}$ are calculated at a cost of $\frac{\deg p(x)-1}{\deg p(x)} \log p$ squarings plus about $(\deg p(x) - 1) \cdot (Hw(x) - 1)$ field multiplications, where $Hw(x)$ stands for the Hamming weight of x. From these intermediate exponentiations, terms of the form $f^{x^i p^j(x)}$ for $0 \le i < \deg p(x)$ and $0 \le j < \varphi(k)$ can be computed efficiently at a cost of some few extra multiplications. In a second phase, a vectorial addition chain is obtained from a valid addition-subtraction sequence that contains all the distinct coefficients $\lambda_{i,j} \in \mathbb{Z}$.*

Assuming that there are a total of s non-zero coefficients $\lambda_{i,j}$, and that the length of the corresponding addition–subtraction sequence is l, then the conversion of it to a vectorial addition chain can be accomplished with $l + s - 1$ steps, by means of the Olivos algorithm [20]. Finally, the powering $f^{d(x)}$ can be obtained by a simultaneous multi-exponentiation involving factors of the form $f^{x^i p^j}$.

A lattice-based refinement

In [9], the authors refined the procedure outlined above by presenting a lattice-based method that finds a multiple d' of d, with r not dividing d', such that $f^{d'}$ can be computed at least as efficiently as f^d.

Let us consider $f \in \mathbb{F}_{p^k}$ with order dividing $\Phi_k(p)$. Since $r(x)d(x) = \Phi_k(p)$, it follows that $f^{r(x)d(x)} = 1$. Hence, to find the multiple $d'(x)$, it suffices to consider a matrix M that includes linear combinations of the monomials, $d(x), xd(x), \ldots, x^{\phi(k)-1}d(x)$.

To this aim, consider a rational matrix M' with dimensions $\deg p \times \varphi(k) \deg p$ such that

$$\begin{bmatrix} d(x) \\ xd(x) \\ \vdots \\ x^{\deg p-1}d(x) \end{bmatrix} = M' \left(\begin{bmatrix} 1 \\ p(x) \\ \vdots \\ p(x)^{\varphi(k)-1} \end{bmatrix} \otimes \begin{bmatrix} 1 \\ x \\ \vdots \\ x^{\deg p-1} \end{bmatrix} \right).$$

---

*A vectorial addition chain is a list of vectors where each vector is the addition of two previous vectors. In the case of a group exponentiation, the last vector contains the final exponent e. Let V be a vector chain, and m be the dimension of every vector. The vector addition chain starts with $V_{i,i} = 1$ for $i = 0, \ldots, m - 1$: $[1, 0, 0, \ldots, 0], [0, 1, 0, \ldots, 0], \ldots, [0, \ldots, 0, 1]$; then the next vector is formed by adding any two previous vectors in the chain, and this process is continued until $V_{j,1} = e$, with $j > m$.
Olivos presented in [20] a procedure that converts a valid addition sequence $\{e_1, \ldots, e_s\}$ of length l to a vectorial addition chain of length $l+s-1$, for $[e_1, \ldots, e_s]$ (See [4] for a historical recount of this algorithm). Since in [20] a procedure that transforms a vectorial addition chain to an addition sequence was also reported, these two structures can be seen as the dual of each other [5, Chapter 9]. The combination of addition–subtraction sequences along with vectorial addition chains lead to efficient computations of the product of simultaneous exponentiations.

Here the symbol '$\otimes$' stands for the Kronecker product and the elements in the rational lattice formed by the matrix M' correspond to $\mathbb{Q}$-linear combinations $d'(x)$ of $d(x)$, $xd(x)$, ..., $x^{\deg r-1}d(x)$.

Thereafter, the Lenstra-Lenstra-Lovász algorithm [19] is applied to M in order to obtain an integer basis for M with small entries. The hope is that short vectorial addition chains can be extracted from linear combinations of these basis elements. By construction, it is guaranteed that one of the solutions found by such a method will match the ones reported in Scott et al. [21] using $d(x)$ as the exponent.

Allowing multiples of $d(x)$ is advantageous, because it provides more flexibility in the choices of the coefficients $\lambda_{i,j}$, which in many instances (as in the case of the pairing-friendly curve KSS-16 to be reviewed here) can yield modest but still noticeable savings in the computation of the hard part of the final exponentiation.

In the next section, this method is explained in detail by applying it to the computation of the final exponentiation step on several pairing-friendly families of curves.

7.2 The Barreto-Naehrig Family of Pairing–Friendly Elliptic Curves

In this section, the main methods for computing the hard part of the final exponentiation for the Barreto-Naehrig family of curves with embedding degree k=12 will be analyzed in detail. Let us recall that Barreto–Naehrig elliptic curves are defined by the equation $E: y^2 = x^3 + b$, with $b \neq 0$. Their embedding degree is $k = 12$, hence, $\phi(k) = 4$. They are parametrized by selecting an arbitrary $x \in \mathbb{Z}$ such that

$$p(x) = 36x^4 + 36x^3 + 24x^2 + 6x + 1, \tag{7.8}$$
$$r(x) = 36x^4 + 36x^3 + 18x^2 + 6x + 1.$$

are both prime.

Furthermore, notice that the twelfth cyclotomic polynomial is given as $\Phi_k(z) = p^4 - p^2 + 1$. Hence the exponent d for the hard part of the final exponentiation is given as $d = \frac{p^4-p^2+1}{r}$.

The rest of this section is organized as follows. We start by discussing the Devegili et al. method, which was proposed in 2007 [6]. Then, we discuss a more elaborated approach that was later presented by Scott et al. in [21]. Finally, we present the Fuentes et al. lattice approach [9], which is arguably the current state-of-the-art procedure for computing the hard part of the final exponentiation.

7.2.1 Devegili et al. Method for Barreto-Naehrig Curves

Devegili et al. showed in [6] that the exponent $d = \frac{p^4-p^2+1}{r}$, corresponding to the hard part of the final exponentiation on Barreto-Naehrig curves, can be expressed in base p as

$$d(x) = -36x^3 + 30x^2 + 18x + 2 + p(x)(-36x^3 - 18x^2 - 12x + 1) \tag{7.9}$$
$$+ p(x)^2(6x^2 + 1) + p(x)^3.$$

Devegili et al. proposed in [6] to decompose Equation (7.9) as

$$\begin{aligned} d(x) = &-36x^3 + 30x^2 + 18x + 2 + p(x)(-36x^3 - 18x^2 - 12x + 1) + p(x)^2(6x^2 + 1) + p(x)^3 \\ = &\, 4 + 9 + (-6x - 5) + (-6x - 5) + (-6x - 5) \cdot (6x^2 + 1) + \\ &\, p(x)\left[(-6x - 5) \cdot (6x^2 + 1) + 2(6x^2 + 1) + (-6x - 5) + 9\right] + \\ &\, p(x)^2(6x^2 + 1) + p(x)^3. \end{aligned}$$

The above expression can be computed efficiently as follows. First, the following series of exponentiations is computed:

$$f \mapsto f^x \mapsto f^{2x} \mapsto f^{2x+2} \mapsto f^{4x+4} \mapsto f^{6x+4} \mapsto f^{6x+5} \mapsto f^{-6x-5}. \tag{7.10}$$

The computation of Equation 7.10 requires 1 exponentiation by x, 3 squarings, and 3 multiplications, plus one inexpensive conjugation, which is a consequence of Remark 7.1. One can then define the terms $a = f^{-6x-5}$ and $b = a \cdot a^p$, at the cost of one extra multiplication and one Frobenius application. Finally, the result f^d is obtained by computing

$$f^{p^3} \cdot \left[f^{p^2} \cdot b \cdot (f^p)^2 \right]^{6x^2+1} \cdot b \cdot (f^p \cdot f)^9 \cdot a \cdot f^4,$$

which requires 2 exponentiations by x, 8 squarings, 11 multiplications, and 3 Frobenius applications.

In total, this method requires 3 exponentiations by x, 11 squarings, and 14 multiplications. Algorithm 7.2 shows an explicit computation of this method.

ALGORITHM 7.2   Final exponentiation by Devegili et al. for BN curves [6].

Input : $f \in \mathbb{F}_{p^k}$, $x \in \mathbb{Z}$

Output: $f^{\frac{\Phi_k(p)}{r}} \in \mathbb{F}_{p^k}$

1   $a = f^x$; $\{\, a \leftarrow f^x \,\}$
2   $b = a^2$; $\{\, b \leftarrow f^{2x} \,\}$
3   $a = b \cdot f^2$; $\{\, a \leftarrow f^{2x+2} \,\}$
4   $a = a^2$; $\{\, a \leftarrow f^{4x+4} \,\}$
5   $a = a \cdot b$; $\{\, a \leftarrow f^{6x+4} \,\}$
6   $a = a \cdot f$; $\{\, a \leftarrow f^{6x+5} \,\}$
7   $a = \bar{a}$; $\{\, a \leftarrow f^{-6x-5} \,\}$
8   $b = a^p$; $\{\, b \leftarrow a^p \,\}$
9   $b = a \cdot b$; $\{\, b \leftarrow a \cdot b \,\}$
10   $a = a \cdot b$; $\{\, a \leftarrow a \cdot b \,\}$
11   $t_0 = f^p$;
12   $t_1 = t_0 \cdot f$;
13   $t_1 = t_1^9$;
14   $a = t_1 \cdot a$; $\{\, a \leftarrow b \cdot (f^p \cdot f)^9 \cdot a \,\}$
15   $t_1 = f^4$;
16   $a = a \cdot t_1$; $\{\, a \leftarrow b \cdot (f^p \cdot f)^9 \cdot a \cdot f^4 \,\}$
17   $t_0 = t_0^2$; $\{\, t_0 \leftarrow f^{p^2} \,\}$
18   $b = b \cdot t_0$; $\{\, b \leftarrow b \cdot (f^p)^2 \,\}$
19   $t_0 = f^{p^2}$;

20   $b = b \cdot t_0$; $\{\, b \leftarrow b \cdot (f^p)^2 \cdot f^{p^2} \,\}$
21   $t_0 = b^x$; $\left\{\, t_0 \leftarrow \left[b \cdot (f^p)^2 \cdot f^{p^2} \right]^x \,\right\}$
22   $t1 = t_0^2$; $\left\{\, t_0 \leftarrow \left[b \cdot (f^p)^2 \cdot f^{p^2} \right]^{2x} \,\right\}$
23   $t_0 = t_1^2$; $\left\{\, t_0 \leftarrow \left[b \cdot (f^p)^2 \cdot f^{p^2} \right]^{4x} \,\right\}$
24   $t_0 = t_0 \cdot t_1$; $\left\{\, t_0 \leftarrow \left[b \cdot (f^p)^2 \cdot f^{p^2} \right]^{6x} \,\right\}$
25   $t_0 = t_0^x$; $\left\{\, t_0 \leftarrow \left[b \cdot (f^p)^2 \cdot f^{p^2} \right]^{6x^2} \,\right\}$
26   $t_0 = t_0 \cdot b$; $\left\{\, t_0 \leftarrow \left[b \cdot (f^p)^2 \cdot f^{p^2} \right]^{6x^2+1} \,\right\}$
27   $a = t_0 \cdot a$;
     $\left\{\, \left[b \cdot (f^p)^2 \cdot f^{p^2} \right]^{6x^2+1} \cdot b \cdot (f^p \cdot f)^9 \cdot a \cdot f^4 \,\right\}$
28   $t_0 = f^{p^3}$;
29   $f = t_0 \cdot a$;
     $\left\{\, f^{p^3} \cdot \left[b \cdot (f^p)^2 \cdot f^{p^2} \right]^{6x^2+1} \cdot b \cdot (f^p \cdot f)^9 \cdot a \cdot f^4 \,\right\}$
30   **return** f;

7.2.2   The Scott et al. Method for Barreto-Naehrig Curves

Notice that the exponent $d = \Phi_k(p)/r$, corresponding to the hard part of the final exponentiation, is a degree-twelve polynomial in the variable x, given as

$$d(x) = 46656x^{12} + 139968x^{11} + 241056x^{10} + 272160x^9 + \tag{7.11}$$
$$225504x^8 + 138672x^7 + 65448x^6 + 23112x^5 + 6264x^4 + 1188x^3 + 174x^2 + 6x + 1.$$

At first glance, it would appear that in order to compute f^d, with $f \in \mathbb{F}_{p^k}$, it is necessary to exponentiate by multiples of powers of x^i, for $i = 1, \ldots, 12$. However, following the method reported in Scott et al. [21], it becomes more efficient to write d in base p as $d(x) = \sum_{i=0}^{3} \lambda_i(x)p(x)^i$, where the coefficients λ_i for $i = 0, \ldots 3$ are polynomials with degree at most 3, given as

$$\lambda_0 = -36x^3 - 30x^2 - 18x - 2, \tag{7.12}$$
$$\lambda_1 = -36x^3 - 18x^2 - 12x + 1,$$
$$\lambda_2 = 6x^2 + 1,$$
$$\lambda_3 = 1.$$

Using the above representation of the exponent $d(x)$ in base p, the hard part of the final exponentiation can be computed as follows. First, three exponentiations of the form f^{x^i}, for $i = 1, \ldots, 3$, must be sequentially computed. Then, the Frobenius operator is applied to pre-compute the following 7 factors

$$f^p, f^{p^2}, f^{p^3}, \left(f^x\right)^p, \left(f^{x^2}\right)^p, \left(f^{x^3}\right)^p, \left(f^{x^2}\right)^{p^2}.$$

Notice that in Equation (7.12) the coefficient 1 appears three times, namely, in the equalities for λ_1, λ_2 and λ_3. Similarly, the coefficients 18 and 36 appear two times each. Those coefficients can be grouped as products of the factors $(f^{x^i})^{p^j}$ previously mentioned. The variables y_i, for $i = 0, \ldots, 6$ are used to group the coefficients $1, 2, 6, 12, 18, 30$, and 36, respectively, as follows,

$$y_0 = f^p \cdot f^{p^2} \cdot f^{p^3}, \qquad\qquad y_1 = \frac{1}{f}, \qquad\qquad y_2 = (f^{x^2})^{p^2}, \tag{7.13}$$

$$y_3 = \frac{1}{(f^x)^p}, \qquad\qquad y_4 = \frac{1}{f^x \cdot (f^{x^2})^p}, \qquad\qquad y_5 = \frac{1}{f^{x^2}},$$

$$y_6 = \frac{1}{f^{x^3} \cdot (f^{x^3})^p}.$$

This leads us to the following multi-exponentiation problem,

$$f^d = y_0 \cdot y_1^2 \cdot y_2^6 \cdot y_3^{12} \cdot y_4^{18} \cdot y_5^{30} \cdot y_6^{36}. \tag{7.14}$$

The cost of extracting all the coefficients y_k from Equation (7.12), so that they can be multi-powered as shown in Equation (7.14), is of 4 multiplications plus one negligible field inversion that can be computed by performing a simple conjugation.

The multi-exponentiation problem of Equation (7.14) can be solved using addition chains along with vectorial addition chains. Let us recall that an addition chain of length l is given as $e_0 = 1, e_1 = 2, \ldots e_l = n$, with the property that $e_i = e_j + e_k$, for some $k \leq j < i$, for all $i = 1, 2, \ldots, l$. A vectorial addition chain can be constructed from an addition chain by means of the Olivos algorithm [20] as explained next (for a more formal description the interested reader is referred to [5, Chapter 9]).

To obtain a vectorial addition chain from a given addition chain, the Olivos procedure sequentially builds a two-dimensional array starting in the upper left corner with the two-by-two array,

$$\begin{array}{cc} 1 & 0 \\ 2 & 2 \end{array}$$

Then, the procedure processes each addition chain element e_i, with $i \geq 2$, by adding at each step a new row in the bottom, as well as two columns in the right. If the element e_i being processed corresponds to the doubling of a previous element e_j, then the line to be added in the bottom is the double of the line j and the two new columns in the right are the i-dimensional canonical vectors $2v_i + v_j$ and $2v_i$. Otherwise, the element e_i corresponds to the addition of two previous elements e_j, e_k. Then, the new line on the bottom is the addition of lines j and k, and the two new columns in the right are the i-dimensional canonical vectors $v_i + v_j$ and $v_i + v_k$.

After having processed all the elements in the addition chain, the vectorial addition chain is obtained by scanning the columns in the Olivos array from right to left. Moreover, the underlined rows and all repeated columns are ignored. This guarantees that the dimension of the vectors so obtained is the length s of the original protosequence. The vectorial addition chain is completed by adding all the s-dimensional canonical vectors.

In the case of Equation (7.14), one first finds a valid addition chain for the so-called protosequence of $s = 7$ elements,

$$\{1, 2, 6, 12, 18, 30, 36\}. \tag{7.15}$$

A valid optimal addition chain of length $l = 7$ for the protosequence (7.15) is

$$\{e_0 = 1, e_1 = 2, \underline{e_2 = 3}, e_3 = 6, e_4 = 12, e_5 = 18, e_6 = 30, e_7 = 36\}. \tag{7.16}$$

Notice that the inserted element $e_2 = 3$ is shown above underlined, whereas the other elements in the chain correspond to the original elements in the protosequence. Since $e_2 = e_0 + e_1$, when processing this element, the Olivos array becomes

1	0	1	0
2	2	0	1
3	2	1	1

Similarly, when processing the element $e_3 = 2e_2$, the Olivos array becomes

1	0	1	0	0	0
2	2	0	1	0	0
3	2	1	1	1	0
6	4	2	2	2	2

After having processed all the elements in the addition chain, the Olivos algorithm generates the array shown in Table 7.1. The vectorial addition chain is then obtained by processing the columns from right to left. For instance, the columns c_{13} and c_{12} produce the vectors $[2, 0, 0, 0, 0, 0, 0]$, $[2, 0, 1, 0, 0, 0, 0]$, respectively. Notice that the values corresponding to the row e_2 were ignored. The complete vectorial addition sequence produced by this procedure is shown in Table 7.2.

TABLE 7.1 Vectorial addition chain construction from the addition chain $[1, 2, 3, 6, 12, 18, 30, 36]$.

	c_0	c_1	c_2	c_3	c_4	c_5	c_6	c_7	c_8	c_9	c_{10}	c_{11}	c_{12}	c_{13}
$e_0 = 1$	1	0	1	0	0	0	0	0	0	0	0	0	0	0
$e1 = 2$	2	2	0	1	0	0	0	0	0	0	0	0	0	0
$e_2 = 3$	3	2	1	1	1	0	0	0	0	0	0	0	0	0
$e_3 = 6$	6	4	2	2	2	2	1	0	1	0	0	0	1	0
$e_4 = 12$	12	8	4	4	4	4	2	2	0	1	1	0	0	0
$e_5 = 18$	18	12	6	6	6	6	3	2	1	1	0	1	0	0
$e_6 = 30$	30	20	10	10	10	10	5	4	1	2	1	1	0	1
$e_7 = 36$	36	24	12	12	12	12	6	4	2	2	1	1	1	1

TABLE 7.2   Vectorial addition sequence for the protosequence $[1, 2, 6, 12, 18, 30, 36]$.

1. $[1, 0, 0, 0, 0, 0, 0]$;	8. $[2, 0, 0, 0, 0, 0, 0]$;	15. $[6, 5, 3, 2, 1, 0, 0]$;
2. $[0, 1, 0, 0, 0, 0, 0]$;	9. $[2, 0, 1, 0, 0, 0, 0]$;	16. $[12, 10, 6, 4, 2, 0, 0]$;
3. $[0, 0, 1, 0, 0, 0, 0]$;	10. $[2, 1, 1, 0, 0, 0, 0]$;	17. $[12, 10, 6, 4, 2, 1, 0]$;
4. $[0, 0, 0, 1, 0, 0, 0]$;	11. $[0, 1, 0, 1, 0, 0, 0]$;	18. $[12, 10, 6, 4, 2, 0, 1]$;
5. $[0, 0, 0, 0, 1, 0, 0]$;	12. $[2, 2, 1, 1, 0, 0, 0]$;	19. $[24, 20, 12, 8, 4, 2, 0]$;
6. $[0, 0, 0, 0, 0, 1, 0]$;	13. $[2, 1, 1, 0, 1, 0, 0]$;	20. $[36, 30, 18, 12, 6, 2, 1]$.
7. $[0, 0, 0, 0, 0, 0, 1]$;	14. $[4, 4, 2, 2, 0, 0, 0]$;	

As shown in Algorithm 7.3, the vectorial addition sequence of Table 7.2 can be directly used to compute the multi-exponentiation $y_0 \cdot y_1^2 \cdot y_2^6 \cdot y_3^{12} \cdot y_4^{18} \cdot y_5^{30} \cdot y_6^{36}$. The cost of this calculation is of nine multiplications plus four squarings, which exactly matches the expected computational cost predicted by the Olivos theorem [20], namely, $l + s - 1 = 7 + 7 - 1 = 13$ field operations.

ALGORITHM 7.3   Multi-exponentiation computation using addition-subtraction vectorial chains.

Input : 7 input elements y_i, for $0 \le i \le 6$
Output: $t_0 = y_0 \cdot y_1^2 \cdot y_2^6 \cdot y_3^{12} \cdot y_4^{18} \cdot y_5^{30} \cdot y_6^{36}$

1 $t_0 = y_6^2$; $\{2, 0, 0, 0, 0, 0, 0\}$
2 $t_0 = t_0 \cdot y_4$; $\{2, 0, 1, 0, 0, 0, 0\}$
3 $t_0 = t_0 \cdot y_5$; $\{2, 1, 1, 0, 0, 0, 0\}$
4 $t_1 = y_3 \cdot y_5$; $\{0, 1, 0, 1, 0, 0, 0\}$
5 $t_1 = t_1 \cdot t_0$; $\{2, 2, 1, 1, 0, 0, 0\}$
6 $t_0 = t_0 \cdot y_2$; $\{2, 1, 1, 0, 1, 0, 0\}$
7 $t_1 = t_1^2$; $\{4, 4, 2, 2, 0, 0, 0\}$
8 $t_1 = t_1 \cdot t_0$; $\{6, 5, 3, 2, 1, 0, 0\}$
9 $t_1 = t_1^2$; $\{12, 10, 6, 4, 2, 0, 0\}$
10 $t_0 = t_1 \cdot y_1$; $\{12, 10, 6, 4, 2, 1, 0\}$
11 $t_1 = t_1 \cdot y_0$; $\{12, 10, 6, 4, 2, 0, 1\}$
12 $t_1 = t_1^2$; $\{24, 20, 12, 8, 4, 2, 0\}$
13 $t_0 = t_0 \cdot t_1$; $\{36, 30, 18, 12, 6, 2, 1\}$
14 **return** t_0;

In total, the hard part of the final exponentiation for the Barreto-Naehrig curve requires 3 exponentiations by x, 13 multiplications, 4 squarings, and 7 Frobenius applications. The complete procedure is shown in Algorithm 7.4.

7.2.3   The Fuentes-Castañeda et al. Method for Barreto-Naehrig Curves

Recall that the exponent $d = \frac{p^4 - p^2 + 1}{r}$, corresponding to the hard part of the final exponentiation on Barreto-Naehrig curves, can be expressed as (cf. with Equation (7.12)),

$$d(x) = -36x^3 - 30x^2 - 18x - 2 + p(x)(-36x^3 - 18x^2 - 12x + 1) \qquad (7.17)$$
$$+ p(x)^2(6x^2 + 1) + p(x)^3.$$

The main algorithmic idea of the Fuentes-Castañeda et al. method as presented in [9] is based on the following observation: Since a fixed power of a pairing is a pairing, instead of raising f to the power d, we can always compute $f^{d'}$, where $d' = d \cdot m$. The only restriction that must be obeyed is to select m such that $r \nmid m$.

ALGORITHM 7.4   Final exponentiation by Scott et al. for BN curves [21].

Input : $f \in \mathbb{F}_{p^k}$, $x \in \mathbb{Z}$

Output: $f^{\frac{\phi_k(p)}{r}} \in \mathbb{F}_{p^k}$

1   $a = f^x$;

2   $b = a^x$; $\{ f^{x^2} \}$

3   $c = b^x$; $\{ f^{x^3} \}$

4   $T_0 = \bar{c} \cdot (\bar{c})^p$; $\left\{ y_6 \leftarrow 1 / \left(f^{x^3} \cdot \left(f^{x^3} \right)^p \right) \right\}$

5   $T_0 \leftarrow T_0^2$; $\{ [y_6]^2 \}$

6   $B \leftarrow (\bar{b})^p$; $\left\{ 1/(f^{x^2})^p \right\}$

7   $B \leftarrow B \cdot \bar{a}$; $\left\{ y_4 \leftarrow 1 / \left(f^x \cdot (f^{x^2})^p \right) \right\}$

8   $T_0 \leftarrow T_0 \cdot B$; $\{ [y_6]^2 \cdot [y_4] \}$

9   $B \leftarrow \bar{b}$; $\left\{ y_5 \leftarrow 1/f^{x^2} \right\}$

10   $T_0 \leftarrow T_0 \cdot B$; $\{ [y_6]^2 \cdot [y_5] \cdot [y_4] \}$

11   $T_1 \leftarrow (\bar{a})^p$; $\{ y_3 \leftarrow (1/f^x)^p \}$

12   $T_1 \leftarrow T_1 \cdot B$; $\{ T_1 \leftarrow y_5 \cdot y_3 \}$

13   $T_1 \leftarrow T_1 \cdot T_0$; $\left\{ [y_6]^2 \cdot [y_5]^2 \cdot [y_4] \cdot [y_3] \right\}$

14   $B \leftarrow (b)^{p^2}$; $\left\{ y_2 \leftarrow (f^{x^2})^{p^2} \right\}$

15   $T_0 \leftarrow T_0 \cdot B$; $\left\{ [y_6]^2 \cdot [y_5] \cdot [y_4] \cdot [y_2] \right\}$

16   $T_1 \leftarrow T_1^2$; $\left\{ [y_6]^4 \cdot [y_5]^4 \cdot [y_4]^2 \cdot [y_3]^2 \right\}$

17   $T_1 \leftarrow T_1 \cdot T_0$; $\left\{ [y_6]^6 \cdot [y_5]^5 \cdot [y_4]^3 \cdot [y_3]^2 \cdot [y_2] \right\}$

18   $T_1 \leftarrow T_1^2$;
  $\left\{ [y_6]^{12} \cdot [y_5]^{10} \cdot [y_4]^6 \cdot [y_3]^4 \cdot [y_2]^2 \right\}$

19   $T_0 \leftarrow T_1 \cdot \bar{f}$;
  $\left\{ [y_6]^{12} \cdot [y_5]^{10} \cdot [y_4]^6 \cdot [y_3]^4 \cdot [y_2]^2 \cdot [y_1] \right\}$

20   $B \leftarrow (f)^p \cdot (f)^{p^2} \cdot (f)^{p^3}$; $\{ y_0 \leftarrow f^p \cdot f^{p^2} \cdot f^{p^3} \}$

21   $T_1 \leftarrow T_1 \cdot B$;
  $\left\{ [y_6]^{12} \cdot [y_5]^{10} \cdot [y_4]^6 \cdot [y_3]^4 \cdot [y_2]^2 \cdot [y_0] \right\}$;

22   $T_0 \leftarrow T_0^2$;
  $\left\{ [y_6]^{24} \cdot [y_5]^{20} \cdot [y_4]^{12} \cdot [y_3]^8 \cdot [y_2]^4 \cdot [y_1]^2 \right\}$

23   $f \leftarrow T_0 \cdot T_1$;
  $\left\{ [y_6]^{36} \cdot [y_5]^{30} \cdot [y_4]^{18} \cdot [y_3]^{12} \cdot [y_2]^6 \cdot [y_1]^2 \cdot [y_0] \right\}$

24   **return** f;

Notice that it is more convenient to interpret the selected value m as a polynomial: $m(x) \bmod (r(x))$. Furthermore, observe that

$$f^{d'(x)} = f^{m(x) \cdot d(x)} \equiv f^{m(x) \cdot d(x) + \phi_k(p(x))} = f^{(m(x) + r(x)) \cdot d(x)}. \tag{7.18}$$

As a consequence, in the case of the Barreto-Naehrig curves, one can write $m(x)$ as $m(x) = a_0 + a_1 x + a_2 x^2 + a_3 x^3$. This implies that $\{d(x), xd(x), x^2 d(x), x^3 d(x)\}$ forms a basis for representing all possible products, $m(x) \cdot d(x)$. In particular, the polynomial $xd(x)$, given as

$$\begin{aligned} xd(x) =& 46656x^{13} + 139968x^{12} + 241056x^{11} + 272160x^{10} + \\ & 225504x^9 + 138672x^8 + 65448x^7 + 23112x^6 + \\ & 6264x^5 + 1188x^4 + 174x^3 + 6x^2 + x, \end{aligned}$$

can be rewritten in base $p(x)$ as

$$\begin{aligned} xd(x) =& 6x^3 + 6x^2 + 4x + 1 \\ & + p(x)(18x^3 + 12x^2 + 7x) \\ & + p(x)^2 (6x^3 + x - 1) \\ & + p(x)^3 x. \end{aligned} \tag{7.19}$$

In the case of the term $x^2 d(x)$, its representation in base $p(x)$ leads to fractional coefficients. Hence, it is more convenient to consider $6x^2 d(x)$ instead, which can be written as

$$\begin{aligned} 6x^2 d(x) =& -1 \\ & + p(x)(-36x^3 - 30x^2 - 18x - 2) \\ & + p(x)^2 (-36x^3 - 18x^2 - 12x + 2) \\ & + p(x)^3 (6x^2 + 1). \end{aligned} \tag{7.20}$$

Similarly, we consider the term $6x^3 d(x)$, which can be written in base $p(x)$ as

$$6x^3 d(x) = -x \qquad\qquad (7.21)$$
$$+ p(x)(6x^3 + 6x^2 + 4x + 1)$$
$$+ p(x)^2(18x^3 + 12x^2 + 8x)$$
$$+ p(x)^3(6x^3 + x - 1).$$

Summarizing, each polynomial, $d(x)$, $xd(x)$, $6x^2 d(x)$, and $6x^3 d(x)$ can be written as a degree-three polynomial in $p(x)$ with coefficients of degree at most three in x. This suggests that each one of the four coefficients above can be mapped to a vector in $\mathbb{Z}^{16}$. For example,

$$d(x) = -36x^3 - 30x^2 - 18x - 2$$
$$+ p(x)(-36x^3 - 18x^2 - 12x + 1)$$
$$+ p(x)^2(6x^2 + 1)$$
$$+ p(x)^3$$

can be mapped to $\mathbb{Z}^{16}$ as

$$d(x) \mapsto [-36, -30, -18, -2, -36, -18, -12, 1, 0, 6, 0, 1, 0, 0, 0, 1].$$

In order to represent the four-element basis $\{d(x),\ xd(x),\ 6x^2 d(x),\ \text{and}\ 6x^3 d(x)\}$, one can define a 4×16 integer matrix M given as

$$\begin{bmatrix} d(x) \\ xd(x) \\ 6x^2 d(x) \\ 6x^3 d(x) \end{bmatrix} = M \left(\begin{bmatrix} 1 \\ p(x) \\ p(x)^2 \\ p(x)^3 \end{bmatrix} \otimes \begin{bmatrix} 1 \\ x \\ x^2 \\ x^3 \end{bmatrix} \right), \qquad\qquad (7.22)$$

where, as mentioned in the previous section, the symbol '$\otimes$' stands for the Kronecker product.

Using Equation (7.12), along with Equations (7.19) and (7.21), one obtains the following matrix defined in the integers:

$$M = \begin{bmatrix} -36 & -30 & -18 & -2 & -36 & -18 & -12 & 1 & 0 & 6 & 0 & 1 & 0 & 0 & 0 & 1 \\ 6 & 6 & 4 & 1 & 18 & 12 & 7 & 0 & 6 & 0 & 1 & -1 & 0 & 0 & 1 & 0 \\ 0 & 0 & 0 & -1 & -36 & -30 & -18 & -2 & -36 & -18 & -12 & 2 & 0 & 6 & 0 & 1 \\ 0 & 0 & -1 & 0 & 6 & 6 & 4 & 1 & 18 & 12 & 8 & 0 & 6 & 0 & 1 & -1 \end{bmatrix}.$$

Notice that the first row in M corresponds to the final exponentiation analized by Scott et al. [21]. Any non-trivial integer linear combination of the rows corresponds to an exponent that produces an element f of the desired order r.

For computational efficiency, a linear combination with coefficients as small as possible is desired. This can be attained by means of the LLL algorithm [19], which can find the shorter vector in the M lattice. After running the LLL algorithm on the matrix M, the following solution was found:

$$\begin{bmatrix} x_0 \\ x_1 \\ x_2 \\ x_3 \end{bmatrix} = \begin{bmatrix} x_0 = \\ x_1 = \\ x_2 = \\ x_3 = \end{bmatrix} \begin{bmatrix} -12 & -6 & -3 & 1 & 6 & 6 & 2 & 0 & 6 & 0 & 0 & -1 & 6 & 6 & 5 & 1 \\ 6 & 6 & 4 & 1 & 18 & 12 & 7 & 0 & 6 & 0 & 1 & -1 & 0 & 0 & 1 & 0 \\ -12 & -6 & -2 & 1 & 0 & 0 & -2 & -1 & -12 & -12 & -8 & -1 & 0 & 6 & 4 & 2 \\ -12 & -12 & -7 & -1 & -6 & 0 & 0 & 1 & 6 & 6 & 2 & 0 & -6 & -6 & -3 & 0 \end{bmatrix}.$$

A visual inspection reveals that the linear combination with smallest coefficients corresponds to the first row of the above solution matrix, namely,

$$\begin{bmatrix} -12 & -6 & -3 & 1 & 6 & 6 & 2 & 0 & 6 & 0 & 0 & -1 & 6 & 6 & 5 & 1 \end{bmatrix}.$$

The above linear combination can be translated to polynomials in the base x as

$$\begin{aligned}
\lambda_0 &= -12x^3 - 6x^2 - 3x + 1, \\
\lambda_1 &= 6x^3 + 6x^2 + 2x, \\
\lambda_2 &= 6x^3 - 1, \\
\lambda_3 &= 6x^3 + 6x^2 + 5x + 1
\end{aligned}$$

where

$$d'(x) = \lambda_0 + \lambda_1 p + \lambda_2 p^2 + \lambda_3 p^3.$$

Applying the method studied in the previous subsection, we now need to compute the simultaneous exponentiation:

$$f^{d'(x)} = y_0^1 \cdot y_1^2 \cdot y_2^3 \cdot y_3^5 \cdot y_4^6 \cdot y_5^{12}.$$

In total, we require 9 Frobenius, 3 exponentiations by x, 14 multiplications, and 3 squarings. However, in the Scott et al. method [21], the cost reported was of just 9 Frobenius, 3 exponentiations by x, 13 multiplications, and 4 squarings.

Unfortunately, none of the basis vectors returned by the LLL algorithm has an advantage over [21]. However, if small integer linear combinations of the short vectors returned by the LLL algorithm are considered, a multiple of d that corresponds to a shorter addition chain could potentially be found. A brute force search of linear combinations of the LLL basis reveals that there exist 18 non-zero vectors with maximal entry 12. Among these vectors the authors of [9, 17] found that the combination, $x_0 - x_2 - x_3$ is equal to:

$$x_0 - x_2 - x_3 = [12, 12, 6, 1, 12, 6, 4, 0, 12, 6, 6, 0, 12, 6, 4, -1],$$

which corresponds to the multiple $d'(x) = \lambda_0 + \lambda_1 p + \lambda_2 p^2 + \lambda_3 p^3 = 2x(6x^2 + 3x + 1)d(x)$, where,

$$\begin{aligned}
\lambda_0(x) &= 1 + 6x + 12x^2 + 12x^3, \\
\lambda_1(x) &= 4x + 6x^2 + 12x^3, \\
\lambda_2(x) &= 6x + 6x^2 + 12x^3, \\
\lambda_3(x) &= -1 + 4x + 6x^2 + 12x^3.
\end{aligned}$$

The resulting multi-exponentiation can be computed more efficiently without using addition chains by applying the following strategy. First, the following exponentiations are computed:

$$f \mapsto f^x \mapsto f^{2x} \mapsto f^{4x} \mapsto f^{6x} \mapsto f^{6x^2} \mapsto f^{12x^2} \mapsto f^{12x^3}.$$

The above computation requires 3 exponentiations by x, 3 squarings, and 1 multiplication. The terms $a = f^{12x^3} \cdot f^{6x^2} \cdot f^{6x}$ and $b = a \cdot (f^{2x})^{-1}$ can be computed using 3 multiplications. Finally, the result $f^{d'}$ is obtained by computing,

$$[a \cdot f^{6x^2} \cdot f] \cdot [b]^p \cdot [a]^{p^2} \cdot [b \cdot f^{-1}]^{p^3},$$

which costs six extra multiplications. In total, this method requires 3 exponentiations by x, 3 squarings, 10 multiplications, and 3 Frobenius applications. The complete procedure just described is shown in Algorithm 7.5.

ALGORITHM 7.5   Final exponentiation by Fuentes-Castañeda et al. for BN curves [9].

Input : $f \in \mathbb{F}_{p^k}$, $x \in \mathbb{Z}$

Output: $f^{\frac{\phi_k(p)}{r}} \in \mathbb{F}_{p^k}$

1 $a = f^x$; $\{\, f^x \,\}$

2 $a = a^2$; $\{\, f^{2x} \,\}$

3 $b = a^2$; $\{\, f^{4x} \,\}$

4 $b = a \cdot b$; $\{\, f^{6x} \,\}$

5 $t = b^x$; $\{\, f^{6x^2} \,\}$

6 $f = f \cdot [\bar{f}]^{p^3}$; $\{\, f \cdot [\frac{1}{f}]^{p^3} \,\}$

7 $f = f \cdot t$; $\{\, f \cdot [\frac{1}{f}]^{p^3} \cdot f^{6x^2} \,\}$

8 $b = b \cdot t$; $\{\, f^{6x^2} \cdot f^{6x} \,\}$

9 $t = t^2$; $\{\, f^{12x^2} \,\}$

10 $t = t^x$; $\{\, f^{12x^3} \,\}$

11 $b = b \cdot t$; $\{\, f^{12x^3} \cdot f^{6x^2} \cdot f^{6x} = A \,\}$

12 $t = b \cdot \bar{a}$; $\{\, A \cdot \frac{1}{f^{2x}} = B \,\}$

13 $f = f \cdot t^{p^3}$; $\{\, [f \cdot f^{6x^2}] \cdot [B \cdot \frac{1}{f}]^{p^3} \,\}$

14 $f = f \cdot t^p$; $\{\, f \cdot f^{6x^2} \cdot [B]^p \cdot [B \cdot \frac{1}{f}]^{p^3} \,\}$

15 $f = f \cdot b$; $\{\, [A \cdot f \cdot f^{6x^2}] \cdot [B]^p \cdot [B \cdot \frac{1}{f}]^{p^3} \,\}$

16 $f = f \cdot b^{p^2}$; $\{\, [A \cdot f \cdot f^{6x^2}] \cdot [B]^p \cdot [A]^{p^2} \cdot [B \cdot \frac{1}{f}]^{p^3} \,\}$

17 **return** f;

7.3   The Kachisa-Schaefer-Scott Family of Pairing-Friendly Elliptic Curves

Kachisa, Schaefer, and Scott proposed in [13] families of pairing-friendly elliptic curves of embedding degrees $k = 16, 18, 32, 36$, and 40. In this section, the hard part of the final exponentiation for the KSS-18 family of curves with embedding degree $k = 18$ will be analyzed in detail. A similar discussion for the KSS-16 and KSS-36 families of curves can be found in [12]. At the end of this section the computational costs of the hard part of the final exponentiation reported in related works are compared.

7.3.1   KSS-18 Curves

KSS-18 curves have embedding degree $k = 18$, which implies $\phi(k) = 6$. They are parametrized by x such that

$$p(x) = (x^8 + 5x^7 + 7x^6 + 37x^5 + 188x^4 + 259x^3 + 343x^2 + 1763x + 2401)/21, \qquad (7.23)$$
$$r(x) = (x^6 + 37x^3 + 343)/343$$

are both prime, with $x \equiv 14 \bmod 42$. The exponent $d = \Phi_k(p)/r$ is a degree-42 polynomial in the variable x.

Following the lattice techniques reported in Fuentes-Castañeda et al. [9], d can be written in base p as $d(x) = \sum_{i=0}^{5} \lambda_i(x)p(x)^i$, where the coefficients λ_i for $i = 0, \ldots, 5$ are polynomials with degree at most 7 given as

$$\lambda_0 = 147x + 108x^2 + 21x^3 + 7x^4 + 5x^5 + x^6, \tag{7.24}$$
$$\lambda_1 = -686 - 505x - 98x^2 - 35x^3 - 25x^4 - 5x^5,$$
$$\lambda_2 = 6 - 133x^2 - 98x^3 - 19x^4 - 7x^5 - 5x^6 - x^7,$$
$$\lambda_3 = 245x + 181x^2 + 35x^3 + 14x^4 + 10x^5 + 2x^6,$$
$$\lambda_4 = -343 - 254x - 49x^2 - 21x^3 - 15x^4 - 3x^5,$$
$$\lambda_5 = 3 + 7x^2 + 5x^3 + x^4.$$

The representation of the exponent d given in Equation (7.24) has the smallest integer coefficients reported in the literature for this curve. The final exponentiation using the representation of the exponent $d'(x)$ in base p can be computed as follows. First, seven exponentiations of the form f^{x^i}, for $i = 1, \ldots, 7$, must be sequentially computed. Thereafter, the Frobenius operator is applied to compute some of the 40 possible factors of the form $(f^{x^i})^{p^j}$, for $i = 0, \ldots, 7$ and $j = 1, \ldots, 5$. In the case of Equation (7.24), only 35 such factors are required, including for example,

$$f^p, f^{p^4}, f^{p^5}, (f^x)^p, (f^{x^2})^{p^2}, (f^x)^{p^4}, (f^{x^2})^{p^5},$$

and so on.

Considering only absolute values, notice that in Equation (7.24), the coefficient 1 appears three times, namely, in the equalities for $\lambda_0, \lambda_2,$ and λ_5. Similarly, the coefficients $2, 3,$ and 5 appear one, two, and three times, respectively. Those coefficients can be grouped as products of the factors $(f^{x^i})^{p^j}$ previously mentioned. The variables $y_0, y_1, y_2,$ and y_3 are used to group the coefficients $1, 2, 3,$ and 5, respectively, as follows:

$$y_0 = \frac{(f^{x^6}) \cdot (f^{x^4})^{p^5}}{(f^{x^7})^{p^2}}, \qquad y_1 = (f^{x^6})^{p^3}, \qquad y_2 = \frac{f^{p^5}}{(f^{x^5})^{p^4}}, \qquad y_3 = \frac{f^{x^5} \cdot (f^{x^3})^{p^5}}{(f^{x^5})^p \cdot (f^{x^6})^{p^2}}.$$

Proceeding in this manner, one can define a total of 24 coefficients y_k for $k = 0, \ldots, 23$. This leads us to the following multi-exponentiation problem,

$$y_0^1 \cdot y_1^2 \cdot y_2^3 \cdot y_3^5 \cdot y_4^6 \cdot y_5^7 \cdot y_6^{10} \cdot y_7^{14} \cdot y_8^{15} \cdot y_9^{19} \cdot y_{10}^{21} \cdot y_{11}^{25} \cdot y_{12}^{35} \cdot y_{13}^{49} \cdot y_{14}^{98} \cdot y_{15}^{108} \cdot y_{16}^{133} \cdot y_{17}^{147} \cdot \tag{7.25}$$
$$y_{18}^{181} \cdot y_{19}^{245} \cdot y_{20}^{254} \cdot y_{21}^{343} \cdot y_{22}^{505} \cdot y_{23}^{686}.$$

It can be readily verified that the cost of extracting all of the coefficients y_k from Equation (7.24), so that they can be multi-powered as shown in Equation (7.25), is of 10 multiplications plus several negligible field inversions that can be computed by simple conjugations.

Considering all the distinct coefficients that appear in Equation (7.24), the multi-exponentiation problem associated to this curve can be solved by finding a valid addition–subtraction chain for the protosequence of $s = 24$ elements:
$\{1, 2, 3, 5, 6, 7, 10, 14, 15, 19, 21, 25, 35, 49, 98, 108, 133, 147, 181, 245, 254, 343, 505, 686\}$.

After applying a specialized algorithm for finding optimal addition–subtraction sequences, the authors of [12] found the following solution:
$\{\, 1, \mathbf{2}, 3, 5, \mathbf{6}, 7, \mathbf{10}, \mathbf{14}, 15, 19, 21, 25, 35, 49, \mathbf{98}, 108, 133, 147, \underline{162}, 181, 245, \underline{248}, 254, 343,$
$505, \mathbf{686} \,\}$

The above 25-step sequence can be computed at a cost of 19 and 6 field multiplications and squarings, respectively. The corresponding vectorial addition–subtraction chain can be found at the extra cost of $s - 1 = 23$ field multiplications. In total, the hard part of the final exponentiation for the curve KSS-18 requires 7 exponentiations by x, 53 multiplications, 6 squarings, and 35 Frobenius applications. The complete procedure just described is shown in Algorithm 7.6.

ALGORITHM 7.6   Final exponentiation for KSS curves with $k = 18$ [1].

Input : $f \in \mathbb{G}_{\Phi_{18}(p)}, x \in \mathbb{Z}$

Output: $f^{\frac{\phi_k(p)}{r}} \in \mathbb{F}_{p^k}$

1 $A = \bar{f}^p;$ $\{ y_{23} \}$

2 $t_0 = A^2;$

3 $B = \bar{f}^{p^4};$ $\{ y_{21} \}$

4 $t_0 = t_0 \cdot B;$

5 $C = (\overline{f^x})^p;$ $\{ y_{22} \}$

6 $t_1 = t_0 \cdot C;$

7 $B = (f^x)^{p^3};$ $\{ y_{19} \}$

8 $t_0 = t_1 \cdot B;$

9 $B = (\overline{f^{x^2}})^p \cdot (\overline{f^{x^3}})^{p^2};$ $\{ y_{14} \}$

10 $t_1 = t_1 \cdot B;$

11 $A = (f)^{p^2};$ $\{ y_4 \}$

12 $B = (\overline{f^x})^{p^4};$ $\{ y_{20} \}$

13 $t_6 = A \cdot B;$

14 $t_0 = t_0 \cdot B;$

15 $A = (\overline{f^{x^5}})^{p^4} \cdot (f)^{p^5};$ $\{ y_2 \}$

16 $t_4 = A \cdot B;$

17 $B = f^x;$ $\{ y_{17} \}$

18 $t_2 = t_0 \cdot B;$

19 $t_0 = t_0 \cdot t1;$

20 $A = (f^{x^2})^{p^3};$ $\{ y_{18} \}$

21 $t_1 = A \cdot C;$

22 $B = (\overline{f^{x^4}})^{p^2};$ $\{ y_9 \}$

23 $t_3 = A \cdot B;$

24 $t_2 = t_1 \cdot t2;$

25 $B = (\overline{f^{x^4}})^{p^4};$ $\{ y_8 \}$

26 $t_5 = t_1 \cdot B;$

27 $B = (\overline{f^{x^2}})^{p^2};$ $\{ y_{16} \}$

28 $t_1 = t_2 \cdot B;$

29 $B = (f^{x^4})^{p^3};$ $\{ y_7 \}$

30 $t_8 = t_2 \cdot B;$

31 $B = f^{x^2};$ $\{ y_{15} \}$

32 $t_2 = t_1 \cdot B;$

33 $B = (\overline{f^{x^4}})^p;$ $\{ y_{11} \}$

34 $t_1 = t_1 \cdot B;$

35 $t_0 = t_2 \cdot t_0;$

36 $B = (f^{x^5})^{p^3};$ $\{ y_6 \}$

37 $t_7 = t_2 \cdot B;$

38 $t_0 = t_0^2;$

39 $B = (\overline{f^{x^2}})^{p^4};$ $\{ y_{13} \}$

40 $t_2 = t_0 \cdot B;$

41 $B = (\overline{f^{x^3}})^p \cdot (f^{x^3})^{p^3};$ $\{ y_{12} \}$

42 $t_0 = t_2 \cdot B;$

43 $t_2 = t_2 \cdot t_8;$

44 $t_1 = t_0 \cdot t_1;$

45 $t_0 = t_0 \cdot t_7;$

46 $t_3 = t_1 \cdot t_3;$

47 $t_1 = t_1 \cdot t_6;$

48 $B = f^{x^3} \cdot (\overline{f^{x^3}})^{p^4};$ $\{ y_{10} \}$

49 $t_6 = t_3 \cdot B;$

50 $A = (f^{x^6})^{p^3};$ $\{ y_1 \}$

51 $t_3 = A \cdot B;$

52 $t_2 = t_6 \cdot t_2;$

53 $B = f^{x^5} \cdot (\overline{f^{x^5}})^p \cdot (\overline{f^{x^6}})^{p^2} \cdot (f^{x^3})^{p^5};$ $\{ y_3 \}$

54 $t_6 = t_6 \cdot B;$

55 $t_2 = t_2 \cdot t_5;$

56 $B = f^{x^6} \cdot (\overline{f^{x^7}})^{p^2} \cdot (f^{x^4})^{p^5};$ $\{ y_0 \}$

57 $t_5 = t_5 \cdot B;$

58 $t_2 = t_2^2;$

59 $B = f^{x^4} \cdot (\overline{f^{x^5}})^{p^2} \cdot (f^{x^2})^{p^5};$ $\{ y_5 \}$

60 $t_2 = t_2 \cdot B;$

61 $t_0 = t_0^2;$

62 $t_0 = t_0 \cdot t_6;$

63 $t_1 = t_2 \cdot t_1;$

64 $t_2 = t_2 \cdot t_5;$

65 $t_1 = t_1^2;$

66 $t_1 = t_1 \cdot t_4;$

67 $t_1 = t_1 \cdot t_0;$

68 $t_0 = t_0 \cdot t_3;$

69 $t_0 = t_0 \cdot t_1;$

70 $t_1 = t_1 \cdot t_2;$

71 $t_0 = t_0^2;$

72 $t_0 = t_0 \cdot t_1;$

---

73 **return** $t_0;$

7.4 Other Families

Curves BW-12

In the particular case of the **cyclotomic families**, given that $r(x) = \Phi_k(x)$, there exists no multiple $d'(x)$ of $d(x) = \Phi_k(p)/r(x)$, such that $g \mapsto g^{d'(z)}$ can be computed at a cheaper cost than $g \mapsto g^{d(z)}$. However, it is possible to save some few multiplications in the field $\mathbb{F}_{p^{12}}$ by using temporary variables. Let $d(z) = \lambda_0 + \lambda_1 p + \lambda_2 p^2 + \lambda_3 p^3$, where

$$
\begin{aligned}
\lambda_0 &= x^5 - 2x^4 + 2x^2 - x + 3, \\
\lambda_1 &= x^4 - 2x^3 + 2x - 1, \\
\lambda_2 &= x^3 - 2x^2 + x, \\
\lambda_3 &= x^2 - 2x + 1.
\end{aligned}
$$

Then, the following elements are pre-computed:

$$
g \to g^{-2} \to g^x \to g^{2x} \to g^{x-2} \to \quad g^{x^2-2x} \to g^{x^3-2x^2} \to g^{x^4-2x^3} \to g^{x^4-2x^3+2x} \to g^{x^5-2x^4+2x^2}.
$$

Thereafter, the powering $g^{d(z)}$ can be computed as

$$
g^{x^5-2x^4+2x^2} \cdot (g^{x-2})^{-1} \cdot g \cdot (g^{x^4-2x^3+2x} \cdot g^{-1})^p \cdot (g^{x^3-2x^2} \cdot g^x)^{p^2} \cdot (g^{x^2-2x} \cdot g)^{p^3},
$$

at a computational cost of 5 exponentiations by x, 3 Frobenius operator applications, 10 multiplications over the field $\mathbb{F}_{p^{12}}$, and 2 squarings in the cyclotomic group $\mathbb{G}_{\Phi_{12}(p)}$. The corresponding procedure is shown in Algorithm 7.7.

ALGORITHM 7.7 Final exponentiation for BW curves with $k = 12$ [1].

Input : $f \in \mathbb{G}_{\Phi_{12}(p)}, \, x \in \mathbb{Z}$

Output: $f^{\frac{\phi_k(p)}{r}} \in \mathbb{F}_{p^k}$

1 $a = \bar{f}^2;$ $\{\, f^{-2} \,\}$

2 $b = f^x;$ $\{\, f^x \,\}$

3 $c = b^2;$ $\{\, f^{2x} \,\}$

4 $a = b \cdot a;$ $\{\, f^{x-2} \,\}$

5 $d = a^x;$ $\{\, f^{x^2-2x} \,\}$

6 $e = d^x;$ $\{\, f^{x^3-2x^2} \,\}$

7 $g = e^x;$ $\{\, f^{x^4-2x^3} \,\}$

8 $g = g \cdot c;$ $\{\, f^{x^4-2x^3+2x} \,\}$

9 $c = g^x;$ $\{\, f^{x^5-2x^4+2x^2} \,\}$

10 $c = c \cdot \bar{a};$ $\{\, [f^{x^5-2x^4+2x^2} \cdot \overline{f^{x-2}}] \,\}$

11 $c = c \cdot f;$ $\{\, [f^{x^5-2x^4+2x^2} \cdot \overline{f^{x-2}} \cdot f] \,\}$

12 $g = g \cdot \bar{f};$ $\{\, [f^{x^4-2x^3+2x} \cdot \bar{f}] \,\}$

13 $e = e \cdot b;$ $\{\, [f^{x^3-2x^2} \cdot f^x] \,\}$

14 $d = d \cdot f;$ $\{\, [f^{x^2-2x} \cdot f] \,\}$

15 $f = c \cdot g^p;$ $\{\, [f^{x^5-2x^4+2x^2} \cdot \overline{f^{x-2}} \cdot f] \cdot [f^{x^4-2x^3+2x} \cdot \bar{f}]^p \,\}$

16 $f = f \cdot e^{p^2};$ $\{\, [f^{x^5-2x^4+2x^2} \cdot \overline{f^{x-2}} \cdot f] \cdot [f^{x^4-2x^3+2x} \cdot \bar{f}]^p \cdot [f^{x^3-2x^2} \cdot f^x]^{p^2} \,\}$

17 $f = f \cdot d^{p^3};$ $\{\, [f^{x^5-2x^4+2x^2} \cdot \overline{f^{x-2}} \cdot f] \cdot [f^{x^4-2x^3+2x} \cdot \bar{f}]^p \cdot [f^{x^3-2x^2} \cdot f^x]^{p^2} \cdot [f^{x^2-2x} \cdot f]^{p^3} \,\}$

18 **return** $f;$

Curves BLS-24

As in the case of the BW-12 curves, there does not exist a multiple $d'(x)$ of $d(x)$ such that the computational cost of the powering $g \mapsto g^{d'(x)}$ is cheaper than the original exponentiation $g \mapsto g^{d(x)}$. Let $d(x) = \lambda_0 + \lambda_1 p + \lambda_2 p^2 + \lambda_3 p^3 + \lambda_4 p^4 + \lambda_5 p^5 + \lambda_6 p^6 + \lambda_7 p^7$, such that,

$$\lambda_7 = x^2 - 2x + 1,$$
$$\lambda_6 = x^3 - 2x^2 + x = x\lambda_7,$$
$$\lambda_5 = x^4 - 2x^3 + x^2 = x\lambda_6,$$
$$\lambda_4 = x^5 - 2x^4 + x^3 = x\lambda_5,$$
$$\lambda_3 = x^6 - 2x^5 + x^4 - x^2 + 2x - 1 = x\lambda_4 - \lambda_7,$$
$$\lambda_2 = x^7 - 2x^6 + x^5 - x^3 + 2x^2 - x = x\lambda_3,$$
$$\lambda_1 = x^8 - 2x^7 + x^6 - x^4 + 2x^3 - x^3 = x\lambda_2,$$
$$\lambda_0 = x^9 - 2x^8 + x^7 - x^5 + 2x^4 - x^3 + 3 = x\lambda_1 + 3.$$

Then, one can proceed to compute the temporary values,

$$g^x \to g^{-2x} \to g^{-2x+1} \to g^{x^2} \to g^{\lambda_7} g^{x\lambda_7} \to g^{x\lambda_6} \to g^{x\lambda_5} \to g^{x\lambda_4} \to g^{x\lambda_4 - \lambda_7} \to g^{x\lambda_3} \to g^{x\lambda_2}$$
$$\to g^{x\lambda_1} \to g^{x\lambda_1 + 3}.$$

The total cost of the hard part of the final exponentiations is of 9 exponentiations to the power x, 7 Frobenius operator applications, 12 multiplications over the extension field $\mathbb{F}_{p^{24}}$, and 2 squarings in the cyclotomic group $\mathbb{G}_{\Phi_{24}(p)}$. The corresponding procedure is shown in Algorithm 7.8, where

$$\lambda_7 = x^2 - 2x + 1, \qquad \lambda_6 = x\lambda_7, \qquad \lambda_5 = x\lambda_6,$$
$$\lambda_4 = x\lambda_5, \qquad \lambda_3 = x\lambda4 - \lambda7, \qquad \lambda_2 = x\lambda_3,$$
$$\lambda_1 = x\lambda_2, \qquad \lambda_0 = x\lambda_1 + 3.$$

7.5   Comparison and Conclusions

Table 7.3 summarizes the computational cost of the hard part of the final exponentiation, as they were reported in [21, 3, 9, 1, 12]. Note that the operation counts are given for only the hard part of the exponentiation, and considering field multiplication and squaring operations only, since the number of exponentiations by x is fixed for each curve and computing p-powers maps is considered negligible.

For a concrete example, let us consider the case of KSS-18 curves, with embedding degree $k = 18$, parametrized with the integer

$$x = -0x1500000150000B7CE,$$

which yields a 192-bit security level as reported in [7]. Using standard towering field techniques as reported in [6], the cost of one multiplication and one cyclotomic squaring (using Karabina's technique as presented in [14]), over the extension field $\mathbb{F}_{p^{18}}$, is approximately equivalent to about 108 and 36 field multiplications in the base field $\mathbb{F}_p$. This implies a ratio of $M \approx 3S$. Then, the total cost to perform the exponentiations f^{x^i} for $i = 1, \ldots, 7$ is around $7 \cdot \lfloor \log_2 x \rfloor = 448$

ALGORITHM 7.8 Final exponentiation for BLS curves with $k = 24$ [1].

Input   : $f \in \mathbb{G}_{\Phi_{24}(p)}, \ x \in \mathbb{Z}$

Output: $f^{\frac{\phi_k(p)}{r}} \in \mathbb{F}_{p^k}$

1 $a = f^x$; $\{ f^x \}$

2 $b = \bar{a}^2$; $\{ f^{-2x} \}$

3 $b = b \cdot f$; $\{ f^{-2x+1} \}$

4 $a = a^x$; $\{ f^{x^2} \}$

5 $c = a \cdot b$; $\{ f^{\lambda_7} = f^{x^2 - 2x + 1} \}$

6 $b = c^x$; $\{ f^{\lambda_6} = f^{x\lambda_7} \}$

7 $a = c^{p^7} \cdot b^{p^6}$; $\{ f^{\lambda_7^{p^7}} \cdot f^{\lambda_6^{p^6}} \}$

8 $b = b^x$; $\{ f^{\lambda_5} = f^{x\lambda_6} \}$

9 $a = a \cdot b^{p^5}$; $\{ f^{\lambda_7^{p^7}} \cdot f^{\lambda_6^{p^6}} \cdot f^{\lambda_5^{p^5}} \}$

10 $b = b^x$; $\{ f^{\lambda_4} = f^{x\lambda_5} \}$

11 $a = a \cdot b^{p^4}$; $\{ f^{\lambda_7^{p^7}} \cdot f^{\lambda_6^{p^6}} \cdot f^{\lambda_5^{p^5}} \cdot f^{\lambda_4^{p^4}} \}$

12 $b = b^x$; $\{ f^{x\lambda_4} \}$

13 $b = b \cdot \bar{c}$; $\{ f^{\lambda_3} = f^{x\lambda_4 - \lambda_7} \}$

14 $a = a \cdot b^{p^3}$; $\{ f^{\lambda_7^{p^7}} \cdot f^{\lambda_6^{p^6}} \cdot f^{\lambda_5^{p^5}} \cdot f^{\lambda_4^{p^4}} \cdot f^{\lambda_3^{p^3}} \}$

15 $b = b^x$; $\{ f^{\lambda_2} = f^{x\lambda_3} \}$

16 $a = a \cdot b^{p^2}$; $\{ f^{\lambda_7^{p^7}} \cdot f^{\lambda_6^{p^6}} \cdot f^{\lambda_5^{p^5}} \cdot f^{\lambda_4^{p^4}} \cdot f^{\lambda_3^{p^3}} \cdot f^{\lambda_2^{p^2}} \}$

17 $b = b^x$; $\{ f^{\lambda_1} = f^{x\lambda_2} \}$

18 $a = a \cdot b^p$; $\{ f^{\lambda_7^{p^7}} \cdot f^{\lambda_6^{p^6}} \cdot f^{\lambda_5^{p^5}} \cdot f^{\lambda_4^{p^4}} \cdot f^{\lambda_3^{p^3}} \cdot f^{\lambda_2^{p^2}} \cdot f^{\lambda_1^{p}} \}$

19 $b = b^x$; $\{ f^{x\lambda_1} \}$

20 $b = b \cdot f^3$; $\{ f^{\lambda_0} = f^{x\lambda_1 + 3} \}$

21 $f = a \cdot b$; $\{ f^{\lambda_7^{p^7}} \cdot f^{\lambda_6^{p^6}} \cdot f^{\lambda_5^{p^5}} \cdot f^{\lambda_4^{p^4}} \cdot f^{\lambda_3^{p^3}} \cdot f^{\lambda_2^{p^2}} \cdot f^{\lambda_1^{p}} \cdot f^{\lambda_0} \}$

22 **return** f;

TABLE 7.3 A comparison of the final exponentiation method cost for several KSS curves.

Curve	Scott et al. [21]	Benger [3]	Fuentes-Castañeda et al. [9]	Guzmán-Trampe et al. [12]	Aranha et al. [1]
BN	13M 4S	13M 4S	10M 3S	—	–
BW-12	—	—	10M 3S	—	10M 2S
BLS-24	—	—	10M 3S	—	12M 2S
KSS-8	31M 6S	27M 6S	26M 7S	—	—
KSS-16	—	83M 20S	—	69M 15S	—
KSS-18	62M 14S	62M 14S	52M 8S	53M 6S	—
KSS-36	—	—	—	177M 69S	—

Note: 'M' denotes a multiplication, and 'S' denotes a squaring in the extension field $\mathbb{F}_{p^k}$. All methods require the same number of exponentiations by x, determined by the curve.

cyclotomic squarings plus $7 \cdot (H(x) - 1) = 112$ multiplications. Using the results reported in Table 7.3, this gives an approximate cost for the hard part of the final exponentiation of $462S + 174M \approx 984S$ for the Scott et al. method, $456S + 164M \approx 948S$ for the Fuentes-Castaneda et al. method, and $454S + 164M \approx 949S$ for the cost presented in Guzmán-Trampe et al. [12].

7.6 SAGE Appendix

7.6.1 BN Curves

This section presents the implementation in SAGE of the final exponentiation, in the family of BN curves to the $\mathbb{G}_T$ group used in the calculation of the Tate family of bilinear pairing functions. The family of BN curves is parametrized in Listing 28, together with the basic functions. Listing 29 shows the Devegili et al. implementation, Listing 30 shows the Scott et al. method, whereas Listing 31 shows the Fuentes-Castañeda et al. method.

7.6.2 BW Curves

This section presents the implementation in SAGE of the final exponentiation in the BW family of curves. The parameters are shown in Listing 32, and the code for the exponentiation is shown in Listing 33.

Listing 28 File `parametersbn.sage`. Definition of parameters and computation of derived parameters for the optimal ate pairing using the Barreto-Naehrig curve $E : Y^2 = X^3 + B$ over $\mathbb{F}_p$, where $p = 36x^4 + 36x^3 + 24x^2 + 6x + 1$, and $r = 36x^4 + 36x^3 + 18x^2 + 6x + 1$.

```
1   z = -(2^62+2^55+1)  #requires beta = -1.  chi = [1, 1]
2
3   print(z)
4   print(ceil(log(abs(6*z + 2),2)))
5   print(ceil(log(abs(z),2)))
6
7   #Parameters of the curve
8   Zx.<x> = PolynomialRing(QQ)
9   k = 12
10
11  px      = 36*x^4 + 36*x^3 + 24*x^2 + 6*x + 1
12  rx      = 36*x^4 + 36*x^3 + 18*x^2 + 6*x + 1
13  tx      = 6*x^2 + 1
14  p       = Integer(px(z))
15  r       = Integer(rx(z))
16  t       = Integer(tx(z))
17
18  beta = -1
19  chi = [1, 1]
20  #gamma = 1
21
22  Fp        = GF(p)
23  K2.<U>    = Fp[]
24  Fp2.<u>   = Fp.extension(U^2 - (beta))
25  K6.<V>    = Fp2[]
26  Fp6.<v>   = Fp2.extension(V^3 -(chi[1]*u+chi[0]*1))
27  K12.<W>   = Fp6[]
28  Fp12.<w>  = Fp6.extension(W^2 - (v))
29  Fp12.is_field=lambda:True
30
31  e = Fp12.random_element()
32
33  Debug = True
34
35
36
37  ###########################################
38  # Final exponentations
39  ###########################################
40
41  def conj(f):
42          (a0,a1) = vector(f)
43          return a0-w*a1
44
45
46
47  # Map-to-Cyclotomic
48  def EasyExpo(f):
49          f = conj(f) * f^(-1)
50          f = f^(p^2)*f
51          return f
```

Listing 29 File `devegilietal.sage`. Final exponentiation for BN curves using Devegili et al. method.

```
1    def FExp_Devegilietal(f):
2            f = EasyExpo(f)
3
4            a = f^abs(z)
5            if z < 0:
6                    a = conj(a)
7            b = a^2
8            a = b * f^2
9            a = a^2
10           a = a * b
11           a = a * f
12           a = conj(a)
13
14           b  = a^p
15           b  = a * b
16           a  = a * b
17           t0 = f^p
18           t1 = t0 * f
19           t1 = t1^9
20           a  = t1 * a
21           t1 = f^4
22           a  = a * t1
23           t0 = t0^2
24           b  = b * t0
25           t0 = (f^(p))^p
26           b  = b * t0
27           t0 = b^abs(z)
28           if z < 0:
29                   t0 = conj(t0)
30           t1 = t0^2
31           t0 = t1^2
32           t0 = t0 * t1
33           t0 = t0^abs(z)
34           if z < 0:
35                   t0 = conj(t0)
36           t0 = t0 * b
37           a  = t0 * a
38           t0 = ((f^p)^p)^p
39           f  = t0 * a
40
41           return f
```

Listing 30 File `scottetal.sage`. Final exponentiation for BN curves using Scott et al. method.

```
1   def FExp_Scottetal(f):
2           f = EasyExpo(f)
3
4           fx1 = f^abs(z)
5           if z < 0:
6                   fx1 = conj(fx1)
7           fx2 = fx1^abs(z)
8           if z < 0:
9                   fx2 = conj(fx2)
10          fx3 = fx2^abs(z)
11          if z < 0:
12                  fx3 = conj(fx3)
13
14          A   = conj(fx3) * (conj(fx3))^p
15          t0  = A^2
16          B   = conj(fx1) * (conj(fx2))^p
17          t0  = t0 * B
18          B   = conj(fx2)
19          t1  = t0 * B
20          A   = conj(fx1)^p
21          #B  = conj(fx2)
22          t0  = A * B
23          t0  = t0 * t1
24          B   = (fx2^p)^p
25          t1  = t1 * B
26          t0  = t0^2
27          t0  = t0 * t1
28          t1  = t0^2
29          B   = conj(f)
30          t0  = t1 * B
31          B   = f^p * (f^p)^p * ((f^p)^p)^p
32          t1  = t1 * B
33          t0  = t0^2
34          t0  = t0 * t1
35
36
37          return t0
```

Listing 31 File `fuentescastanedaetal.sage`.   Final exponentiation for BN curves using Fuentes-Castañeda et al. method.

```
1  def FExp_FuentesCastanedaetal(f):
2          f = EasyExpo(f)
3
4          a = f^abs(z)
5          if z < 0:
6                  a = conj(a)
7          a = a^2
8          b = a^2
9          b = a * b
10         t = b^abs(z)
11         if z < 0:
12                 t = conj(t)
13
14         f = f * ((conj(f)^p)^p)^p
15         f = f * t
16
17         b = b * t
18         t = t^2
19         t = t^abs(z)
20         if z < 0:
21                 t = conj(t)
22         b = b * t
23         t = b * conj(a)
24         f = f * ((t^p)^p)^p
25         f = f * t^p
26         f = f * b
27         f = f * (b^p)^p
28
29
30         return f
```

Listing 32 File `parametersbw12.sage`. Definition of parameters and computation of derived parameters for the optimal ate pairing using the Brezing-Weng curve with $k = 12$ $E : Y^2 = X^3 + B$ over $\mathbb{F}_p$, where $p = \frac{1}{3}(x+1)(x^{\frac{k}{3}} - x^{\frac{k}{6}} + 1) + x$, and $r = \Phi_k(x)$.

```
1   z = int("0x10008010000804000",16)
2   print(z)
3
4   #Parameters of the curve
5   Zx.<x> = PolynomialRing(QQ)
6   k = 12
7
8   px        = (x^6 - 2*x^5 + 2*x^3 -2*x +1)//3 + x
9   rx        = x^4 - x^2 + 1
10  tx        = x + 1
11  p         = Integer(px(z))
12  r         = Integer(rx(z))
13  t         = Integer(tx(z))
14
15  beta = 2
16  chi = [4, 1]
17  gamma = 5
18
19  Fp        = GF(p)
20  K2.<U>    = Fp[]
21  Fp2.<u>   = Fp.extension(U^2 - (beta))
22  K6.<V>    = Fp2[]
23  Fp6.<v>   = Fp2.extension(V^3 -(chi[1]*u+chi[0]*1))
24  K12.<W>   = Fp6[]
25  Fp12.<w>  = Fp6.extension(W^2 - (gamma*v))
26  Fp12.is_field=lambda:True
27
28  e = Fp12.random_element()
```

Listing 33 File `finalexpobw12.sage`. Final exponentiation for BW:$k = 12$ curves using Fuentes-Castaneda et al. method.

```
1   # Fuentes-Castaneda et al. BW:K=12
2   def FExp_BW12(f):
3           f = EasyExpo(f)
4
5           a = conj(f^2)
6           b = f^abs(z)
7           if z < 0:
8                   b = conj(b)
9           c = b^2
10          a = b * a
11          d = a^abs(z)
12          if z < 0:
13                  d = conj(d)
14          e = d^abs(z)
15          if z < 0:
16                  e = conj(e)
17          g = e^abs(z)
18          if z < 0:
19                  g = conj(g)
20          g = g * c
21          c = g^abs(z)
22          if z < 0:
23                  c = conj(c)
24
25          c = c * conj(a)
26          c = c * f
27          g = g * conj(f)
28          e = e * b
29          d = d * f
30          f = c * g^p
31          f = f * (e^p)^p
32          f = f * ((d^p)^p)^p
33
34          return f
```

References

[1] Diego F. Aranha, Laura Fuentes-Castañeda, Edward Knapp, Alfred Menezes, and Francisco Rodríguez-Henríquez. Implementing pairings at the 192-bit security level. In M. Abdalla and T. Lange, editors, *Pairing-Based Cryptography – Pairing 2012*, volume 7708 of *Lecture Notes in Computer Science*, pp. 177–195. Springer, Heidelberg, 2013.

[2] Paulo S. L. M. Barreto, Hae Yong Kim, Ben Lynn, and Michael Scott. Efficient algorithms for pairing-based cryptosystems. In M. Yung, editor, *Advances in Cryptology – CRYPTO 2002*, volume 2442 of *Lecture Notes in Computer Science*, pp. 354–368. Springer, Heidelberg, 2002.

[3] Naomi Benger. *Cryptographic Pairings: Efficiency and DLP Security*. PhD thesis, Dublin City University, 2010.

[4] Daniel J. Bernstein. Pippenger's exponentiation algorithm. Unpublished manuscript, available at http://cr.yp.to/papers.html#pippenger, 2001.

[5] Henri Cohen and Gerhard Frey, editors. *Handbook of Elliptic and Hyperelliptic Curve Cryptography*, volume 34 of *Discrete Mathematics and Its Applications*. Chapman & Hall/CRC, 2006.

[6] Augusto Jun Devegili, Michael Scott, and Ricardo Dahab. Implementing cryptographic pairings over Barreto-Naehrig curves (invited talk). In T. Takagi et al., editors, *Pairing-Based Cryptography – Pairing 2007*, volume 4575 of *Lecture Notes in Computer Science*, pp. 197–207. Springer, Heidelberg, 2007.

[7] Luis J. Dominguez Perez, Ezekiel J. Kachisa, and Michael Scott. Implementing cryptographic pairings: A Magma tutorial. Cryptology ePrint Archive, Report 2009/072, 2009. http://eprint.iacr.org/2009/072.

[8] Sylvain Duquesne and Loubna Ghammam. Memory-saving computation of the pairing final exponentiation on BN curves. Cryptology ePrint Archive, Report 2015/192, 2015. http://eprint.iacr.org/2015/192.

[9] Laura Fuentes-Castañeda, Edward Knapp, and Francisco Rodríguez-Henríquez. Faster hashing to $\mathbb{G}_2$. In A. Miri and S. Vaudenay, editors, *Selected Areas in Cryptography – SAC 2011*, volume 7118 of *Lecture Notes in Computer Science*, pp. 412–430. Springer, Heidelberg, 2012.

[10] Robert Granger, Dan Page, and Nigel P. Smart. High security pairing-based cryptography revisited. In F. Hess, S. Pauli, and M. E. Pohst, editors, *Algorithmic Number Theory (ANTS-VII)*, volume 4076 of *Lecture Notes in Computer Science*, pp. 480–494. Springer, 2006.

[11] Robert Granger and Michael Scott. Faster squaring in the cyclotomic subgroup of sixth degree extensions. In P. Q. Nguyen and D. Pointcheval, editors, *Public Key Cryptography – PKC 2010*, volume 6056 of *Lecture Notes in Computer Science*, pp. 209–223. Springer, Heidelberg, 2010.

[12] Juan E. Guzmán-Trampe, Nareli Cruz Cortés, Luis J. Dominguez Perez, Daniel Ortiz Arroyo, and Francisco Rodríguez-Henríquez. Low-cost addition-subtraction sequences for the final exponentiation in pairings. *Finite Fields and Their Applications*, 29:1–17, 2014.

[13] Ezekiel J. Kachisa, Edward F. Schaefer, and Michael Scott. Constructing Brezing-Weng pairing-friendly elliptic curves using elements in the cyclotomic field. In S. D. Galbraith and K. G. Paterson, editors, *Pairing-Based Cryptography – Pairing 2008*, volume 5209 of *Lecture Notes in Computer Science*, pp. 126–135. Springer, Heidelberg, 2008.

[14] Koray Karabina. Squaring in cyclotomic subgroups. *Mathematics of Computation*, 82(281):555–579, 2013.

[15] Taechan Kim, Sungwook Kim, and Jung Hee Cheon. Accelerating the final exponentiation in the computation of the tate pairings. Cryptology ePrint Archive, Report 2012/119, 2012. http://eprint.iacr.org/2012/119.

[16] Taechan Kim, Sungwook Kim, and Jung Hee Cheon. On the final exponentiation in Tate pairing computations. *IEEE Transactions on Information Theory*, 59(6):4033–4041, 2013.

[17] Edward Knapp. *On the Efficiency and Security of Cryptographic Pairings*. PhD thesis, University of Waterloo, 2010.

[18] Neal Koblitz and Alfred Menezes. Pairing-based cryptography at high security levels (invited paper). In N. P. Smart, editor, *Cryptography and Coding*, volume 3796 of *Lecture Notes in Computer Science*, pp. 13–36. Springer, Heidelberg, 2005.

[19] Arjen K. Lenstra, Hendrik W. Lenstra Jr., and László Lovász. Factoring polynomials with rational coefficients. *Mathematische Annalen*, 261(4):515–534, 1982.

[20] Jorge Olivos. On vectorial addition chains. *Journal of Algorithms*, 2(1):13–21, 1981.

[21] Michael Scott, Naomi Benger, Manuel Charlemagne, Luis J. Dominguez Perez, and Ezekiel J. Kachisa. On the final exponentiation for calculating pairings on ordinary elliptic curves. In H. Shacham and B. Waters, editors, *Pairing-Based Cryptography – Pairing 2009*, volume 5671 of *Lecture Notes in Computer Science*, pp. 78–88. Springer, Heidelberg, 2009.

[22] Thomas Unterluggauer and Erich Wenger. Efficient pairings and ECC for embedded systems. In L. Batina and M. Robshaw, editors, *Cryptographic Hardware and Embedded Systems – CHES 2014*, volume 8731 of *Lecture Notes in Computer Science*, pp. 298–315. Springer, Heidelberg, 2014.

8

Hashing into Elliptic Curves

Eduardo Ochoa-Jiménez
CINVESTAV-IPN

Francisco Rodríguez-Henríquez
CINVESTAV-IPN

Mehdi Tibouchi
NTT Secure Platform Laboratories

This chapter discusses the general problem of hashing into elliptic curves, particularly in the context of pairing-based cryptography. Our goal is to obtain practical, efficient, and secure algorithms for hashing values to elliptic curve subgroups $\mathbb{G}_1$ and $\mathbb{G}_2$.

Indeed, numerous pairing-based protocols involve hashing to either of these groups. For example, in the Boneh–Franklin identity-based encryption scheme [10] already discussed in Chapter 1, the public key for identity ID $\in \{0,1\}^*$ is an element P_{ID} in the group $\mathbb{G}_1$ (effectively an elliptic curve point) obtained as the image of ID under a hash function $\mathcal{H}: \{0,1\}^* \to \mathbb{G}_1$. A $\mathbb{G}_1$- or $\mathbb{G}_2$-valued hash function is also needed in many other pairing-based cryptosystems including IBE and HIBE schemes [4, 29, 32], signature and identity-based signature schemes [9, 11, 12, 17, 53], and identity-based signcryption schemes [14, 39].

In all of those cases, the hash functions are modeled as random oracles [8] in security proofs. However, it is not immediately clear how such a hash function can be instantiated in practice. Indeed, random oracles to groups like $(\mathbb{Z}/p\mathbb{Z})^*$ can be easily constructed from random oracles to fixed-length bit strings, for which conventional cryptographic hash functions usually provide acceptable substitutes. On the other hand, constructing random oracles to a elliptic curves, even from random oracles to bit strings, appears difficult in general, and some of the more

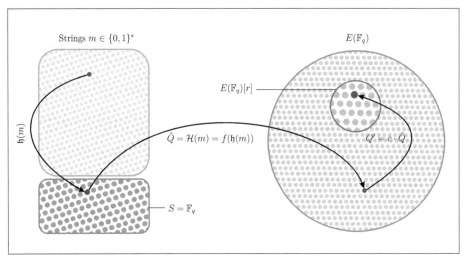

FIGURE 8.1   Hashing into pairing-friendly elliptic curve subgroups.

obvious instantiations actually break security completely. We therefore discuss how it can be done correctly, both from a theoretical and a very concrete standpoint.

The general approach taken in this chapter to construct secure hash functions to the subgroups $\mathbb{G}_1$ and $\mathbb{G}_2$ of a pairing-friendly elliptic curve E is illustrated in Figure 8.1, and consists of three main steps. The first step takes an arbitry message m to some element $\mathfrak{h}(\mathsf{m})$ of a set S that is "easy to hash to", in the sense that the function $\mathfrak{h}\colon \{0,1\}^* \to S$ can be easily obtained from a traditional cryptographic hash function like SHA-2 or SHA-3. In the main case of interest, the set S is the base field $\mathbb{F}_q$, although we may have $S = \mathbb{F}_q \times \mathbb{F}_q$, $S = \mathbb{F}_q \times \mathbb{Z}/r\mathbb{Z}$ and so on in some other settings. The second step maps the resulting value $\mathfrak{h}(\mathsf{m})$ to a point $\tilde{Q} = \mathcal{H}(\mathsf{m}) = f(\mathfrak{h}(\mathsf{m}))$ in the elliptic curve group $E(\mathbb{F}_q)$ using a map $f\colon S \to E(\mathbb{F}_q)$ called an encoding function. For the overall construction to be secure, the map f should satisfy a number of statistical properties related to the so-called indifferentiability of the hash function $\mathcal{H}$ from a random oracle, and we will also see that it should preferably be implemented in constant time. Finally, the last step takes the point $\tilde{Q}$ (which can lie anywhere on the curve) and maps it to a point Q' in the group $\mathbb{G}_1$ or $\mathbb{G}_2$. This is typically achieved by carrying out the scalar multiplication by the cofactor c of the elliptic curve (which is normally relatively small for the group $\mathbb{G}_1$, but much larger in the case of $\mathbb{G}_2$).

The chapter is organized as follows. In §8.1, we show that naive approaches for constructing hash functions to elliptic curves can be totally insecure. In §8.2, we discuss a classical algorithm called try-and-increment, which has good security properties from a theoretical standpoint, but can cause difficulties in practical implementations due to timing side-channel attacks. Arguably more robust, state-of-the-art approaches are then presented in in §8.3 and §8.4. They make it possible to construct constant-time hash functions to pairing-friendly elliptic curves, such as Barreto–Naehrig curves (a case which we discuss in more detail). Due to large cofactors, hashing to $\mathbb{G}_2$ requires particular care to be efficiently implemented; we discuss this point in §8.5, where we describe how the rich endomorphism structure associated to pairing-friendly elliptic curves lets us speed up the relevant scalar multiplication. At that point, the algorithmic description of hash functions to pairing groups will be complete, and we will turn to concrete implementation results in §8.6, on both Intel and ARM platforms. Finally, in Appendix 8.7, we include some SAGE source code implementing the main algorithms presented in this chapter, which can be downloaded from `http://sandia.cs.cinvestav.mx/Site/GuidePBCcode`.

8.1 The Trivial Encoding: Totally Insecure

To gain a sense of why the construction of hash functions to elliptic curves requires some care, we first show how a naive construction can completely break the security of a protocol that uses it.

A naive construction

We would like to construct a hash function $\mathcal{H} : \{0,1\}^* \to \mathbb{G}$ to an elliptic curve group $\mathbb{G}$, which we can assume is cyclic of order r and generated by a given point G. The simplest, most naive way to do so is probably to start from an integer-valued hash function $\mathfrak{h} : \{0,1\}^* \to \mathbb{Z}/r\mathbb{Z}$ (for which reasonable instantiations are easy to come by) and to define $\mathcal{H}$ as:

$$\mathcal{H}(\mathsf{m}) = [\mathfrak{h}(\mathsf{m})]G. \tag{8.1}$$

This is, however, a bad idea on multiple levels.

On the one hand, it is easy to see why this will typically break security *proofs* in the random oracle model. Indeed, at some point in a random oracle model security reduction, the simulator will typically want to "program" the random oracle by setting some of its outputs to specific values. In this case, it will want to set the value $\mathcal{H}(\mathsf{m})$ for some input m to a certain elliptic curve point P. However, if $\mathcal{H}$ is defined as in (8.1), the simulator should actually program the integer-valued random oracle $\mathfrak{h}$ to satisfy $[\mathfrak{h}(\mathsf{m})]G = P$. In other words, it should set $\mathfrak{h}(\mathsf{m})$ to the discrete logarithm of P with respect to G. But this discrete logarithm isn't usually known to the simulator, and it cannot be computed efficiently, therefore, the security reduction breaks down.

On the other hand, it is often not clear how this problem translates into an actual security weakness for a protocol using the hash function $\mathcal{H}$: one could think that it is mostly an artifact of the security proof. Nevertheless, a construction like (8.1) leaks the discrete logarithm of $\mathcal{H}(\mathsf{m})$ whenever m is known, which certainly feels uncomfortable from a security standpoint. We demonstrate below that this discomfort is entirely warranted, by showing that the Boneh–Lynn–Shacham signature scheme [12], already presented in Chapter 1, becomes completely insecure if the hash function involved is instantiated as in (8.1).

BLS signatures

Proposed by Boneh, Lynn, and Shacham in 2001 [12], the BLS signature scheme remains the efficient scheme that achieves the shortest signature length to this day: about 160 bits at the 80-bit security level. Recall from Chapter 1 that public parameters are a bilinear pairing $e \colon \mathbb{G}_1 \times \mathbb{G}_2 \to \mathbb{G}_T$ between groups of order r, generators G_1, G_2 of $\mathbb{G}_1$ and $\mathbb{G}_2$, and a hash function $\mathcal{H} \colon \{0,1\}^*$ modeled as a random oracle. The secret key is a random element $x \in \mathbb{Z}/r\mathbb{Z}$, the public key is $P = [x]G_2$, and a signature on a message $\mathsf{m} \in \{0,1\}^*$ is obtained as $S = [x]\mathcal{H}(\mathsf{m})$. To check that S is correct, a verifier then simply tests whether $e(\mathcal{H}(\mathsf{m}), P) = e(S, G_2)$.

Boneh, Lynn, and Shacham proved that this scheme is secure (in the usual sense of existential unforgeability under chosen message attacks) under the Computational Diffie–Hellman assumption when $\mathcal{H}$ is modeled as a random oracle.

Now consider the case when $\mathcal{H}$ is instantiated as in (8.1). Then, the signature on a message m can be written as:

$$S = [x]\mathcal{H}(\mathsf{m}) = \left[x\mathfrak{h}(\mathsf{m})\right]G_2 = [\mathfrak{h}(\mathsf{m})]P$$

and hence, one can forge a signature on any message using only publicly available data! There is no security left at all when using the trivial hash function construction.

A slightly less naive variant of the trivial construction consists in defining $\mathcal{H}$ as:

$$\mathcal{H}(\mathsf{m}) = [\mathfrak{h}(\mathsf{m})]Q \tag{8.2}$$

where $Q \in \mathbb{G}_2$ is an auxiliary public point distinct from the generator G_2 and whose discrete logarithm α with respect to G_2 is not published. Using this alternate construction for $\mathcal{H}$ thwarts the key-only attack described above against BLS signatures. However, the scheme remains far from secure. Indeed, the signature on a message m can be written as:

$$S = \big[x\mathfrak{h}(\mathsf{m}) \big] Q = \big[\alpha x\mathfrak{h}(\mathsf{m}) \big] G = [\mathfrak{h}(\mathsf{m})] \alpha P.$$

Now suppose an attacker knows a valid signature S_0 on some message m_0. Then the signature S on an arbitrary m is simply:

$$S = \left[\frac{\mathfrak{h}(\mathsf{m})}{\mathfrak{h}(\mathsf{m}_0)} \right] [\mathfrak{h}(\mathsf{m}_0)][\alpha] P = \left[\frac{\mathfrak{h}(\mathsf{m})}{\mathfrak{h}(\mathsf{m}_0)} \right] S_0,$$

where the division is computed in $\mathbb{Z}/r\mathbb{Z}$. Thus, even with this slightly less naive construction, knowing a single valid signature is enough to produce forgeries on arbitrary messages: again, a complete security breakdown.

8.2 Hashing by Random Trial

A classical construction of a hash function to elliptic curves that does work (and one variant of which is suggested by Boneh, Lynn, and Shacham in the original short signatures paper [12]) is the so-called "try-and-increment" algorithm. We present this algorithm here, as well as some of the limitations that explain why different constructions may be preferable.

8.2.1 The Try-and-Increment Algorithm

Consider an elliptic curve E over a finite field $\mathbb{F}_q$ of odd characteristic, defined by the Weierstrass equation:

$$E \colon y^2 = x^3 + ax^2 + bx + c \tag{8.3}$$

for some $a, b, c \in \mathbb{F}_q$. A probabilistic way to find a point on $E(\mathbb{F}_q)$ is to pick a random $x \in \mathbb{F}_q$, check whether $t = x^3 + ax^2 + bx + c$ is a square in $\mathbb{F}_q$, and if so, set $y = \pm\sqrt{t}$ and return (x, y). If t is not a square, then x is not the abscissa of a point on the curve: then one can pick another x and try again.

It is an easy consequence of the Hasse bound [30] on the number of points on $E(\mathbb{F}_q)$ (namely, $|\#E(\mathbb{F}_q) - q - 1| \leq 2\sqrt{q}$) that the success probability of a single trial is very close to $1/2$. Indeed, if we denote by χ_q the non-trivial quadratic character of $\mathbb{F}_q^*$ (see Remark 2.14), extended by 0 to $\mathbb{F}_q$, we have:

$$\#E(\mathbb{F}_q) = 1 + \sum_{x \in \mathbb{F}_q} \big(1 + \chi_q(x^3 + ax^2 + bx + c) \big) = q + 1 + \sum_{x \in \mathbb{F}_q} \chi_q(x^3 + ax^2 + bx + c).$$

On the other hand, the success probability ϖ of a single iteration of this point construction algorithm is the proportion of $x \in \mathbb{F}_q$ such that $\chi_q(x^3 + ax^2 + bx + c) = 1$ or 0, namely:

$$\varpi = \frac{\alpha}{2q} + \frac{1}{q} \sum_{x \in \mathbb{F}_q} \frac{1 + \chi_q(x^3 + ax^2 + bx + c)}{2}$$

where $\alpha \in \{0, 1, 2, 3\}$ is the number of roots of the polynomial $x^3 + ax^2 + bx + c$ in $\mathbb{F}_q$. This gives:

$$\varpi = \frac{1}{2} + \frac{\#E(\mathbb{F}_q) - q - 1 + \alpha}{2q} = \frac{1}{2} + O\left(\frac{1}{\sqrt{q}} \right).$$

Now this point construction algorithm can be turned into a hash function based on an $\mathbb{F}_q$-valued random oracle $\mathfrak{h} : \{0,1\}^* \to \mathbb{F}_q$. To hash a message m, the idea is to pick the x-coordinate as, essentially, $\mathfrak{h}(\mathsf{m})$ (which amounts to picking it at random once) and carry out the point construction above. However, since one should also be able to retry in case the first x-coordinate that is tried out is not the abscissa of an actual curve point, we rather let $x \leftarrow \mathfrak{h}(c\|m)$, where c is a fixed-length counter initially set to 0 and incremented in case of a failure. Since there is a choice of sign to make when taking the square root of $t = x^3 + ax^2 + bx + c$, we also modify $\mathfrak{h}$ to output an extra bit for that purpose: $\mathfrak{h} : \{0,1\}^* \to \mathbb{F}_q \times \{0,1\}$. This is the try-and-increment algorithm, described more precisely in Algorithm 8.1 (and called MapToGroup in [12]*). The failure probability after up to ℓ iterations is about $2^{-\ell}$ by the previous computations, so choosing the length of the counter c to be large enough for up to $\ell \approx 128$ iterations, say, is enough to ensure that the algorithm succeeds except with negligible probability.

Boneh, Lynn, and Shacham proved that this construction can replace the random oracle $\mathcal{H} : \{0,1\}^* \to E(\mathbb{F}_q)$ in BLS signatures without compromising security. In fact, it is not hard to see that it is *indifferentiable* from such a random oracle, in the sense of Maurer, Renner, and Holenstein [40]: This ensures that this construction can be plugged into almost all protocols (with some caveats [45]) requiring a random oracle $\mathcal{H} : \{0,1\}^* \to E(\mathbb{F}_q)$ while preserving random oracle security proofs.

Nevertheless, there are various reasons why Algorithm 8.1 is not a completely satisfactory construction for hash functions to elliptic curves. There is arguably a certain lack of mathematical elegance in the underlying idea of picking x-coordinates at random until a correct one is found, especially as the length of the counter, and hence the maximum number of trials, has to be fixed (to prevent collisions). More importantly, this may have adverse consequences for the security of physical devices implementing a protocol using this construction: for example, since the number of iterations in the algorithm depends on the input m, an adversary can obtain information on m by measuring the running time or the power consumption of a physical implementation.

ALGORITHM 8.1   The try-and-increment algorithm.

Input   : the message $\mathsf{m} \in \{0,1\}^*$ to be hashed
Output: the resulting point (x,y) on the curve $E : y^2 = x^3 + ax^2 + bx + c$

```
1  c ← 0                          // c is represented as a ⌈log₂ ℓ⌉-bit bit string
2  (x, b) ← 𝔥(c‖m)               // 𝔥 is a random oracle to 𝔽_q × {0,1}
3  t ← x³ + ax² + bx + c
4  if t is a square in 𝔽_q then
5  │    y ← (−1)ᵇ · √t   // define √· as the smaller square root wrt some ordering
6  │    return (x, y)
7  else
8  │    c ← c + 1
9  │    if c < ℓ then goto step 2
10 end
11 return ⊥
```

---

*Boneh et al. were in fact concerned with hashing to a supersingular curve of characteristic 3 of the form $y^2 = x^3 + 2x \pm 1$. In this case, it was later observed by Barreto and Kim [5] that picking y at random and solving the resulting Artin–Schreier equation for x was actually much more efficient, as that equation can be seen as a linear system over $\mathbb{F}_3$. But the basic principle of trying a value and incrementing a counter in case of failure remains the same. Moreover, pairing-friendly curves of small characteristic should not be used anymore: see the discussion in Chapter 9!

Alice (Passport)		Bob (Reader)
	$\xleftarrow{\quad s \quad}$	$s \xleftarrow{\$} \{0,1\}^k$
$G \leftarrow \mathcal{H}(s\|\pi)$		$G \leftarrow \mathcal{H}(s\|\pi)$
$r_A \xleftarrow{\$} \mathbb{Z}_p$		$r_B \xleftarrow{\$} \mathbb{Z}_p$
$A \leftarrow [r_A]G$	$\xrightarrow{\quad A \quad}$	
	$\xleftarrow{\quad B \quad}$	$B \leftarrow [r_B]G$
$K \leftarrow [r_A]B$		$K \leftarrow [r_B]A$

FIGURE 8.2   A randomized variant of the SPEKE protocol.

8.2.2   The Issue of Timing Attacks

A concrete situation in which this varying running time can be a serious issue is the case of embedded devices (especially e-passports) implementing an elliptic curve-based Password-Authenticated Key Exchange (PAKE) protocol.

PAKE is a method for two parties sharing a common low-entropy secret (such as a four-digit PIN, or a self-picked alphabetic password) to derive a high-entropy session key for secure communication in an authenticated way. One of the main security requirements is, informally, that an attacker should not be able to gain any information about the password, except through a brute force online dictionary attack (i.e., impersonating one of the parties in the protocol and attempting to authenticate with each password, one password at a time), which can be prevented in practice by latency, smart card blocking, and other operational measures. In particular, a PAKE protocol should be considered broken if a *passive* adversary can learn any information about the password.

Now consider the PAKE protocol described in Figure 8.2, which is essentially Jablon's Simple Password-base Exponential Key Exchange (SPEKE) [34] implemented over an elliptic curve, except with a random salt as suggested in [35]. The public parameters are an elliptic curve group $\mathbb{G}$ of prime order p and a hash function $\mathcal{H}\colon \{0,1\}^* \to \mathbb{G}$. The two parties share a common password π, and derive a high-entropy $K \in \mathbb{G}$ using Diffie–Hellman key agreement in $\mathbb{G}$ but with a variable generator $G \in \mathbb{G}$ computed by hashing the password.

But if the hash function $\mathcal{H}$ is instantiated by the try-and-increment construction and an eavesdropper is able to measure the running time of one of the parties, he will find different running times or different power traces depending on how many trials it takes to find a suitable x-coordinate in the computation of $\mathcal{H}(s\|\pi)$. Since it takes a single iteration with probability close to $1/2$, an execution of the protocol provides at least one bit of information about π to the adversary (and about $-\sum_{k \geq 1} 2^{-k} \log_2(2^{-k}) = 2$ bits on average).

This leads to a so-called "partition attack", conceptually similar to those described by Boyd et al. in [13]: The adversary can count the number of iterations needed to compute $\mathcal{H}(s\|\pi_0)$ for each password π_0 in the password dictionary, keeping only the π_0's for which this number of iterations matches the side-channel measurement. This reduces the search space by a factor of at least 2 (and more typically 4) for each execution of the protocol, as the running times for different values of s are independent. As a result, the eavesdropper can typically reduce his search space to a single password after at most a few dozen executions of the protocol!

A rather inefficient countermeasure that can be considered is to run all ℓ iterations of the try-and-increment algorithm every time. However, even that is probably insufficient to thwart the attack: indeed, the usual algorithm (using quadratic reciprocity) for testing whether an element of $\mathbb{F}_q$ is a square, as is done in Step 4 of Algorithm 8.1, also has different running times depending on its input. This can provide information to the adversary as well, unless this part is somehow tweaked to run in constant time,[*] which seems difficult to do short of computing the quadratic character with an exponentiation and making the algorithm prohibitively slow with ℓ exponentiations every time. In principle, padding the quadratic reciprocity-based algorithm with dummy operations might provide a less computationally expensive solution, but implementing such a countermeasure securely seems quite daunting. A construction that naturally runs in constant time would certainly be preferable.

8.3 Encoding Functions to Elliptic Curves

8.3.1 Main Idea

A natural way to construct a constant-time hash function to an elliptic curve E would be to use, as a building block, a suitable function $f\colon \mathbb{F}_q \to E(\mathbb{F}_q)$ that can be efficiently computed in constant time. Then, combining f with a hash function $\mathfrak{h}\colon \{0,1\}^* \to \mathbb{F}_q$, we can hope to obtain a well-behaved hash function to $E(\mathbb{F}_q)$.

Of course, not all such functions f are appropriate: for example, when $q = p$ is prime, the trivial encoding described in §8.1 is essentially of that form, with $f\colon u \mapsto [\hat{u}]G$ (and $u \mapsto \hat{u}$ any lifting of $\mathbb{F}_p$ to $\mathbb{Z}$).

On the other hand, if f is a bijection between $\mathbb{F}_q$ and $E(\mathbb{F}_q)$ whose inverse is also efficiently computable, then the following construction:

$$\mathcal{H}(\mathsf{m}) = f\big(\mathfrak{h}(\mathsf{m})\big) \tag{8.4}$$

is well-behaved, in the sense that if $\mathfrak{h}$ is modeled as a random oracle to $\mathbb{F}_q$, then $\mathcal{H}$ can replace a random oracle to $E(\mathbb{F}_q)$ in any protocol while preserving proofs of security in the random oracle model. Indeed, contrary to what happens in the case of the trivial encoding (where programming the random oracle would require computing discrete logarithm), a simulator can easily choose a value $\mathcal{H}(\mathsf{m}_0) = P_0$ by setting $\mathfrak{h}(\mathsf{m}_0) = f^{-1}(P_0)$. More generally, such a construction is, again, indifferentiable from a random oracle to $E(\mathbb{F}_q)$ (and even *reset indifferentiable* in the sense of Ristenpart et al. [45]).

The same holds if f induces a bijection from $\mathbb{F}_q \setminus T$ to $E(\mathbb{F}_q) \setminus W$ for some finite or negligibly small sets of points T, W.

More generally, we will be considering cases where f is not necessarily an efficiently invertible bijection but only a so-called *samplable* mapping, in the sense that for each $P \in E(\mathbb{F}_q)$, one can compute a random element of $f^{-1}(P)$ in probabilistic polynomial time.

8.3.2 The Boneh–Franklin Encoding

It was actually one of the first papers requiring hashing to elliptic curves, namely Boneh and Franklin's construction [10] of identity-based encryption from the Weil pairing, that introduced the first practical example of a hash function of the form (8.4). Boneh and Franklin used elliptic curves of a very special form:

$$E\colon y^2 = x^3 + b \tag{8.5}$$

over a field $\mathbb{F}_q$ such that $q \equiv 2 \pmod 3$. In $\mathbb{F}_q$, $u \mapsto u^3$ is clearly a bijection, and thus each element has a unique cube root. This makes it possible, following Boneh and Franklin, to define

---

[*]By constant time, we mean "whose running time does not depend on the input" (once the choice of parameters like E and $\mathbb{F}_q$ is fixed), and not $O(1)$ time in the sense of complexity theory.

a function f as:

$$f \colon \mathbb{F}_q \to E(\mathbb{F}_q) \tag{8.6}$$

$$u \mapsto \left(\left(u^2 - b \right)^{1/3}, u \right). \tag{8.7}$$

In other words, instead of picking the x-coordinate and trying to deduce the y-coordinate by taking a square root (which may not exist) as before, we first choose the y-coordinate and deduce the x-coordinate by taking a cube root (which always exists).

Obviously, the function f is a bijection from $\mathbb{F}_q$ to all the finite points of $E(\mathbb{F}_q)$. In particular, this implies that $\#E(\mathbb{F}_q) = 1 + \#\mathbb{F}_q = q + 1$; thus, E is supersingular (and hence comes with an efficient symmetric pairing). This also means that f satisfies the conditions mentioned in the previous section; therefore, construction (8.4) can replace the random oracle $\mathcal{H}$ required by the Boneh–Franklin IBE scheme, or any other protocol proved secure in the random oracle model. And it can also easily be computed in constant time: It suffices to compute the cube root as an exponentiation to a fixed power α such that $3\alpha \equiv 1 \pmod{q-1}$.

Note that in fact, the group $\mathbb{G}$ considered by Boneh and Franklin isn't $E(\mathbb{F}_q)$ itself, but a subgroup $\mathbb{G} \subset E(\mathbb{F}_q)$ of prime order. More precisely, the cardinality q of the base field is chosen of the form $6r - 1$ for some prime $r \neq 2, 3$. Then $E(\mathbb{F}_q)$ has a unique subgroup $\mathbb{G}$ of order r (the curve has cofactor 6), which is the group actually used in the scheme. Hashing to $\mathbb{G}$ rather than $E(\mathbb{F}_q)$ is then easy:

$$\mathcal{H}(\mathsf{m}) = f'\big(\mathfrak{h}(\mathsf{m})\big) \quad \text{where} \quad f'(u) = [6]f(u). \tag{8.8}$$

The encoding f' defined in that way isn't injective but it is samplable: indeed, to compute a random preimage of some point $P \in \mathbb{G}$, we can simply compute the six points Q_i such that $[6]Q_i = P$, and return $f^{-1}(Q_i)$ for a random index i. Using that observation, Boneh and Franklin prove that construction (8.8) can replace the random oracle to $\mathbb{G}$ in their IBE scheme. More generally, it is easy to see that it is indifferentiable from a random oracle in the sense of Maurer et al. [40].

8.3.3 Beyond Supersingular Curves

The previous example suggests that a sensible first step towards constructing well-behaved constant-time hash functions to elliptic curves is to first obtain mappings $f \colon \mathbb{F}_q \to E(\mathbb{F}_q)$ that are computable in deterministic polynomial time and samplable, and admit constant-time implementations. We will refer to such mappings as *encoding functions* or simply *encodings*. Note that despite what the name might suggest, there is no assumption of injectivity for those mappings.

It turns out that constructing encodings to elliptic curves beyond special cases such as Eq. (8.6) is far from an easy task. In fact, Schoof mentioned the presumably easier problem of constructing a *single* non-identity point on a general elliptic curve over a finite field as open in his 1985 paper on point counting [46], and little progress was made on this problem before the 2000s. Nevertheless, we now know how to construct encodings to essentially all elliptic curves thanks to the work of numerous researchers.

We now present the two most important constructions, due to Shallue and van de Woestijne on the one hand, and Icart on the other.

8.3.4 The Shallue–van de Woestijne Approach

In a paper presented at ANTS in 2006, Shallue and van de Woestijne [50] proposed a general construction of an encoding function that applies to all elliptic curves over finite fields of odd characterstic.

Consider the general Weierstrass equation for an elliptic curve in odd characteristic (possibly including 3):

$$E: y^2 = x^3 + ax^2 + bx + c.$$

Let further $g(x) = x^3 + ax^2 + bx + c \in \mathbb{F}_q[x]$. It is possible to construct an encoding function to $E(\mathbb{F}_q)$ from a rational curve on the three-dimensional variety:

$$V: y^2 = g(x_1)g(x_2)g(x_3)$$

(which, geometrically, is the quotient of $E \times E \times E$ by $(\mathbb{Z}/2\mathbb{Z})^2$, where each non-trivial element acts by $[-1]$ on two components and by the identity on the third one). Indeed, if $\phi \colon \mathbb{A}^1 \to V$, $t \mapsto \big(x_1(t), x_2(t), x_3(t), y(t)\big)$ is such a rational curve (i.e., a rational map of the affine line $\mathbb{A}^1$, parametrized by t, to the variety V), then for any $u \in \mathbb{F}_q$ that is not a pole of ϕ, at least one of $g(x_i(u))$ for $i = 1, 2, 3$ is a quadratic residue (because the product of three quadratic nonresidues is not a square, and hence cannot be equal to $y(u)^2$). This yields a well-defined point $\big(x_j(u), \sqrt{g(x_j(u))}\big) \in E(\mathbb{F}_q)$ where j is the first index such that $g(x_j(u))$ is a square, and thus we obtain the required encoding function.

Then, Shallue and van de Woestijne show how to construct such a rational curve ϕ (and in fact a large number of them). They first obtain an explicit rational map $\psi : S \to V$, where S is the surface of equation:

$$S: y^2 \cdot \big(u^2 + uv + v^2 + a(u+v) + b\big) = -g(u),$$

which can also be written, by completing the square with respect to v, as:

$$\left[y\big(v + \tfrac{1}{2}u + \tfrac{1}{2}a\big)\right]^2 + \left[\tfrac{3}{4}u^2 + \tfrac{1}{2}au + b - \tfrac{1}{4}a^2\right]y^2 = -g(u).$$

Now observe that for any fixed $u \in \mathbb{F}_q$, the previous equation defines a curve of genus 0 in the (v, y)-plane. More precisely, it can be written as:

$$z^2 + \alpha y^2 = -g(u) \tag{8.9}$$

with $z = y(v + \tfrac{1}{2}u + \tfrac{1}{2}a)$ and $\alpha = \tfrac{3}{4}u^2 + \tfrac{1}{2}au + b - \tfrac{1}{4}a^2$. This is a non-degenerate conic as soon as α and $g(u)$ are both non-zero (which happens for all $u \in \mathbb{F}_q$ except at most 5), and then admits a rational parametrization, yielding a rational curve $\mathbb{A}^1 \to S$. Composing with ψ, we get the required rational curve on V, and hence an encoding, provided that $q > 5$.

8.3.5 Icart's Approach

In [33], Icart introduced an encoding function based on a very different idea, namely, trying to adapt the Boneh–Franklin encoding discussed in §8.3.2 to the case of an ordinary elliptic curve. More precisely, consider again an elliptic curve E given by a short Weierstrass equation:

$$E: y^2 = x^3 + ax + b$$

over a field $\mathbb{F}_q$ of odd characteristic with $q \equiv 2 \pmod 3$. The idea is again to reduce the equation to a binomial cubic, which can be solved directly in $\mathbb{F}_q$ (where $u \mapsto u^3$ is a bijection).

Unlike the simple case considered by Boneh and Franklin, this cannot be done by picking y as a constant: doing so results in a trinomial cubic which does not always have a root in $\mathbb{F}_q$. Icart's idea is to set $y = ux + v$ for two parameters u, v to be chosen later. This gives:

$$x^3 - u^2x^2 + (a - 2uv)x + b - v^2 = 0$$

$$f \colon \mathbb{F}_q \longrightarrow E(\mathbb{F}_q)$$

$$u \longmapsto \left(\left(v^2 - b - \frac{u^6}{27} \right)^{1/3} + \frac{u^2}{3} \; ; \; ux + v \right)$$

where $v = (a - 3u^4)/(6u)$. By convention, $f(0) = O$, the identity element.

FIGURE 8.3   Icart's encoding to $E \colon y^2 = x^3 + ax + b$ over $\mathbb{F}_q$ with $q \equiv 2 \pmod 3$.

and after completing the cube:

$$\left(x - \frac{u^2}{3} \right)^3 + \left(a - 2uv - \frac{u^4}{3} \right) x = v^2 - b - \frac{u^6}{27},$$

Thus, by setting $v = (a - 3u^4)/(6u)$, it is possible to cancel the term of degree 1 and obtain a binomial cubic equation:

$$\left(x - \frac{u^2}{3} \right)^3 = v^2 - b - \frac{u^6}{27},$$

which is easy to solve for x in $\mathbb{F}_q$. This gives Icart's encoding, described in Figure 8.3.

This encoding applies to a more restricted setting than the Shallue–van de Woestijne encodings, due to the requirement that $q \equiv 2 \pmod 3$, but it has the advantage of being very easy to describe and implement in constant time.

8.4   Constant-Time Hashing to Pairing-Friendly Curves

8.4.1   From Encodings to Hash Functions

In the previous section, we have described several constructions of encoding functions to elliptic curves. It is not clear, however, that they solve our initial problem of hashing to elliptic curve groups. There are two issues at play: the first is the lack of indifferentiability, and the second is the fact that we want to map to the subgroup $\mathbb{G}_1$ or $\mathbb{G}_2$ rather than the whole curve.

The issue of indifferentiability

The basic construction of a hash function $\mathcal{H} \colon \{0,1\}^* \to E(\mathbb{F}_q)$ from an $\mathbb{F}_q$-valued random oracle $\mathfrak{h} \colon \{0,1\}^* \to \mathbb{F}_q$ and an encoding $f \colon \mathbb{F}_q \to E(\mathbb{F}_q)$, as suggested in §8.3, is simply:

$$\mathcal{H}(\mathrm{m}) = f\big(\mathfrak{h}(\mathrm{m})\big). \tag{8.10}$$

However, unlike what happens for the Boneh–Franklin encoding, the resulting hash function $\mathcal{H}$ does not necessarily have strong security properties.

Consider the case when f is Icart's encoding, for example (most other encodings are similar). One can then prove some limited security properties on $\mathcal{H}$, such as that $\mathcal{H}$ is one-way if $\mathfrak{h}$ is [33, Lemma 5]. However, unlike the Boneh–Franklin encoding, f is not a surjective or "almost" surjective function to the target group $E(\mathbb{F}_q)$. Indeed, in his original paper [33], Icart could only show that the image $f(\mathbb{F}_q)$ satisfies $\#f(\mathbb{F}_q) \gtrsim (1/4) \cdot \#E(\mathbb{F}_q)$, and conjectured that, in fact, $\#f(\mathbb{F}_q) \approx (5/8) \cdot \#E(\mathbb{F}_q)$ (a conjecture which was later proved in [23, 24]). As a result, the hash function $\mathcal{H}$ constructed from f using formula (8.10) is easily distinguished from a random oracle!

To see this, note that since f is an algebraic function, we can efficiently compute $f^{-1}(P)$ for any $P \in E(\mathbb{F}_q)$ by solving a polynomial equation over $\mathbb{F}_q$. In particular, it is possible to decide efficiently whether P is in the image of f or not. Therefore, we can construct a distinguisher $\mathcal{D}$ between $\mathcal{H}_0 = \mathcal{H}$ and a random oracle $\mathcal{H}_1$ to $E(\mathbb{F}_q)$ as follows. $\mathcal{D}$ is given as input $P = \mathcal{H}_b(\mathsf{m}) \in E(\mathbb{F}_q)$ for some message m and a random bit $b \in \{0, 1\}$. It answers with a guess of the bit b, as $b = 0$ if P is in $f(\mathbb{F}_q)$ and $b = 1$ otherwise. Then $\mathcal{D}$ has a constant positive advantage. Indeed, it answers correctly with probability 1 if $P \notin f(\mathbb{F}_q)$, and with probability $1/2$ otherwise, hence it has a non-negligible advantage in the distinguishing game. Thus, clearly, construction (8.10) does not behave like a random oracle when f is Icart's encoding (or most other encodings), and cannot replace a random oracle in a generic way.

In many protocols requiring a hash function to an elliptic curve group, this is actually not much of a problem, and an encoding with an image size that is a constant fraction of $\#E(\mathbb{F}_q)$ is often good enough. The reason is that, in a random oracle proof of security, the simulator programs the random oracle by setting the hash of some message m to a value P, but that point P itself can usually be anything depending on some randomness. So the simulator might typically want to set $\mathcal{H}(\mathsf{m})$ to $P = [r]G$ for some random r, say. Now if $\mathcal{H}$ is defined in the protocol using a construction like (8.10), the simulator would pick a random r and set $\mathfrak{h}(\mathsf{m})$ to one of the preimages $u \in f^{-1}(P)$ if $P \in f(\mathbb{F}_q)$. If, however, P is not in the image of f, the simulator would pick another random r and try again.

Nevertheless, it seems difficult to give formal sufficient conditions on a protocol for it to remain secure when the elliptic curve-valued random oracle is replaced by a construction like (8.10). One can actually find protocols that are secure in the random oracle model, but in which using that construction instead breaks security completely [16].

Therefore, it would be desirable to obtain from the encodings discussed thus far a construction that does satisfy the *indifferentiability* property mentioned in §8.2, and can thus be used as a plug-in replacement for elliptic curve-valued random oracles in a very large class of protocols. The problem was solved by Brier et al. [16] in the case of Icart's function, and by Farashahi et al. [22] in general. They prove that the following construction achieves indifferentiability from a random oracle:

$$\mathcal{H}(\mathsf{m}) = f\big(\mathfrak{h}_1(\mathsf{m})\big) + f\big(\mathfrak{h}_2(\mathsf{m})\big) \tag{8.11}$$

where $\mathfrak{h}_1$ and $\mathfrak{h}_2$ are modeled as random oracles $\{0,1\}^* \to \mathbb{F}_q$ (and the addition is the usual group operation in $E(\mathbb{F}_q)$). As a result, to obtain an efficient indifferentiable hash function construction, it suffices to know how to compute a function of the form (8.10) efficiently: do it twice, add the results together, and you get indifferentiability. Therefore, and since in many cases it is sufficient by itself, the form (8.10) is what the rest of this chapter will focus on.

Hashing to subgroups

Most of the discussion so far has focused on the problem of hashing to the whole group $E(\mathbb{F}_q)$ of points on the elliptic curve E. But this is not in fact what we need for pairing-based cryptography: The groups we would like to hash to are pairing groups $\mathbb{G}_1$ and $\mathbb{G}_2$, which are *subgroups* of an elliptic curve group. Let us review how hashing to those subgroups works.

Hashing to $\mathbb{G}_1$

This case is simpler. Consider a pairing-friendly curve $E/\mathbb{F}_p$ over a prime field. The group $\mathbb{G}_1$ is just the group $E(\mathbb{F}_p)[r]$ of r-torsion points in $E(\mathbb{F}_p)$, for some large prime divisor r of $\#E(\mathbb{F}_p)$. If we denote by c the cofactor of E, i.e., the integer such that $\#E(\mathbb{F}_p) = c \cdot r$, then c is always coprime to r, and $\mathbb{G}_1$ can thus be obtained as the image of the homomorphism $[c]$ of multiplication by c in $E(\mathbb{F}_p)$.

Now suppose we are given some well-behaved hash function $\mathcal{H} \colon \{0,1\}^* \to \#E(\mathbb{F}_p)$. Then we can construct a map $\mathcal{H}_1 \colon \{0,1\}^* \to \mathbb{G}_1$ by defining $\mathcal{H}_1(\mathsf{m}) = [c]\mathcal{H}(\mathsf{m})$, and it turns out that $\mathcal{H}_1$ is still a well-behaved hash function. For example, Brier et al. [16] show that if $\mathcal{H}$ is indifferentiable from a random oracle, then so is $\mathcal{H}_1$. This results from the fact that the muliplication-by-c homomorphism is efficiently computable, regular (all elements of $\mathbb{G}_1$ have the same number of preimages, namely c), and efficiently samplable (we can sample a uniformly distributed preimage of an element in $\mathbb{G}_1$ by adding to it a random element of order c in $E(\mathbb{F}_p)$, which we can, for example, generate as $[r]P$ for P, a uniformly sampled random point).

As a result, to hash to $\mathbb{G}_1$ efficiently, we simply need an efficient way of computing some curve-valued hash function $\mathcal{H}$, and an efficient way of evaluating the multiplication-by-c map. The latter is typically quite cheap.

Hashing to $\mathbb{G}_2$

This case is more complicated in general. Indeed, generally speaking, $\mathbb{G}_2$ is one specific subgroup of order r in the group $E(\mathbb{F}_{p^k})[r]$ of r-torsion points of E over the embedding field $\mathbb{F}_{p^k}$. But since $E(\mathbb{F}_{p^k})[r]$ is isomorphic to $(\mathbb{Z}/r\mathbb{Z})^2$, there are many subgroups of order r, and one cannot just multiply by some cofactor to map into $\mathbb{G}_2$. The approach used to hash to $\mathbb{G}_2$ will differ according to which of the three *pairing types*, in the sense of Galbraith, Paterson, and Smart [28], we are working with (we also refer Chapter 3 for a discussion of Type I, Type II, and Type III pairings).

For Type I pairings, the distortion map provides an efficiently computable isomorphism from $\mathbb{G}_1$ to $\mathbb{G}_2$. Therefore, we can simply hash to $\mathbb{G}_1$ as above and compose with the distortion map to obtain a hash function to $\mathbb{G}_2$.

For Type II pairings, $\mathbb{G}_2$ is not the image of any efficiently computable homomorphism, and as a result, there is in fact no way of efficiently hashing to that group. *One cannot instantiate protocols that require hashing to $\mathbb{G}_2$ in the Type II setting.* In rare cases where one really needs both the ability to hash to $\mathbb{G}_2$ and the existence of a one-way isomorphism $\mathbb{G}_2 \to \mathbb{G}_1$, a possible workaround is to replace $\mathbb{G}_2$ with the entire group $E(\mathbb{F}_{p^k})[r]$ of order r^2, which we can hash to as above using the multiplication by $\#E(\mathbb{F}_{p^k})/r^2$ in $E(\mathbb{F}_{p^k})$. This is usually called the Type IV pairing setting [20, 18]. The cofactor multiplication in that case is quite costly, so it may be interesting to optimize it. However, it is usually possible to convert such protocols to the significantly more efficient Type III setting [19], so we will not consider that case further in the rest of this chapter.

For Type III pairings, as was seen in Chapter 3, $\mathbb{G}_2 \subset E(\mathbb{F}_{p^k})[r]$ is the eigenspace of the Frobenius endomorphism π associated with the eigenvalue q, and the complementary subspace of $\mathbb{G}_1$ (which is the eigenspace for the eigenvalue 1). As a result, $\mathbb{G}_2$ is the image of the efficient endomorphism $\frac{\pi - \mathrm{Id}}{q - 1}$ of $E(\mathbb{F}_{p^k})[r]$. Therefore, we can hash to $\mathbb{G}_2$ by cofactor multiplication to get into $E(\mathbb{F}_{p^k})[r]$, composed with that endomorphism. In practice, however, it is much more preferable to represent $\mathbb{G}_2$ as a subgroup in the degree-d twist E' of E over a lower degree extension $\mathbb{F}_{p^{k/d}}$. Doing so, $\mathbb{G}_2$ simply appears as the subgroup $E'(\mathbb{F}_{p^{k/d}})[r]$ of r-torsion points on that curve, and hashing can be done exactly as in the case of $\mathbb{G}_1$. Contrary to the case of $\mathbb{G}_1$, however, the cofactor in this case is usually quite large, and it is thus a major concern to make it as fast as possible. This is one of the main issues discussed in the coming sections.

8.4.2   Case Study: The Barreto–Naehrig Elliptic Curves

In this subsection we discuss how to apply the the Shallue and van de Woestijne encoding described in §8.3.4 along with the hashing techniques discussed in §8.4, in order to construct points that belong to the groups of a popular instantiation of type III pairings, namely, bilinear pairings implemented over the Barreto–Naehrig elliptic curves.

As already studied in Chapter 7, Barreto–Naehrig elliptic curves are defined by the equation $E : y^2 = x^3 + b$, with $b \neq 0$. Their embedding degree is $k = 12$, hence, $\phi(k) = 4$. They are

---

ALGORITHM 8.2 Constant-time hash function to $\mathbb{G}_1$ on Barreto–Naehrig curves [25].

Input : $t \in \mathbb{F}_p^*$, parameter $b \in \mathbb{F}_p$ of $E/\mathbb{F}_p : y^2 = x^3 + b$
Output: A point Point $P = (x, y) \in \mathbb{G}_1$

1 **Precomputation:**
2 $sqrt_3 \leftarrow \sqrt{-3}$;
3 $j \leftarrow (-1 + sqrt_3)/2$;
4 **Main Computation:**
5 $w \leftarrow sqrt_3 \cdot \frac{t}{1+b+t^2}$;
6 $x_1 \leftarrow j - t \cdot w$;
7 $x_2 \leftarrow -1 - x_1$;
8 $x_3 \leftarrow 1 + 1/w^2$;
9 $\alpha \leftarrow \chi_q(x_1^3 + b)$;   // Using Euler's Criterion: $\left(x_1^3 + b\right)^{\frac{p-1}{2}}$.
10 $\beta \leftarrow \chi_q(x_2^3 + b)$;   // Using Euler's Criterion: $\left(x_2^3 + b\right)^{\frac{p-1}{2}}$.
11 $i \leftarrow [(\alpha - 1) \cdot \beta \mod 3] + 1$;
12 **return** $P \leftarrow (x_i, \chi_q(t) \cdot \sqrt{x_i^3 + b})$;   // Using Euler's Criterion: $t^{\frac{p-1}{2}}$.

---

parametrized by selecting an arbitrary $x \in \mathbb{Z}$ such that

$$p(x) = 36x^4 + 36x^3 + 24x^2 + 6x + 1; \qquad (8.12)$$
$$r(x) = 36x^4 + 36x^3 + 18x^2 + 6x + 1;$$

are both prime.

Constant-time hashing to $\mathbb{G}_1$

In Latincrypt 2012, Fouque and Tibouchi [25] presented a specialization of the procedure proposed by Shallue and van de Woestijne [50] applied to Barreto–Naehrig curves, which are defined over the finite field $\mathbb{F}_p$, with $p \equiv 7 \mod 12$, or $p \equiv 1 \mod 12$. The mapping covers a 9/16 fraction of the prime group size $r = \#E(F_p)$. In a nutshell, the procedure proposed in [25] consists in the following.

Let t be an arbitrary nonzero element in the base field $\mathbb{F}_p^*$ such that $x_1, x_2, x_3 \in \mathbb{F}_p^*$ are defined as

$$x_1 = \frac{-1 + \sqrt{-3}}{2} - \frac{\sqrt{-3} \cdot t^2}{1 + b + t^2},$$
$$x_2 = \frac{-1 - \sqrt{-3}}{2} + \frac{\sqrt{-3} \cdot t^2}{1 + b + t^2},$$
$$x_3 = 1 - \frac{(1 + b + t^2)^2}{3t^2}.$$

The Shallue–van de Woestijne encoding applied to the Barreto–Naehrig curves of the form $E : y^2 = g(x) = x^3 + b$ is given by the following projection:

$$f : \mathbb{F}_p^* \rightarrow E(\mathbb{F}_p)$$
$$t \mapsto \left(x_i, \chi_p(t) \cdot \sqrt{g(x_i)}\right),$$

where the index $i \in \{1, 2, 3\}$ is the smallest integer such that $g(x_i)$ is a square in $\mathbb{F}_p$ and the function $\chi_p : \{-1, 0, 1\}$, computes the non-trivial quadratic character over the field $\mathbb{F}_p^*$, also known as quadratic residuosity test (Definition 2.29). The procedure just outlined is presented in Algorithm 8.2.

ALGORITHM 8.3   Blind factor version of the Hash function to $\mathbb{G}_1$ on Barreto–Naehrig curves [25].

Input : $t \in \mathbb{F}_p^*$, parameter $b \in \mathbb{F}_p$ of $E/\mathbb{F}_p : y^2 = x^3 + b$
Output: A point Point $P = (x, y) \in \mathbb{G}_1$

1 **Precomputation:**
2 $sqrt_3 \leftarrow \sqrt{-3}$;
3 $j \leftarrow (-1 + sqrt_3)/2$;
4 **Main Computation:**
5 $w \leftarrow sqrt_3 \cdot \frac{t}{1+b+t^2}$;
6 $x_1 \leftarrow j - t \cdot w$;
7 $x_2 \leftarrow -1 - x_1$;
8 $x_3 \leftarrow 1 + 1/w^2$;
9 $r_1, r_2, r_3 \xleftarrow{\$} \mathbb{F}_p^*$;
10 $\alpha \leftarrow \chi_q(r_1^2 \cdot (x_1^3 + b))$;   // Using Algorithm 2.3
11 $\beta \leftarrow \chi_q(r_1^2 \cdot (x_2^3 + b))$;   // Using Algorithm 2.3
12 $i \leftarrow [(\alpha - 1) \cdot \beta \mod 3] + 1$;
13 **return** $P \leftarrow (x_i, \chi_q(r_3^2 \cdot t) \cdot \sqrt{x_i^3 + b})$;   // Using Algorithm 2.3

Remark 8.1   The Barreto–Naehrig curve $\mathbb{G}_1$ subgroup. Notice that the Barreto–Naehrig curves are exceptional in the sense that the subgroup $\mathbb{G}_1$ is exactly the same as $E(\mathbb{F}_p)$. In other words, for this case, the cofactor c is equal to one, and therefore, the procedure presented in Algorithm 8.2 effectively completes the hashing to $\mathbb{G}_1$ as illustrated in Figure 8.1.

Remark 8.2   Implementation aspects. All the computations of Algorithm 8.2 are performed over the base field $\mathbb{F}_p$ at a cost of two field inversions, three quadratic character tests, one square root, and few field multiplications. Notice that the values $\sqrt{-3}$ and $\frac{-1+\sqrt{-3}}{2}$ are precomputed offline in Steps 2–3. Moreover, when p is chosen such that $p \equiv 3 \mod 4$, the square root $\sqrt{x_i^3 + b}$ (line 13) can be computed by the power $(x_i^3 + b)^{\frac{q+1}{4}}$.

In order to ensure a constant-time behavior, the quadratic residuosity test of a field element a can be computed by performing the exponentiation $a^{\frac{p-1}{2}}$. Alternatively, one can invoke the procedure discussed in Chapter 2 and shown in Algorithm 2.3.

Algorithm 2.3 performs the quadratic residuosity test by recursively applying Gauss' law of quadratic reciprocity at a computational cost similar to computing the greatest common divisor of a and p. Unfortunately, it is difficult to implement Algorithm 2.3 in constant-time. That is why the authors of [25] suggested using blinding techniques in order to thwart potential timing attacks. This variant was adopted in several papers such as [21, 52] and implemented in Algorithm 8.3.

Remark 8.3   On the security of Algorithm 8.3. Strictly speaking, the blinding factor protection of Algorithm 8.3 is not *provably* secure against timing attacks. This is because even if the blinding factors are uniformly distributed in the base field, and kept unknown to the adversary, the input of the algorithm computing the quadratic character χ_q, is not uniformly distributed in all of $\mathbb{F}_p^*$, but only among its quadratic residues or quadratic nonresidues. Practically speaking, this is not very significant: Very little secret information can leak in that way. Moreover, if the quadratic residuosity test is performed in constant-time, then no information will be leaked at all. Nevertheless, it is always possible to achieve provable protection through additional blinding. For example, one can randomly multiply by a blind factor that is a known

ALGORITHM 8.4 Deterministic construction of points in $E'(\mathbb{F}_{p^2})$ for Barreto–Naehrig curves.

Input : $t \in \mathbb{F}_p^*$, parameter $B = b_0 + b_1 u \in \mathbb{F}_{p^2}$ of $E'/\mathbb{F}_{p^2} : Y^2 = X^3 + B$
Output: Point $Q = (x, y) \in E'(\mathbb{F}_{p^2})$

1 **Precomputations:**
2 $sqrt_3 \leftarrow \sqrt{-3}$
3 $j \leftarrow (-1 + sqrt_3)/2$

4 **Calculations:**
5 $a_0 \leftarrow 1 + b_0 + t^2$
6 $a_1 \leftarrow b_1$
7 $A \leftarrow 1/A$ // with $A = a_0 + a_1 u \in \mathbb{F}_{p^2}$
8 $c \leftarrow sqrt_3 \cdot t$
9 $W \leftarrow (c \cdot a_0) + (c \cdot a_1)u$ // with $W = w_0 + w_1 u \in \mathbb{F}_{p^2}$.
10 $a_0 \leftarrow w_0 \cdot t$
11 $a_1 \leftarrow w_1 \cdot t$
12 $X_1 \leftarrow (j - a_0) - a_1 u$ // with $X_1 = x_{1,0} + x_{1,1} u \in \mathbb{F}_{p^2}$
13 $X_2 \leftarrow (-x_{1,0} - 1) - x_{1,1} u$
14 $X_3 \leftarrow 1/W^2$ // with $X_3 = x_{3,0} + x_{3,1} u \in \mathbb{F}_{p^2}$
15 $X_3 \leftarrow (1 + x_{3,0}) + x_{3,1} u$
16 $\alpha \leftarrow \chi_{p^2}(X_1^3 + B)$
17 $\beta \leftarrow \chi_{p^2}(X_2^3 + B)$
18 $i \leftarrow [(\alpha - 1) \cdot \beta \mod 3] + 1$
19 **return** $Q \leftarrow (X_i, \chi_p(t) \cdot \sqrt{X_i^3 + B})$

square/non-square with probability $1/2$, and then adjust the output accordingly.

Deterministic construction of points in $E'(\mathbb{F}_{q^2})$ for Barreto–Naehrig curves

The Barreto–Naehrig family of elliptic curves has an embedding degree of $k = 12$, and an associated twist curve E' with degree $d = 6$. This defines the group $\mathbb{G}_2$ as

$$\mathbb{G}_2 = E'(\mathbb{F}_{p^{k/d}})[r] = E'(\mathbb{F}_{p^2})[r].$$

As already mentioned, the encoding described by Fouque and Tibouchi in [25] applies over finite fields $\mathbb{F}_p$, where $p \equiv 7 \mod 12$ or $p \equiv 1 \mod 12$. In the case of the Barreto–Naehrig curves, one observes that since $p \equiv 7 \mod 12$, then $p^2 \equiv 1 \mod 12$. As a result, the encoding presented in [25] can be applied as it is in order to find random points over $E'/\mathbb{F}_{p^2}$, except that several computations must be performed over the quadratic field extension $\mathbb{F}_{p^2}$. The corresponding procedure is shown in Algorithm 8.4.

Notice that all the operations of Algorithm 8.4 are performed over the base field $\mathbb{F}_p$ and its quadratic extension $\mathbb{F}_{p^2} = \mathbb{F}_p[u]/u^2 - \beta$, where $\beta = -1$ is not a square over $\mathbb{F}_p$. In particular, the steps 16 and 17 of Algorithm 8.4 must compute in constant-time the quadratic character $\chi_{p^2}(\cdot)$ over the extension field $\mathbb{F}_{p^2}$ (Definition 2.29). To this end, one can use the procedure described in [1], which is an improvement over the work made by Bach and Huber [3]. The authors of [1] proposed to compute the quadratic character over the quadratic field extension $\mathbb{F}_{p^2}$ by descending the computation to the base field $\mathbb{F}_p$, using the following identity:

$$\chi_{p^2}(a) = a^{\frac{p^2-1}{2}} = (a \cdot a^p)^{\frac{p-1}{2}} \tag{8.13}$$
$$= (a \cdot \bar{a})^{\frac{p-1}{2}} = \chi_p(a \cdot \bar{a}) = \chi_p(|a|)$$

where $\bar{a} = a_0 - a_1 u$ and $|a|$ is the conjugate and the norm of a, respectively (Definition 2.27). The above computation can be carried out at a cost of two squarings, one addition, and the computation of the quadractic character $\chi_p(|a|)$. As before, $\chi_p(|a|)$ can be computed either by performing one exponentiation over the base field $\mathbb{F}_p$, or alternatively, by applying Algorithm 2.3 along with blinding techniques.

Furthermore, in line 19 of Algorithm 8.4 one needs to extract a square root over the quadratic field $\mathbb{F}_{p^2}$. This operation can be efficiently computed using the complex method proposed by Scott in [47], which was already reviewed in Algorithm 5.18 of Chapter 5.

Remark 8.4 Notice that in line 19 of Algorithm 8.4, the procedure proposed in [25] guarantees that the term $X_i^3 + B$ is always a quadratic residue over the field $\mathbb{F}_{p^2}$. Hence, one can safely omit the first quadratic residuosity test performed in Algorithm 5.18 of Chapter 5.

8.5   Efficient Hashing to $\mathbb{G}_2$

In this section we are interested in the efficient computation of the third mapping shown in Figure 8.1, namely, the computation of the scalar multiplication, $Q' = [c]\tilde{Q}$.

Let $E'(\mathbb{F}_{p^{k/d}})$ be an Abelian group of order $\#E'(\mathbb{F}_{p^{k/d}}) = c \cdot r$, where c is a composite integer known as the *cofactor* of the twist elliptic curve E'. As we have seen, hashing to $\mathbb{G}_2$ can be done by deterministically selecting a random point $\tilde{Q}$ in $E'(\mathbb{F}_{p^{k/d}})$, and then by virtue of the Lagrange's theorem for groups already presented in Chapter 5, we have that

$$\{[c]\tilde{Q} \mid \tilde{Q} \in E'(\mathbb{F}_{p^{k/d}})\} = \{Q' \in E'(\mathbb{F}_{p^{k/d}}) \mid [r]Q' = \mathcal{O}\}. \tag{8.14}$$

However, since in most pairing-friendly elliptic curves the cofactor c in the group $\mathbb{G}_2$ has a considerably large size, which is certainly much larger than the prime order r, a direct computation of such scalar multiplication will be quite costly.

In the rest of this section, we describe a method which for several families of pairing-friendly elliptic curves, allows us to compute the scalar multiplication $Q' = [c]\tilde{Q}$ on a time-computational complexity of $O(1/\varphi(k)\log c)$. The following material closely follows the discussion presented in [49, 26, 37].

8.5.1   Problem Statement

Let us recall that an asymmetric pairing $e \colon \mathbb{G}_2 \times \mathbb{G}_1 \to \mathbb{G}_T$ is a mapping where the groups $\mathbb{G}_1, \mathbb{G}_2$ are defined as $\mathbb{G}_1 = E(\mathbb{F}_p)[r]$ and $\mathbb{G}_2 = E'(\mathbb{F}_{p^{k/d}})[r]$, where k is the embedding degree of the elliptic curve $E/\mathbb{F}_p$, r is a large prime divisor of $\#E(\mathbb{F}_p)$, and $\tilde{E}$ is the degree-d twist of E over $\mathbb{F}_{p^{k/d}}$ with $r \mid \#E'(\mathbb{F}_{p^{k/d}})$. Given the isomorphism $\phi \colon E'(\mathbb{F}_q) \to E(\mathbb{F}_{p^k})$, let $\pi \colon E(\mathbb{F}_{p^k}) \to E(\mathbb{F}_{p^k})$ denote the Frobenius endomorphism of E. Taking a random point $\tilde{Q}$ that belongs to the twist curve, followed by a scalar multiplication by the cofactor c, gives us an $r-$torsion point $Q' = [c] \cdot \tilde{Q}$ in $E'(\mathbb{F}_{p^{k/d}})[r]$. Notice that the resulting $Q' \in \mathbb{G}_2$ has the property that $\pi(Q') = [p]Q'$, where π is the Frobenius endomorphism.

8.5.2   Determining the Order of the Twist

Hess, Smart, and Vercauteren showed in [31] the existence of a unique non-trivial twist $\tilde{E}$ of E over $\mathbb{F}_q$, with $q = p^{k/d}$ such that r divides $\#\tilde{E}(\mathbb{F}_q)$. If $d = 2$, then $\#\tilde{E}(\mathbb{F}_q) = q + 1 + \hat{t}$, where $\hat{t}$ is the trace of the q-power Frobenius of E. Using the Weil theorem, the order of that twist can be found by first determining the trace $\hat{t}$ of the q-power Frobenius of E from the trace t of the p-power Frobenius of E.

TABLE 8.1 Possible values for the trace $\tilde{t}$ of the q-power Frobenius of a degree-d twist $\tilde{E}$ of E.

d	2	3	4	6
$\tilde{t}$	$-\hat{t}$	$(\pm 3\hat{f} - \hat{t})/2$	$\pm\hat{f}$	$(\pm 3\hat{f} + \hat{t})/2$

The interested reader can check out reference [41], where it is shown that the trace t_m of the p^m-power Frobenius of E for an arbitrary m can be determined using the recursion $t_0 = 2$, $t_1 = t$, and $t_{i+1} = t \cdot t_i - p \cdot t_{i-1}$ for all $i > 1$.

Having computed the trace $\hat{t}$ of the q-power Frobenius of E, the possible values for the trace $\tilde{t}$ of the q-power Frobenius of $\tilde{E}$ over $\mathbb{F}_q$ are shown in Table 8.1 [31].

Recall that for any pairing-friendly elliptic curve E, given its discriminant D, there exists $\hat{f} \in \mathbb{Z}$ such that $\hat{t}^2 - 4q = D\hat{f}^2$.

Once the order of the twist has been determined, the cofactor c can be computed as

$$c = \frac{q + 1 - \tilde{t}}{r}. \tag{8.15}$$

8.5.3 The Scott et al. Method

Scott et al. observed in [49] that the endomorphism of $\psi : E'(\mathbb{F}_{p^{k/d}}) \to E'(\mathbb{F}_{p^{k/d}})$, defined as $\psi = \phi^{-1} \circ \pi \circ \phi$, can be used to speed up the computation of $\tilde{Q} \mapsto [c]\tilde{Q}$, by opportunistically using the fact that for all $\tilde{Q} \in E'(\mathbb{F}_{p^{k/d}})$, the endomorphism ψ satisfies the following identity [27]:

$$\psi^2(\tilde{Q}) - [t]\psi(\tilde{Q}) + [p]\tilde{Q} = \mathcal{O}. \tag{8.16}$$

This way, Scott et al. represented the cofactor c as a polynomial in base p, and then, by means of Equation (8.16), the cofactor c can be expressed as a polynomial in ψ with coefficients g_i less than p. For parameterized pairing-friendly elliptic curves, the acceleration in the computation of $Q \mapsto [c]Q$ is often dramatic.

Example 8.1 The MNT pairing-friendly family of elliptic curves have embedding degree $k = 6$ and are parameterized by x so that

$$p(x) = x^2 + 1,$$
$$r(x) = x^2 - x + 1$$

are both prime. It was shown in [49] that

$$[c(x)]P = [x^4 + x^3 + 3x^2]P = [p^2 + (x+1)p - x - 2]P$$
$$= \psi([2x]P) + \psi^2([2x]P).$$

Since the cost of computing the endomorphism ψ can usually be neglected, this approach has a computational cost of two scalar multiplications by the parameter x, plus one point addition.

Example 8.2 For the Barreto–Naehrig curves, the cofactor of the twist is given as,

$$c(x) = 36x^4 + 36x^3 + 30x^2 + 6x + 1.$$

Using the Scott et al. method presented in [49] the scalar multiplication $[c]Q$ becomes

$$[c(x)]\tilde{Q} = [36x^4 + 36x^3 + 30x^2 + 6x + 1]\tilde{Q} = \psi([6x^2]\tilde{Q}) + [6x^2]\tilde{Q} + \psi(\tilde{Q}) - \psi^2(\tilde{Q}).$$

Since the cost of computing the endomorphism ψ can usually be neglected, this approach has a computational cost of two scalar multiplications by the parameter x, plus three point additions.

8.5.4   The Fuentes et al. Method

Observe that a multiple c' of the cofactor c such that $c' \not\equiv 0 \pmod{r}$ will also hash correctly to the group $\mathbb{G}_2$, since the point $[c']\tilde{Q}$ is also in $E(\mathbb{F}_{p^{k/d}})[r]$. The method presented in [26, 37] is based in the following theorem.

THEOREM 8.1   *Since $p \equiv 1 \mod d$ and $\tilde{E}(\mathbb{F}_q)$ is a cyclic group, then there exists a polynomial $h(z) = h_0 + h_1 z + \cdots + h_{\varphi(k)-1} z^{\varphi(k)-1} \in \mathbb{Z}[z]$ such that $[h(\psi)]P$ is a multiple of $[c]P$ for all $P \in \tilde{E}(\mathbb{F}_q)$ and $|h_i|^{\varphi(k)} \leq \#\tilde{E}(\mathbb{F}_q)/r$ for all i.*

Theorem 8.1 was proved in [26] by means of the following two auxiliary lemmas.

LEMMA 8.1   *Let d be the degree of the twist curve E', if $p \equiv 1 \mod d$, then $\psi(\tilde{Q}) \in E'(\mathbb{F}_{p^{k/d}})$, for all $\tilde{Q} \in E'(\mathbb{F}_{p^{k/d}})$.*

The above lemma proves that the endomorphism $\psi \colon \tilde{E} \to \tilde{E}$, defined over $\mathbb{F}_{q^d}$, fixes $\tilde{E}(\mathbb{F}_q)$ as a set. The next lemma shows the effect of ψ on elements in $\tilde{E}(\mathbb{F}_q)$.

LEMMA 8.2   *Let $t^2 - 4p = Df^2$ and $\tilde{t}^2 - 4q = D\tilde{f}^2$, for some value of f and $\tilde{f}$, where $q = p^{k/d}$ and D is the discriminant of the curve E. Also, let $\tilde{n} = \#E'(\mathbb{F}_{p^{k/d}})$. If the following conditions are satisfied,*

- *$p \equiv 1 \mod d$,*
- *$gcd(\tilde{f}, \tilde{n}) = 1$,*
- *$E'(\mathbb{F}_{p^{k/d}})$ is cyclic,*

then $\psi(\tilde{Q}) = [a]\tilde{Q}$ for all $\tilde{Q} \in E'(\mathbb{F}_{p^{k/d}})$, where,

$$a = (t \pm f(\tilde{t} - 2)/\tilde{f})/2.$$

Once the value of a such that $[a]\tilde{Q} = \psi(\tilde{Q})$ has been computed, it is necessary to find the polynomial $h \in \mathbb{Z}[w]$, with the smallest coefficients, such that $h(a) \equiv 0 \mod c$. To this end, one needs to consider a matrix M, with rows representing the polynomials $h_i(w) = w^i - a^i$, such that $h_i(a) \equiv 0 \mod c$. Hence, any linear combination of the rows of the matrix M will correspond with a polynomial $h'(w)$ that satisfies the above condition.

Since the Frobenius endomorphism π acting over $E(\mathbb{F}_{p^k})$ has order k, and since ψ is an endomorphism that acts on the cyclic group $E'(\mathbb{F}_{p^{k/d}})$, then ψ operating over $E'(\mathbb{F}_{p^{k/d}})$ also has order k. Furthermore, since the integer number a satisfies the congruence $\Phi_k(a) \equiv 0 \mod \tilde{n}$, where Φ_k is the k-th cyclotomic polynomial, then the polynomials $h(w) = w^i - a^i$ with $i \geq \varphi(k)$ can be written as linear combinations of $w - a, \ldots, w^{\varphi(k)-1} - a^{\varphi(k)-1} \mod c$, where $\varphi(\cdot)$ is the Euler's totient function. Because of the aforementioned argument, only the polynomials with degree less than $\varphi(k)$ are considered, as shown in Figure 8.4. As it can be seen in Figure 8.4

$$M = \begin{pmatrix} & a^0 & a^1 & a^2 & \cdots & a^{\varphi(k)-1} \\ & c & 0 & 0 & \cdots & 0 \\ & -a & 1 & 0 & \cdots & 0 \\ & -a^2 & 0 & 1 & \cdots & 0 \\ & \vdots & \vdots & \vdots & \ddots & \\ & -a^{\varphi(k)-1} & 0 & 0 & \cdots & 1 \end{pmatrix} \quad \longrightarrow \quad \begin{matrix} c & \equiv & 0 & \mod c \\ -a+a & \equiv & 0 & \mod c \\ -a^2+a^2 & \equiv & 0 & \mod c \\ \vdots & \vdots & \vdots & \vdots \\ -a^{\varphi(k)-1}+a^{\varphi(k)-1} & \equiv & 0 & \mod c \end{matrix}$$

FIGURE 8.4   Matrix M with rows representing polynomials $h_i(w)$ such that $h_i(a) \equiv 0 \mod c$.

the strategy proposed in [26, 37] is analogous to the one used for the final exponentiation in Chapter 7, which was in turn inspired from the definition of optimal pairings as given in [51].

In this case, the rows of the matrix M can be seen as vectors, which form a lattice basis. Now, the Lenstra–Lenstra–Lovász algorithm [38] can be applied to M in order to obtain an integer basis for M with small entries. According to the Minkowski's theorem [42], a vector v that represents a linear combination of the basis of the lattice L, will be found. This solution will correspond to the polynomial h with coefficients smaller than $|c|^{1/\varphi(k)}$.

In the rest of this Section, explicit equations for computing the hash to the group $\mathbb{G}_2$ on several families of pairing-friendly curves will be described. We cover the following families: the Barreto–Naehrig (BN) curves [7], the Brezing–Weng curves with embedding degree $k = 12$, (BW-12) [15], the Kachisa–Schaefer–Scott curves with embedding degree $k = 8$ (KSS-8) and $k = 18$ (KSS-18) [36], and the Barreto–Lynn–Scott curves with embedding degree $k = 24$ (BLS-24) [6].

Barreto–Naehrig curves

For the Barreto–Naehrig elliptic curves, the group order $\tilde{n} = \#E'(\mathbb{F}_{p^2})$ and the trace of the twist E' over $\mathbb{F}_{p^2}$, are parametrized as follows:

$$\tilde{n} = (36x^4 + 36x^3 + 18x^2 + 6x + 1)(36x^4 + 36x^3 + 30x^2 + 6x + 1),$$
$$\tilde{t} = 36x^4 + 1$$

where $\tilde{n}(x) = r(x)c(x)$. Using Lemma 8.2, we find that

$$a(x) = -\frac{1}{5}(3456x^7 + 6696x^6 + 7488x^5 + 4932x^4 + 2112x^3 + 588x^2 + 106x + 6).$$

It is interesting to notice that $a(x) \equiv p(x) \mod r(x)$ and $\psi(Q') = [a(x)]Q' = [p(x)]Q'$ for all $Q' \in \tilde{E}(\mathbb{F}_{p^2})[r]$. Note that $a(x) \equiv p(x) \pmod r$ and thus $\psi Q = [a(x)]Q = [p(x)]Q$ for all $Q \in \tilde{E}(\mathbb{F}_q)[r]$.

Following the strategy mentioned above, the lattice L can be built, and then reducing $-a(x)^i$ modulo $c(x)$, one obtains

$$\begin{bmatrix} c(x) & 0 & 0 & 0 \\ -a(x) & 1 & 0 & 0 \\ -a(x)^2 & 0 & 1 & 0 \\ -a(x)^3 & 0 & 0 & 1 \end{bmatrix} \rightarrow \begin{bmatrix} 36x^4 + 36x^3 + 30x^2 + 6x + 1 & 0 & 0 & 0 \\ 48/5x^3 + 6x^2 + 4x - 2/5 & 1 & 0 & 0 \\ 36/5x^3 + 6x^2 + 6x + 1/5 & 0 & 1 & 0 \\ 12x^3 + 12x^2 + 8x + 1 & 0 & 0 & 1 \end{bmatrix}.$$

From this lattice, one finds the polynomial $h(x) = x + 3xz + xz^2 + z^3$. Working modulo $\tilde{n}(x)$, we have that

$$h(a) = -(18x^3 + 12x^2 + 3x + 1)c(x),$$

and since $\gcd(18x^3 + 12x^2 + 3x + 1, r(x)) = 1$, the following map is a homomorphism of $\tilde{E}(\mathbb{F}_q)$ with image $\tilde{E}(\mathbb{F}_q)[r]$:

$$Q \mapsto [x]Q + \psi([3x]Q) + \psi^2([x]Q) + \psi^3(Q).$$

We can compute $Q \mapsto [x]Q \mapsto [2x]Q \mapsto [3x]Q$ using one doubling, one addition, and one multiply-by-x. Given Q, $[x]Q$, $[3x]Q$, we can compute $[h(a)]Q$ using three ψ-maps, and three additions. In total, we require one doubling, four additions, one multiply-by-x, and three ψ-maps.

8.5.5 KSS-8 Curves

KSS-8 curves [36] have embedding degree $k = 8$ and are parameterized by x such that

$$r = r(x) = \frac{1}{450}(x^4 - 8x^2 + 25),$$

$$p = p(x) = \frac{1}{180}(x^6 + 2x^5 - 3x^4 + 8x^3 - 15x^2 - 82x + 125)$$

are both prime. Define q as $q = p^{k/d} = p^2$. Then, there exists a twist $\tilde{E}(\mathbb{F}_q)$ of degree four and order given as

$$\tilde{n}(x) = \frac{1}{72}(x^8 + 4x^7 + 6x^6 + 36x^5 + 34x^4 - 84x^3 + 486x^2 + 620x + 193)r(x).$$

Define the cofactor c in polynomial form as $c(x) = \tilde{n}(x)/r(x)$. After some work, it is found that the endomorphism ψ is such that $\psi Q = [a]Q$ for all $Q \in \tilde{E}(\mathbb{F}_q)$, where

$$a = \frac{1}{184258800}\big(-52523x^{11} - 174115x^{10} + 267585x^9 - 193271x^8$$
$$- 325290x^7 + 15093190x^6 - 29000446x^5 - 108207518x^4$$
$$+ 235138881x^3 + 284917001x^2 - 811361295x - 362511175\big).$$

At this point, one strives to find a short basis for the lattice generated by the matrix

$$\begin{bmatrix} c(x) & 0 & 0 & 0 \\ -a(x) & 1 & 0 & 0 \\ -a(x)^2 & 0 & 1 & 0 \\ -a(x)^3 & 0 & 0 & 1 \end{bmatrix}.$$

The solution discovered from this matrix corresponds to

$$h(a) = \frac{1}{75}(x^2 - 25)c(x) = \lambda_0 + \lambda_1 a + \lambda_2 a^2 + \lambda_3 a^3$$

of c such that $\lambda = (\lambda_0, \lambda_1, \lambda_2, \lambda_3) = (-x^2 - x, x - 3, 2x + 6, -2x - 4)$.

For an element $Q \in \tilde{E}(\mathbb{F}_q)$. Then, one can compute $[h(a)]Q$ at the following computational cost. First, the computation $Q \mapsto [x]Q \mapsto [x+1]Q \mapsto [x^2 + x]Q$ and $Q \mapsto [2]Q \mapsto [4]Q$ requires one point addition, two point doublings, and two scalar multiplications by x. Afterwards, one can compute

$$\lambda_0 Q = -[x^2 + x]Q,$$
$$\lambda_1 Q = [x+1]Q - [4]Q,$$
$$\lambda_2 Q = [2(x+1)]Q + [4]Q \text{ and}$$
$$\lambda_3 Q = -[2(x+1)]Q - [2]Q.$$

The above computation requires three more point additions and one more point doubling. Finally, the operation,

$$h(a)Q = [\lambda_0]Q + \psi([\lambda_1]Q) + \psi^2([\lambda_2]Q) + \psi^3([\lambda_3]Q)$$

requires three more point additions and three ψ maps.

In total, hashing to $\mathbb{G}_2$ in this family of curves has a cost of seven, three, two, and three point additions, point doublings, scalar multiplications by the parameter x, and ψ maps, respectively.

8.5.6 KSS-18 Curves

KSS-18 curves [36] have embedding degree $k = 18$ and a twist of order $d = 6$. These curves are parameterized by x such that

$$r = r(x) = \frac{1}{343}(x^6 + 37x^3 + 343),$$

$$p = p(x) = \frac{1}{21}(x^8 + 5x^7 + 7x^6 + 37x^5 + 188x^4$$
$$+ 259x^3 + 343x^2 + 1763x + 2401)$$

are both prime. We find that

$$c(x) = \frac{1}{27}\big(x^{18} + 15x^{17} + 96x^{16} + 409x^{15} + 1791x^{14} + 7929x^{13} + 27539x^{12}$$
$$+ 81660x^{11} + 256908x^{10} + 757927x^9 + 1803684x^8$$
$$+ 4055484x^7 + 9658007x^6 + 19465362x^5 + 30860595x^4$$
$$+ 50075833x^3 + 82554234x^2 + 88845918x + 40301641\big).$$

Constructing our lattice, one can obtain the vector corresponding to the multiple

$$h(a) = -\frac{3}{343}x(8x^3 + 147)c(x) = \lambda_0 + \lambda_1 a + \lambda_2 a^2 + \lambda_3 a^3 + \lambda_2 a^4 + \lambda_3 a^5$$

of $c(x)$, where

$$\lambda_0 = 5x + 18,$$
$$\lambda_1 = x^3 + 3x^2 + 1,$$
$$\lambda_2 = -3x^2 - 8x,$$
$$\lambda_3 = 3x + 1,$$
$$\lambda_4 = -x^2 - 2 \text{ and}$$
$$\lambda_5 = x^2 + 5x.$$

With the help of the addition chain $\{1, 2, 3, 5, 8, \underline{10}, 18\}$, one can compute $Q \mapsto [h(a)]Q$ using sixteen additions, two doublings, three scalar multiplications by the parameter x's, and five ψ maps.

BW-12 curves

For the BW-12 curves, the cofactor c can be parametrized as

$$c(x) = \frac{1}{9}x^8 - \frac{4}{9}x^7 + \frac{5}{9}x^6 - \frac{4}{9}x^4 + \frac{2}{3}x^3 - \frac{4}{9}x^2 - \frac{4}{9}x + \frac{13}{9},$$

and for this particular case, it is more efficient to compute the hashing to $\mathbb{G}_2$ operation using the Scott et al. method, where

$$h(a) = (x^3 - x^2 - x + 4) + (x^3 - x^2 - x + 1)a + (-x^2 + 2x - 1)a^2.$$

Hence,

$$\tilde{Q} \mapsto [x^3 - x^2 - x + 4]\tilde{Q} + \psi([x^3 - x^2 - x + 1]\tilde{Q}) + \psi^2([-x^2 + 2x - 1]\tilde{Q}),$$

which can be computed at a cost of 6 point additions, 2 point doublings, 3 scalar multiplications by the parameter x, and three applications of the endomorphism ψ.

TABLE 8.2   Cost summary of hashing to $\mathbb{G}_2$ using the Scott et al. and the Fuentes et al. methods.

Curve	Scott et al. [49]	Fuentes et al. [26]
BN	4A 2D 2Z 3ψ	4A 1D 1Z 3ψ
KSS-8	22A 5D 5Z 2ψ	7A 3D 2Z 3ψ
KSS-18	59A 5D 7Z 4ψ	16A 2D 3Z 5ψ
BW-12	6A 3D 3Z 3ψ	-
BLS-24	21A 4D 8Z 6ψ	-

BLS-24 curves

As in the case of the previous family of curves, the most efficient method to compute the hashing to $\mathbb{G}_2$ for the BLS-24 family of curves is to follow the approach proposed by Scott et al. in [49]. For this family of curves, the cofactor is parametrized as follows:

$$c(z) = \frac{1}{81} \cdot x^{32} - 8x^{31} + 28x^{30} - 56x^{29} + 67x^{28} - 32x^{27} - 56x^{26} +$$
$$160x^{25} - 203x^{24} + 44x^{23} + 4x^{22} - 44x^{21} + 170x^{20} - 124x^{19} + 44x^{18} - 4x^{17} +$$
$$2x^{16} + 20x^{15} - 46x^{14} + 20x^{13} + 5x^{12} + 8x^{11} - 14x^{10} + 16x^9 - 101x^8 + 100x^7 +$$
$$70x^6 - 128x^5 + 70x^4 - 56x^3 - 44x^2 + 40x + 100,$$

and one can find that

$$h(a) = 3c(x) = \lambda_0 + \lambda_1 a + \lambda_2 a^2 + \lambda_3 a^3 + \lambda_4 a^4 + \lambda_5 a^5 + \lambda_6 a^6,$$

where

$$\lambda_0 = -2x^8 + 4x^7 - 3x^5 + 3x^4 - 2x^3 - 2x^2 + x + 4,$$
$$\lambda_1 = x^5 - x^4 - 2x^3 + 2x^2 + x - 1,$$
$$\lambda_2 = x^5 - x^4 - x + 1,$$
$$\lambda_3 = x^5 - x^4 - x + 1,$$
$$\lambda_4 = -3x^4 + x^3 + 4x^2 + x - 3,$$
$$\lambda_5 = 3x^3 - 3x^2 - 3x + 3 \text{ and}$$
$$\lambda_6 = -x^2 + 2x - 1.$$

Using the method described in [49], the associated computational cost is of 21 point additions, 4 point doublings, 8 scalar multiplications by the parameter x, and 6 applications of the endomorphism ψ.

8.5.7   Comparison

In Table 8.2, there is summary of the costs associated with the hashing to $\mathbb{G}_2$ operation by using the Scott et al. and the Fuentes et al. strategies. "A", "D" stand for point addition and point doubling, respectively. Moreover, "X" and "ψ" denote a scalar multiplication by the curve parameter x and the application of the mapping $\psi(\cdot)$, respectively. Notice that by far, the most costly operation reported in Table 8.2 is the scalar multiplication by the scalar x. Let us consider the case of the Barreto–Naehrig curves. In order to achieve a 128-bit security level, the bit-size of the parameter x must be $\log_2(x) \approx 64$. Assuming that x has a Hamming weight of 3 and the the addition-and-double method for the scalar multiplication has been used, then the cost of computing $[x]Q$ for $Q \in \mathbb{G}_2$ is about $63D + 2A$. Therefore, the method proposed by Fuentes et al. in [26] is approximately twice as fast as the Scott et al. method of [49], as shown in Table 8.3.

TABLE 8.3 Cost summary of hashing to $\mathbb{G}_2$ on Barreto–Naehrig curves.

Scott et al.	Fuentes et al.
$\approx 126\,D + 4\,A$	$\approx 63\,D + 2\,A$

8.6 Practical Algorithms and Implementation

8.6.1 Intel Processor

Tables 8.6 and 8.7 report the timings (in 10^3 clock cycles) achieved by our software implementation of all the required building blocks for computing the hash functions to the Barreto–Naehrig curve subgroups $\mathbb{G}_1$ and $\mathbb{G}_2$. Our library was written in the C and C++ languages and compiled with gcc 4.9.2. It was run on a Haswell Intel core i7-4700MQ processor running at 2.4 GHz, with both the Turbo-Boost and Hyper-threading technologies disabled.

We used two values of the Barreto–Naehrig parameter x, namely, $x = -(2^{62} + 2^{55} + 1)$, which is a standard choice recommended in [44] and used in many pairing libraries, such as [2, 43, 52]. We also report the timings for the parameter choice: $x = -(2^{62} + 2^{47} + 2^{38} + 2^{37} + 2^{14} + 2^7 + 1)$, which is the value recommended in [48] to avoid subgroup attacks in $\mathbb{G}_T$. It is noticed that in [52] slightly faster timings were reported.

8.6.2 ARM Processor

Tables 8.4 and 8.5 report the timings (in 10^3 clock cycles) achieved by our software implementation of all the required building blocks for computing the hash functions to the Barreto–Naehrig curve subgroups $\mathbb{G}_1$ and $\mathbb{G}_2$. Our library was written in the C language and compiled using the android-ndk-r10b *Native Development Kit*, and executed on an ARM Exynos 5 Cortex-A15 platform running at 1.7GHz. Our library makes use of the NEON technology. Once again, we used the two values for the x parameters chosen for the Intel processor in the preceding section.

TABLE 8.4 Cost of the hashing to $\mathbb{G}_1$ operation on Barreto–Naehrig curves at the 128-bit security level.

	x parameter used to define the BN curve	
Operation	$x = -(2^{62} + 2^{55} + 1)$	$x = -(2^{62} + 2^{47} + 2^{38} + 2^{37} + 2^{14} + 2^7 + 1)$
SHA256	8.67	8.67
Algorithm 8.2	1047.80	1357.71
Algorithm 8.3	655.32	709.40
Hash to $\mathbb{G}_1$ with Algorithm 8.2	1056.43	1366.23
Hash to $\mathbb{G}_1$ with Algorithm 8.3	664.02	718.12

Note: Timings are given in 10^3 clock cycles as measured on an ARM Exynos 5 Cortex-A15 1.7GHz.

TABLE 8.5 Cost of the hashing to $\mathbb{G}_2$ operation on Barreto–Naehrig curves at the 128-bit security level.

	x parameter used to define the BN curve	
Operation	$x = -(2^{62} + 2^{55} + 1)$	$x = -(2^{62} + 2^{47} + 2^{38} + 2^{37} + 2^{14} + 2^7 + 1)$
SHA256	8.67	8.67
$[c]\bar{Q}$	754.10	806.06
Algorithm 8.4 (constant-time)	1337.21	1717.81
Algorithm 8.4 (with blinding)	809.01	863.29
Hash to $\mathbb{G}_2$ (constant-time)	2099.95	2530.20
Hash to $\mathbb{G}_2$ (with blinding)	1570.75	1680.37

Note: Timings are given in 10^3 clock cycles as measured on an ARM Exynos 5 Cortex-A15 1.7GHz.

TABLE 8.6   Cost of the hashing to $\mathbb{G}_1$ operation on Barreto–Naehrig curves at the 128-bit security level.

	x parameter used to define the BN curve	
Operation	$x = -(2^{62} + 2^{55} + 1)$	$x = -(2^{62} + 2^{47} + 2^{38} + 2^{37} + 2^{14} + 2^7 + 1)$
SHA256	1.81	1.81
Algorithm 8.2	122.81	156.48
Algorithm 8.3	95.83	104.83
Hash to $\mathbb{G}_1$ with Algorithm 8.2	124.62	158.29
Hash to $\mathbb{G}_1$ with Algorithm 8.3	97.64	106.64

Note: Timings are given in 10^3 clock cycles as measured on a Haswell Intel core I7-4700MQ 2.4GHz.

TABLE 8.7   Cost of the hashing to $\mathbb{G}_2$ operation on Barreto–Naehrig curves at the 128-bit security level.

	x parameter used to define the BN curve	
Operation	$x = -(2^{62} + 2^{55} + 1)$	$x = -(2^{62} + 2^{47} + 2^{38} + 2^{37} + 2^{14} + 2^7 + 1)$
SHA256	1.81	1.81
$[c]\tilde{Q}$	161.63	175.02
Algorithm 8.4 (constant-time)	186.71	236.94
Algorithm 8.4 (with blinding)	134.03	153.12
Hash to $\mathbb{G}_2$ (constant-time)	344.82	408.24
Hash to $\mathbb{G}_2$ (with bliding)	293.29	322.21

Note: Timings are given in 10^3 clock cycles as measured on a Haswell Intel core I7-4700MQ 2.4GHz.

8.7   SAGE Appendix

8.7.1   Barreto–Naehrig Curves

This section presents a SAGE implementation of the hashing to the groups $\mathbb{G}_1$ and $\mathbb{G}_2$ procedures, as defined in the Barreto–Naehrig curves.

     The family of Barreto–Naehrig curves are parameterized by the equations shown in Listing 34, where the field characteristic p, the order r of the elliptic curve defined over $\mathbb{F}_p$, and the trace of Frobenius t are defined. Note that the parameter x is chosen in such a way that the values $p(x)$ and $r(x)$ are prime numbers. In this example we decided to use $x = -(2^{62} + 2^{55} + 1)$, mainly because it is the preferred parameter for efficient implementations of bilinear pairings in many pairing libraries, such as [2, 43, 52].

     Once the parameters of the curve have been defined, we construct the finite field towering required for the pairing computations. In Listing 35, the base field $\mathbb{F}_p$, as well as its quadratic extension $\mathbb{F}_{p^2} = \mathbb{F}_p[i]/(i^2 - \beta)$ with $\beta = -1$ are defined. The Barreto–Naehrig curve E and its twist E' are defined as $E/\mathbb{F}_p : y^2 = x^3 + b$ and $E'/\mathbb{F}_{p^2} : Y^2 = X^3 + B$, with $B = b/\xi \in \mathbb{F}_{p^2}$.

     Several precomputations will be useful in subsequent calculations, such as the Pre-Frobenius function to be used for calculating the map $\psi^i : \phi \circ \pi^i \circ \phi^{-1}$, where π is the Frobenius endomorphism and $\phi : E'(\mathbb{F}_{p^2}) \to E(\mathbb{F}_{p^{12}})$. Some other constants used in the hash encodings are also precomputed, as shown in Listing 36.

     Now we have the necessary ingredients for computing the hash function to the group $\mathbb{G}_1$, by applying the method proposed by Fouque and Tibouchi [25]. Furthermore, we also verify that the point obtained has the right order, as shown in Listing 37.

     As for the hash function calculation to $\mathbb{G}_2$, Listing 38 shows the auxiliary functions required to compute square roots over $\mathbb{F}_{p^2}$ in a deterministic fashion, as well as functions that allow us to compute the projection ψ.

     At this point, one can build a random point in the twist curve E' using the method proposed

---

Listing 34 Definition of parameters for the Barreto-Naehrig curves.

```
1  reset
2  ################################################
3  #          Elliptic curve parameters          #
4  ################################################
5
6  x = -(2^62+2^55+1)
7  px = 36*x^4 + 36*x^3 + 24*x^2 + 6*x + 1
8  rx = 36*x^4 + 36*x^3 + 18*x^2 + 6*x + 1
9  tx = 6*x^2 + 1
10
11 p = Integer(px)
12 r = Integer(rx)
13 t = Integer(tx)
14
15 is_prime(p)
16 is_prime(r)
```

Listing 35 Definition of the base and quadratic fields and the Barreto-Naehrig curve $E/\mathbb{F}_p$: $y^2 = x^3 + b$ and $E'/\mathbb{F}_{p^2} : Y^2 = X^3 + b/\xi$.

```
1  ################################################
2  #  Finite field and Elliptic curve definitions  #
3  ################################################
4  Fp = GF(p)
5  beta = Fp(-1)
6  K2.<U>    = Fp[]
7  Fp2.<i>   = Fp.extension(U^2 - (beta))
8
9  b = Fp(2)
10 xi = Fp2([1,1])
11 b_xi = b/xi
12 #        Elliptic curve E/Fp: y^2 = x^3 + b       #
13 E = EllipticCurve(Fp, [0,0,0,0,b])
14 #        Etwist E'/Fp2: Y^2 = X^3 + b/xi          #
15 Etwist = EllipticCurve(Fp2, [0,0,0,0,b_xi])
```

---

by Fouque and Tibouchi in [25]. Note that several computations of Listing 39 now take place in the quadratic extension field $\mathbb{F}_{p^2}$.

Once a random point has been built using the above function, the next step is to map it to an r-torsion point that belongs to the subgroup $\mathbb{G}_2$. To this end, a scalar multiplication is carried out by the cofactor c defined as $c = \#E'/r$. As we have seen, this can be efficiently computed using the method proposed by Fuentes et al. in [26], as shown in Listing 40.

One can now define the hashing to $\mathbb{G}_2$ function and perform the corresponding sobriety test shown in the Sage code 41.

A more efficient way to perform the hashing computations on Barreto–Naehrig curves, which was coded somewhat in a similar fashion as one would do it for a C implementation, can be downloaded from http://sandia.cs.cinvestav.mx/Site/GuidePBCcode.

Listing 36 Precomputations of the pre-Frobenius constants and some other values used in the hashing to $\mathbb{G}_1$ and $\mathbb{G}_2$ functions.

```
1  ##################################################
2  #            Prefrobenius computation            #
3  ##################################################
4  def preFrobenius():
5      gamp_2 = xi^((p-1)/3)
6      gamp_3 = xi^((p-1)/2)
7
8      gamp2_2 = gamp_2*gamp_2^p
9      gamp2_3 = gamp_3*gamp_3^p
10
11     gamp3_2 = gamp_2*gamp2_2^p
12     gamp3_3 = gamp_3*gamp2_3^p
13
14     return(gamp_2, gamp_3, gamp2_2,
15         gamp2_3, gamp3_2, gamp3_3)
16
17 ##################################################
18 #                Precomputations                 #
19 ##################################################
20 sqrt_3 = Fp(-3)^((p+1)/4)
21 j = (-1 + sqrt_3)/2
22 gp_2, gp_3, gp2_2, gp2_3, gp3_2, gp3_3 = preFrobenius()
```

Listing 37 Hash function to the group $\mathbb{G}_1$ and sobriety test.

```
1  ##################################################
2  #        Hash function to G1 (map-to-point)      #
3  ##################################################
4  def Hash_to_G1(t):
5      w = sqrt_3*t/(1 + b + t^2)
6      x = [0, 0, 0]
7      x[0] = j - t*w
8      x[1] = -1 - (j-t*w)
9      x[2] = 1 + 1/w^2
10     alpha = (x[0]^3 + b)^((p-1)/2)
11     beta = (x[1]^3 + b)^((p-1)/2)
12     i = (Integer(3+(alpha-1)*beta)).mod(3);
13     y = (x[i]^3 + b)^((p+1)/4)
14     y = t^((p-1)/2)*y
15     return(E((x[i], y)))
16
17 ##################################################
18 #                Hash to G1 Test                 #
19 ##################################################
20 l = 0
21 for i in range(0,10):
22     P = Hash_to_G1(Fp.random_element())
23     if r*P != E(0):
24         l = l+1
```

Listing 38 Square root in $\mathbb{F}_{p^2}$ using the complex method and $\psi(\cdot)$ functions.

```
1   #################################################
2   #      Square root in Fp2 with Complex Method   #
3   #################################################
4   def Sqrt_CM(a):
5       b = vector(a)
6       x = (Fp(b[0])^2 - beta*Fp(b[1])^2)^((p+1)/4)
7       aux = (b[0] + x)/2
8       if aux.is_square()==false:
9           aux = (b[0] - x)/2
10      x = aux^((p+1)/4)
11      y = b[1]/(2*x)
12      return(Fp2([x,y]))
13
14  #################################################
15  #           \psi and \psi^2 function            #
16  #################################################
17  def conj(a):
18      b = vector(a)
19      return Fp2([b[0],-b[1]])
20  ##psi and psi3
21  def psi(P, gp_2, gp_3):
22      x = conj(P[0])*gp_2
23      y = conj(P[1])*gp_3
24      return(Etwist([x, y]))
25
26  def psi2(P, gp2_2, gp2_3):
27      x = P[0]*gp2_2
28      y = P[1]*gp2_3
29      return(Etwist([x, y]))
```

8.7.2 KSS-18 Curves

This section presents the implementation in SAGE of the hashing to the groups $\mathbb{G}_1$ and $\mathbb{G}_2$, this time using the pairing-friendly family of the KSS-18 curves. This family of curves is also parameterized, hence at the beginning of the implementation, it is necessary to define the field characteristic p, the prime group order r of the elliptic curve defined over $\mathbb{F}_p$, and the Frobenius trace t. In this example, the parameter choice $x = 2^{64} - 2^{61} + 2^{56} - 2^{13} - 2^7$ was picked. This choice defines a 506-bit prime p. The SAGE Listing 42 shows the parameter definition for the KSS-18 elliptic curves.

Once the curve parameters have been defined, the finite fields where the KSS-18 elliptic curve lies must be defined. Hence, we define the base field $\mathbb{F}_p$ and its cubic extension $\mathbb{F}_{p^3} = \mathbb{F}_p[i]/(i^3 - \beta)$ with $\beta = -6$. The KSS-18 elliptic curve E and its twist E' are then defined as $E/\mathbb{F}_p : y^2 = x^3 + b$ and $E'/\mathbb{F}_{p^3} : Y^2 = X^3 + B$, with $B = b/\xi \in \mathbb{F}_{p^3}$. The corresponding definitions are shown in Listing 43.

In Listing 44, several precomputations that will be useful in the subsequent calculations are performed. We precompute some necesary values for calculating square roots in the cubic extension field when $p \equiv 5 \mod 8$, and pre-Frobenius values used in the calculation of ψ^i : $\phi \circ \pi^i \circ \phi^{-1}$, where π is the Frobenius endomorphism and $\phi : E'(\mathbb{F}_{p^3}) \to E(\mathbb{F}_{p^{18}})$.

Now we have the necessary elements for the calculation of the hash function to the group $\mathbb{G}_1$ using the method proposed by Fouque and Tibouchi [25]. As a sobriety check, we also verify that the point obtained has the correct order r, as shown in Listing 45.

For the remaining part, one still needs to code several auxiliary functions that are required for computing square roots over the cubic extension $\mathbb{F}_{p^3}$ in a deterministic fashion. Moreover, the

---

Listing 39 Residuosity quadratic test of an element in $\mathbb{F}_{p^2}$ and deterministic point construction in the curve $E'/\mathbb{F}_{p^2} : Y^2 = X^3 + b/\xi$.

```
1  ###################################################
2  #    deterministic point construction in Etwist   #
3  ###################################################
4  def RQT(a):
5      b = vector(a)
6      c = Fp(b[0]^2 + b[1]^2)^((p-1)/2)
7      if c == 1:
8          return 1
9      return -1
10
11 def deterministic_point_Etwist(t):
12     b = vector(b_xi)
13     a0 = Fp(1 + b[0] + t^2)
14     a1 = b[1]
15     A = 1/Fp2([a0,a1])
16     (a0,a1) = vector(A)
17     c = sqrt_3*t
18     w = [a0*c, a1*c]
19     x = [0,0,0]
20     x[0] = Fp2([j - t*w[0], -t*w[1]])
21     x[1] = -1 - x[0]
22     x[2] = 1 + 1/Fp2(w)^2
23     alpha = RQT(x[0]^3 + b_xi)
24     beta = RQT(x[1]^3 + b_xi)
25     i = (Integer(3+(alpha-1)*beta)).mod(3)
26     y = t^((p-1)/2)*Sqrt_CM(x[i]^3 + b_xi)
27     return(Etwist(x[i], y))
```

---

projection ψ also requires several building blocks. The SAGE code 46 shows these calculations. We note that the square root procedure was done following the Atkin algorithm described in [1], which extracts square roots when the characteristic p satisfies, $p \equiv 5 \mod 8$.

Next, the calculation of a random point that belongs to the twist elliptic curve group $E'(\mathbb{F}_{p^3})$ is carried out by means of the method proposed by Fouque and Tibouchi [25], as shown in the SAGE code 47. Note that most computations now take place in the cubic extension field $\mathbb{F}_{p^3}$.

Once a random point in the twist elliptic curve has been obtained, it is necessary to ensure that this point has the correct order r. To this end, a scalar multiplication by the cofactor c, defined as $c = \#E'/r$, must be performed. As we have seen, this can be efficiently accomplished using the method proposed by Fuentes et al. [26]. The corresponding procedure is shown in Listing 48. Finally, the hash function to $\mathbb{G}_2$ and its corresponding sobriety test are shown in Listing 49.

A more efficient way to perform the hashing computations on KSS-18 curves, which was coded in a somewhat similar fashion to what one would do for a C implementation, can be downloaded from: http://sandia.cs.cinvestav.mx/Site/GuidePBCcode.

Listing 40 Scalar multiplication by the cofactor c that produces a point in E' with order r.

```
1   ##################################################
2   #      Scalar multiplicaction by the cofactor      #
3   ##################################################
4   def Mult_cofactor(P):
5       Px = x*P
6       P2x = 2*Px
7       P3x = P2x + Px
8       psiP3x = psi(P3x, gp_2, gp_3)
9       psi2Px = psi2(Px, gp2_2, gp2_3)
10      psi3P = psi(P, gp3_2, gp3_3)
11      Q = Px + psiP3x + psi2Px + psi3P
12      return(Q)
```

Listing 41 Definition of the hashing to $\mathbb{G}_2$ function and test.

```
1   ##################################################
2   #              Hash function to G2              #
3   ##################################################
4   def Hash_to_G2(t):
5       P = deterministic_point_Etwist(t)
6       Q = Mult_cofactor(P)
7       return(Q)
8
9   ##################################################
10  #              Hash to G2 test                  #
11  ##################################################
12  l = 0
13  for i in range(1,10):
14      P = Hash_to_G2(Fp.random_element())
15      if r*P != Etwist(0):
16          l = l+1
```

Listing 42 Parameter definition for the KSS-18 elliptic curve.

```
1   reset
2   ##################################################
3   #         Elliptic curve parameters             #
4   ##################################################
5
6   x = 2^64-2^61+2^56-2^13-2^7
7   px = (x^8+5*x^7+7*x^6+37*x^5+188*x^4
8         +259*x^3+343*x^2+1763*x+2401)/21
9   rx = (x^6+37*x^3+343)/343
10  tx = (x^4+16*x+7)/7
11
12  p = Integer(px)
13  r = Integer(rx)
14  t = Integer(tx)
15
16  is_prime(p)
17  is_prime(r)
```

Listing 43 Definitions of the base and cubic extension field and the KSS-18 curve $E/\mathbb{F}_p : y^2 = x^3 + b$ and $E'/\mathbb{F}_{p^3} : Y^2 = X^3 + b/\xi$.

```
1   #################################################
2   #   Finite field and Elliptic curve definitions   #
3   #################################################
4   Fp = GF(p)
5   beta = Fp(-6)
6   K3.<U>   = Fp[]
7   Fp3.<i> = Fp.extension(U^3 - (beta))
8
9   b = Fp(13)
10  xi = Fp3([0,1,0])
11  b_xi = b/xi
12  #         Elliptic curve E/Fp: y^2 = x^3 + b        #
13  E = EllipticCurve(Fp, [0, 0,0,0,b])
14  #           Etwist E'/Fp3: Y^2 = X^3 + b/xi           #
15  Etwist = EllipticCurve(Fp3, [0, 0,0,0,b_xi])
```

Listing 44 Precomputation of the Pre-Frobenius constants, and some values used in the hash functions to $\mathbb{G}_1$ and $\mathbb{G}_2$, and the square root calculation in $\mathbb{F}_{p^3}$.

```
1   #################################################
2   #       Prefrobenius used in psi^i computation       #
3   #################################################
4   landa_1 = beta^((p-1)/3)
5   landa_2 = landa_1^2
6
7   def mul_landa(a, lan1, lan2):
8       b = vector(a)
9       return Fp3([b[0], b[1]*lan1, b[2]*lan2])
10
11  def preFrobenius():
12      gamp_1 = xi^((p-1)/6)
13      gamp_2 = gamp_1^2
14      gamp_3 = gamp_2*gamp_1
15      gamp2_2 = gamp_2*mul_landa(gamp_2,landa_1,landa_2)
16      gamp2_3 = gamp_3*mul_landa(gamp_3,landa_1,landa_2)
17      gamp3_2 = gamp2_2*mul_landa(gamp_2,landa_2,landa_1)
18      gamp3_3 = gamp2_3*mul_landa(gamp_3,landa_2,landa_1)
19      gamp4_2 = gamp3_2*gamp_2
20      gamp4_3 = gamp3_3*gamp_3
21      gamp5_2 = gamp4_2*mul_landa(gamp_2,landa_1,landa_2)
22      gamp5_3 = gamp4_3*mul_landa(gamp_3,landa_1,landa_2)
23
24      return(gamp_2,gamp_3,gamp2_2,gamp2_3,gamp3_2,
25          gamp3_3,gamp4_2,gamp4_3,gamp5_2,gamp5_3)
26
27  #################################################
28  #                 Precomputations                 #
29  #################################################
30  sqrt_3 = Fp(-3).sqrt()
31  j = (-1 + sqrt_3)/2
32  p_5 = (p-5) / 8;
33  e = p_5 + p*(p*p_5 + 5*p_5 +3)
34  t_A = Fp(2)^e
35
36  gp_2,gp_3,gp2_2,gp2_3,gp3_2,gp3_3,
37          gp4_2,gp4_3,gp5_2,gp5_3 = preFrobenius()
```

Listing 45 Hash to the group $\mathbb{G}_1$ function and sobriety test.

```
1   ##################################################
2   #         Hash function to G1 (map-to-point)     #
3   ##################################################
4   def Hash_to_G1(t):
5       w = sqrt_3*t/(1 + b + t^2)
6       x = [0,0,0]
7       x[0] = j - t*w
8       x[1] = -1 - (j-t*w)
9       x[2] = 1 + 1/w^2
10      alpha = (x[0]^3 + b)^((p-1)/2)
11      beta = (x[1]^3 + b)^((p-1)/2)
12      i = (Integer(3+(alpha-1)*beta)).mod(3)
13      y = t^((p-1)/2)*(x[i]^3 + b).sqrt()
14      return(E((x[i], y)))
15
16  ##################################################
17  #                Hash to G1 Test                 #
18  ##################################################
19  l = 0
20  cof = int(E.order()/r)
21  for i in range(1,10):
22      P = Hash_to_G1(Fp.random_element())
23      if (cof*r)*P != E(0):
24          l = l+1
25  l
```

Listing 46 Square root computation over $\mathbb{F}_{p^3}$ using the Atkin method for $p \equiv 5 \mod 8$, and $\psi(\cdot)$ function computations.

```
1   ##################################################
2   #       Square root in Fp3 with p = 5 mod 8      #
3   ##################################################
4   def Sqrt_5_mod_8(a):
5       a1 =  a^e
6       b = t_A*a1
7       i = 2*(a*b)*b
8       x = (a*b)*(i-1)
9       return x
10
11  ##################################################
12  #              \psi^i  functions                 #
13  ##################################################
14  ## psi and psi4
15  def psi(P, g_2, g_3):
16      x = mul_landa(P[0], landa_1, landa_2)
17      y = mul_landa(P[1], landa_1, landa_2)
18      return(Etwist([x*g_2, y*g_3]))
19  ## psi2 and psi5
20  def psi2(P, g2_2, g2_3):
21      x = mul_landa(P[0], landa_2, landa_1)
22      y = mul_landa(P[1], landa_2, landa_1)
23      return(Etwist([x*g2_2, y*g2_3]))
24  def psi3(P, g3_2, g3_3):
25      return(Etwist([P[0]*g3_2, P[1]*g3_3]))
```

Listing 47 Residuosity quadratic test over $\mathbb{F}_{p^3}$ and deterministic point construction in the twist curve $E'/\mathbb{F}_{p^3} : Y^2 = X^3 + b/\xi$.

```
1   ###################################################
2   #     deterministic point construction in Etwist  #
3   ###################################################
4   def QRT(a):
5       v0 = a * mul_landa(a, landa_1, landa_2)
6       v0 = v0 * mul_landa(a, landa_2, landa_1)
7       v0 = vector(v0)
8       c = v0[0]^((p-1)/2)
9       if c == 1:
10          return 1
11      return -1
12
13  def deterministic_point_Etwist(t):
14      b = vector(b_xi)
15      a0 = Fp(1 + b[0] + t^2)
16      a1 = b[1]
17      a2 = b[2]
18      A = 1/Fp3([a0,a1, a2])
19      (a0,a1, a2) = vector(A)
20      c = sqrt_3*t
21      w = [a0*c, a1*c, a2*c]
22      x = [0,0,0]
23      x[0] = Fp3([j - t*w[0], -t*w[1], -t*w[2]])
24      x[1] = -1 - x[0]
25      x[2] = 1 + 1/Fp3(w)^2
26      alpha = QRT(x[0]^3 + b_xi)
27      beta = QRT(x[1]^3 + b_xi)
28      i = (Integer(3+(alpha-1)*beta)).mod(3)
29      y = t^((p-1)/2)*Sqrt_5_mod_8(x[i]^3 + b_xi)
30      return(Etwist(x[i], y))
```

Listing 48 Scalar multiplication by the cofactor c that produces a point in E' with order r.

```
1   ###################################################
2   #     Scalar multiplicaction by the cofactor      #
3   ###################################################
4   def Mult_cofactor(P):
5       Qx0_ = -P
6       Qx1  = x*P
7       Qx1_ = -Qx1
8       Qx2  = x*Qx1
9       Qx2_ = -Qx2
10      Qx3  = x*Qx2
11
12      t1 = P
13      t2 = psi2(Qx1_,gp2_2,gp2_3)
14      t3 = Qx1+psi2(Qx1,gp5_2,gp5_3)
15      t4 = psi3(Qx1,gp3_2,gp3_3)+psi(Qx2,gp_2,gp_3)+
16           psi2(Qx2_,gp2_2,gp2_3)
17      t5 = psi(Qx0_,gp4_2,gp4_3)
18      t6 = psi(P,gp_2,gp_3)+psi3(P,gp3_2,gp3_3)+
19           psi(Qx2_,gp4_2,gp4_3)+psi2(Qx2,gp5_2,gp5_3)+
20           psi(Qx3,gp_2,gp_3)
21
22      t2 = t2 + t1
23      t1 = t1 + t1
24      t1 = t1 + t3
25      t1 = t1 + t2
26      t4 = t4 + t2
27      t5 = t5 + t1
28      t4 = t4 + t1
29      t5 = t5 + t4
30      t4 = t4 + t6
31      t5 = t5 + t5
32      t5 = t5 + t4
33      return(t5)
```

Listing 49 Definition of the hashing to $\mathbb{G}_2$ function and sobriety test.

```
1   ###################################################
2   #              Hash function to G2                #
3   ###################################################
4   def Hash_to_G2(t):
5       P = deterministic_point_Etwist(t)
6       Q = Mult_cofactor(P)
7       return(Q)
8
9   ###################################################
10  #              Hash to G2 test                    #
11  ###################################################
12  l = 0
13  for i in range(1,100):
14      P = Hash_to_G2(Fp.random_element())
15      if r*P != Etwist(0):
16          l = l+1
17  l
```

References

[1] Gora Adj and Francisco Rodríguez-Henríquez. Square root computation over even extension fields. *IEEE Transactions on Computers*, 63(11):2829–2841, 2014.

[2] Diego F. Aranha, Koray Karabina, Patrick Longa, Catherine H. Gebotys, and Julio López. Faster explicit formulas for computing pairings over ordinary curves. In K. G. Paterson, editor, *Advances in Cryptology – EUROCRYPT 2011*, volume 6632 of *Lecture Notes in Computer Science*, pp. 48–68. Springer, Heidelberg, 2011.

[3] Eric Bach and Klaus Huber. Note on taking square-roots modulo N. *IEEE Transactions on Information Theory*, 45(2):807–809, 1999.

[4] Joonsang Baek and Yuliang Zheng. Identity-based threshold decryption. In F. Bao, R. Deng, and J. Zhou, editors, *Public Key Cryptography – PKC 2004*, volume 2947 of *Lecture Notes in Computer Science*, pp. 262–276. Springer, Heidelberg, 2004.

[5] Paulo S. L. M. Barreto and Hae Yong Kim. Fast hashing onto elliptic curves over fields of characteristic 3. Cryptology ePrint Archive, Report 2001/098, 2001. http://eprint.iacr.org/2001/098.

[6] Paulo S. L. M. Barreto, Ben Lynn, and Michael Scott. On the selection of pairing-friendly groups. In M. Matsui and R. J. Zuccherato, editors, *Selected Areas in Cryptography (SAC 2003)*, volume 3006 of *Lecture Notes in Computer Science*, pp. 17–25. Springer, Heidelberg, 2004.

[7] Paulo S. L. M. Barreto and Michael Naehrig. Pairing-friendly elliptic curves of prime order. In B. Preneel and S. Tavares, editors, *Selected Areas in Cryptography (SAC 2005)*, volume 3897 of *Lecture Notes in Computer Science*, pp. 319–331. Springer, Heidelberg, 2006.

[8] Mihir Bellare and Phillip Rogaway. Random oracles are practical: A paradigm for designing efficient protocols. In V. Ashby, editor, *1st ACM Conference on Computer and Communications Security*, pp. 62–73. ACM Press, 1993.

[9] Alexandra Boldyreva. Threshold signatures, multisignatures and blind signatures based on the gap-Diffie-Hellman-group signature scheme. In Y. Desmedt, editor, *Public Key Cryptography – PKC 2003*, volume 2567 of *Lecture Notes in Computer Science*, pp. 31–46. Springer, Heidelberg, 2003.

[10] Dan Boneh and Matthew K. Franklin. Identity-based encryption from the Weil pairing. In J. Kilian, editor, *Advances in Cryptology – CRYPTO 2001*, volume 2139 of *Lecture Notes in Computer Science*, pp. 213–229. Springer, Heidelberg, 2001.

[11] Dan Boneh, Craig Gentry, Ben Lynn, and Hovav Shacham. Aggregate and verifiably encrypted signatures from bilinear maps. In E. Biham, editor, *Advances in Cryptology – EUROCRYPT 2003*, volume 2656 of *Lecture Notes in Computer Science*, pp. 416–432. Springer, Heidelberg, 2003.

[12] Dan Boneh, Ben Lynn, and Hovav Shacham. Short signatures from the Weil pairing. In C. Boyd, editor, *Advances in Cryptology – ASIACRYPT 2001*, volume 2248 of *Lecture Notes in Computer Science*, pp. 514–532. Springer, Heidelberg, 2001.

[13] Colin Boyd, Paul Montague, and Khanh Quoc Nguyen. Elliptic curve based password authenticated key exchange protocols. In V. Varadharajan and Y. Mu, editors, *Information Security and Privacy (ACISP 2001)*, volume 2119 of *Lecture Notes in Computer Science*, pp. 487–501. Springer, Heidelberg, 2001.

[14] Xavier Boyen. Multipurpose identity-based signcryption (a swiss army knife for identity-based cryptography). In D. Boneh, editor, *Advances in Cryptology – CRYPTO 2003*, volume 2729 of *Lecture Notes in Computer Science*, pp. 383–399. Springer, Heidelberg, 2003.

[15] Friederike Brezing and Annegret Weng. Elliptic curves suitable for pairing based cryptography. *Designs, Codes and Cryptography*, 37(1):133–141, 2005.

[16] Eric Brier, Jean-Sébastien Coron, Thomas Icart, David Madore, Hugues Randriam, and Mehdi Tibouchi. Efficient indifferentiable hashing into ordinary elliptic curves. In T. Rabin, editor, *Advances in Cryptology – CRYPTO 2010*, volume 6223 of *Lecture Notes in Computer Science*, pp. 237–254. Springer, Heidelberg, 2010.

[17] Jae Choon Cha and Jung Hee Cheon. An identity-based signature from gap Diffie-Hellman groups. In Y. Desmedt, editor, *Public Key Cryptography – PKC 2003*, volume 2567 of *Lecture Notes in Computer Science*, pp. 18–30. Springer, Heidelberg, 2003.

[18] Sanjit Chatterjee, Darrel Hankerson, and Alfred Menezes. On the efficiency and security of pairing-based protocols in the Type 1 and Type 4 settings. In M. A. Hasan and T. Helleseth, editors, *Arithmetic of Finite Fields (WAIFI 2010)*, volume 6087 of *Lecture Notes in Computer Science*, pp. 114–134. Springer, 2010.

[19] Sanjit Chatterjee and Alfred Menezes. On cryptographic protocols employing asymmetric pairings – The role of Ψ revisited. *Discrete Applied Mathematics*, 159(13):1311–1322, 2011.

[20] Liqun Chen, Zhaohui Cheng, and Nigel P. Smart. Identity-based key agreement protocols from pairings. *International Journal of Information Security*, 6(4):213–241, 2007.

[21] Chitchanok Chuengsatiansup, Michael Naehrig, Pance Ribarski, and Peter Schwabe. PandA: Pairings and arithmetic. In Z. Cao and F. Zhang, editors, *Pairing-Based Cryptography – Pairing 2013*, volume 8365 of *Lecture Notes in Computer Science*, pp. 229–250. Springer, Heidelberg, 2014.

[22] Reza Rezaeian Farashahi, Pierre-Alain Fouque, Igor Shparlinski, Mehdi Tibouchi, and José Felipe Voloch. Indifferentiable deterministic hashing to elliptic and hyperelliptic curves. *Mathematics of Computation*, 82(281):491–512, 2013.

[23] Reza Rezaeian Farashahi, Igor E. Shparlinski, and José Felipe Voloch. On hashing into elliptic curves. *Journal of Mathematical Cryptology*, 3(4):353–360, 2009.

[24] Pierre-Alain Fouque and Mehdi Tibouchi. Estimating the size of the image of deterministic hash functions to elliptic curves. In M. Abdalla and P. S. L. M. Barreto, editors, *Progress in Cryptology – LATINCRYPT 2010*, volume 6212 of *Lecture Notes in Computer Science*, pp. 81–91. Springer, Heidelberg, 2010.

[25] Pierre-Alain Fouque and Mehdi Tibouchi. Indifferentiable hashing to Barreto-Naehrig curves. In A. Hevia and G. Neven, editors, *Progress in Cryptology – LATINCRYPT 2012*, volume 7533 of *Lecture Notes in Computer Science*, pp. 1–17. Springer, Heidelberg, 2012.

[26] Laura Fuentes-Castañeda, Edward Knapp, and Francisco Rodríguez-Henríquez. Faster hashing to $\mathbb{G}_2$. In A. Miri and S. Vaudenay, editors, *Selected Areas in Cryptography – SAC 2011*, volume 7118 of *Lecture Notes in Computer Science*, pp. 412–430. Springer, Heidelberg, 2012.

[27] Steven D. Galbraith, Xibin Lin, and Michael Scott. Endomorphisms for faster elliptic curve cryptography on a large class of curves. *Journal of Cryptology*, 24(3):446–469, 2011.

[28] Steven D. Galbraith, Kenneth G. Paterson, and Nigel P. Smart. Pairings for cryptographers. *Discrete Applied Mathematics*, 156(16):3113–3121, 2008.

[29] Craig Gentry and Alice Silverberg. Hierarchical ID-based cryptography. In Y. Zheng, editor, *Advances in Cryptology – ASIACRYPT 2002*, volume 2501 of *Lecture Notes in Computer Science*, pp. 548–566. Springer, Heidelberg, 2002.

[30] Helmut Hasse. Zur Theorie der abstrakten elliptischen Funktionenkörper III. Die Struktur des Meromorphismenrings; die Riemannsche Vermutung. *Journal für die reine und angewandte Mathematik*, 175:193–208, 1936.

[31] Florian Hess, Nigel P. Smart, and Frederik Vercauteren. The Eta pairing revisited. *IEEE Transactions on Information Theory*, 52(10):4595–4602, 2006.

[32] Jeremy Horwitz and Ben Lynn. Toward hierarchical identity-based encryption. In L. R. Knudsen, editor, *Advances in Cryptology – EUROCRYPT 2002*, volume 2332 of *Lecture Notes in Computer Science*, pp. 466–481. Springer, Heidelberg, 2002.

[33] Thomas Icart. How to hash into elliptic curves. In S. Halevi, editor, *Advances in Cryptology – CRYPTO 2009*, volume 5677 of *Lecture Notes in Computer Science*, pp. 303–316. Springer, Heidelberg, 2009.

[34] David P. Jablon. Strong password-only authenticated key exchange. *SIGCOMM Computer Communication Review*, 26(5):5–26, 1996.

[35] David P. Jablon. Extended password key exchange protocols immune to dictionary attacks. In *6th IEEE International Workshops on Enabling Technologies: Infrastructure for Collaborative Enterprises (WETICE 1997)*, pp. 248–255. IEEE Computer Society, 1997.

[36] Ezekiel J. Kachisa, Edward F. Schaefer, and Michael Scott. Constructing Brezing-Weng pairing-friendly elliptic curves using elements in the cyclotomic field. In S. D. Galbraith and K. G. Paterson, editors, *Pairing-Based Cryptography – Pairing 2008*, volume 5209 of *Lecture Notes in Computer Science*, pp. 126–135. Springer, Heidelberg, 2008.

[37] Edward Knapp. *On the Efficiency and Security of Cryptographic Pairings*. PhD thesis, University of Waterloo, 2010.

[38] Arjen K. Lenstra, Hendrik W. Lenstra Jr., and László Lovász. Factoring polynomials with rational coefficients. *Mathematische Annalen*, 261(4):515–534, 1982.

[39] Benoît Libert and Jean-Jacques Quisquater. Efficient signcryption with key privacy from gap Diffie-Hellman groups. In F. Bao, R. Deng, and J. Zhou, editors, *Public Key Cryptography – PKC 2004*, volume 2947 of *Lecture Notes in Computer Science*, pp. 187–200. Springer, Heidelberg, 2004.

[40] Ueli M. Maurer, Renato Renner, and Clemens Holenstein. Indifferentiability, impossibility results on reductions, and applications to the random oracle methodology. In M. Naor, editor, *Theory of Cryptography Conference (TCC 2004)*, volume 2951 of *Lecture Notes in Computer Science*, pp. 21–39. Springer, Heidelberg, 2004.

[41] Alfred Menezes. *Elliptic Curve Public Key Cryptosystems*. Kluwer Academic Publishers, 1993.

[42] Hermann Minkowski. *Geometrie der Zahlen*. Leipzig und Berlin, Druck ung Verlag von B.G. Teubner, 1910.

[43] Shigeo Mitsunari. A fast implementation of the optimal Ate pairing over BN curve on Intel Haswell processor. Cryptology ePrint Archive, Report 2013/362, 2013. http://eprint.iacr.org/2013/362.

[44] Geovandro C. C. F. Pereira, Marcos A. Simplício, Jr., Michael Naehrig, and Paulo S. L. M. Barreto. A family of implementation-friendly BN elliptic curves. *Journal of Systems and Software*, 84(8):1319–1326, 2011.

[45] Thomas Ristenpart, Hovav Shacham, and Thomas Shrimpton. Careful with composition: Limitations of the indifferentiability framework. In K. G. Paterson, editor, *Advances in Cryptology – EUROCRYPT 2011*, volume 6632 of *Lecture Notes in Computer Science*, pp. 487–506. Springer, Heidelberg, 2011.

[46] René Schoof. Elliptic curves over finite fields and the computation of square roots mod *p*. *Mathematics of Computation*, 44(170):483–494, 1985.

[47] Michael Scott. Implementing cryptographic pairings. In T. Takagi et al., editors, *Pairing-Based Cryptography – Pairing 2007*, volume 4575 of *Lecture Notes in Computer Science*, pp. 177–196. Springer, Heidelberg, 2007.

[48] Michael Scott. Unbalancing pairing-based key exchange protocols. Cryptology ePrint Archive, Report 2013/688, 2013. `http://eprint.iacr.org/2013/688`.

[49] Michael Scott, Naomi Benger, Manuel Charlemagne, Luis J. Dominguez Perez, and Ezekiel J. Kachisa. Fast hashing to G_2 on pairing-friendly curves. In H. Shacham and B. Waters, editors, *Pairing-Based Cryptography – Pairing 2009*, volume 5671 of *Lecture Notes in Computer Science*, pp. 102–113. Springer, Heidelberg, 2009.

[50] Andrew Shallue and Christiaan van de Woestijne. Construction of rational points on elliptic curves over finite fields. In F. Hess, S. Pauli, and M. E. Pohst, editors, *Algorithmic Number Theory (ANTS-VII)*, volume 4076 of *Lecture Notes in Computer Science*, pp. 510–524. Springer, 2006.

[51] Frederik Vercauteren. Optimal pairings. *IEEE Transactions on Information Theory*, 56(1):455–461, 2009.

[52] Eric Zavattoni, Luis J. Dominguez Perez, Shigeo Mitsunari, Ana H. Sánchez-Ramírez, Tadanori Teruya, and Francisco Rodríguez-Henríquez. Software implementation of an attribute-based encryption scheme. *IEEE Transactions on Computers*, 64(5):1429–1441, 2015.

[53] Fangguo Zhang and Kwangjo Kim. ID-based blind signature and ring signature from pairings. In Y. Zheng, editor, *Advances in Cryptology – ASIACRYPT 2002*, volume 2501 of *Lecture Notes in Computer Science*, pp. 533–547. Springer, Heidelberg, 2002.

9

Discrete Logarithms

Aurore Guillevic
INRIA-Saclay and École Polytechnique/LIX

François Morain
*École Polytechnique/LIX and CNRS and
INRIA-Saclay*

The Discrete Logarithm Problem (DLP) is one of the most used mathematical problems in asymmetric cryptography design, the other one being the integer factorization. It is intrinsically related to the Diffie-Hellman problem (DHP). DLP can be stated in various groups. It must be hard in well-chosen groups, so that secure-enough cryptosystems can be built. In this chapter, we present the DLP, the various cryptographic problems based on it, the commonly used groups, and the major algorithms available at the moment to compute discrete logarithms in such groups. We also list the groups that must be avoided for security reasons.

Our computational model will be that of classical computers. It is to be noted that in the quantum model, DLP can be solved in polynomial time for cyclic groups [108].

9.1   Setting and First Properties

9.1.1   General Setting

Let G be a finite cyclic group of order N generated by g. The group law on G is defined multiplicatively and noted $\circ$, the neutral element is noted 1_G, and the inverse of an element a will be noted $1/a = a^{-1}$. The *discrete logarithm problem* is the following:

   DLP: *given $h \in G$, find an integer n, $0 \le n < N$ such that $h = g^n$.*

Along the years, and motivated by precise algorithms or cryptographic implementations, variants of this basic problem appeared. For instance:

   Interval-DLP (IDLP): *given $h \in G$, find an integer n, $a \le n < b$ such that $h = g^n$.*

We may also need to compute multiple instances of the DLP:

Batch-DLP (BDLP): *given $\{h_1, \ldots, h_k\} \subset G$, find integers n_i's, $0 \leq n_i < N$ such that $h_i = g^{n_i}$.*

A variant of this is **Delayed target DLP** where we can precompute many logs before receiving the actual target. This was used in the *logjam attack* where logarithms could be computed in real time during an SSL connection [9].

We may relate them to Diffie-Hellman problems, such as

Computational DH problem (DHPor CDH): *given (g, g^a, g^b), compute g^{ab}.*

Decisional DH problem (DDHP): *given (g, g^a, g^b, g^c), do we have $c = ab \bmod N$?*

We use $A \leq B$ to indicate that A is easier than B (i.e., there is polynomial time reduction from B to A). The first easy result is the following:

PROPOSITION 9.1 DDHP $\leq$ DHP $\leq$ DLP.

In some cases, partial or heuristic reciprocals have been given, most notably in [86, 87]. With more and more applications and implementations available of different DH bases protocols, more problems arose, notably related to static variants. We refer to [77] for a survey.

9.1.2 The Pohlig-Hellman Reduction

In this section, we reduce the cost of DLP in a group of size N to several DLPs whose overall cost is dominated by that of the DLP for the largest $p \mid N$.

PROPOSITION 9.2 *Let G be a finite cyclic group of order N whose factorization into primes is known,*

$$N = \prod_{i=1}^{r} p_i^{\alpha_i}$$

where the p_i's are all distinct. Then DLP in G can be solved using the DLPs on all subgroups of order $p_i^{\alpha_i}$.

Proof. Solving $g^x = a$ is equivalent to finding $x \bmod N$, i.e., $x \bmod p_i^{\alpha_i}$ for all i, using the Chinese remaindering theorem.

Suppose $p^\alpha \parallel N$ (which means $p^\alpha \mid N$ but $p^{\alpha+1} \nmid N$) and $m = N/p^\alpha$. Then $b = a^m$ is in the cyclic group of order p^α generated by $h = g^m$. We can find the log of b in this group, which yields $x \bmod p^\alpha$. From this, we have reduced DLP to r DLPs in smaller cyclic groups.

How do we proceed? First compute $c_1 = b^{p^{\alpha-1}}$ and $h_1 = h^{p^{\alpha-1}}$. Both elements belong to the cyclic subgroup of order p of G, generated by h_1. Writing $y = x \bmod p^\alpha = y_0 + y_1 p + \cdots + y_{\alpha-1}p^{\alpha-1}$ with $0 \leq y_i < p$, we see that $y = \log_h(b) \bmod p^\alpha$. We compute

$$h_1^y = h_1^{y_0}$$

so that y_0 is the discrete logarithm of c in base h_1. Write

$$c_2 = b^{p^{\alpha-2}} = h_1^{(y_0+y_1 p)p^{\alpha-2}}$$

or

$$c_2 h_1^{-y_0 p^{\alpha-2}} = h_1^{y_1}.$$

In this way, we recover y_1 by computing the discrete logarithm of the left hand side w.r.t. h_1 again. $\qquad\square$

We have shown that DLP in a cyclic group of order p^α can be replaced by α solutions of DLP in a cyclic group of order p and some group operations. If p is small, all these steps will be easily solved by table lookup. Otherwise, the methods presented in the next section will apply and give a good complexity.

A direct cryptographic consequence is that for cryptographic use, N must have at least one large prime factor.

9.1.3 A Tour of Possible Groups

Easy groups

Let us say a group is *easy* for DL if DLP can be solved in polynomial time for this group. DLP is easy in $(\mathbb{Z}/N\mathbb{Z}, +)$, since $h = ng \bmod N$ is solvable in polynomial time (Euclid). This list was recently enlarged to cases where the discrete log can be computed in quasi-polynomial time [17].

As for algebraic curves, supersingular elliptic curves were shown to be somewhat weaker in [89]; the same is true for hyperelliptic curves [49, 37]. Elliptic curves of trace 1 (also called *anomalous curves*) were shown to be easy in [109, 104]; the result was extended to hyperelliptic anomalous curves in [99].

Not-so-easy groups

Relatively easy groups are those for which subexponential methods exist: finite fields (of medium or large characteristic), algebraic curves of very large genus, class groups of number fields.

Probably difficult groups, for which we know of nothing but exponential methods, include elliptic curves (see [50] for a recent survey) and curves of genus 2.

9.2 Generic Algorithms

We start with algorithms solving DLP in a generic way, which amounts to saying that we use group operations only. We concentrate on generic groups. For particular cases, as elliptic curves, we refer to [53] for optimized algorithms. We emphasize that generic methods are the only known methods for solving the DLP over ordinary elliptic curves.

The chosen setting is that of a group $G = \langle g \rangle$ of prime order N, following the use of the Pohlig-Hellman reduction. Enumerating all possible powers of g is an $O(N)$ process, and this is enough when N is small. Other methods include Shanks's and Pollard's, each achieving a $O(\sqrt{N})$ time complexity, but different properties as determinism or space. We summarize this in Table 9.1. The baby-steps giant-steps (BSGS) method and its variants are deterministic, whereas the other methods are probabilistic. It is interesting to note that Nechaev and Shoup have proven that a lower bound on generic DLP (algorithms that use group operations only) is precisely $O(\sqrt{N})$ (see for instance [111]).

Due to the large time complexity, it is desirable to design distributed versions with a gain of p in time when p processors are used. This will be described method by method.

TABLE 9.1 Properties of generic DL algorithms.

Algorithm	group	interval	time	space
BSGS	✓	✓	$O(\sqrt{N})$	$O(\sqrt{N})$
RHO	✓	–	$O(\sqrt{N})$	$O(\log N)$
Kangaroo	✓	✓	$O(\sqrt{N})$	$O(\log N)$

TABLE 9.2   Table of constants C such that the complexity is $C\sqrt{N}$.

Algorithm	Average-case time	Worst-case time
BSGS	1.5	2.0
BSGS optimized for av. case	1.414	2.121
IBSGS	1.333	2.0
Grumpy giants	1.25?	≤ 3
RHO with dist. pts	$1.253(1 + o(1))$	∞

9.2.1   Shanks's Baby-Steps Giant-Steps Algorithm

Shanks's idea, presented in the context of class groups [107], became a fundamental tool for operating in generic groups, for order or discrete logarithm computations. All variants of it have a time complexity $O(\sqrt{N})$ group operations for an $O(\sqrt{N})$ storage of group elements. They all differ by the corresponding constants and scenarii in which they are used. From [53], we extract Table 9.2.

The goal of this subsection is to present the original algorithm together with some of its more recent variants. We refer to the literature for more material and analyses.

Original algorithm and analysis

The standard algorithm runs as follows. Write the unknown n in base u for some integer u that will be precised later on:

$$n = cu + d, 0 \leq d < u, \quad 0 \leq c < N/u.$$

We rewrite our equation as

$$g^n = h \Leftrightarrow h \circ (g^{-u})^c = g^d.$$

The algorithm is given in Figure 9.1. It consists of evaluating the right-hand side for all possible d by increment of 1 (baby steps), and then computing all left-hand sides by increment of u (giant steps), until a match is found.

The number of group operations is easily seen to be $C_o = u + N/u$ in the worst case, minimized for $u = \sqrt{N}$, leading to a deterministic complexity of $2\sqrt{N}$ group operations. On average, c will be of order $N/(2u)$, so that the average cost is $(1 + 1/2)u$, which gives us the first line in Table 9.2.

Step 1 requires u insertions in the set $\mathcal{B}$ and Step 2 requires N/u membership tests in the worst case. This explains why we should find a convenient data structure for $\mathcal{B}$ that has the smallest time for both operations. This calls for $\mathcal{B}$ to be represented by a hash table of some sort. The cost of these set operations will be $(u + N/u)O(1)$, again minimized by $u = \sqrt{N}$.

Remarks.

Variants exist when inversion has a small cost compared to multiplication, leading to writing $n = cu + d$ where $-c/2 \leq d < c/2$, thereby gaining in the constant in front of $\sqrt{N}$ (see [53] for a synthetic view). Cases where the distribution of parameters is non-uniform are studied in [26].

All kinds of trade-offs are possible if low memory is available. Moreover, different variants of BSGS exist when one wants to run through the possible steps in various orders (see [110]). If no bound is known on N, there are slower incremental algorithms that will find the answer; see [113] and [110].

As a final comment, BSGS is easy to distribute among processors with a shared memory.

Solving IDLP

BSGS works if we have a bound on N only. It also works when x is known to belong to some interval $[a, b[\subset [0, N[$. The process amounts to a translation of x to $[0, b - a[$ of length $b - a$ and therefore BSGS will require time $O(\sqrt{b - a})$.

ALGORITHM 9.1 Baby-steps giant-steps.

Function *BSGS(G, g, N, h)*

    Input   : $G \supset \langle g \rangle$, g of order N; $h \in \langle g \rangle$

    Output: $0 \leq n < N$, $g^n = h$

    $u \leftarrow \lceil \sqrt{N} \rceil$

    // Step 1 (baby steps)

    initialize a table $\mathcal{B}$ for storing u pairs (element of G, integer $< N$)

    store$(\mathcal{B}, (1_G, 0))$

    $H \leftarrow g$; store$(\mathcal{B}, (H, 1))$

    for $d := 2$ **to** $u - 1$ **do**

        $H \leftarrow H \circ g$

        store$(\mathcal{B}, (H, d))$

    end

    // Step 2 (giant steps)

    $H \leftarrow H \circ g$

    $f \leftarrow 1/H = g^{-u}$

    $H \leftarrow h$

    for $c := 0$ **to** N/u **do**

        // $H = h \circ f^c$

        if $\exists (H', d) \in \mathcal{B}$ *such that* $H = H'$ **then**

            // $H = h \circ f^c = g^d$ hence $n = cu + d$

            return $cu + d$;

        end

        $H \leftarrow H \circ f$

    end

Optimizing BSGS on average

On average, we could anticipate c to be around $N/2$, so that we may want to optimize the mean number of operations of BSGS, or $C_m = u + (N/2)/u$, leading to $u = \sqrt{N/2}$ and $C_m = \sqrt{2N}$, decreasing the memory used by the same quantity. The number of set operations also decreases. This gives us the line for BSGS optimized for average-case. Algorithm 9.1 is usable *mutatis mutandis*.

Interleaving baby steps and giant steps

Pollard [98] proposed a variant of the BSGS algorithm interleaving baby steps and giant steps, in order to decrease the average cost of BSGS. The idea is the following: If $x = cu + d$, we may find c and d after $\max(c, d)$ steps using the following algorithm. The rationale for the choice of u will be explained next.

First of all, remark that any integer n in $[0, N[$ may be written as $cu + d$ where $0 \leq c, d < u$. Algorithm 9.2 performs $2 \max(c, d)$ group operations. We need to evaluate the average value of this quantity over the domain $[0, u[\times [0, u[$, which is equivalent to computing

$$\int_{x=0}^{1} \int_{y=0}^{1} \max(x, y) \, dx \, dy.$$

Fixing x, we see that $\max(x, y) = x$ for $y \leq x$ and y otherwise. Therefore the double integral is

$$\int_{x=0}^{1} \left(\int_{y=0}^{x} x \, dy + \int_{y=x}^{1} y \, dy \right) dx = \frac{2}{3}.$$

ALGORITHM 9.2   Interleaved Baby steps-giant steps.

Function *IBSGS(G, g, N, h)*

    Input : $G \supset \langle g \rangle$, g of order N; $h \in \langle g \rangle$

    Output: $0 \le n < N$, $g^n = h$

    $u \leftarrow \lceil \sqrt{N} \rceil$

    initialize two tables $\mathcal{B}$ and $\mathcal{G}$ for storing u pairs (element of G, integer $< N$)

    $H \leftarrow 1_G$; store($\mathcal{B}$, $(1_G, 0)$); store($\mathcal{G}$, $(1_G, 0)$)

    $F \leftarrow h$

    $f \leftarrow g^{-u} = 1/g^u$

    for $i := 1$ **to** u **do**

        $H \leftarrow H \circ g$

        if $\exists (H', c) \in \mathcal{G}$ *such that* $H = H'$ **then**

            // $H = g^i = H' = h \circ g^{-uc}$ hence $n = cu + i$

            return $cu + i$;

        end

        store($\mathcal{B}$, (H, i))

        $F \leftarrow F \circ f$

        if $\exists (H', d) \in \mathcal{B}$ *such that* $F = H'$ **then**

            // $F = h \circ g^{-ui} = H' = g^d$ hence $n = iu + d$

            return $iu + d$;

        end

        store($\mathcal{G}$, (F, i))

    end

We have proven that the mean time for this algorithm is $4/3\sqrt{N}$, hence a constant that is smaller than $\sqrt{2}$ for the original algorithm.

Grumpy giants

In [23], the authors designed a new variant of BSGS to decrease the average case running time again. They gave a heuristic analysis of it. This was precised and generalized to other variants (such as using negation) in [53]. We follow the presentation therein. For $u = \lceil \sqrt{N}/2 \rceil$, the algorithm computes the three sets of cardinality L that will be found later:

$$\mathcal{B} = \{ g^i \text{ for } 0 \le i < L \},$$

$$\mathcal{G}_1 = \{ h \circ (g^{ju}) \text{ for } 0 \le j < L \},$$

$$\mathcal{G}_2 = \{ : h^2 \circ g^{-k(u+1)} \text{ for } 0 \le k < L \},$$

and waits for collisions between any of the two sets, in an interleaved manner. The algorithm succeeds when one of the following sets contains the discrete logarithm we are looking for:

$$
\begin{aligned}
\mathcal{L}_L \ &= \ \{ i - ju \pmod{N}, 0 \le i, j < L \} \\
&\cup \ \{ 2^{-1}(i + k(u+1)) \pmod{N}, 0 \le i, k < L \} \\
&\cup \ \{ ju + k(u+1) \pmod{N}, 0 \le j, k < L \}.
\end{aligned}
$$

ALGORITHM 9.3 Two grumpy giants and a baby.

Function *Grumpy(G, g, N, h)*

    Input : $G \supset \langle g \rangle$, g of order N; $h \in \langle g \rangle$

    Output: $0 \leq n < N$, $g^n = h$

    $u \leftarrow \lceil \sqrt{N} \rceil$

    initialize three tables $\mathcal{B}$, $\mathcal{G}_1$ and $\mathcal{G}_2$ for storing u pairs (element of G, integer $< N$)

    $H \leftarrow 1_G$; store($\mathcal{B}$, $(H, 0)$)

    $F_1 \leftarrow h$; store($\mathcal{G}_1$, $(F_1, 0)$)

    $F_2 \leftarrow h^2$; store($\mathcal{G}_2$, $(F_2, 0)$)

    $f_1 \leftarrow g^u$

    $f_2 \leftarrow 1/(f_1 \circ g) = 1/g^{u+1}$

    for $i := 1$ **to** L **do**

        $H \leftarrow H \circ g$

        for $j := 1$ **to** 2 **do**

            if $\exists (H', c) \in \mathcal{G}_j$ *such that* $H = H'$ **then**

                // $H = g^i = H' = h^j \circ f_j^c$

                return $(-Expo(u, j)c + i)/j$ (mod N);

            end

        end

        store($\mathcal{B}$, (H, i))

        for $j := 1$ **to** 2 **do**

            $F_j \leftarrow F_j \circ f_j$

            if $\exists (H', d) \in \mathcal{B}$ *such that* $F_j = H'$ **then**

                // $F_j = h^j \circ g^{ui} = H' = g^d$

                return $(-Expo(u, j)i + d)/j$ (mod N)

            end

            $j' \leftarrow 3 - j$

            if $\exists (H', c) \in \mathcal{G}_{j'}$ *such that* $F_j = H'$ **then**

                // $F_j = h^j \circ f_j^i = H' = h^{j'} \circ f_{j'}^c$

                return $(Expo(u, j')c - Expo(u, j)i) + d)/(j - j')$ (mod N)

            end

            store($\mathcal{G}_j$, (F_j, i))

        end

    end

For ease of exposition of Algorithm 9.3, we define $Expo(u, j)$ to be the exponent of g in case of $\mathcal{G}_j$ for $j = 1..2$. Precisely, a member of $\mathcal{G}_j$ is

$$h^j \circ f_j = h^j \circ g^{Expo(u,j)},$$

with $Expo(u, j) = (-1)^{j-1}(u + j - 1)$.

It is conjectured that u is optimal and that L can be taken as $O(\sqrt{N})$. Experiments were carried out to support this claim in [23, 53]. Moreover [53] contains an analysis of the algorithm, leading to $1.25\sqrt{N}$ as total group operations.

9.2.2 The RHO Method

The idea of Pollard was to design an algorithm solving DLP for which the memory requirement would be smaller than that of BSGS. Extending his RHO method of factoring, he came up with the idea of RHO for computing discrete logarithms [97].

A basic model

Let E be a finite set of cardinality m and suppose we draw uniformly n elements from E with replacement. The probability that all n elements are distinct is

THEOREM 9.1

$$\text{Proba} = \frac{1}{m} \prod_{k=1}^{n-1} \left(1 - \frac{k}{m} \right).$$

Taking logarithms, and assuming $n \ll m$, we get

$$\log \text{Proba} \approx \log(n/m) - \frac{n(n-1)}{2m}.$$

This means that taking $n = O(\sqrt{m})$ will give a somewhat large value for this probability.

We can derive from this a very simple algorithm for computing discrete logarithms, presented as Algorithm 9.4. Its time complexity would be $O(\sqrt{m} \log m)$ on average, together with a space $O(\sqrt{m})$, which is no better than BSGS.

ALGORITHM 9.4   Naive DLP algorithm.

Function *NaiveDL(G, g, N, h)*
    Input   : $G \supset \langle g \rangle$, g of order N; $h \in \langle g \rangle$
    Output: $0 \leq n < N$, $g^n = h$
    initialize a table $\mathcal{L}$ for storing u triplets (element of G, two integers $< N$)
    repeat
        draw u and v at random modulo N
        $H \leftarrow g^u \circ h^v$
        if $\exists (H', u', v') \in \mathcal{L}$ *such that* $H = H'$ **then**
            // $H = g^u \circ h^v = g^{u'} \circ h^{v'}$ hence $n(v - v') = u' - u$
            if $v - v'$ *is invertible modulo* N **then**
                return $(u' - u)/(v - v') \bmod N$
            end
        end
        else
            store($\mathcal{L}$, (H, u, v))
        end
    until *a collision is found*

If we assume that N is prime, the only case where $v - v'$ is non-invertible is that of $v = v'$. In that case, we hit a useless relation between g and h that is discarded.

Our basic model is highly distributable. Unfortunately, the memory problem is still there. It does not solve the space problem, so that we have to replace this by deterministic random walks, as explained now.

Functional digraphs

Consider E of cardinality m as above and let $f : E \to E$ be a function on E. Consider the sequence $X_{n+1} = f(X_n)$ for some starting point $X_0 \in E$. The *functional digraph* of X is built

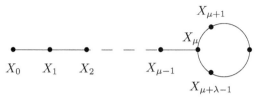

FIGURE 9.1   Functional digraph.

with vertices X_i's; an edge is put between X_i and X_j if $f(X_i) = X_j$. Since E is finite, the graph has two parts, as indicated in Figure 9.1.

Since E is finite, the sequence X must end up looping. The first part of the sequence is the set of X_i's that are reached only once and there are μ of them; the second part forms a loop containing λ distinct elements.

Examples. 1) $E = G$ is a finite group, we use $f(x) = ax$, and $x_0 = a$, (x_n) purely is periodic, i.e., $\mu = 0$, and $\lambda = \mathrm{ord}_G(a)$.
2) Take $E_m = \mathbb{Z}/11\mathbb{Z}$ and $f : x \mapsto x^2 + 1 \bmod 11$: We give the complete graph for all possible starting points in Figure 9.2. The shape of it is quite typical: a cycle and trees plugged on the structure.

By Theorem 9.1, λ and μ cannot be too large on average, since $n = \lambda + \mu$. A convenient source for all asymptotic complexities of various parameters of the graph can be found in [46]. In particular:

THEOREM 9.2   *When* $m \to \infty$

$$\overline{\lambda} \sim \overline{\mu} \sim \sqrt{\frac{\pi m}{8}} \approx 0.627\sqrt{m}.$$

Finding λ and μ is more easily done using the notion of *epact*.

PROPOSITION 9.3   *There exists a unique $e > 0$ (epact) s.t. $\mu \le e < \lambda + \mu$ and $X_{2e} = X_e$. It is the smallest non-zero multiple of λ that is $\ge \mu$: If $\mu = 0$, $e = \lambda$ and if $\mu > 0$, $e = \lceil \frac{\mu}{\lambda} \rceil \lambda$.*

Proof. The equation $X_i = X_j$ with $i < j$ only if i and j are larger than or equal to μ. Moreover, λ must divide $j - i$. If we put $i = e$ and $j = 2e$, then $\mu \le e$ and $\lambda \mid e$. There exists a single multiple of λ in any interval of length λ, which gives unicity for e. When $\mu = 0$, it is clear that the smallest $e > 0$ must be λ. When $\mu > 0$, the given candidate satisfies all the properties.   □

From [46], we extract

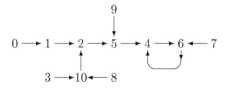

FIGURE 9.2   The functional digraph of $f : x \mapsto x^2 + 1 \bmod 11$.

THEOREM 9.3 $\bar{e} \sim \sqrt{\frac{\pi^5 m}{288}} \approx 1.03\sqrt{m}$.

which means that finding the epact costs $O(\sqrt{m})$ with a constant not too large compared to the actual values of μ and λ. Note that in most cryptographic applications, the collision $x_{2e} = x_e$ will be enough to solve our problem.

From a practical point of view, a nice and short algorithm by Floyd can be used to recover the epact and is given as Algorithm 9.5. We need $3e$ evaluations of f and e comparisons. Ideas for decreasing the number of evaluations are given in [31] (see also [91] when applied to integer factorization).

ALGORITHM 9.5 Floyd's algorithm.

Function *epact(f, x_0)*
> **Input** : A function f, a starting point x_0
> **Output**: The epact of (x_n) defined by $x_{n+1} = f(x_n)$
> $x \leftarrow x_0;\ y \leftarrow x_0;\ e \leftarrow 0$
> **repeat**
>> $e \leftarrow e + 1$
>> $x \leftarrow f(x)$
>> $y \leftarrow f(f(y))$
>
> **until** $x = y$
> **return** e.

More parameters can be studied and their asymptotic values computed. Again, we refer to [46], from which we extract the following complements.

THEOREM 9.4 *The expected values of some of the parameters related to the functional graph G are*

- *the number of components is $\frac{1}{2}\log m$;*
- *the component size containing a node $\nu \in G$ is $2m/3$;*
- *the tree size containing ν is $m/3$ (maximal tree rooted on a circle containing ν);*
- *the number of cyclic nodes is $\sqrt{\pi m/2}$ (a node is cyclic if it belongs to a cycle).*

A way to understand these results is to imagine that there is a giant component that contains almost all nodes.

Discrete logarithms

The idea of Pollard is to build a function f from G to G appearing to be random, in the sense that the epact of f is $c\sqrt{N}$ for some small constant c. This can be realized via multiplications by random points and/or perhaps squarings in G.

Building on [105], Teske [114] has suggested the following: precompute r random elements $z_i = g^{\gamma_i} \circ h^{\delta_i}$ for $1 \leq i \leq r$ for some random exponents. Then use some hash function $\mathcal{H} : G \to \{1, \ldots, r\}$. Finally, define $f(y) = y \circ z_{\mathcal{H}(y)}$. The advantage of this choice is that we can represent any iterate x_i of f as

$$x_i = g^{c_i} \circ h^{d_i},$$

where (c_i) and (d_i) are two integer sequences. When e is found:

$$g^{c_{2e}} \circ h^{d_{2e}} = g^{c_e} \circ h^{d_e},$$

or

$$g^{c_{2e} - c_e} = h^{d_e - d_{2e}},$$

i.e.,

$$n(c_{2e} - c_e) \equiv (d_e - d_{2e}) \bmod N.$$

With high probability, $c_{2e} - c_e$ is invertible modulo N and we get the logarithm of h. When we hit a collision and it is trivial, it is no use continuing the algorithm.

Experimentally, $r = 20$ is enough to have a large mixing of points. Under a plausible model, this leads to a $O(\sqrt{N})$ method (see [114]). We give the algorithm in Algorithm 9.6. As an example, if G contains integers, we may simply use $\mathcal{H}(x) = 1 + (x \bmod r)$.

ALGORITHM 9.6   RHO algorithm.

Function *RHO(G, g, N, h, $\mathcal{H}$, $(z_i, \gamma_i, \delta_i)$)*

   Input : $\mathcal{H} : G \to \{1, \ldots, r\}$; $(z_i)_{1 \leq i \leq r}$ random powers $z_i = g^{\gamma_i} \circ h^{\delta_i}$ of G

   Output: $0 \leq n < N$, $g^n = h$

   if $h = 1_G$ then

    | return 0

   end

   // invariant: $x = g^{u_x} \circ h^{v_x}$

   $x \leftarrow h$; $u_x \leftarrow 0$; $v_x \leftarrow 1$

   $y \leftarrow x$; $u_y \leftarrow u_x$; $v_y \leftarrow v_x$

   repeat

    | $(x, u_x, v_x) \leftarrow$ Iterate$(G, N, \mathcal{H}, (z_i, \gamma_i, \delta_i), x, u_x, v_x)$

    | $(y, u_y, v_y) \leftarrow$ Iterate$(G, N, \mathcal{H}, (z_i, \gamma_i, \delta_i), y, u_y, v_y)$

    | $(y, u_y, v_y) \leftarrow$ Iterate$(G, N, \mathcal{H}, (z_i, \gamma_i, \delta_i), y, u_y, v_y)$

   until $x = y$

   // $g^{u_x} \circ h^{v_x} = g^{u_y} \circ h^{v_y}$

   if $v_x - v_y$ *is invertible modulo N* then

    | return $(u_y - u_x)/(v_x - v_y) \pmod{N}$;

   end

   else

    | return *Failure.*

   end

ALGORITHM 9.7   RHO iteration algorithm.

Function *Iterate(G, N, $\mathcal{H}$, $(z_i, \gamma_i, \delta_i)$, x, u_x, v_x)*

   Input : $\mathcal{H} : G \to \{1, \ldots, r\}$; $(z_i)_{1 \leq i \leq r}$ random powers $z_i = g^{\gamma_i} \circ h^{\delta_i}$ of G;

       $x = g^{u_x} \circ h^{v_x}$

   Output: $f(x, u_x, v_x) = (w, u_w, v_w)$ such that $w = g^{u_w} \circ h^{v_w}$

   $i \leftarrow \mathcal{H}(x)$

   return $(x \circ z_i, u_x + \gamma_i \pmod{N}, v_x + \delta_i \pmod{N})$.

Parallel RHO

How would one program a parallel version of RHO? We have to modify the algorithm. First of all, we cannot use the notion of epact any more. If we start p processors on finding the epact of their own sequence, we would not gain anything, since all epacts are of the same (asymptotic) size. We need to share computations.

FIGURE 9.3   Paths.

The idea is to launch p processors on the same graph with the same iteration function and wait for a collision. Since we cannot store all points, we content ourselves with *distinguished elements*, i.e., elements having a special form, uniformly with some probability θ over G. (For integers, one can simply decide that a distinguished integer is 0 modulo a prescribed power of 2.) Each processor starts its own path from a random value in $\langle g \rangle$ and each time it encounters a distinguished element, it compares it with shared distinguished elements already found and when a useful collision is found, the program stops. The idea is that two paths colliding at some point will eventually lead to the same distinguished element, that will be found a little while later (see Figure 9.3). Typically, if $\theta < 1$ is the proportion of distinguished elements, the time to reach one of these will be $1/\theta$. Remembering properties of functional digraphs, this probability should satisfy $1/\theta < c\sqrt{N}$ for some constant $c > 0$.

In view of Theorem 9.4, the method succeeds since there is a giant component in which the processors have a large probability to run in. At worst, we would need $O(\log m)$ of these to be sure to have at least two processors in the same component.

There are many fine points that must be dealt with in an actual implementation. For ease of reading, we first introduce a function that computes a *distinguished path*, starting from a point and iterating until a distinguished element is reached, at which point is it returned.

ALGORITHM 9.8   Finding a distinguished path.

Function *DistinguishedPath(f, x_0)*
    Input   : A function f, a starting point x_0
    Output: The first distinguished element found starting at x_0,
    $x \leftarrow x_0$; **repeat**
       | $x \leftarrow f(x)$
    until x *is distinguished*
    return x.

At this point, the master can decide to continue from this distinguished element, or start a new path. One of the main problems we can encounter is that a processor be trapped in a (small) cycle. By the properties of random digraph, a typical path should be of length $O(\sqrt{N})$; if θ is small enough, the probability to enter a cycle will be small. However, in some applications, small cycles exist. Therefore, we need some cycle detection algorithm, best implemented using a bound on the number of elements found. Modifying Algorithm 9.8 can be done easily, for instance, giving up on paths with length $> 20/\theta$ as suggested in [118]. The expected running time is $\sqrt{\pi N/2}/p + 1/\theta$ group operations.

Note that in many circumstances, we can use an automorphism in G, and we take this into account for speeding up the parallel RHO process, despite some technical problems that arise (short cycles). See [44], and more recently [72, 24].

Other improvements are discussed in [36] for prime fields, with the aim of reducing the cost of the evaluation of the iteration function.

9.2.3 The Kangaroo Method

This method was designed to solve Interval-DLP with a space complexity as small as that of RHO, assuming the discrete logarithm we are looking for belongs to $[0, \ell]$ with $\ell \leq N$. We would like to obtain an algorithm whose running time is $O(\sqrt{\ell})$ instead of $O(\sqrt{N})$.

The idea is to have two processes, traditionally called *tame kangaroo* and *wild kangaroo*. The tame kangaroo follows a random path starting from $g^{\ell/2}$ and adding random integers to the exponent, while the wild kangaroo starts from $h = g^n$ and uses the same deterministic random function. We use a sequence of integer increments $(\delta_i)_{1 \leq i \leq r}$ whose mean size is m. Then, we iterate: $f(x) = x \circ g^{\delta_{\mathcal{H}(x)}}$. Both kangaroos can be written $T = g^{d_T}$ and $W = h \circ g^{d_W}$ for two integer sequences d_T and d_W that are updated when computing f.

When hitting a distinguished element, it is stored in a list depending on its character (tame or wild). When a collision occurs, the discrete logarithm is found. The analysis is heuristic along the following way. The original positions of K_T and K_W can be either

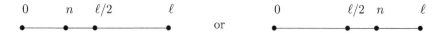

$$0 \quad n \quad \ell/2 \quad \ell \qquad \text{or} \qquad 0 \quad \ell/2 \quad n \quad \ell$$

In either case, we have a back kangaroo (B) and a front kangaroo (F) heading right. They are at mean mutual distance $\ell/4$ at the beginning. Since the average distance between two points is m, B needs $\ell/(4m)$ jumps to reach the initial position of F. After that, B needs m jumps to reach a point already reached by F. The total number of jumps is therefore $2(\ell/(4m) + m)$, which is minimized for $m = \sqrt{\ell}/2$, leading to a $2\sqrt{\ell}$ cost. A more precise analysis is given in [118]. The reader can find details as Algorithm 9.9.

ALGORITHM 9.9   Sequential kangaroos.

Function *Kangaroo(G, g, N, h, ℓ)*
    Input : $G \supset \langle g \rangle$, g of order N; $h \in \langle g \rangle$
    Output: $0 \leq n < \ell$, $g^n = h$
    $m \leftarrow \lceil \sqrt{\ell}/2 \rceil$
    compute positive increments $(\delta_i)_{1 \leq i \leq r}$ of mean m
    initialize two tables $\mathcal{T}$ and $\mathcal{W}$ for storing pairs (element of G, integer $< N$)
    $T \leftarrow g^{\ell/2}$; $d_T \leftarrow \ell/2$
    $W \leftarrow h$; $d_W \leftarrow 0$
    while *true* **do**
        $(T, d_T) \leftarrow f((\delta_i), T, d_T)$
        if $\exists (W', d') \in \mathcal{W}$ *such that* $W' = T$ **then**
            // $T = g^{d_T}$, $W' = h \circ g^{d'}$
            return $(d_T - d') \pmod{N}$
        end
        $(W, d_W) \leftarrow f((\delta_i), W, d_W)$
        if $\exists (T', d') \in \mathcal{T}$ *such that* $T' = W$ **then**
            // $T' = g^{d'}$, $W = h \circ g^{d_W}$
            return $(d' - d_W) \pmod{N}$
        end
    end

A close algorithm that uses another model of analysis (though still heuristic) is that of Gaudry-Schost [54], improving on work by Gaudry and Harley. This algorithm is generalized to any dimension (e.g., solving $g^x = g^{a_1 n_1 + a_2 n_2 + \cdots + a_d n_d}$ for given (a_i)'s) and improved in [51] (see also [52] for the use of equivalence classes).

Parallel kangaroos

The idea, as for parallel RHO, is to start p kangaroos that will discover and store distinguished elements. Following [98], we assume $p = 4p'$, and select $u = 2p' + 1$, $v = 2p' - 1$, so that $p = u + v$. Increments of the jumps will be $(uvs_1, \ldots, uvs_k)$ for small s_i's, insisting on the mean to be $\approx \sqrt{\ell/(uv)}$. The i-th tame kangaroo will start at $g^{\ell/2+iv}$ for $0 \le i < u$; wild kangaroo W_i will start from $h \circ g^{iu}$, $0 \le i < v$. A collision will be $\ell/2 + iv = n + ju \mod (uv)$ and the solution is unique. This prevents kangaroos of the same herd from colliding. The final running time is effectively divided by p.

9.2.4   Solving Batch-DLP

Using BSGS

If we know in advance that we have to solve the DLP for k instances, then Step 1 of BSGS is unchanged (but with another u) and Step 2 is performed at most kN/u times. This implies that we can minimize the total cost $u + kN/u$ using $u = \sqrt{kN}$, a gain of $\sqrt{k}$ compared to applying the algorithm k times. Mixing this with other tricks already mentioned is easy.

Parallel methods

The work of [45] was analyzed in [80]: A batch of k discrete logarithms in a group of order N reduces to an average $\Theta(k^{-1/2} N^{1/2})$ group operations for $k \ll N^{1/4}$; each DL costs $\Theta(N^{3/8})$. [80] also defines some problems related to DLP. The method was further studied in [60].

A more systematic way to consider the problem is the following; see [23] and its follow-up [22], where interval-batch-DLP is also considered and solved with the same ideas. We consider again a table of random steps not involving any target in its definition. The idea is to build a table T of distinguished elements found by random walks starting at random elements g^x. If T is the table size and W the length of the walk, then TW elements will be encountered. When given a target h, a random walk of length W will encounter one of the elements in T, solving DLP for h. The probability that none of the points in the new walk encounters any of the first is

$$\left(1 - \frac{1}{N}\right)^{TW^2}.$$

Taking logarithms, this is close to TW^2/N, so that a reasonable chance of success is for $W \approx \alpha\sqrt{N/T}$ for some constant α. Using this, the probability can be written $\exp(-\alpha^2)$, favoring rather large αs, therefore enabling (and favoring) parallel work too.

Extending this algorithm to the finding of k targets leads to a total cost of

$$O(TW) + kO(W) = O((T + k)W) = O(\sqrt{TN} + k\sqrt{N/T}),$$

and this is dominated by kW for $k > T$. If we want to optimize the cost as a function of k, we see that $T = k$ is minimal. For $T = N^{1/3}$, we get $W = N^{1/3}$ for each walk and $TW = N^{2/3}$ for the precomputation phase.

For a real implementation, we can choose $t = \lceil \log_2 N/3 \rceil$. If G contains integers, define x to be distinguished if $x \equiv 0 \mod 2^t$. With this choice, we need to store 2^t elements; 2^{2t} operations are needed for the precomputation phase, and 2^t for each of the 2^t target.

9.3 Finite Fields and Such

9.3.1 Introduction

There exist dedicated algorithms to compute DL in finite fields that exploit the structure of the finite field. These algorithms are sub-exponential in the size of the field (but not in the involved subgroup order). Moreover, in 2014 two variants of a quasi-polynomial-time algorithm were proposed, for a class of finite fields such as $\mathbb{F}_{2^n}$ and $\mathbb{F}_{3^n}$, where n is composite. As a consequence, any pairing-friendly curve defined over a finite field of small characteristic should definitely be avoided. We explain in the following the main idea of these algorithms known as *index calculus methods*. Three main variants apply to three different types of finite fields (small, medium, and large characteristic, as explained below). We give the specializations of each variant and the range of finite fields to which they apply. Moreover when the prime defining the finite field is of special form (e.g., given by a polynomial), special variants provide an even lower complexity to compute DL.

Interest for pairing-based cryptography

Studying the complexity of DL computations in finite fields is essential for pairing-based cryptosystem designers. The pairing target group for any algebraic curve defined over a finite field is a subgroup in an extension of that finite field. Here are the most common finite fields that arise with pairing-friendly curves, in increasing order of hardness of DL computation:

1. $\mathbb{F}_{2^{4n}}$ and $\mathbb{F}_{3^{6m}}$ for supersingular curves defined over $\mathbb{F}_{2^n}$, $\mathbb{F}_{3^m}$, resp.

2. Small extension $\mathbb{F}_{p^n}$ of prime field with p of special form, given by a polynomial. This is the case for curves in families such as MNT [90], BLS [20], BN [21] curves, and any family obtained with the Brezing-Weng method [32].

3. Small extension of prime field $\mathbb{F}_p$ with p, without any special form, e.g., $\mathbb{F}_{p^2}, \mathbb{F}_{p^6}$ for supersingular curves in large characteristic, and any curves generated with the Cocks-Pinch or Dupont-Enge-Morain methods.

The first class: Supersingular curves defined over a small characteristic finite field also correspond to a variant of the index calculus method where computing a DL is much easier. The two other classes are each parted in medium and large characteristic. This is explained in the next section.

Small, medium, and large characteristic

The finite fields are commonly divided into three cases, depending on the size of the prime p (the finite field characteristic) compared to the extension degree n, with $Q = p^n$. Each case has its own index calculus variant, and the most appropriate variant that applies qualifies the characteristic (as small, large, or medium):

- Small characteristic: One uses the function field sieve algorithm, and the quasi-polynomial-time algorithm when the extension degree is suitable for that (i.e., smooth enough);

- Medium characteristic: one uses the NFS-HD algorithm. This is the High Degree variant of the Number Field Sieve (NFS) algorithm. The elements involved in the relation collection are of higher degree compared to the regular NFS algorithm.

- Large characteristic: one uses the Number Field Sieve algorithm.

Each variant (QPA, FFS, NFS-HD, and NFS) has a different asymptotic complexity. The asymptotic complexities are stated with the L-notation. This comes from the smoothness probability of integers. The explanation will be provided in Section 9.3.2. The L-notation is defined as follows.

DEFINITION 9.1   Let Q be a positive integer. The L-notation is defined by

$$L_Q[\alpha, c] = \exp\left(\big(c + o(1)\big)(\log Q)^\alpha (\log \log Q)^{1-\alpha}\right) \quad \text{with } \alpha \in [0, 1] \text{ and } c > 0 \ .$$

The α parameter measures the gap between polynomial time: $L_Q[\alpha = 0, c] = (\log Q)^c$, and exponential time: $L_Q[\alpha = 1, c] = Q^c$. When c is implicit, or obvious from the context, one simply writes $L_Q[\alpha]$. When the complexity relates to an algorithm for a prime field $\mathbb{F}_p$, one writes $L_p[\alpha, c]$.

Main historical steps

Here is a brief history of DL computations in finite fields. The DLP for cryptographic use was first stated in prime fields, without pairing context [43]. Binary fields were used for efficiency reasons (they provide a better arithmetic). Finite fields are not generic groups and there exist subexponential time algorithms to compute discrete logarithms in finite fields. A long time ago, Kraitchik [78, pp. 119–123], [79, pp. 69–70, 216–267] introduced the *index calculus* method (from the French *calcul d'indice*) to compute discrete logarithms. His work was rediscovered in the cryptographic community in the 1970s. The first algorithms to compute discrete logarithms in prime fields are attributed to Adleman and Western–Miller [6, 119]. These algorithms had a complexity of $L_p[1/2, c]$. However Coppersmith showed as early as 1984 [40] that the DLP is much easier in binary fields than in prime fields of the same size. He obtained an $L_Q[1/3, c]$ running-time complexity for his algorithm. He also showed how fast his algorithm can compute discrete logarithms in $\mathbb{F}_{2^{127}}$ (his algorithm was better for fields whose extension degree is very close to a power of 2, in his record $127 = 2^7 - 1$). Later, Adleman and Huang generalized the Coppersmith's method to other small characteristic finite fields and named it the Function Field Sieve (FFS) algorithm [7, 8]. The asymptotic complexity was $L_Q[1/3, (\frac{32}{9})^{1/3} \approx 1.526]$. In 1986, the state-of-the-art for computing DL in prime fields was the Coppersmith, Odlyzko, and Schroeppel (COS) algorithm [42], in time $L_p[1/2, 1]$. Then in 1993, Gordon designed the Number Field Sieve algorithm for prime fields [55] and reached the same class of sub-exponential asymptotic complexity as the FFS algorithm: $L[1/3, c]$ but with a larger constant $c = 9^{1/3} \approx 2.080$. Between the two extremes ($\mathbb{F}_{2^n}$ and $\mathbb{F}_p$) is the medium characteristic case. In 2006, a huge work was done by Joux, Lercier, Smart, and Vercauteren [67] to propose algorithms to compute the DLP in $L_Q[1/3, c]$ for any finite field. Until 2013, the complexity formulas were frozen at this point:

- a complexity of $L_Q[1/3, (\frac{32}{9})^{1/3} \approx 1.526]$ for small characteristic finite fields with the Function Field Sieve algorithm [8];
- a complexity of $L_Q[1/3, (\frac{128}{9})^{1/3} \approx 2.423]$ for medium characteristic finite fields with the Number Field Sieve–High Degree algorithm [67];
- a complexity of $L_Q[1/3, (\frac{64}{9})^{1/3} \approx 1.923]$ for large characteristic finite fields with the Number Field Sieve algorithm [106, 67, 85].

It was not known until recently whether small characteristic fields could be a gap weaker than prime fields. However, computing power was regularly increasing and in 2012, Hayashi, Shimoyama, Shinohara, and Takagi were able to compute a discrete logarithm record in $\mathbb{F}_{3^{6 \cdot 97}}$, corresponding to a 923-bit field [58], with the Function Field Sieve. Then in December 2012 and January 2013, Joux released two preprints later published in [63] with a $L_Q[1/4]$ algorithm, together with record-breaking discrete logarithms in $\mathbb{F}_{2^{4080}}$ and $\mathbb{F}_{2^{6168}}$. This was improved by various researchers. In 2014, Barbulescu, Gaudry, Joux, and Thomé [17] on one side and Granger,

Kleinjung, and Zumbrägel [56] on the other side proposed two versions of a *quasi polynomial-time* algorithm (QPA) to solve the DLP in small characteristic finite fields. All the techniques that allowed this breakdown are not applicable to medium and large characteristic so far.

9.3.2 Index-Calculus Methods

We saw in Section 9.2 that the best complexity is obtained by balancing parameters in the algorithms. Over finite fields, FFS and NFS algorithms also reach sub-exponential complexity through balancing parameters.

Throughout this section, we follow the clear presentation of index-calculus methods made in [88]. We present in Algorithm 9.10 the basic index-calculus method for computing DL in a prime field.

ALGORITHM 9.10   Index calculus in $(\mathbb{Z}/p\mathbb{Z})^*$.

Function *IndexCalculus($\mathbb{F}_p^*$, g, h)*

   Input   : $\mathbb{F}_p^* \supset \langle g \rangle$, g of order dividing $p-1$

   Output: $0 \leq x < p-1$, $g^x = h$

   Phase 1: fix $\#\mathcal{B}$ and find the logarithms modulo $p-1$ of small primes in $\mathcal{B} = \{p_1, p_2, \ldots, p_{\#\mathcal{B}} \leq B\}$:

   Choose integers $t_i \in [1, \ldots, p-1]$ s.t. g^{t_i} as an integer splits completely in small primes of $\mathcal{B}$:

$$g^{t_i} \bmod p = \prod_{b=1}^{\#\mathcal{B}} p_b^{\alpha_{b,i}}$$

so that taking the logarithm to the base g gives a *relation*:

$$t_i = \sum_{b=1}^{\#\mathcal{B}} \alpha_{b,i} \log_g p_b \bmod (p-1)$$

   Phase 2: When enough relations are collected, solve the system to get $\{\log_g p_b\}_{1 \leq b \leq \#\mathcal{B}}$.

   Phase 3: Compute the individual discrete logarithm of h in base g.

   Look for t s.t. $hg^t \bmod p$ as an integer factors into small primes of $\mathcal{B}$:

$$hg^t \bmod p = \prod_{b=1}^{\#\mathcal{B}} p_b^{\alpha_b} \Leftrightarrow x + t \equiv \sum_{b=1}^{\#\mathcal{B}} \alpha_b \log_g p_b \bmod (p-1)$$

   return x.

Example for a tiny prime p

Let $p = 1019$ and $g = 2$. We need to find the logarithms modulo 2 and 509, since $p-1 = 2 \cdot 509$. We first locate some smooth values of $g^b \bmod p$:

$$2^{909} = 2 \cdot 3^2 \cdot 5, 2^{10} = 5, 2^{848} = 3^3 \cdot 5, 2^{960} = 2^2 \cdot 3.$$

The system to be solved is

$$\begin{pmatrix} 1 & 2 & 1 \\ 10 & 0 & 0 \\ 0 & 3 & 1 \\ 2 & 1 & 0 \end{pmatrix} \cdot X = \begin{pmatrix} 909 \\ 10 \\ 848 \\ 960 \end{pmatrix} \bmod 1018.$$

Solving modulo 2 and 509 separately and recombining by the Chinese remainder theorem, we find

$$\log_2 2 = 1, \log_2 3 = 958, \log_2 5 = 10.$$

Note that solving modulo 2 can be replaced by computations of Legendre symbols.

Consider computing $\log_2 314$. We find that

$$h \cdot g^{372} \equiv 2^4 \cdot 5^2 \bmod p$$

from which $\log_g h = 4 + 2 \cdot 10 - 372 \bmod 1018$ or $\log_2(314) = 670$. Had we used rational reconstruction (a classical trick for DL target solution), we would have found

$$h \cdot g^{409} \equiv 2/3 \bmod p$$

from which the log of t follows from

$$\log_g h + 409 \equiv 1 - \log_2 3 \bmod 1018.$$

A first complexity analysis

We will prove that:

THEOREM 9.5   *The asymptotic heuristic running-time of Algorithm 9.10 is $L_p[1/2, 2]$ with $\#\mathcal{B} \approx B = L_p[1/2, 1/2]$.*

The running-time of algorithm 9.10 (pp. 9–17) is related to smoothness probabilities of integers in Phase 1 and 3, and linear algebra in phase 2. We will use the L-notation formula given in 9.1. To estimate the smoothness probability, we need the result of Corollary 9.1 from Theorem 9.6.

THEOREM 9.6 (Canfield–Erdős–Pomerance [34])   *Let $\psi(x, y)$ be the number of natural numbers smaller or equal to x which are y-smooth. If $x \geq 10$ and $y \geq \log x$, then it holds that*

$$\psi(x, y) = x u^{-u(1+o(1))} \quad \text{with } u = \frac{\log x}{\log y}, \tag{9.1}$$

where the limit implicit in the $o(1)$ is for $x \to \infty$.

The Canfield–Erdős–Pomerance [34] theorem provides a useful result to measure smoothness probability:

COROLLARY 9.1 (B-smoothness probability)   *For an integer S bounded by $L_Q[\alpha_S, \sigma]$ and a smoothness bound $B = L_Q[\alpha_B, \beta]$ with $\alpha_B < \alpha_S$, the probability that S is B-smooth is*

$$Pr[S \text{ is } B\text{-smooth}] = L_Q \left[\alpha_S - \alpha_B, -(\alpha_S - \alpha_B)\frac{\sigma}{\beta} \right]. \tag{9.2}$$

We will also need these formulas:

$$L_Q[\alpha, c_1]L_Q[\alpha, c_2] = L_Q[\alpha, c_1 + c_2] \text{ and } L_Q[\alpha, c_1]^{c_2} = L_Q[\alpha, c_1 c_2] \ . \tag{9.3}$$

Proof. (of Theorem 9.5.)

In our context, we want to find a smooth decomposition of the least integer r ($|r| < p$) s.t. $r = g^{b_i} \pmod{p}$. We need to estimate the probability of an integer smaller than p to be B-smooth. We write the formula, then balance parameters to ensure the optimal cost of this algorithm. Let us write the smoothness-bound B in sub-exponential form: $B = L_p[\alpha_B, \beta]$. Any prime in $\mathcal{B}$ is smaller than B. The probability of r bounded by $p = L_p[1,1]$ to be B-smooth is

$$\Pr[r \text{ is } B\text{-smooth}] = L_p\left[1 - \alpha_B, (1 - \alpha_B)\frac{1}{\beta}\right]^{1+o(1)} \ .$$

To complete Phase 2, we need enough relations to get a square matrix and solve the system, in other words, more than $\#\mathcal{B}$. Since the number of prime numbers $\leq B$ is $B/\log B$ for $B \to \infty$, we approximate $\#\mathcal{B} \approx B$. The number of iterations over g^{t_i} and smoothness tests to be made to get enough relations is

$$\text{number of tests} = \frac{\text{number of relations}}{B\text{-smoothness probability}} = \frac{B}{\Pr} = L_p[\alpha_B, \beta]L_p[1 - \alpha_B, (1 - \alpha_B)/\beta] \ .$$

The dominating α-parameter will be $\max(\alpha_B, 1 - \alpha_B)$ and is minimal for $\alpha_B = 1/2$. The number of tests is then $L_p[1/2, \beta + \frac{1}{2\beta}]$ (thanks to Equation (9.3)).

Now we compute the running time to gather the relations: This is the above quantity times the cost of a smoothing step. At the beginning of index calculus methods, a trial division was used, so one B-smooth test costs at most $\#\mathcal{B} = B = L_p[1/2, \beta]$ divisions. The total relation collection running time is then

$$L_p[1/2, 2\beta + 1/(2\beta)] \ .$$

The linear algebra phase finds the kernel of a matrix of dimension B. It has running-time of $B^\omega \approx L_p[1/2, \omega\beta]$ (ω is a constant, equal to 3 for classical Gauss, and nowadays we use iterative methods, of complexity $B^{2+o(1)}$; see Section 9.3.3). The total running time of the first two steps is

$$L_p[1/2, 2\beta + 1/(2\beta)] + L_p[1/2, 3\beta] \ .$$

The minimum of $\beta \mapsto 2\beta + \frac{1}{2\beta}$ is 2, for $\beta = 1/2$ (take the derivative of the function $x \mapsto 2x + \frac{1}{2x}$ to obtain its minimum, for $x > 0$). We conclude that the total cost of the first two phases is dominated by $L_p[1/2, 2]$: the relation collection phase.

The last phase uses the same process as finding one relation in Phase 1. It needs $1/\Pr[r = g^t \pmod{p}$ is B-smooth] tries of cost B each, hence $B/\Pr = L_p[1/2, \beta + \frac{1}{2\beta}] = L_p[1/2, 3/2]$. $\qquad \square$

COS algorithm, Gaussian integer variant

Coppersmith, Odlyzko, and Schroeppel proposed in [42] a better algorithm by modifying the relation collection phase. They proposed to *sieve* over elements of well-chosen form, of size $\sim \sqrt{p}$ instead of p. They also proposed to use sparse matrix linear algebra, to improve the second phase of the algorithm. Computing the kernel of a sparse square matrix of size B has a running-time of $O(B^{2+o(1)})$ with their modified Lanczos algorithm. We present now their Gaussian Integer variant, so that the Gordon Number Field Sieve will be clearer in Section 9.3.4. We consider a generator g of $\mathbb{F}_p^*$ and want to compute the discrete logarithm x of h in base g, in the subgroup of $\mathbb{F}_p^*$ of prime order ℓ, with $\ell \mid p - 1$.

The idea of [42] is to change the relation collection: In the former case, iterating over g^{t_i} and taking the smallest integer $r \equiv g^{t_i} \pmod{p}$ always produces r of the same size as p. In this version, another iteration is made. The idea is to produce elements r much smaller than p, to improve their smoothness probability. In the previous index calculus, we made relations between integers (we lifted $g^{t_i} \bmod p$ to $r \in \mathbb{Z}$). Here one side will consider integers, the second side will treat algebraic integers. Let A be a small negative integer which is a quadratic residue modulo p. Preferably, $A \in \{-1, -2, -3, -7, -11, -19, -43, -67, -163\}$ so that $\mathbb{Q}[\sqrt{A}]$ is a unique factorization domain. For ease of presentation, we assume that $p \equiv 1 \bmod 4$ and take $A = -1$. Our algebraic side will be the Gaussian integer ring $\mathbb{Z}[i]$. Now let $0 < U, V < \sqrt{p}$ such that $p = U^2 + V^2$ (computed via the rational reconstruction method, for example). The element U/V is a root of $x^2 + 1$ modulo p. For an analogy with the number field sieve, one can define $f = x^2 + 1$ for the first (algebraic) side and $g = U - xV$ for the second (rational) side. The two polynomials have a common root U/V modulo p. We define a map from $\mathbb{Z}[i]$ to $\mathbb{F}_p$:

$$\rho: \quad \begin{array}{ccc} \mathbb{Z}[i] & \to & \mathbb{F}_p \\ i & \mapsto & UV^{-1} \bmod p \end{array} \tag{9.4}$$
$$\text{hence} \qquad a - bi \; \mapsto \; V^{-1}(aV - bU) \bmod p \; .$$

Now we sieve over pairs (a, b) on the rational side, looking for a B-smooth decomposition of the integer $aV - bU$, as in Algorithm 9.10. What will be the second member of a relation? Here comes the algebraic side. We consider the elements $a - bi \in \mathbb{Z}[i]$ (with the same pairs (a, b)) and iterate over them such that $a - bi$, as an ideal of $\mathbb{Z}[i]$, factors into prime ideals $\mathfrak{p}$ of $\mathbb{Z}[i]$ of norm $\mathcal{N}_{\mathbb{Z}[i]/\mathbb{Z}}(\mathfrak{p})$ smaller than B. (For example: $1 + 3i = (1 + i)(2 + i)$ with $\mathcal{N}(1 + 3i) = 1^2 + 3^2 = 10 = 2 \cdot 5 = \mathcal{N}(1 + i)\mathcal{N}(2 + i)$). Here is the magic: Since $\mathcal{N}_{\mathbb{Z}[i]/\mathbb{Z}}(a - bi) = a^2 + b^2$ and we sieve over small $0 < a < E$, $-E < b < E$, the norm of $a - bi$ will be bounded by E^2 and the product of the norms of the prime ideals $\mathfrak{p}$ in the factorization, which is equal to $a^2 + b^2$, will be bounded by E^2 as well. At this point, we end up with pairs (a, b) such that

$$aV - bU = \prod_{p_b \leq B} p_b^{s_j}, \quad \text{and} \quad a - bi = \prod_{\mathcal{N}(\mathfrak{p}_{b'}) \leq B} \mathfrak{p}_{b'}^{t_j} \; .$$

Then we use the map ρ to show up an equality, then get a relation. We have $\rho(a - bi) = a - bUV^{-1} = V^{-1}(aV - bU)$ so up to a factor V (which is constant along the pairs (a, b)), we have:

$$aV - bU = \prod_{p_b \leq B} p_b^{s_j} = V \prod_{\mathcal{N}(\mathfrak{p}_{b'}) \leq B} \rho(\mathfrak{p}_{b'})^{t_j} \; . \tag{9.5}$$

We don't need to know explicitly the value of $\rho(\mathfrak{p}_{b'})$ in $\mathbb{F}_p$. We simply consider it as an element of the basis $\mathcal{B}$ of small elements: $\mathcal{B} = \{V\} \cup \{p_b \leq B\} \cup \{\rho(\mathfrak{p}_{b'}) : \mathcal{N}_{\mathbb{Z}[i]/\mathbb{Z}}(\mathfrak{p}_{b'}) \leq B\}$. In the COS algorithm, the matrix is indeed two times larger than in the basic version of Algorithm 9.10 ($2B$ instead of B), but with norms $a^2 + b^2$ and $Va - bU$ much smaller; we can decrease the smoothness bound B. Taking the logarithm of Equation (9.5), we obtain an equation between logarithms of elements in $\mathcal{B}$:

$$\sum_{p_b \leq B} s_j \log p_b = \log V + \sum_{\mathcal{N}_{\mathbb{Z}[i]/\mathbb{Z}}(\mathfrak{p}_{b'}) \leq B} t_j \log \rho(\mathfrak{p}_{b'}). \tag{9.6}$$

The optimal choice of parameters is $E = B = L_p[1/2, 1/2]$, so that both sieving and linear algebra cost $L_p[1/2, 1]$, better than the previous $L_p[1/2, 2]$ thanks to the much better smoothness probabilities of the elements considered. Phase 2 of the algorithm computes the kernel of a large sparse matrix of more than $2B$ rows. Its expected running time is $(2B)^2$, hence again $L_p[1/2, 1]$. The expected running time of the individual logarithm computation (Phase 3) is $L_p[1/2, 1/2]$ ([42, §7]).

9.3.3 Linear Algebra

Before diving into the explanation of the most powerful algorithms for computing DLs, we make a pause and give some ideas on a technical but crucial problem: linear algebra and how we solve the systems arising in DL computations.

At the beginning of index calculus methods, the only available method was the ordinary Gauss algorithm, whose complexity is cubic in the number of rows (or columns) N of the matrix M. For small matrices, this is enough, but remember that entries in the matrix are elements of some finite field $\mathbb{F}_\ell$ for some large ℓ, in contrast with matrices we encounter in integer factorization, which are boolean. Fill-in in DL matrices is therefore extremely costly and we should be careful in doing this.

DL-matrices are very sparse, as factorization matrices. The coefficients are small integers (in absolute value). Special methods have been designed for all these matrices. Generally, some form of sparse Gaussian elimination is done in order to reduce the size of the matrix prior to the use of more sophisticated methods to be described below. The idea is to perform elimination using a sparse structure and minimize fill-in as long as possible. The general term for these is *filtering*, and it was optimized and made necessary due to the use of many large primes in recent years (see [30] for recent progress in the field).

Adaptation of numerical methods was done: The Lanczos iterative algorithm could be generalized to the finite field case. A new class of iterative methods was invented by Wiedemann [120]. Both classes have a complexity of $O(N^{2+\varepsilon})$ where we assume that our matrix has size $N^{1+\varepsilon}$. The core of the computation is the determination of the so-called Krylov subspaces, namely $Vect\langle M^i \cdot b \rangle$ for some fixed vector b. Applying M to b costs $O(N^{1+\varepsilon})$, and N iterations are needed. The advantage of such methods is also to be able to handle sparse structures for M.

Variants operating on blocks of vectors were designed by Coppersmith [39], for integer factorization as for DLP. Despite the use of such methods, space and time become quite a problem in recent records. The natural idea is to try to distribute the computations over clusters or larger networks. The only method that can be partly distributed is the block Wiedemann algorithm. A good reference for this part is Thomé's thesis [117]. All his work is incorporated and available in the CADO-NFS package [112], including the many refinements to distribute the computations over the world and some special tricks related to Schirokauer maps (see [70] for an independent work on the same topic). Record DL computations now handle (sparse) matrices with several millions of rows and columns.

9.3.4 The Number Field Sieve (NFS)

Many improvements for computing discrete logarithms first concerned prime fields and were adapted from improvements on integer factorization methods. Until 1993, the state-of-the-art algorithm for computing discrete logarithms in prime fields was the Coppersmith, Odlyzko, and Schroeppel (COS) algorithm [42] in $L_p[1/2, 1]$. In some cases, integer factorization was easier because the modulus had a special form. The equivalent for DL computation in prime fields is when p has a special form, given by a polynomial P of very small coefficients, $p = 2^{127} - 1$ for example. In that case, one can define an algebraic side with this polynomial P. By construction, P will have a root m modulo p of size $\sim p^{1/\deg P}$, hence the polynomial of the rational side ($g = U - Vx$ in the COS algorithm) will have coefficients of size $m \sim p^{1/\deg P}$. However, a generic method was found to reduce the coefficient size of the polynomials when one of the degrees increases. In 1993, Gordon [55] proposed the first version of the NFS–DL algorithm for prime fields $\mathbb{F}_p$ with asymptotic complexity $L_p[1/3, 9^{1/3}]$. Gordon's $L_p[1/3]$ algorithm is interesting for very large values of p that were not yet targets for discrete logarithm computations in the 1990s. Buhler, H. Lenstra, and Pomerance [33] estimated the crossover point at between 100 and 150 decimal digits, i.e., between 330 and 500 bits.

In Gordon's algorithm, Phase 1 and Phase 3 are modified. The Phase 2 is still a large sparse matrix kernel computation. We explain the polynomial selection method and the sieving phase. We also explain why the Phase 3 (individual logarithm computation) needs important modifications. The hurried reader can skip the proof of Theorem 9.7.

Polynomial selection with the base-m method

The polynomial selection of [55] is an analogy for prime fields of the method [33] for integer factorization. We will build a polynomial f of degree $d > 1$ and a polynomial g of degree 1 such that they have a common root m modulo p, and have coefficients of size $\sim p^{1/d}$. Set $m = [p^{1/d}]$ and write p to the base-m:

$$p = c_d m^d + c_{d-1} m^{d-1} + \ldots + c_0,$$

where $0 \leq c_i < m$. Then set

$$f = c_d x^d + c_{d-1} x^{d-1} + \ldots + c_0 \quad \text{and} \quad g = x - m .$$

Under the condition $p > 2^{d^2}$, f will be monic [33, Prop. 3.2]. These two polynomials have a common root m modulo p, hence a similar map than in Equation (9.4) is available:

$$
\begin{array}{rccc}
\rho: & \mathbb{Z}[x]/(f(x)) = \mathbb{Z}[\alpha_f] & \to & \mathbb{F}_p \\
& \alpha_f & \mapsto & m \bmod p \\
\text{hence} & a - b\alpha_f & \mapsto & a - bm \bmod p .
\end{array}
\tag{9.7}
$$

Relation collection

The new technicalities concern factorization of ideals $a - b\alpha_f$ into prime ideals of $\mathbb{Z}[\alpha_f]$. This is not as simple as for $\mathbb{Z}[i]$: $\mathbb{Z}[\alpha_f]$ may not be a unique factorization domain, moreover what is called *bad ideals* can appear in the factorization. To end up with good relations, one stores only pairs (a, b) such that $a - b\alpha_f$ factors into *good* prime ideals of degree one, and of norm bounded by B. The sieve on the rational side is as in the COS algorithm.

Individual discrete logarithm computation

The other new issue is the individual logarithm computation. Since the sieving space and the factor basis $\mathcal{B}$ are much smaller ($L_p[1/3, \beta]$ instead of $L_p[1/2, \beta]$), there are much fewer known logarithms. It is hopeless to compute $g^t h$ with t at random until a smooth factorization is found, because now the smoothness bound is too small. A strategy in two phases was proposed by Joux and Lercier: Fix a larger smoothness bound $B_1 = L_p[2/3, \beta_1]$. First find a B_1-smoothness decomposition of $g^t h$. Secondly, treat separately each prime factor less than B_1 but larger than B to obtain a B-smooth decomposition. Finally retrieve the individual logarithm of h in base g. Each step has a cost $L_Q[1/3, c']$ with c' smaller than the constant $c = 1.923$ of the two dominating steps (relation collection and linear algebra).

Asymptotic complexity

We present how to obtain the expected heuristic running-time of $L_p[1/3, (\frac{64}{9})^{1/3}]$ to compute DL in $\mathbb{F}_p$ with the base-m method and a few improvements to the original Gordon algorithm. The impatient reader can admit the result of Theorem 9.7 and skip this section.

THEOREM 9.7   *The running-time of the NFS-DL algorithm with base-m method is*

$$L_p\left[1/3, \left(\frac{64}{9}\right)^{1/3} \approx 1.923\right],$$

TABLE 9.3 Optimal value $\delta(\frac{\log p}{\log\log p})^{1/3}$ with $\delta = 1.44$ for p of 100 to 300 decimal digits. One takes $d = \lceil x \rceil$ or $d = \lfloor x \rfloor$ in practice.

$\log_{10} p$	40	60	80	100	120	140	160	180	200	220	240	260	280	300
$\lceil\log_2 B\rceil$ (bits)	18b	21b	24b	27b	29b	31b	33b	35b	36b	38b	39b	41b	42b	43b
$\delta\left(\frac{\log p}{\log\log p}\right)^{1/3}$	3.93	4.37	4.72	5.02	5.27	5.50	5.71	5.90	6.08	6.24	6.39	6.54	6.68	6.81

obtained for a smoothness bound $B = L_p[1/3, \beta]$ with $\beta = (8/9)^{1/3} \approx 0.96$, a sieving bound $E = B$ (s.t. $|a|, |b| < E$), and a degree of f to be $d = \lceil\delta(\frac{\log p}{\log\log p})^{1/3}\rceil$ with $\delta = 3^{1/3} = 1.44$.

We present in Table 9.3 the optimal values of B (in bits) and d with $\beta \approx 0.96$ and $\delta \approx 1.44$ for p from 100 to 300 decimal digits (dd).

Proof. (of Theorem 9.7.) One of the key-ingredients is to set an optimal degree d for f. So let

$$d = \delta\left(\frac{\log p}{\log\log p}\right)^{1/3} \quad \text{so that } m = p^{1/d} = L_p\left[2/3, \frac{1}{\delta}\right] . \tag{9.8}$$

(Compute $\log m = \frac{1}{d}\log p = \frac{1}{\delta}\left(\frac{\log\log p}{\log p}\right)^{1/3}\log p = \frac{1}{\delta}\log^{2/3}p\log^{1/3}\log p$). We will compute the optimal value of δ under the given constraints later. The aim is to get a bound on the norms of the elements $a - b\alpha_f$ and $a - bm$ of size $L_p[2/3, \cdot]$ and a smoothness bound $L_p[1/3, \cdot]$, so that the smoothness probability will be $L_p[1/3, \cdot]$. A smoothness test is done with the Elliptic Curve Method (ECM). The cost of the test depends on the smoothness bound B and the total size of the integer tested. We first show that the cost of an ECM B-smoothness test with $B = L_p[1/3, \beta]$, of an integer of size $L_p[2/3, \eta]$ is $L_p[1/6, \sqrt{2\beta/3}]$, hence is negligible compared to any $L_p[1/3, \cdot]$. The cost of this ECM test depends on the size of the smoothness bound:

$$\text{cost of an ECM test} = L_B[1/2, \sqrt{2}] .$$

Writing $\log B = \beta\log^{1/3}p\log^{2/3}\log p$, we compute $\log L_B[1/2, \sqrt{2}] = \sqrt{2}(\log B\log\log B)^{1/2}$, cancel the negligible terms, and get the result.

We denote the infinity norm of a polynomial to be the largest coefficient in absolute value:

$$\|f\|_\infty = \max_{0 \le i \le \deg f} |f_i| . \tag{9.9}$$

We have

$$\|f\|_\infty, \|g\|_\infty \le m = L_p\left[\frac{2}{3}, \frac{1}{\delta}\right] .$$

In Phase 1, we sieve over pairs (a, b) satisfying $0 < a < E, -E < b < E$ and $\gcd(a, b) = 1$, so the sieving space is of order E^2. We know that we can sieve over no more than $L_p[1/3, \cdot]$ pairs to be able to balance the three phases of the algorithm. So let $E = L_p[1/3, \epsilon]$ pairs, with ϵ to be optimized later. The sieving space is $E^2 = L_p[1/3, 2\epsilon]$. Since the cost of a B-smoothness test with ECM is negligible compared to $L_p[1/3, \epsilon]$, we conclude that the running time of the sieving phase is $E^2 = L_p[1/3, 2\epsilon]$.

We need at least B relations to get a square matrix, and the linear algebra cost will be $B^2 = L_p[1/3, 2\beta]$. To balance the cost of the sieving phase and the linear algebra phase, we set

$$E^2 = B^2, \text{ hence } \epsilon = \beta$$

and we replace ϵ by β in the following computations.

What is the norm bound for $a - b\alpha_f$? We need it to estimate its probability to be B-smooth. The norm is computed as the *resultant* (denoted Res) of the element $a - b\alpha_f$ as a polynomial $a - bx$ in x, and the polynomial f. Then we bound the norm. The norm is

$$
\begin{aligned}
\mathcal{N}(a - b\alpha_f) &= \operatorname{Res}(f(x), a - bx) \\
&= a^d + f_{d-1}a^{d-1}b + \ldots + ab^{d-1}f_1 + b^d f_0 = \sum_{i=0}^{d} a^i b^{d-i} f_i \\
&\leq (\deg f + 1)\|f\|_\infty E^{\deg f} = (d+1)p^{1/d}E^d
\end{aligned}
$$

with $d+1$ negligible, $p^{1/d} = m = L_p[2/3, 1/\delta]$, and $E^d = B^d = L_p[2/3, \beta\delta]$ (first compute $d \log B$ to get the result).

$$
\mathcal{N}(a - b\alpha_f) \leq L_p[2/3, 1/\delta + \delta\beta] \ .
$$

Then for the g-side we compute

$$
\begin{aligned}
a - bm &\leq Ep^{1/d} = L_p[1/3, \beta]L_p[2/3, \tfrac{1}{\delta}] \\
&\leq L_p[2/3, \tfrac{1}{\delta}]
\end{aligned}
$$

since the $L_p[1/3, \beta]$ term is negligible.

We make the usual heuristic assumption that the norm of $a - b\alpha_f$ follows the same smoothness probability as a random integer of the same size. Moreover, we assume that the probability of the norm of $a - b\alpha_f$ and $a - bm$ to be B-smooth at the same time is the same as the probability of their product (bounded by $L_p[2/3, 2/\delta + \delta\beta]$) to be B-smooth.

Finally we apply Corollary 9.1 and get

$$
\Pr[\mathcal{N}(a - b\alpha_f) \text{ and } a - bm \text{ are } B\text{-smooth}] = 1/L_p\left[\frac{1}{3}, \frac{1}{3}\left(\frac{2}{\delta\beta} + \delta\right)\right] \ .
$$

How many relations do we have ? We multiply this smoothness probability Pr by the sieving space E^2 and obtain

$$
\text{number of pairs } (a, b) \text{ tested} = \frac{\text{number of relations}}{B\text{-smoothness probability}} = \frac{B}{\Pr} = L_p\left[1/3, \beta + \frac{1}{3}\left(\frac{2}{\delta\beta} + \delta\right)\right] \ .
$$

Since B^2 pairs were tested, we obtain the equation

$$
2\beta = \beta + \frac{1}{3}\left(\frac{2}{\delta\beta} + \delta\right) \Leftrightarrow \beta = \frac{1}{3}\left(\frac{2}{\delta\beta} + \delta\right) \ . \tag{9.10}
$$

We want to minimize the linear algebra and sieving phases, hence we minimize $\beta > 0$ through finding the minimum of the function $x \mapsto \frac{1}{3}(\frac{2}{\beta x} + x)$ (by computing its derivative): This is $2/3\sqrt{2/\beta}$, obtained with $\delta = x = \sqrt{2/\beta}$. We end up by solving Equation (9.10): $\beta = 2/3\sqrt{2/\beta} \Leftrightarrow \beta = (8/9)^{1/3}$. Since the running time of Phase 1 and Phase 2 is $L_p[1/3, 2\beta]$, we obtain $2\beta = (64/9)^{1/3}$ as expected. The optimal degree of the polynomial f is $d = \delta(\frac{\log p}{\log\log p})^{1/3}$ with $\delta = 3^{1/3} \approx 1.44$.

$\square$

9.3.5   Number Field Sieve: Refinements

Norm approximation

We can approximate the norm of an element b in a number field $K_f = \mathbb{Q}[x]/(f(x))$ by the two following bounds.

The Kalkbrener bound [71, Corollary 2] is the following:

$$
|\operatorname{Res}(f, b)| \leq \kappa(\deg f, \deg b) \cdot \|f\|_\infty^{\deg b}\|b\|_\infty^{\deg f}, \tag{9.11}
$$

where $\kappa(n, m) = \binom{n+m}{n}\binom{n+m-1}{n}$, and $\|f\|_\infty = \max_{0 \leq i \leq \deg f} |f_i|$ is the absolute value of the greatest coefficient. An upper bound for $\kappa(n, m)$ is $(n + m)!$.

Bistritz and Lifshitz proved the other following bound [25, Theorem 7]:

$$|\operatorname{Res}(f, \phi)| \leq \|f\|_2^n \|\phi\|_2^m \leq (m + 1)^{n/2}(n + 1)^{m/2}\|f\|_\infty^n \|\phi\|_\infty^m . \quad (9.12)$$

When the degree of the involved polynomials is negligible, we can approximate $|\operatorname{Res}(f, \phi)|$ by $O(\|f\|_\infty^n \|\phi\|_\infty^m)$. This simpler bound will be used to bound the norm of elements $\phi = a - bx$ and $\phi = \sum_{i=0}^{t-1} a_i x^i$ in a number field defined by a polynomial f.

Rational reconstruction and LLL

Throughout the polynomial selections, two algorithms are extensively used: the Rational Reconstruction algorithm and the Lenstra/Lenstra/Lovasz (LLL) algorithm. Given an integer y and a prime p, the Rational Reconstruction algorithm computes a quotient u/v such that $u/v \equiv y$ mod p and $|u|, |v| < p$.

The Lenstra–Lenstra–Lovász algorithm (LLL) [83] computes a short vector in a lattice. Given a lattice $\mathcal{L}$ of $\mathbb{Z}^n$ defined by a basis given in an $n \times n$ matrix L, and parameters $\frac{1}{4} < \delta < 1$, $\frac{1}{2} < \eta < \sqrt{\delta}$, the LLL algorithm outputs a (η, δ)-*reduced basis* of the lattice. The coefficients of the first (shortest) vector are bounded by

$$(\delta - \eta^2)^{\frac{n-1}{4}} \det(L)^{1/n} .$$

With (η, δ) close to $(0.5, 0.999)$ (as in NTL or Magma), the approximation factor $C = (\delta - \eta^2)^{\frac{n-1}{4}}$ is bounded by 1.075^{n-1} (see [35, §2.4.2]). A very fast software implementation of the LLL algorithm is available with the `fplll` library [10].

Improvements of the polynomials

Joux and Lercier proposed in [65] another polynomial selection method that is an improvement of the base-m method. In this case, the polynomials are no longer monic but have smaller coefficients, bounded by $p^{\frac{1}{d+1}}$ instead of $p^{\frac{1}{d}}$. The size of the coefficients is spread over one more coefficient (the leading coefficient). Comeine and Semaev analyzed the complexity of their algorithm in [38]. The asymptotic complexity does not change because the gain is hidden in the $o(1)$ term. Their improvement is very important in practice, however.

For a given pair of polynomials (f, g) obtained with the Joux-Lercier method, one can improve their quality. We want the polynomials to have as many roots as possible modulo small primes, in order to get many relations in the relation collection step. The quality of the polynomials was studied my B. A. Murphy [93, 92]. The α value and the Murphy's $\mathbb{E}$ value measure the root properties of a pair of polynomials. The aim is to improve the root properties while keeping the coefficients of reasonable size. A recent work on this subject can be found in Shi Bai's PhD thesis [11] and in [12]. The sieving can be speeded up in practice by choosing coefficients of the polynomial f whose size increases while the monomial degree decreases, and redefining the sieving space as $-E\sqrt{s} < a < E\sqrt{s}$, $0 < b < E/\sqrt{s}$ [74, 75].

9.3.6 Large Characteristic Non-Prime Fields

In 2006, Joux, Lercier, Smart, and Vercauteren [67] provided a polynomial construction that permitted us to conclude that for any finite field $\mathbb{F}_q$, there exists a $L_q[1/3, c]$ algorithm to compute DL: They proposed a method for medium-characteristic finite fields (known as the JLSV$_1$ method) and for large-characteristic finite fields. Another method (gJL: generalization of Joux-Lercier for prime fields) was independently proposed by Matyukhin [85] and Barbulescu [13, §8.3] and achieved the same asymptotic complexity as for prime fields: $L_Q[1/3, (\frac{64}{9})^{1/3}]$, with $Q = p^n$. The two polynomials are of degree $d + 1$ and $d \geq n$, for a parameter d that depends on $\log Q$ as for the prime case. Note that the optimal choice of d for the gJL method is $d = \delta \left(\frac{\log p}{\log \log p} \right)^{1/3}$ with $\delta = 3^{1/3}/2 \approx 0.72$ instead of $\delta = 3^{1/3}$ for the NFS-DL algorithm in prime fields. This is not surprising: In this case the sum of polynomial degrees $\deg f + \deg g$ is $2d + 1$ instead of $d + 1$.

Recent results (2015) on DL record computation in non-prime finite fields with the NFS algorithm

In 2015, Barbulescu, Gaudry, Guillevic, and Morain published a DL record computation in a 595-bit quadratic extension $\mathbb{F}_{p^2}$, where p is a generic 298-bit prime [16]. They designed a new polynomial selection method called the Conjugation. The polynomials allowed a factor-two speed-up in the relation collection step, thanks to an order-two Galois automorphism.

The area is moving and later in 2015, Sarkar and Singh in [378] proposed a variant that combines the gJL and the Conjugation method. The asymptotic complexity is the same: $L_Q[1/3, (\frac{64}{9})^{1/3}]$ and the polynomials might provide slightly smaller norm values in practice.

We lack one last piece of information to be able to recommend parameter sizes for a given level of security in large characteristic fields $\mathbb{F}_Q$: record computations with one of these methods.

9.3.7   Medium Characteristic Fields

This range of finite fields is where the discrete logarithm is more difficult to compute at the moment. Moreover, the records published are for quite small sizes of finite fields only: $\mathbb{F}_{p^3}$ of 120dd (400 bits) in [67], and in [121].

There was a big issue in obtaining an $L[1/3]$ algorithm as for the FFS algorithm in small characteristic and the NFS algorithm in large characteristic. The solution proposed in [67] introduces a modification in the relation collection. The small elements are of higher degree $t - 1 \geq 2$, where t is of the form $\frac{n}{t} = \frac{1}{c_t} \left(\frac{\log Q}{\log \log Q} \right)^{1/3}$. To obtain similar asymptotic formulas in $L_Q[1/3, \cdot]$ as for prime fields (see Section 9.3.4), one sets the total number of elements considered in the relation collection to be E^2. With these settings,

- the total number of elements $\phi = \sum_{i=0}^{t-1} a_i x^i$ considered in the relation collection is $\|\phi\|_\infty^t = E^2$, so that $\|\phi\|_\infty = E^{2/t}$;

- the norm of the elements ϕ on the f side is

$$|\mathcal{N}_f(\phi)| \leq \mathrm{Res}(\phi, f) \leq \kappa(t-1, \deg f) \|\phi\|_\infty^{\deg f} \|f\|_\infty^{t-1} = \kappa(t-1, \deg f) E^{2 \deg f/t} \|f\|_\infty^{t-1}.$$

The two polynomials defined for the 120dd record computation in [67] were $f = x^3 + x^2 - 2x - 1$ and $g = f + p$. This method is not designed for scaling well and the authors propose a variant where the two polynomials have a balanced coefficient size of $p^{1/2}$ each. This polynomial selection method, combined with the relation collection over elements of degree $t - 1$, provides an asymptotic complexity of $L_Q[1/3, (\frac{128}{9})^{1/3}] \simeq 2.42$. This asymptotic complexity went down in 2015 in [16] to $L_Q[1/3, (\frac{96}{9})^{1/3}] \simeq 2.201$. These two asymptotic complexities were improved in [19, 96] to $L_Q[1/3, 2.39]$ and $L_Q[1/3, 2.156]$, respectively, by using a *multiple number field sieve* variant that we explain in the next paragraph. This variant has not been implemented for any finite field yet.

Multiple number field sieve

This paragraph is about refinements in the algorithm that slightly improve the asymptotic complexity but whose practical gain is not known, and not certain yet.

The idea is to use additional number fields and hope to generate more relations per polynomial (the $a - bx$ elements in Gordon's algorithm, for example). There are again two versions: one asymmetric where one number field is preferred, say f_0, and additional number fields f_i are considered. For each element $a - bx$, one tests the smoothness of the image of $a - bx$ first in the number field defined by f_0, and if successful, then in all of the number fields defined by the f_i, to generate relations. This is used with a polynomial selection that produces the first polynomial much better than the second. This is the case for the base-m method and was studied by Coppersmith [41]. The same machinery applies to the generalized Joux-Lercier method [85, 96]. In these two cases, the asymptotic complexity is $L_Q[1/3, 1.90]$ instead of $L_Q[1/3, 1.923]$ (where $Q = p^n$). This MNFS version also applies to the conjugation method [96] and the complexity is $L_Q[1/3, 2.156]$ instead of $L_Q[1/3, 2.201]$.

The second symmetric version tests the smoothness of the image of $a - bx$ in all the pairs of possible number fields defined by f_i, f_j (for $i < j$). It applies to the NFS algorithm used with the JLSV$_1$ polynomial selection method, in medium characteristic. The complexity is $L_Q[1/3, 2.39]$ instead of $L_Q[1/3, 2.42]$.

Unfortunately, none of these methods were implemented (yet), even for a reasonable finite field size, hence we cannot realize how much in practice the smaller c constant improves the running-time. There was a similar unknown in 1993. The cross-over point between the COS algorithm in $L_p[1/3]$ and Gordon's algorithm in $L_p[1/3]$ was estimated in 1995 at about 150 decimal digits. In the MNFS algorithm, the constant is slightly reduced but no one knows the size of Q for which, "in practice," an MNFS variant will be faster than a regular NFS algorithm.

Special-NFS, Tower-NFS, and pairing-friendly families of curves

This paragraph lists the improvements to the NFS algorithm dedicated to fields $\mathbb{F}_{p^n}$ where the prime p is of special form, i.e., given by a polynomial evaluated at a given value. This is the case for embedding fields of pairing-friendly curves in families, such as the Barreto-Naehrig curves.

When the prime p defining the finite field is of a special form, there exist better algorithms, such as the Special NFS (SNFS) for prime fields. Joux and Pierrot [69] proved a complexity of

$$L_Q\left[\frac{1}{3}, \left(\frac{\deg P + 1}{\deg P}\frac{64}{9}\right)^{1/3}\right]$$

with P the polynomial defining the prime p. We have $\deg P = 4$, for example, for a BN curve [21]. Note that when $\deg P = 2$ as for MNT curves, the complexity $L_Q[1/3, (\frac{96}{9})^{1/3}]$ is the same as the conjugation method complexity. This promising method has not yet been implemented.

Very recently, [18] proposed another construction named the Tower-NFS that would be the best choice for finite field target groups, where $\deg P \geq 4$. One of the numerous difficult technicalities of this variant is the degree of the polynomials: a multiple of n. Handling such polynomials and elements of high degree throughout the algorithm, in particular in the relation collection phase, is a very difficult task, and is not implemented at the moment.

A clear rigorous recommendation of parameter sizes is not possible at the moment for pairing target groups, since the area is not fixed. Theoretical algorithms are published but real-life implementations and records over reasonable-size finite fields (more than 512 bits) to estimate their running-time are not available.

We can clearly say that all these theoretical propositions shall be seriously taken into consideration. The constant c in the $L_Q[1/3, c]$ asymptotic formula is decreasing, and might reach the $(\frac{64}{9})^{1/3}$ value of prime fields one day. A generic recommendation in the large characteristic case, based on a $L_Q[1/3, (64/9)^{1/3}]$ asymptotic complexity, seems reasonable.

9.3.8   Small Characteristic: From the Function Field Sieve (FFS) to the Quasi-Polynomial-Time Algorithm (QPA)

The main difference between a small-characteristic and a large-characteristic field is the Frobenius map $\pi : x \mapsto x^p$, where p is the characteristic of the field. This map is intensively used to obtain multiple relations from an initial one, at a negligible cost. For prime fields, the Frobenius map is the identity, so we cannot gain anything with it. In small-characteristic fields, the use of the Frobenius map is one of the key ingredients that provides a much better asymptotic complexity. This should be combined with a specific field representation that allows a very fast evaluation of the Frobenius map.

For large-characteristic finite fields $\mathbb{F}_{p^n}$, the Frobenius map of order n can provide a speed-up of a factor up to n if the polynomial selection provides a compatible representation, i.e., if a is B-smooth, then we want $\pi_p(a)$ to be B-smooth as well.

Example 9.1 (Systematic equations in $\mathbb{F}_{2^{127}}$)   Blake, Fuji-Hara, Mullin, and Vanstone in [27] targeted the finite field $\mathbb{F}_{2^{127}}$. They used the representation $\mathbb{F}_{2^{127}} = \mathbb{F}_2[x]/(x^{127} + x + 1)$ to implement Adleman's algorithm. The polynomial $f(x) = x^{127} + x + 1$ is primitive so that they have chosen x as a generator of $\mathbb{F}_{2^{127}}^*$. They observed that $x^{2^7-1} = x^{127} = x + 1 \bmod f(x)$, $x^{2^7} \equiv x^2 + x \bmod f(x)$ and $x^{2^i} \equiv x^{2^{i-6}} + x^{2^{i-7}} = x^{2^{i-7}}(1 + x^{2^{i-7}}) \bmod f(x)$ for any $i \geq 7$. Moreover, $(1 + x^{2^i}) = (1 + x)^{2^i} \equiv (x^{127})^{2^i}$ so that $\log_x(1 + x^{2^i}) = 127 \cdot 2^i$. Combining these equations, they obtain logarithms for free.

A second notable difference between small-characteristic fields and prime fields is the cost of factorization. In Algorithm 9.10, a preimage in $\mathbb{N}$ of elements in $\mathbb{F}_p$ is factorized. In small characteristic, elements of $\mathbb{F}_{2^n}$ are lifted to polynomials of $\mathbb{F}_2[x]$, then factorized. Over finite fields, there exists polynomial time algorithms to factor polynomials, hence the time needed per smooth test is much lower.

Brief history

Figure 9.4 shows the records in small-characteristic finite fields with the Function Field Sieve (FFS) and its various improvements, especially from 2012. Most of the records were announced on the number theory list. [*] Finite fields of characteristic 2 and 3 and composite extension degree such as target groups of pairing-friendly (hyper-)elliptic curves must definitively be avoided, since fields of even more than 3072 bits were already reached in 2014.

The Waterloo algorithm

In 1984, Blake, Fuji-Hara, Mullin, and Vanstone proposed a dedicated implementation of Adleman's algorithm to $\mathbb{F}_{2^{127}}$ [27, 28]. They introduced the idea of *systematic equations* (using the Frobenius map) and *initial splitting* (this name was introduced later). Their idea works for finite fields of characteristic two and extension degree close to a power of 2 (e.g., 127). The asymptotic complexity of their method was $L_Q[1/2]$. In the same year, Blake, Mullin, and Vanstone [28] proposed an improved algorithm known as the Waterloo algorithm. Odlyzko computed the asymptotic complexity of this algorithm in [94]. The asymptotic complexity needs the estimation of the probability that a random polynomial over $\mathbb{F}_q$ of degree m factors entirely into polynomials of degree at most b, i.e., is b-smooth. Odlyzko in [94, Equation (4.5), p. 14]

---

[*]https://listserv.nodak.edu/cgi-bin/wa.exe?A0=NMBRTHRY

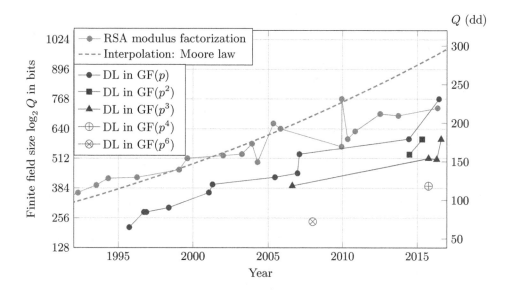

FIGURE 9.4 Records of DL computation in fields $\mathbb{F}_{2^n}, \mathbb{F}_{3^n}, \mathbb{F}_{r^n}$ of small characteristic, with n prime or composite. All the fields $\mathbb{F}_{2^n}, \mathbb{F}_{3^n}$ with n composite are target fields of supersingular pairing-friendly (hyper-)elliptic curves.

gave the following estimation.

$$p(m,n) = \exp\left((1 + o(1))\frac{n}{m}\log_e\frac{m}{n}\right) \text{ for } n^{1/100} \leq m \leq n^{99/100}. \tag{9.13}$$

The *initial splitting* idea is still used nowadays, combined with the QPA algorithm. Given a random element $a(x)$ of $\mathrm{GF}(2^n)$ represented by a degree $n-1$ polynomial over $\mathrm{GF}(2)$ modulo an irreducible degree n polynomial $f(x)$, the algorithm computes the extended Euclidean algorithm to compute the GCD of $f(x)$ and $a(x)$. At each iteration, the following equation holds [27, §2]:

$$s_i(x)a(x) + t_i(x)f(x) = r_i(x). \tag{9.14}$$

Reducing this equation modulo $f(x)$, one obtains $a(x) \equiv r_i(x)/s_i(x) \bmod f(x)$. The degree of $r_i(x)$ decreases while the degree of $s_i(x)$ increases. By stopping the extended Euclidean algorithm at the state i where $\deg r_i(x), \deg s_i(x) \leq \lfloor n/2 \rfloor$, one obtains the initial splitting of $a(x)$ of degree $n-1$ into two polynomials $r_i(x), s_i(x)$ of degree at most $\lfloor n/2 \rfloor$.

Odlyzko computed the asymptotic complexity of the Waterloo algorithm to be [94, Equation (4.17), p. 19] $L_Q[1/2, (2\log_e(2))^{1/2} \approx 1.1774]$.

Coppersmith's $L_Q[1/3]$ algorithm and FFS algorithm

Building on the idea of systematic equations, Coppersmith [40] gave the first $L_Q[1/3, c]$ algorithm for DL computations over $\mathbb{F}_{2^n}$ (with $Q = 2^n$). He found $(32/9)^{1/3} \leq c \leq 4^{1/3}$ and did a record computation for $\mathbb{F}^*_{2^{127}}$. In 1994, Adleman [7] generalized this work to the case of any small characteristic, and this is now called the Function Field Sieve (FFS). This gave a $L_Q[1/3, (\frac{64}{9})^{\frac{1}{3}}]$ for Q of small characteristic, with function field (in place of number field for prime fields). Later, Adleman-Huang improved that to $L_Q[1/3, (\frac{32}{9})^{\frac{1}{3}}]$ for Q of small characteristic [8], however, this was slower than Coppersmith for $\mathbb{F}_{2^n}$.

Outside of the pairing-based cryptography context, the research and the records are focused on prime extensions degrees. In 2002 Thomé increased the Coppersmith record up to $\mathrm{GF}(2^{607})$

[115, 116]. During the same time, Joux and Lercier implemented FFS for $GF(2^{521})$ in [64]. Continuing the record series, in 2005, Joux and Lercier recomputed a record in $GF(2^{607})$ and went slightly further with a record in $GF(2^{613})$. They also investigated the use of FFS for larger characteristic finite fields in [66]. In 2013, a record of the CARAMEL group in $GF(2^{1089})$ with FFS was announced by Bouvier on the NMBRTHRY list [14] and published in [15]. The actual record is held by Kleinjung, in $GF(2^{1279})$ [76].

Since 2000, examples of supersingular pairing-friendly elliptic curves of cryptographic size arise. Two curves are well studied in characteristic 2 and 3 for the various speed-up they provide, in particular, in hardware. Supersingular curves of almost prime order in small characteristic are very rare. No ordinary pairing-friendly curves were ever known in small characteristic. The embedding degree for supersingular curves is at most 4 in characteristic 2 and at most 6 in characteristic 3. The first cryptanalysts exploited this composite degree extension to improve the FFS algorithm. In 2010, the record in characteristic three was in $GF(3^{6 \cdot 71})$ of 676 bits [59]. In 2012, due to increasing computer power and the prequel of the use of the additional structure provided by the composite extension degree of the finite field, a DL record-breaking in $GF(3^{6 \cdot 97})$ (a 923-bit finite field) was made possible [47]. This announcement had a quite important effect over the community at that time, probably because the broken curve was the one used in the initial paper on short signatures from pairings [29]. The targeted finite field $\mathbb{F}_{3^{582}}$ was the target field of a pairing-friendly elliptic curve in characteristic 3, considered safe for 80-bit security implementations.

The real start of mathematical improvements occurred at Christmas 2012: Joux proposed a conjectured heuristic $L_Q[1/4]$ algorithm [63] and announced two records [62] of much larger size. In finite fields that can be represented as Kummer extensions, using the Frobenius gives many relations at one time, hence speeding-up the relation-collection phase. As we can see in Figure 9.4, records in prime extensions n of $\mathbb{F}_2$ do not grow as extraordinary as composite extensions coming from pairing-friendly curves.

The Quasi-Polynomial-time Algorithm (QPA)

In 2013 [48] an improved *descent* phase was proposed, one that provided a quasi-polynomial-time algorithm (QPA). Two variants of the algorithm were published [17, 56]. The polynomial selection differs and induces differences in the algorithm. We are at the "beginning of the end"— much work is still needed for a complete implementation of this algorithm. In particular, the descent phase is still costly in memory requirements.

The two versions of the QPA algorithm intensively exploit the Frobenius map, to obtain many relations for free. It works when the extension degree n is composite and satisfies some properties.

As a conclusion, we list the last records published. In 2014 Adj, Menezes, Oliveira, and Rodrígues-Henríquez published a discrete logarithm record in $GF(3^{6 \cdot 137})$ and $GF(3^{6 \cdot 163})$, corresponding to a 1303-bit and a 1551-bit finite field [5]. In 2014, Joux and Pierrot published a record in $GF(3^{5 \cdot 479})$ corresponding to a 3796-bit field [68]. In 2014, Granger, Kleinjung, and Zumbragel announced a record in $GF(2^{9234})$ [57].

Recent improvements in finite fields of composite extension degree (Spring 2016)

There were major theoretical improvements in finite fields $\mathbb{F}_{p^n}$ where n is composite, in 2015 and 2016. This paragraph tries to summarize the news. Two preprints by Kim on one side and Barbulescu on the other side evolved to a common paper at CRYPTO'16 [73]. In parallel, Sarkar and Singh combined Kim–Barbulescu's work with their own techniques [103, 101, 102]. We need to mention Jeong and Kim's work [61] to complete the list or recent preprints on the subject. This one paper and four preprints exploit the extension degree that should be composite. They each propose an improved polynomial selection step that allows us to reduce the size of the norms of the elements involved in the relation collection. Since the improvement is notable, it reduces

TABLE 9.4 Estimate of security levels according to NFS variants.

$\log_2 p^n$	Conj $L_{p^n}[\frac{1}{3}, 2.20]$	Joux–Pierrot $d = 4, L_{p^n}[\frac{1}{3}, 2.07]$	$-$ $L_{p^n}[\frac{1}{3}, 1.923]$	Conj Ext. TNFS $L_{p^n}[\frac{1}{3}, 1.747]$	Special Ext. TNFS $L_{p^n}[\frac{1}{3}, 1.526]$
3072	$2^{159-\delta_1}$	$2^{149-\delta_2}$	$2^{139-\delta_3}$	$2^{126-\delta_4}$	$2^{110-\delta_5}$
3584	$2^{169-\delta_1}$	$2^{159-\delta_2}$	$2^{148-\delta_3}$	$2^{134-\delta_4}$	$2^{117-\delta_5}$
4096	$2^{179-\delta_1}$	$2^{169-\delta_2}$	$2^{156-\delta_3}$	$2^{142-\delta_4}$	$2^{124-\delta_5}$
4608	$2^{188-\delta_1}$	$2^{177-\delta_2}$	$2^{164-\delta_3}$	$2^{149-\delta_4}$	$2^{130-\delta_5}$
5120	$2^{197-\delta_1}$	$2^{185-\delta_2}$	$2^{172-\delta_3}$	$2^{156-\delta_4}$	$2^{136-\delta_5}$
5632	$2^{204-\delta_1}$	$2^{192-\delta_2}$	$2^{179-\delta_3}$	$2^{162-\delta_4}$	$2^{142-\delta_5}$
6144	$2^{212-\delta_1}$	$2^{199-\delta_2}$	$2^{185-\delta_3}$	$2^{168-\delta_4}$	$2^{147-\delta_5}$

Note: The numbers should be read as follows: a 3072-bit finite field, which is the embedding field of a BN curve whose p is of special form and n is composite will provide approximately a security level of $2^{110-\delta_{\text{BN}}}$, where δ_{BN} depends on the curve and on the implementation of the special extended NFS variant.

the asymptotic complexity of the NFS algorithm. These papers exploit the finite field structure and contruct a degree-n extension as a tower of three levels: a base field $\mathbb{F}_p$ as first level, a second level $\mathbb{F}_{p^\eta}$, and a third level $\mathbb{F}_{p^{\eta\kappa}} = \mathbb{F}_{p^n}$. The extension degree n should be composite and the divisor η of quite small size. The two (or multiple) number fields will exploit this structure as well. This setting provides the following new asymptotic complexities for medium-characteristic fields:

1. $L_{p^n}[1/3, (48/9)^{1/3} \approx 1.747]$ when n is composite, $n = \eta\kappa$, $\kappa = \left(\frac{1}{12^{1/3}} + o(1)\right)\left(\frac{\log Q}{\log \log Q}\right)^{1/3}$, and p is generic;

2. $L_{p^n}[1/3, (32/9)^{1/3} \approx 1.526]$ when n is composite and p has a special form.

The generic case where n is prime is not affected by these new improvements. We summarize in Table 9.4 the new theoretical security of a pairing-friendly curve where (1) n is composite and (2) n is composite and p of special form, for p^n of 3072 bits.

9.3.9   How to Choose Real-Size Finite Field Parameters

At some point, to design a cryptosystem, we want to translate an asymptotic complexity to a size recommendation for a given security level, usually equivalent to an AES level of security: 128, 192, or 256 bits. In other words, we would like that for a given finite field $\mathbb{F}_q$ of given size, the running-time required to break an instance of DLP is equivalent to 2^{128}, 2^{192}, or 2^{256} group operations. For the DLP in a generic group, we saw in Section 9.2 that the expected time is in $O(\sqrt{N})$, with N the prime-order subgroup considered. A group of size $2n$ bits (where only generic attacks apply) is enough to achieve an n-bit security level.

We present in Table 9.5 the usual key-length recommendations from `http://www.keylength. com`. The NIST recommendations are the less-conservative ones. A modulus of length 3072 is recommended to achieve a security level equivalent to a 128-bit symmetric key. The ECRYPT II recommendations are slightly larger: 3248 bit modulus are suggested.

We explain here where these key sizes come from. The running-time complexity of the most efficient attacks on discrete logarithm computation and factorization are considered and balanced to fit the last records. In practice, we calibrate the asymptotic complexity (we set the constant hidden in the $O()$ notation) so that it matches the largest DL record computations.

TABLE 9.5   Cryptographic key length recommendations, August 2015.

Method	Date	Sym-metric	Asymmetric	Discrete Log Key	Discrete Log Group	Elliptic curve	Hash function
Lenstra / Verheul [84]	2076	129	6790–5888	230	6790	245	257
Lenstra Updated [82]	2090	128	4440–6974	256	4440	256	256
ECRYPT II (EU) [1]	2031–2040	128	3248	256	3248	256	256
NIST (US) [4]	> 2030	128	3072	256	3072	256	256
ANSSI (France) [3]	2021–2030	128	2048	200	2048	256	256
NSA (US) [2]	–	128	–	–	–	256	256
RFC3766 [95]	–	128	3253	256	3253	242	–

Note: All key sizes are provided in bits. These are the minimal sizes for security.

For prime fields $\mathbb{F}_p$ with no special form of the prime p, the asymptotic formula of NFS-DL is $L_p[1/3, (\frac{64}{9})^{1/3}]$, and we consider its logarithm in base 2:

$$\log_2 L[\alpha, c](n) = \big(c + o(1)\big)\, n^\alpha \log_2^{1-\alpha}(n \ln 2) \qquad (9.15)$$

with $n = \log_2 N$. The last record was a DL computation in a prime field of 180dd or 596 bits, `https://listserv.nodak.edu/cgi-bin/wa.exe?A2=ind1406&L=NMBRTHRY&F=&S=&P=3161`.

Figure 9.5 presents the records of DL computation in prime fields, the records of RSA modulus factorization, and an interpolation according to [81, §3] by a Moore law doubling every nine months.

To estimate the required modulus size, we compute the logarithm in base 2 of the L-notation (9.15) and translate it such that $\log_2 L[c, \alpha](598) \approx 60$ (with 180dd=598bits). We obtain $\log_2 L[c, \alpha](598) = 68.5$ so we set $a = -8.5$. We obtain $\log_2 L[c, \alpha](3072) - 8.5 = 130$ so we can safely deduce that a 3072-bit prime field with a generic safe prime is enough to provide a 128-bit security level.

Conservative recommendations

To avoid dedicated attacks, and specific NFS variants, common-sense advice would be to avoid the curves with too much structure in the parameters. Here is a list of points to take into account.

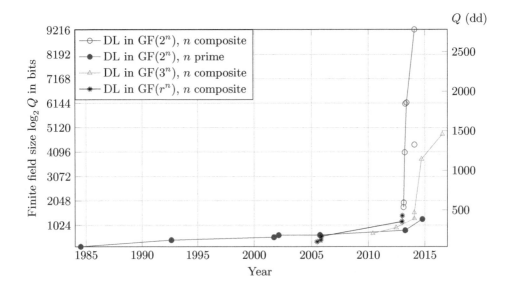

FIGURE 9.5   Records of DL computation in prime fields and RSA modulus factorization.

- Use a generic curve constructed with the Cocks-Pinch or Dupont-Enge Morain methods;
- Use a curve in a family with a non-special form seed, i.e., the prime $p = P(x_0)$ is such that x_0 has no special form (e.g., $x_0 \neq 2^{63} + 1$);
- Use a curve with low-degree polynomials defining the parameters, e.g., degree 2 (MNT and Galbraith-McKee-Valença curves) or degree 4 (Freeman curves);
- Use a curve whose discriminant D is large (e.g., constructed with the Cocks-Pinch or Dupont-Enge-Morain method, or an MNT, a Galbraith-McKee-Valença, or a Freeman curve);
- Use a prime embedding degree.

9.3.10 Discrete Logarithm Algorithms in Pairing-Friendly Target Finite Fields $\mathbb{F}_{p^n}$

Given a finite field $\mathbb{F}_{p^n}$ which contains the target group of a cryptographic pairing, different NFS-based algorithms can be applied to compute discrete logarithms, depending on the structure of the finite field. Two criteria should be taken into account: whether n is prime, and whether the characteristic p has a special form given by a polynomial of degree greater than two.

1. If n is prime:

   (a) and p has no special form (e.g., supersingular curves where $k = 2$, MNT curves where $n = 3$, and any curves constructed with the Cocks-Pinch or Dupont-Enge-Morain methods), then only the generic NFS algorithms apply.

      i. In a large characteristic, the generalized Joux–Lercier method of asymptotic complexity $L_Q[1/3, 1.923]$ (and $L_Q[1/3, 1.90]$ in the multiple-NFS version) applies.

      ii. In a medium characteristic, the conjugation method of asymptotic complexity $L_Q[1/3, 2.20]$ applies. The multiple-NFS version has an asymptotic complexity of $L_Q[1/3, 2.15]$. The finite field size does not need to be enlarged for now.

   In practice for large sizes of finite fields, the Sarkar-Singh method that interpolates between the GJL and the conjugation methods provides smaller norms. In this case, the key size should be enlarged by maybe 10% but not significantly since the asymptotic complexity is not lower than the complexity of NFS in a prime field: $L_Q[1/3, 1.923]$.

   (b) If p is given by a polynomial degree of at least three, i.e., $p = P(u)$ where $\deg(P) \geq 3$, then the Joux–Pierrot method applies. In a medium characteristic, the asymptotic complexity tends to $L_Q[1/3, 1.923]$ for large $\deg(P)$. In a large characteristic, the pairing-friendly curves ($k = 2, 3, 4, 6$ for instance) are such that $\deg(P) = 2$ only.

2. If n is composite, then the extended tower-NFS technique, first introduced by Kim then improved by Barbulescu, Kim, Sarkar and Singh, and Jeong, applies.

   (a) If p has no special form or is given by a polynomial of degree at most 2 (MNT curves of embedding degree 4 and 6, Cocks-Pinch and Dupont-Enge-Morain methods), then the asymptotic complexity is $L_Q[1/3, 1.74]$ so asymptotically, the finite field size should be enlarged by a factor $4/3$.

   (b) If p has a special form, then the asymptotic complexity is $L_Q[1/3, 1.56]$ and asymptotically, the finite field size should be doubled.

References

[1] D.SPA.20. *ECRYPT2 Yearly Report on Algorithms and Keysizes (2011-2012)*. European Network of Excellence in Cryptology II, September 2012.

[2] NSA Suite B. *Fact Sheet Suite B Cryptography*. National Security Agency, U.S.A., September 2014.

[3] RGS-B1. *Mécanismes cryptographiques - Règles et recommandations concernant le choix et le dimensionnement des mécanismes cryptographiques*. Agence Nationale de la Sécurité des Systèmes d'Information, France, February 2014. version 2.03.

[4] SP-800-57. *Recommendation for Key Management – Part 1: General*. National Institute of Standards and Technology, U.S. Department of Commerce, July 2012.

[5] Gora Adj, Alfred Menezes, Thomaz Oliveira, and Francisco Rodríguez-Henríquez. Computing discrete logarithms in $\mathbb{F}_{3^{6 \cdot 137}}$ and $\mathbb{F}_{3^{6 \cdot 163}}$ using Magma. In Ç. K. Koç, S. Mesnager, and E. Savas, editors, *Arithmetic of Finite Fields (WAIFI 2014)*, volume 9061 of *Lecture Notes in Computer Science*, pp. 3–22. Springer, 2014.

[6] Leonard Adleman. A subexponential algorithm for the discrete logarithm problem with applications to cryptography. In *20th Annual Symposium on Foundations of Computer Science*, pp. 55–60. IEEE Computer Society Press, 1979.

[7] Leonard Adleman. The function field sieve. In L. M. Adleman and M.-D. Huang, editors, *Algorithmic Number Theory (ANTS-I)*, volume 877 of *Lecture Notes in Computer Science*, pp. 141–154. Springer, 1994.

[8] Leonard M. Adleman and Ming-Deh A. Huang. Function field sieve method for discrete logarithms over finite fields. *Information and Computation*, 151(1/2):5–16, 1999.

[9] David Adrian, Karthikeyan Bhargavan, Zakir Durumeric, Pierrick Gaudry, Matthew Green, J. Alex Halderman, Nadia Heninger, Drew Springall, Emmanuel Thomé, Luke Valenta, Benjamin VanderSloot, Eric Wustrow, Santiago Zanella-Béguelin, and Paul Zimmermann. Imperfect forward secrecy: How Diffie-Hellman fails in practice. In I. Ray, N. Li, and C. Kruegel, editors, *22nd ACM Conference on Computer and Communications Security*, pp. 5–17. ACM Press, 2015.

[10] M. Albrecht, S. Bai, D. Cadé, X. Pujol, and D. Stehlé. fplll-4.0, a floating-point LLL implementation. Available at http://perso.ens-lyon.fr/damien.stehle.

[11] Shi Bai. *Polynomial Selection for the Number Field Sieve*. PhD thesis, Australian National University, 2011. http://maths.anu.edu.au/~brent/pd/Bai-thesis.pdf.

[12] Shi Bai, Richard Brent, and Emmanuel Thomé. Root optimization of polynomials in the number field sieve. *Mathematics of Computation*, 84(295):2447–2457, 2015.

[13] Razvan Barbulescu. *Algorithmes de logarithmes discrets dans les corps finis*. PhD thesis, Université de Lorraine, 2013. https://tel.archives-ouvertes.fr/tel-00925228.

[14] Razvan Barbulescu, Cyril Bouvier, Jérémie Detrey, Pierrick Gaudry, Hamza Jeljeli, Emmanuel Thomé, Marion Videau, and Paul Zimmermann. Discrete logarithm in GF(2^{809}) with ffs, April 2013. Announcement available at the NMBRTHRY archives, item 004534.

[15] Razvan Barbulescu, Cyril Bouvier, Jérémie Detrey, Pierrick Gaudry, Hamza Jeljeli, Emmanuel Thomé, Marion Videau, and Paul Zimmermann. Discrete logarithm in GF(2809) with FFS. In H. Krawczyk, editor, *PKC 2014: 17th International Conference on Theory and Practice of Public Key Cryptography*, volume 8383 of *Lecture Notes in Computer Science*, pp. 221–238. Springer, Heidelberg, 2014.

[16] Razvan Barbulescu, Pierrick Gaudry, Aurore Guillevic, and François Morain. Improving NFS for the discrete logarithm problem in non-prime finite fields. In E. Oswald and M. Fischlin, editors, *Advances in Cryptology – EUROCRYPT 2015, Part I*, volume 9056 of *Lecture Notes in Computer Science*, pp. 129–155. Springer, Heidelberg, 2015.

[17] Razvan Barbulescu, Pierrick Gaudry, Antoine Joux, and Emmanuel Thomé. A heuristic quasi-polynomial algorithm for discrete logarithm in finite fields of small characteristic. In P. Q. Nguyen and E. Oswald, editors, *Advances in Cryptology – EUROCRYPT 2014*, volume 8441 of *Lecture Notes in Computer Science*, pp. 1–16. Springer, Heidelberg, 2014.

[18] Razvan Barbulescu, Pierrick Gaudry, and Thorsten Kleinjung. The tower number field sieve. In T. Iwata and J. H. Cheon, editors, *Advances in Cryptology – ASIACRYPT 2015, Part II*, volume 9453 of *Lecture Notes in Computer Science*, pp. 31–55. Springer, Heidelberg, 2015.

[19] Razvan Barbulescu and Cécile Pierrot. The Multiple Number Field Sieve for Medium and High Characteristic Finite Fields. *LMS Journal of Computation and Mathematics*, 17:230–246, 2014.

[20] Paulo S. L. M. Barreto, Ben Lynn, and Michael Scott. Constructing elliptic curves with prescribed embedding degrees. In S. Cimato, C. Galdi, and G. Persiano, editors, *Security in Communication Networks (SCN 2002)*, volume 2576 of *Lecture Notes in Computer Science*, pp. 257–267. Springer, Heidelberg, 2003.

[21] Paulo S. L. M. Barreto and Michael Naehrig. Pairing-friendly elliptic curves of prime order. In B. Preneel and S. Tavares, editors, *Selected Areas in Cryptography (SAC 2005)*, volume 3897 of *Lecture Notes in Computer Science*, pp. 319–331. Springer, Heidelberg, 2006.

[22] Daniel J. Bernstein and Tanja Lange. Computing small discrete logarithms faster. In S. D. Galbraith and M. Nandi, editors, *Progress in Cryptology – INDOCRYPT 2012*, volume 7668 of *Lecture Notes in Computer Science*, pp. 317–338. Springer, Heidelberg, 2012.

[23] Daniel J. Bernstein and Tanja Lange. Non-uniform cracks in the concrete: The power of free precomputation. In K. Sako and P. Sarkar, editors, *Advances in Cryptology – ASIACRYPT 2013, Part II*, volume 8270 of *Lecture Notes in Computer Science*, pp. 321–340. Springer, Heidelberg, 2013.

[24] Daniel J. Bernstein, Tanja Lange, and Peter Schwabe. On the correct use of the negation map in the Pollard rho method. In D. Catalano et al., editors, *Public Key Cryptography – PKC 2011*, volume 6571 of *Lecture Notes in Computer Science*, pp. 128–146. Springer, Heidelberg, 2011.

[25] Yuval Bistritz and Alexander Lifshitz. Bounds for resultants of univariate and bivariate polynomials. *Linear Algebra and its Applications*, 432(8):1995–2005, 2009.

[26] Simon R. Blackburn and Edlyn Teske. Baby-step giant-step algorithms for non-uniform distributions. In W. Bosma, editor, *Algorithmic Number Theory (ANTS-IV)*, volume 1838 of *Lecture Notes in Computer Science*, pp. 153–168. Springer, 2000.

[27] Ian F. Blake, Ryoh Fuji-Hara, Ronald C. Mullin, and Scott A. Vanstone. Computing logarithms in finite fields of characteristic two. *SIAM Journal on Algebraic Discrete Methods*, 5(2):276–285, 1984.

[28] Ian F. Blake, Ronald C. Mullin, and Scott A. Vanstone. Computing logarithms in $gf(2^n)$. In G. R. Blakley and D. Chaum, editors, *Advances in Cryptology, Proceedings of CRYPTO '84*, volume 196 of *Lecture Notes in Computer Science*, pp. 73–82. Springer, Heidelberg, 1984.

[29] Dan Boneh, Ben Lynn, and Hovav Shacham. Short signatures from the Weil pairing. In C. Boyd, editor, *Advances in Cryptology – ASIACRYPT 2001*, volume 2248 of *Lecture Notes in Computer Science*, pp. 514–532. Springer, Heidelberg, 2001.

[30] Cyril Bouvier. *Algorithmes pour la factorisation d'entiers et le calcul de logarithme discret*. PhD thesis, Université de Lorraine, 2015. https://tel.archives-ouvertes.fr/tel-01167281.

[31] Richard P. Brent. An improved Monte Carlo factorization algorithm. *BIT*, 20:176–184, 1980.

[32] Friederike Brezing and Annegret Weng. Elliptic curves suitable for pairing based cryptography. *Designs, Codes and Cryptography*, 37(1):133–141, 2005.

[33] Joe P. Buhler, Hendrik W. Lenstra Jr., and Carl Pomerance. Factoring integers with the number field sieve. In A. K. Lenstra and H. W. Lenstra Jr., editors, *The Development of the Number Field Sieve*, volume 1554 of *Lecture Notes in Mathematics*, pp. 50–94. Springer, 1993.

[34] Earl R. Canfield, Paul Erdős, and Carl Pomerance. On a problem of Oppenheim concerning "factorisatio numerorum". *Journal of Number Theory*, 17(1):1–28, 1983.

[35] Yuanmi Chen. *Réduction de réseaux et sécurité concrète du chiffrement complètement homomorphe*. PhD thesis, Université Paris 7 Denis Diderot, 2013. http://www.di.ens.fr/~ychen/research/these.pdf.

[36] Jung Hee Cheon, Jin Hong, and Minkyu Kim. Speeding up the pollard rho method on prime fields. In J. Pieprzyk, editor, *Advances in Cryptology – ASIACRYPT 2008*, volume 5350 of *Lecture Notes in Computer Science*, pp. 471–488. Springer, Heidelberg, 2008.

[37] Young Ju Choie, Eun Kyung Jeong, and Eun Jeong Lee. Supersingular hyperelliptic curves of genus 2 over finite fields. *Journal of Applied Mathematics and Computation*, 163(2):565–576, 2005.

[38] An Commeine and Igor Semaev. An algorithm to solve the discrete logarithm problem with the number field sieve. In M. Yung, Y. Dodis, A. Kiayias, and T. Malkin, editors, *Public Key Cryptography – PKC 2006*, volume 3958 of *Lecture Notes in Computer Science*, pp. 174–190. Springer, Heidelberg, 2006.

[39] D. Coppersmith. Solving linear equations over GF(2) via block Wiedemann algorithm. *Mathematics of Computation*, 62(205):333–350, 1994.

[40] Don Coppersmith. Fast evaluation of logarithms in fields of characteristic two. *IEEE Transactions on Information Theory*, 30(4):587–594, 1984.

[41] Don Coppersmith. Modifications to the number field sieve. *Journal of Cryptology*, 6(3):169–180, 1993.

[42] Don Coppersmith, Andrew M. Odlyzko, and Richard Schroeppel. Discrete logarithms in GF(p). *Algorithmica*, 1(1):1–15, 1986.

[43] Whitfield Diffie and Martin E. Hellman. New directions in cryptography. *IEEE Transactions on Information Theory*, 22(6):644–654, 1976.

[44] Iwan M. Duursma, Pierrick Gaudry, and François Morain. Speeding up the discrete log computation on curves with automorphisms. In K.-Y. Lam, E. Okamoto, and C. Xing, editors, *Advances in Cryptology – ASIACRYPT '99*, volume 1716 of *Lecture Notes in Computer Science*, pp. 103–121. Springer, Heidelberg, 1999.

[45] A. E. Escott, J. C. Sager, A. P. L. Selkirk, and D. Tsapakidis. Attacking elliptic curve cryptosystems using the parallel Pollard rho method. *CryptoBytes*, 4, 1999.

[46] Philippe Flajolet and Andrew M. Odlyzko. Random mapping statistics. In J.-J. Quisquater and J. Vandewalle, editors, *Advances in Cryptology – EURO-*

CRYPT '89, volume 434 of *Lecture Notes in Computer Science*, pp. 329–354. Springer, Heidelberg, 1990.

[47] Fujitsu Laboratories, NICT, and Kyushu University. DL record in $\mathbb{F}_{3^{6 \cdot 97}}$ of 923 bits (278 dd). NICT press release, June 18, 2012. http://www.nict.go.jp/en/press/2012/06/18en-1.html.

[48] Steven Galbraith. Quasi-polynomial-time algorithm for discrete logarithm in finite fields of small/medium characteristic. The Elliptic Curve Cryptography blog, June 2013. https://ellipticnews.wordpress.com/2013/06/21.

[49] Steven D. Galbraith. Supersingular curves in cryptography. In C. Boyd, editor, *Advances in Cryptology – ASIACRYPT 2001*, volume 2248 of *Lecture Notes in Computer Science*, pp. 495–513. Springer, Heidelberg, 2001.

[50] Steven D. Galbraith and Pierrick Gaudry. Recent progress on the elliptic curve discrete logarithm problem. Cryptology ePrint Archive, Report 2015/1022, 2015. http://eprint.iacr.org/.

[51] Steven D. Galbraith and Raminder S. Ruprai. An improvement to the Gaudry-Schost algorithm for multidimensional discrete logarithm problems. In M. G. Parker, editor, *Cryptography and Coding*, volume 5921 of *Lecture Notes in Computer Science*, pp. 368–382. Springer, Heidelberg, 2009.

[52] Steven D. Galbraith and Raminder S. Ruprai. Using equivalence classes to accelerate solving the discrete logarithm problem in a short interval. In P. Q. Nguyen and D. Pointcheval, editors, *Public Key Cryptography – PKC 2010*, volume 6056 of *Lecture Notes in Computer Science*, pp. 368–383. Springer, Heidelberg, 2010.

[53] Steven D. Galbraith, Ping Wang, and Fangguo Zhang. Computing elliptic curve discrete logarithms with improved baby-step giant-step algorithm. Cryptology ePrint Archive, Report 2015/605, 2015. http://eprint.iacr.org/2015/605.

[54] Pierrick Gaudry and Éric Schost. A low-memory parallel version of Matsuo, Chao, and Tsujii's algorithm. In D. A. Buell, editor, *Algorithmic Number Theory (ANTS-VI)*, volume 3076 of *Lecture Notes in Computer Science*, pp. 208–222. Springer, 2004.

[55] Daniel M. Gordon. Discrete logarithms in GF(p) using the number field sieve. *SIAM Journal on Discrete Mathematics*, 6(1):124–138, 1993.

[56] Robert Granger, Thorsten Kleinjung, and Jens Zumbrägel. Breaking '128-bit secure' supersingular binary curves - (or how to solve discrete logarithms in $F_{2^{4 \cdot 1223}}$ and $F_{2^{12 \cdot 367}}$). In J. A. Garay and R. Gennaro, editors, *Advances in Cryptology – CRYPTO 2014, Part II*, volume 8617 of *Lecture Notes in Computer Science*, pp. 126–145. Springer, Heidelberg, 2014.

[57] Robert Granger, Thorsten Kleinjung, and Jens Zumbragel. Discrete logarithms in GF(2^{9234}), 2014. Announcement available at the NMBRTHRY archives, item 004666.

[58] Takuya Hayashi, Takeshi Shimoyama, Naoyuki Shinohara, and Tsuyoshi Takagi. Breaking pairing-based cryptosystems using η_T pairing over GF(3^{97}). In X. Wang and K. Sako, editors, *Advances in Cryptology – ASIACRYPT 2012*, volume 7658 of *Lecture Notes in Computer Science*, pp. 43–60. Springer, Heidelberg, 2012.

[59] Takuya Hayashi, Naoyuki Shinohara, Lihua Wang, Shin'ichiro Matsuo, Masaaki Shirase, and Tsuyoshi Takagi. Solving a 676-bit discrete logarithm problem in GF(3^{6n}). In P. Q. Nguyen and D. Pointcheval, editors, *Public Key Cryptography – PKC 2010*, volume 6056 of *Lecture Notes in Computer Science*, pp. 351–367. Springer, Heidelberg, 2010.

[60] Yvonne Hitchcock, Paul Montague, Gary Carter, and Ed Dawson. The efficiency of solving multiple discrete logarithm problems and the implications for the security

of fixed elliptic curves. *International Journal of Information Security*, 3(2):86–98, 2004.

[61] Jinhyuck Jeong and Taechan Kim. Extended tower number field sieve with application to finite fields of arbitrary composite extension degree. Cryptology ePrint Archive, Report 2016/526, 2016. http://eprint.iacr.org/.

[62] Antoine Joux. Faster index calculus for the medium prime case application to 1175-bit and 1425-bit finite fields. In T. Johansson and P. Q. Nguyen, editors, *Advances in Cryptology – EUROCRYPT 2013*, volume 7881 of *Lecture Notes in Computer Science*, pp. 177–193. Springer, Heidelberg, 2013.

[63] Antoine Joux. A new index calculus algorithm with complexity $L(1/4 + o(1))$ in small characteristic. In T. Lange, K. Lauter, and P. Lisonek, editors, *Selected Areas in Cryptography – SAC 2013*, volume 8282 of *Lecture Notes in Computer Science*, pp. 355–379. Springer, Heidelberg, 2014.

[64] Antoine Joux and Reynald Lercier. The function field sieve is quite special. In C. Fieker and D. R. Kohel, editors, *Algorithmic Number Theory (ANTS-V)*, volume 2369 of *Lecture Notes in Computer Science*, pp. 431–445. Springer, 2002.

[65] Antoine Joux and Reynald Lercier. Improvements to the general number field sieve for discrete logarithms in prime fields. A comparison with the Gaussian integer method. *Mathematics of Computation*, 72(242):953–967, 2003.

[66] Antoine Joux and Reynald Lercier. The function field sieve in the medium prime case. In S. Vaudenay, editor, *Advances in Cryptology – EUROCRYPT 2006*, volume 4004 of *Lecture Notes in Computer Science*, pp. 254–270. Springer, Heidelberg, 2006.

[67] Antoine Joux, Reynald Lercier, Nigel Smart, and Frederik Vercauteren. The number field sieve in the medium prime case. In C. Dwork, editor, *Advances in Cryptology – CRYPTO 2006*, volume 4117 of *Lecture Notes in Computer Science*, pp. 326–344. Springer, Heidelberg, 2006.

[68] Antoine Joux and Cécile Pierrot. Improving the polynomial time precomputation of frobenius representation discrete logarithm algorithms - simplified setting for small characteristic finite fields. In P. Sarkar and T. Iwata, editors, *Advances in Cryptology – ASIACRYPT 2014, Part I*, volume 8873 of *Lecture Notes in Computer Science*, pp. 378–397. Springer, Heidelberg, 2014.

[69] Antoine Joux and Cécile Pierrot. The special number field sieve in $\mathbb{F}_{p^n}$ - application to pairing-friendly constructions. In Z. Cao and F. Zhang, editors, *Pairing-Based Cryptography – Pairing 2013*, volume 8365 of *Lecture Notes in Computer Science*, pp. 45–61. Springer, Heidelberg, 2014.

[70] Antoine Joux and Cécile Pierrot. Nearly sparse linear algebra. Cryptology ePrint Archive, Report 2015/930, 2015. http://eprint.iacr.org/.

[71] Michael Kalkbrener. An upper bound on the number of monomials in determinants of sparse matrices with symbolic entries. *Mathematica Pannonica*, 8:73–82, 1997.

[72] Minkyu Kim, Jung Hee Cheon, and Jin Hong. Subset-restricted random walks for Pollard rho method on $\mathbb{F}_{p^m}$. In S. Jarecki and G. Tsudik, editors, *Public Key Cryptography – PKC 2009*, volume 5443 of *Lecture Notes in Computer Science*, pp. 54–67. Springer, Heidelberg, 2009.

[73] Taechan Kim and Razvan Barbulescu. Extended Tower Number Field Sieve: A New Complexity for Medium Prime Case. In M. Robshaw and J. Katz, editors, *CRYPTO 2016*, LNCS. Springer, 2016. To appear, preprint available at http://eprint.iacr.org/2015/1027.

[74] Thorsten Kleinjung. On polynomial selection for the general number field sieve. *Mathematics of Computation*, 75(256):2037–2047, 2006.

[75] Thorsten Kleinjung. Polynomial selection. Invited talk at the CADO-NFS workshop, Nancy, France, October 2008. slides available at `http://cado.gforge.inria.fr/workshop/slides/kleinjung.pdf`.

[76] Thorsten Kleinjung. Discrete logarithms in $GF(2^{1279})$, October 2014. Announcement available at the NMBRTHRY archives, item 004751.

[77] Neal Koblitz and Alfred Menezes. Another look at non-standard discrete log and Diffie-Hellman problems. *Journal of Mathematical Cryptology*, 2(4):311–326, 2008.

[78] Maurice Kraitchik. *Théorie des Nombres*. Gauthier–Villars, 1922.

[79] Maurice Kraitchik. *Recherches sur la Théorie des Nombres*. Gauthier–Villars, 1924.

[80] Fabian Kuhn and René Struik. Random walks revisited: Extensions of Pollard's rho algorithm for computing multiple discrete logarithms. In S. Vaudenay and A. M. Youssef, editors, *Selected Areas in Cryptography (SAC 2001)*, volume 2259 of *Lecture Notes in Computer Science*, pp. 212–229. Springer, Heidelberg, 2001.

[81] Arjen K. Lenstra. Unbelievable security: Matching AES security using public key systems (invited talk). In C. Boyd, editor, *Advances in Cryptology – ASIACRYPT 2001*, volume 2248 of *Lecture Notes in Computer Science*, pp. 67–86. Springer, Heidelberg, 2001.

[82] Arjen K. Lenstra. Key lengths. In H. Bidgoli, editor, *Handbook of Information Security*, volume 3, pp. 617–635. John Wiley & Sons, 2006.

[83] Arjen K. Lenstra, Hendrik W. Lenstra Jr., and László Lovász. Factoring polynomials with rational coefficients. *Mathematische Annalen*, 261(4):515–534, 1982.

[84] Arjen K. Lenstra and Eric R. Verheul. Selecting cryptographic key sizes. *Journal of Cryptology*, 14(4):255–293, 2001.

[85] D. V. Matyukhin. Effective version of the number field sieve for discrete logarithms in the field $GF(p^k)$ (in Russian). *Trudy po Discretnoi Matematike*, 9:121–151, 2006.

[86] Ueli M. Maurer and Stefan Wolf. The relationship between breaking the Diffie-Hellman protocol and computing discrete logarithms. *SIAM Journal on Computing*, 28(5):1689–1721, 1999.

[87] Ueli M. Maurer and Stefan Wolf. The Diffie-Hellman protocol. *Designs, Codes and Cryptography*, 19(2/3):147–171, 2000.

[88] Kevin S. McCurley. The discrete logarithm problem. In C. Pomerance, editor, *Cryptology and Computational Number Theory*, volume 42 of *Proceedings of Symposia in Applied Mathematics*, pp. 49–74. AMS, 1990.

[89] Alfred J. Menezes, Tatsuaki Okamoto, and Scott A. Vanstone. Reducing elliptic curves logarithms to logarithms in a finite field. *IEEE Transactions on Information Theory*, 39(5):1639–1646, 1993.

[90] Atsuko Miyaji, Masaki Nakabayashi, and Shunzo Takano. Characterization of elliptic curve traces under FR-reduction. In D. Won, editor, *Information Security and Cryptology – ICISC 2000*, volume 2015 of *Lecture Notes in Computer Science*, pp. 90–108. Springer, Heidelberg, 2001.

[91] Peter L. Montgomery. Speeding the Pollard and elliptic curve methods of factorization. *Mathematics of Computation*, 48(177):243–264, 1987.

[92] B. A. Murphy. *Polynomial Selection for the Number Field Sieve Integer Factorisation Algorithm*. PhD thesis, Australian National University, 1999. `http://maths-people.anu.edu.au/~brent/pd/Murphy-thesis.pdf`.

[93] Brian A. Murphy. Modelling the yield of number field sieve polynomials. In J. P. Buhler, editor, *Algorithmic Number Theory: Third International Symposiun, ANTS-III Portland, Oregon, USA, June 21–25, 1998 Proceedings*, Lecture Notes in Computer Science, pp. 137–150. Springer Berlin Heidelberg, 1998.

[94]   Andrew M. Odlyzko. Discrete logarithms in finite fields and their cryptographic signif-
       icance. In T. Beth, N. Cot, and I. Ingemarsson, editors, *Advances in Cryptology –
       EUROCRYPT '84*, volume 209 of *Lecture Notes in Computer Science*, pp. 224–314.
       Springer, Heidelberg, 1985.

[95]   Hilarie Orman and Paul Hoffman. Determining strengths for public keys used for ex-
       changing symmetric keys. Request for Comments RFC 3766, Internet Engineering
       Task Force (IETF), 2004.

[96]   Cécile Pierrot. The multiple number field sieve with conjugation and generalized joux-
       lercier methods. In E. Oswald and M. Fischlin, editors, *Advances in Cryptology –
       EUROCRYPT 2015, Part I*, volume 9056 of *Lecture Notes in Computer Science*,
       pp. 156–170. Springer, Heidelberg, 2015.

[97]   John M. Pollard. Monte Carlo methods for index computation (mod p). *Mathematics
       of Computation*, 32(143):918–924, 1978.

[98]   John M. Pollard. Kangaroos, monopoly and discrete logarithms. *Journal of Cryptol-
       ogy*, 13(4):437–447, 2000.

[99]   Hans-Georg Rück. On the discrete logarithm in the divisor class group of curves.
       Mathematics of Computation, 68(226):805–806, 1999.

[100] Palash Sarkar and Shashank Singh. New Complexity Trade-Offs for the (Multiple)
       Number Field Sieve Algorithm in Non-Prime Fields. Cryptology ePrint Archive,
       Report 2015/944, 2015. http://eprint.iacr.org/2015/944.

[101] Palash Sarkar and Shashank Singh. A general polynomial selection method and new
       asymptotic complexities for the tower number field sieve algorithm. Cryptology
       ePrint Archive, Report 2016/485, 2016. http://eprint.iacr.org/.

[102] Palash Sarkar and Shashank Singh. A generalisation of the conjugation method for
       polynomial selection for the extended tower number field sieve algorithm. Cryp-
       tology ePrint Archive, Report 2016/537, 2016. http://eprint.iacr.org/.

[103] Palash Sarkar and Shashank Singh. Tower number field sieve variant of a recent
       polynomial selection method. Cryptology ePrint Archive, Report 2016/401, 2016.
       http://eprint.iacr.org/.

[104] Takakazu Satoh and Kiyomichi Araki. Fermat quotients and the polynomial time
       discrete log algorithm for anomalous elliptic curves. *Commentarii Math. Univ. St.
       Pauli*, 47(1):81–92, 1998.

[105] Jürgen Sattler and Claus-Peter Schnorr. Generating random walks in groups. *Ann.
       Univ. Sci. Budapest. Sect. Comput.*, 6:65–79, 1985.

[106] Oliver Schirokauer. Discrete logarithms and local units. *Philosophical Transactions
       of the Royal Society*, 345(1676):409–423, 1993.

[107] Daniel Shanks. Class number, a theory of factorization, and genera. In D. J. Lewis,
       editor, *1969 Number Theory Institute*, volume 20 of *Proceedings of Symposia in
       Applied Mathematics*, pp. 415–440. AMS, 1971.

[108] Peter W. Shor. Polynomial-time algorithms for prime factorization and discrete log-
       arithms on a quantum computer. *SIAM Journal on Computing*, 26(5):1484–1509,
       1997.

[109] Nigel P. Smart. The discrete logarithm problem on elliptic curves of trace one. *Journal
       of Cryptology*, 12(3):193–196, 1999.

[110] Andreas Stein and Edlyn Teske. Optimized baby step–giant step methods. *J. Ra-
       manujan Math. Soc.*, 20(1):27–58, 2005.

[111] Douglas R. Stinson. *Cryptography: Theory and Practice*. Discrete Mathematics and
       Its Applications. Chapman and Hall/CRC, 3rd edition, 2006.

[112] The CADO-NFS Development Team. CADO-NFS, an implementation of the number
       field sieve algorithm, 2015. Release 2.2.0.

[113] David C. Terr. A modification of Shanks' baby-step giant-step algorithm. *Mathematics of Computation*, 69(230):767–773, 2000.

[114] Edlyn Teske. Speeding up Pollard's rho method for computing discrete logarithms. In J. P. Buhler, editor, *Algorithmic Number Theory (ANTS-III)*, volume 1423 of *Lecture Notes in Computer Science*, pp. 541–554. Springer, 1998.

[115] Emmanuel Thomé. Computation of discrete logarithms in $F_{2^6 07}$. In C. Boyd, editor, *Advances in Cryptology – ASIACRYPT 2001*, volume 2248 of *Lecture Notes in Computer Science*, pp. 107–124. Springer, Heidelberg, 2001.

[116] Emmanuel Thomé. Discrete logarithms in GF(2^{607}), February 2002. Announcement available at the NMBRTHRY archives, item 001894.

[117] Emmanuel Thomé. *Algorithmes de calcul de logarithme discret dans les corps finis.* Thèse, École polytechnique, 2003. `https://tel.archives-ouvertes.fr/tel-00007532`.

[118] Paul C. van Oorschot and Michael J. Wiener. Parallel collision search with cryptanalytic applications. *Journal of Cryptology*, 12(1):1–28, 1999.

[119] A. E. Western and J. C. P. Miller. *Tables of Indices and Primitive Roots*, volume 9 of *Royal Society Mathematical Tables*. Cambridge University Press, 1968.

[120] D. H. Wiedemann. Solving sparse linear equations over finite fields. *IEEE Transactions on Information Theory*, IT–32(1):54–62, 1986.

[121] Pavol Zajac. *Discrete Logarithm Problem in Degree Six Finite Fields*. PhD thesis, Slovak University of Technology, 2008. `http://www.kaivt.elf.stuba.sk/kaivt/Vyskum/XTRDL`.

10

Choosing Parameters

Sylvain Duquesne
Université de Rennes I

Nadia El Mrabet
EMSE

Safia Haloui
LIASD

Damien Robert
INRIA

Franck Rondepierre
Oberthur Technologies

10.1 Which Parameters?

In this section, we explain how to construct or choose the parameters necessary to implement a pairing. We recall that in order to define a pairing, we need

- a finite field $\mathbb{F}_p$, where p is a prime number,
- an elliptic curve E defined over $\mathbb{F}_p$,
- a prime number r dividing $\mathrm{card}(E(\mathbb{F}_p))$,
- the embedding degree k, i.e., the smallest integer such that r divides $(p^k - 1)$,
- the set of points of r-torsion $E[r] = \mathbb{Z}/r\mathbb{Z} \times \mathbb{Z}/r\mathbb{Z}$ subdivided as $\mathbb{G}_1$ and $\mathbb{G}_2$. For the implementation of the Tate pairing, $\mathbb{G}_1$ and $\mathbb{G}_2$ are defined by $\mathbb{G}_1 = E(\mathbb{F}_p)[r]$ and $\mathbb{G}_2 = E(\mathbb{F}_{p^k})[r] \setminus rE(\mathbb{F}_p)$. For the (optimal) Ate, (optimal) twisted Ate we have that $\mathbb{G}_1 = E(\mathbb{F}_p)[r] \cap \mathrm{Ker}(\pi_p - [1])$ and $\mathbb{G}_2 = E(\mathbb{F}_{p^k})[r] \cap \mathrm{Ker}(\pi_p - [p])$, where π_p is the Frobenius endomorphism on E.
- A rational function $f_{\delta,P}(Q)$ or $f_{\delta,Q}(P)$, where δ is an integer defined by the pairing. We denote by $f_{\delta,X}$ the normalized function on the curve with divisor $\mathrm{div}(f_{\delta,X}) = \delta[Q] - [\delta Q] - (\delta - 1)[0_E]$, with $X = P$ or $X = Q$. This rational function evaluated in the second point of $E[r]$ is computed using the Miller algorithm [29]. Such functions are the core of all known pairings. They are computed thanks to the Miller loop (de-

TABLE 10.1   Sizes of curves parameters and corresponding embedding degrees to obtain commonly desired levels of security.

Security level in bits	Subgroup size r in bits	Extension field size p^k in bits	Embedding degree k $\rho \approx 1$	$\rho \approx 2$
80	160	960 − 1280	6 − 8	3 − 4
112	224	2200 − 3600	10 − 16	5 − 8
128	256	3000 − 5000	12 − 20	6 − 10
192	384	8000 − 10000	20 − 26	10 − 13
256	512	14000 − 18000	28 − 36	14 − 18

scribed in Section 10.7.4), which is an adaptation of the classical scalar multiplication algorithm.

Remark 10.1   We choose here to describe the construction of a pairing over $\mathbb{F}_p$. It is possible to define a pairing over a finite field $\mathbb{F}_q$, with q a power of a prime number. But, according to the recent records of solving the discrete logarithm problems [5], pairings over finite fields of characteristic 2 and 3 are no longer secure. Moreover, according to our knowledge, it is difficult to construct a pairing defined over a finite field $\mathbb{F}_q$ for q, a power of a prime number different from 2 and 3. There is no construction of such a pairing in the literature.

We will construct the corresponding parameters in the following order:

- First we fix the security level we want to achieve.
- Then, we first fix k, the embedding degree of the elliptic curve relative to r.
- Once k is chosen, we can use the stop-and-go shop from the article [17] to construct a suitable elliptic curve; see Chapter 4. In this step, we will carefully choose the bit size of r and p.
- We then have to construct the suitable subgroups of order r, $\mathbb{G}_1 \subseteq E(\mathbb{F}_p)$ and $\mathbb{G}_2 \subseteq E(\mathbb{F}_{p^k})$.
- The last step is the choice of the pairing algorithm: Tate, Ate, twisted Ate, or an optimal version of those.

10.2   Security Level

The first choice corresponds to the security level. Let p be a prime number, E an elliptic curve defined over $\mathbb{F}_p$. Let r denote a large prime factor of $\mathrm{Card}(E(\mathbb{F}_p))$. Let k be the embedding degree of E relatively to r. A pairing implementation involves computations in subgroups of order r of $E(\mathbb{F}_p)$, $E(\mathbb{F}_{p^k})$, and $\mathbb{F}_{p^k}^\star$. The mathematical problem that the security of pairing-based cryptosystems relies on is the discrete logarithm problem. In Chapter 9, the discrete logarithm problem is presented together with the existing algorithms to solve it. The size of the subgroups involved in a pairing computation must be large enough to ensure that the discrete logarithm is hard. This condition implies lower bounds on the bit size of r and p^k, for a given security level. Table 10.1, taken from [17], gives the minimal size of the parameters for a given security level.

According to this Table 10.1, the security level gives us the minimal size for r, and range for the size of p^k, and k relative to ρ. The value ρ is defined as $\rho = \frac{\log(p)}{\log(r)}$. In order to save bandwidth during the calculation we are looking for ρ as small as possible (in the ideal case, $\rho \approx 1$). Indeed, if ρ is greater than 2, than $\mathbb{F}_p$ is twice as large as necessary. As a consequence, the computations over $\mathbb{F}_p$ and its extensions are more expensive for a given security level, compared to a curve with $\rho \approx 1$.

TABLE 10.2   Comparison of security level estimates for BN curves, according to the available NFS variants.

$\log_2(p)$	n	$\log_2 p^n$	Joux–Pierrot $L_{p^n}[1/3, 2.07]$	$-$ $L_{p^n}[1/3, 1.92]$	ExTNFS $L_{p^n}[1/3, 1.74]$	Special ExTNFS $L_{p^n}[1/3, 1.53]$
256	12	3072	$\approx 2^{149-\delta_1}$	$\approx 2^{139-\delta_2}$	$\approx 2^{126-\delta_3}$	$\approx 2^{110-\delta_4}$
384	12	4608	$\approx 2^{177-\delta_1}$	$\approx 2^{164-\delta_2}$	$\approx 2^{149-\delta_3}$	$\approx 2^{130-\delta_4}$
448	12	5376	$\approx 2^{189-\delta_1}$	$\approx 2^{175-\delta_2}$	$\approx 2^{159-\delta_3}$	$\approx 2^{139-\delta_4}$
512	12	6144	$\approx 2^{199-\delta_1}$	$\approx 2^{185-\delta_2}$	$\approx 2^{168-\delta_3}$	$\approx 2^{147-\delta_4}$

Note: There is no variant in that case whose complexity is $L_{p^n}[1/3, 1.92]$, the values are given only for comparison.

Example 10.1   We assume that we want to implement a pairing for the AES 128-bit security level. The divisor of $\mathrm{Card}(E(\mathbb{F}_p))$, r, must be composed by at least 256 bits. The size in bits of p^k must be in the range $[3000; 5000]$.

Remark 10.2   Table 10.1 is extracted from the article [17] published in 2010. Currently, several works are performed on the resolution of the discrete logarithm problem over pairing-friendly elliptic curves. For instance, at the present day, a preprint presents new records on the number field sieve algorithm in finite fields $\mathbb{F}_{p^n}$ [24]. Those results would imply a revaluation of the minimal bit size of r and p^k. In the sequel, we use published results for determining the bit size of r and p^k. Obviously, our method can be applied for future bounds that can be achieved by new records of the discrete logarithm problem. For instance, when published, the results in [24] will give new bounds. An approximation of those new bounds is presented in Table 10.2. Be careful, this is a rough and theoretical estimate. This only says that a new low bound on the size of BN-like $\mathbb{F}_{p^{12}}$ to achieve a 128-bit security level would be at least 5376 bits (p of 448 bits). The website devoted to this book will keep updating the security recommendations.

Remark 10.3   To obtain a key size according to a security level, a value δ_i is required, which is not known. The order of magnitude of this δ_i is usually of a dozen. A 4608-bit field (corresponding to Barreto-Naehrig curves to p of 384 bits) might not be large enough to achieve a 128-bit security level [20].

10.3   The Embedding Degree

After the security level, the next parameter we have to find is the embedding degree k. Table 10.1 provides us with potential values of k.

In practice, the value k is often chosen smooth. This property allows the extension field $\mathbb{F}_{p^k}$ to be constructed using tower field extensions. The interest in using tower field extensions is based on an optimization of the arithmetic. In particular, the multiplication over $\mathbb{F}_{p^k}$ can be constructed using intermediate multiplications on the floor of the tower field extension. The most efficient multiplication over $\mathbb{F}'_{p^k}$ is obtained with the use of Karatsuba multiplication. This is possible when k is a power of 2. An alternative efficient arithmetic can be obtained by using Karatsuba and Toom Cook multiplication, which can be done when k is a product of powers of 2 and 3. An important trick when computing a pairing is the elimination of denominators. The denominators are the equations of vertical lines during the computation of the Miller algorithm. As explained in Chapter 3, Section 3.2.3, when 2 divides k, the computation during the Miller algorithm can be improved by the elimination of the denominators during the final exponentiation of the pairing. Obviously, this trick is interesting for pairings including a final exponentiation. Fortunately, the most efficient implementations are obtained for pairings with a final exponentiation. The elimination of the denominator is possible for any degree of twists, not only even ones. In practice, the computation of pairings admitting an odd degree of twist are less efficient than computation of pairings with an even degree of twists [30]. Another vector of optimization is the possibility to use a twisted elliptic curve to $E(\mathbb{F}_{p^k})$. As presented in Chapter 2, Section 2.3.6, there exists a morphism between $E(\mathbb{F}_{p^k})$ and its twisted elliptic curve $E'(\mathbb{F}_{p^{k/d}})$, where d is the degree of the twist and $d \in \{2, 3, 4, 6\}$. The bigger d is, the better it is. Indeed, all the operations over $E(\mathbb{F}_{p^k})$ can be performed over $E(\mathbb{F}_{p^{k/d}})$ by using the map into $E(\mathbb{F}_{p^k})$.

To sum up, the embedding degree should be

- even in order to allow the denominator elimination,
- smooth for an efficient arithmetic over finite fields,
- a product of powers of 2 and 3, admitting 6 as a divisor.

Example 10.2 At the AES 128-bits security level, it is recommended to take k between 12 and 20 for curves with $\rho \approx 1$. The products of powers of 2 and 3 in this range are 12, 16, and 18. For $k = 12$, it is possible to achieve $\rho = 1$ (by using the Barreto-Naehrig family). For $k = 16$ or 18, all the known constructions give $\rho > 1$. The elliptic curve with $k = 12$ should be chosen with a twist of degree 6. The same holds for the curve with $k = 18$. For the curve with $k = 16$, the degree of the twist should be 4. Since the arithmetic over $\mathbb{F}_{p^2}$ is more efficient for the same size p than the arithmetic over $\mathbb{F}_{p^3}$, the value $k = 12$ seems the most accurate for the AES 128-bits security level.

10.4 Construction of the Elliptic Curve

The pairing-friendly elliptic curves that are the most interesting for implementation purposes are obtained from families. The definition of a family of pairing-friendly elliptic curves was introduced by Freeman, Scott, and Teske [17], and is recalled in Chapter 4. A family of pairing-friendly elliptic curves with embedding degree k is given by a triple $(p(x), r(x), t(x))$ of polynomials with coefficients in $\mathbb{Q}$. In this representation, $p(x)$ is the characteristic of the finite field, $r(x)$ a prime factor of $\mathrm{Card}(E(\mathbb{F}_p))$, and $t(x)$ is the trace of the elliptic curve. If x_0 is an integer such that $p(x_0)$ and $r(x_0)$ are prime numbers, then there exists an elliptic curve with embedding degree k and parameters $(p(x_0), r(x_0), t(x_0))$. Such a curve can be efficiently constructed, provided by the fact that the discriminant D of the family is not too large (D is the positive integer defined by $4q(x) - t(x)^2 = Dy(x)^2$, for some $y(x) \in \mathbb{Q}[x]$).

An extensive survey about the known families of pairing-friendly elliptic curves and their constructions is given in [17]. Chapter 4 recalls the most used families. The ρ-value $\rho = \deg q(x)/\deg r(x)$ and the embedding degree of the chosen family should fulfill the conditions described in Sections 10.2 and 10.3. Moreover, in order to ensure that the constructed curve has twists of order greater than 2, the discriminant D of the family should be equal to 1 (for twists of degree 4) or to 3 (for twists of degree 6).

Now, we consider a fixed family $(p(x), r(x), t(x))$ and search for an integer x_0 that gives a prime value for $p(x)$ and $r(x)$. The integer x_0 is implied in the exponent in the Miller loop, in the final exponentiation, and it can have a great impact on the $\mathbb{F}_{p^k}$ arithmetic (see Section 10.7.7 for the example of BN curves). For this reason, x_0 should be chosen as sparse as possible in order to improve the efficiency of the pairing computation.

TABLE 10.3   Cost of Miller's algorithm for various models of elliptic curve.

Model	Doubling	Addition	Mixed addition
Huff	$(k11)\mathbf{m6sS}_k\mathbf{S}_k$	$(k15)\mathbf{mM}_k$	$(k13)\mathbf{mM}_k$
Jacobi quartic	$(k9)\mathbf{m8scM}_k$	–	$(k16)\mathbf{msM}_k$
Edwards	$(k6)\mathbf{m5scM}_k\mathbf{S}_k$	$(k14)\mathbf{mcM}_k$	$(k12)\mathbf{mcM}_k$
Weierstrass J.	$(k1)\mathbf{m11scM}_k\mathbf{S}_k$	–	$(k6)\mathbf{m6sM}_k$

TABLE 10.4   Cost of one step in Miller's algorithm for even embedding degree.

Degree of twist and coordinates	Doubling	Mixed addition
$d = 2$ and $k = 2$ Jacobian	$10\mathbf{s3m1a1S}_k\mathbf{1M}_k$	$6\mathbf{s6m}k\mathbf{m1M}_k$
$d = 2$ and $k > 2$ Jacobian	$11\mathbf{s}(k1)\mathbf{m1a1S1M}_k$	$6\mathbf{s6m}k\mathbf{m1M}_k$
$d = 4$ Jacobian	$(2k/d2)\mathbf{s8s1d}_a\mathbf{1S}_k\mathbf{1M}_k$	$((2k/d)9)\mathbf{m5s1M}_k$
$d = 6$ Projective	$(2k/d)\mathbf{m5s1d}_b\mathbf{1S}_k\mathbf{1M}_k$	$(2k/d9)\mathbf{m2s1M}_k$

Once we have prime values for $p(x)$ and $r(x)$, we have to construct the equation of the elliptic curve. This can be done thanks to the Complex Multiplication (CM for short) method. There exist several models for elliptic curves, but the most efficient computation of pairings are obtained using Weierstrass model: $E : y = x^3 axb$. Table 10.3 recalls the cost of pairing computations for other elliptic curve models. We denote by m (resp. s) the cost of a multiplication (resp. a square) over $\mathbb{F}_p$ and by M_λ (resp. S_λ) the cost of a multiplication (resp. a square) over $\mathbb{F}_{p^\lambda}$. We denote by J. Weierstrasss the short Weiestrass model of an elliptic curve, given in Jacobian coordinates. If $D = 1$ or 3, then it is not necessary to use the CM method. Indeed, any elliptic curve with discriminant $D = 1$ (resp. $D = 3$) has an equation of the shape $E : y = x^3 ax$ (resp. $E : y = x^3 b$) and conversely, any curve having this shape of equation is either the desired curve, or one of its twists.

We begin with a brief discussion on the parameters a and b. Two cases are possible for the parameter a with implications on the pairing computation, either it is 0 or not. When $a = 0$, the pairing computations are the most efficient in projective coordinates, as described in Chapter 3, Section 3.3.3. When $a \neq 0$, the most accurate choice is $a = -3$. In this case, the equation during the Miller computation admits a little optimization compared to $a \neq \{0, -3\}$. This optimization is a factorization in the doubling formulas over E, and in the related line in the Miller loop. The parameter b has occured in the doubling formulas over E. Therefore, it should be chosen such that the multiplication by b is as cheap as possible.

The Jacobian coordinates often provide the most efficient formula for pairing computations. The only exception is when the curve admits a twist of degree 6, in this case, the projective coordinates provide the most efficient implementation. In Table 10.4 we summarize the algebraic complexity of Miller's algorithm, considering the degree of the twist and the system of coordinates.

When choosing the elliptic curve, one must take into consideration the subgroup security problem [6]. Indeed, if the elliptic curve admits at least one subgroup with order smaller than r, then the discrete logarithm problem is easier to solve in this subgroup. The attack path consists then in providing to the pairing computation a point of the elliptic curve belonging to the wrong subgroup. The countermeasure consists of constructing elliptic curves such that the subgroup involved in the pairing computation has orders greater than r.

10.5   Construction of $\mathbb{G}_1$ and $\mathbb{G}_2$

In this section, we present an explicit method to construct the subgroups $\mathbb{G}_1$ and $\mathbb{G}_2$ involved in the optimal Ate pairing computations, as it is one of the most efficient pairing. Furthermore, the same construction is efficient for the computation of the Tate or twisted Ate pairing.

10.5.1   The Subgroup $\mathbb{G}_1$

For the optimal Ate pairing, the elements of $\mathbb{G}_1$ are the points P of order r in $E(\mathbb{F}_{p^k})$, which are eigenvectors with eigenvalues 1 of the Frobenius map (acting as an endomorphism on the $\mathbb{F}_r$-vector space $E(\mathbb{F}_{p^k})[r]$). We can then describe $\mathbb{G}_1$ as $\mathbb{G}_1 = E(\mathbb{F}_{p^k})[r] \cap \mathrm{Ker}(\pi_p - [1])$, which can be written $\mathbb{G}_1 = \{P \in E(\mathbb{F}_{p^k}), [r]P = 0_E \text{ and } \pi_p(P) = P\}$.

Since $\pi_p(P) = P$ if and only if $P \in E(\mathbb{F}_p)$, we are just searching for points of order r in $E(\mathbb{F}_p)$. Such a point can be constructed in the following way:

- choose a random point P in $E(\mathbb{F}_p)$,
- compute $P' = \frac{\mathrm{Card}(E(\mathbb{F}_p))}{r} \times P$,
- if $P' = 0_E$, choose another P,
- if $P' \neq 0_E$, then P' is a generator of $\mathbb{G}_1$. We denote $P = P'$.

Remark 10.4   The construction of $\mathbb{G}_1$ is the same for every possible pairing.

10.5.2   The Subgroup G_2 with Twist of Degree d over $\mathbb{F}_{p^k}$

For the optimal Ate pairing, the elements of $\mathbb{G}_2$ are the points Q of order r in $E(\mathbb{F}_{p^k})$ which are eigenvectors with eigenvalues p of the Frobenius. We can then describe $\mathbb{G}_2$ as $\mathbb{G}_2 = E(\mathbb{F}_{p^k})[r] \cap \mathrm{Ker}(\pi_p - [p])$, which can be written $\mathbb{G}_2 = \{Q \in E(\mathbb{F}_{p^k}), [r]Q = 0_E \text{ and } \pi_p(Q) = [p]Q\}$.

When $E(\mathbb{F}_{p^k})$ admit a twist of degree d, there is an efficient way of constructing $\mathbb{G}_2$ [23].

In Chapter 2, the definition of a twisted elliptic curve is given. Twists are useful as they allow a simplified representation of $\mathbb{G}_2$ if d divides k.

DEFINITION 10.1   Let E and E' be two elliptic curves defined over $\mathbb{F}_q$, for q a power of a prime number p. Then the curve E' is a twist of degree d of E if there exists an isomorphism Ψ_d defined over $\mathbb{F}_{q^d}$ from E' into E, and such that d is minimal.

The possible number of twists for a given elliptic curve is bounded. It depends on the group of endomorphisms of the elliptic curve E. Theorem 2.12 gives the classification of the potential twists. In Chapter 2, Section 2.3.6, they describe the possible equations and cardinality of twisted elliptic curves. The possible twists are of degree $d = 2$, 3, 4, or 6.

We can compute the cardinal of a twisted elliptic curve according to the degree of the twist as described in Section 2.3.6.

We can then provide an efficient representation for $\mathbb{G}_2$. Let E be an elliptic curve admitting a twist of degree d. Let $z = \gcd(k, d)$ and $e = k/z$. Since $k > 1$ is the embedding degree of E relative to r, we have that r divides $\mathrm{Card}(E(\mathbb{F}_{p^e}))$ but r^2 does not. As $r > 6$, there is a unique degree z twist E' such that r divides $\mathrm{Card}(E'(\mathbb{F}_{p^e}))$ [23]. The cardinality of a twisted elliptic curve can be computed as illustrated in Property 2.5.

PROPOSITION 10.1 ([23])   *Let $\mathbb{G}_2'$ be the unique subgroup of order r of $E'(\mathbb{F}_{p^e})$ and let $\Phi_z : E' \to E$ be the twisting isomorphism, then*

$$\mathbb{G}_2 = \Phi_z(\mathbb{G}_2').$$

As a consequence, in order to construct $\mathbb{G}_2$, we first construct a point of order r on the twisted elliptic curve $E'(\mathbb{F}_{p^{k/d}})$. Then, we use the map between $E'(\mathbb{F}_{p^{k/d}})$ and $E(\mathbb{F}_{p^k})$ to find a generator of $\mathbb{G}_2$. As who can do more can do less, the group $\mathbb{G}_2$ generated in this way is suitable for the computation of the Weil or Tate pairing.

10.5.3 The Subgroup $\mathbb{G}_2$ without Twist over $\mathbb{F}_{p^k}$

For the optimal Ate pairing, elements in $\mathbb{G}_2$ are elements of $E(\mathbb{F}_{p^k})$ of order r and eigenvectors with eigenvalues p of the Frobenius. The subgroup $\mathbb{G}_2$ can be described as $\mathbb{G}_2 = \{Q \in E(\mathbb{F}_{p^k}), [r]Q = 0_E$ and $\pi_p(Q) = [p]Q\}$.

The first step is to find a point of order r in $E(\mathbb{F}_{p^k})$ independent from the point $P \in \mathbb{G}_1$.

- Choose a random point Q in $E(\mathbb{F}_{p^k})$.
- Compute $Q' = \dfrac{\mathrm{Card}(E(\mathbb{F}_{p^k}))}{r}Q$.
- If $Q' = 0_E$, choose another point Q,
- If $Q' \neq 0_E$ and all the coordinates of Q' are in $\mathbb{F}_p$, choose another point Q,
- Else, Q' is a valuable candidate.

Remark 10.5 The point Q' constructed above is a suitable candidate for the computation of the Tate pairing.

Now we have to construct a p-eigenvector of the Frobenius. We present below a general construction that does not use a twist. We know that $E[r] \cong \mathbb{Z}/r\mathbb{Z} \times \mathbb{Z}/r\mathbb{Z}$, consequently we can define a basis of $E[r]$ to be (P, Q').

By construction, $P \in E(\mathbb{F}_p)$ and $Q' \in E(\mathbb{F}_{p^k})$. In order to construct the matrix of the Frobenius on this basis we have to compute the Frobenius of P and Q'. First of all, $\pi_p(P) = P$ by construction. In the basis (P, Q') of $E[r]$, knowing that $Q' \in E[r]$ we have that $\pi_p(Q') = \alpha P \beta Q'$, for α and β integers. We have $\#E(\mathbb{F}_p) = p1 - t$, where t is the trace of the Frobenius. Modulo the cardinal of the elliptic curve, then we have that $\beta = p$. Indeed, the trace of the matrix should be $t = 1\beta$ and we have that $t \equiv 1p \mod (\#E(\mathbb{F}_p))$. Consequently, the matrix of the Frobenius in the basis (P, Q') is the following:

$$\begin{bmatrix} 1 & \alpha \\ 0 & p \end{bmatrix}.$$

In order to obtain the value of α, we can use the fact that $\pi_p(Q') = [\alpha]P[p]Q'$ to write $\pi_p(Q') - [p]Q' = [\alpha]P$. With the knowledge of P and Q', finding α is equivalent to solving the discrete logarithm problem. But, we cannot, as we are working over a subgroup where the discrete logarithm problem is hard. However, we can consider that P is in fact αP. By abuse of notation, we will now consider that P is constructed as $\pi_p(Q') - [p]Q'$. The point P constructed like that is of order r and we verify that $\pi_p(P) = P$. Furthermore, the couple (P, Q') is still a basis of $E(\mathbb{F}_{p^k})$. In this basis the matrix of the Frobenius is

$$\begin{bmatrix} 1 & 1 \\ 0 & p \end{bmatrix}.$$

Now, we have to construct a p-eigenvector of the Frobenius. We are looking for $Q = c_1 P c_2 Q'$ such that $\pi_p(Q) = pQ$. By simplification, we obtain the following equations:

$$\begin{aligned}
Q &= c_1 P c_2 Q' \\
\pi_p(Q) &= \pi_p(c_1 P c_2 Q') \\
&= c_1 \pi_p(P) c_2 \pi_p Q' \quad \text{(because } \pi_p \text{ is a morphism of the group)} \\
&= c_1 P c_2 (P p Q') \\
&= (c_1 c_2) P (c_2 p) Q'.
\end{aligned}$$

Modulo r, we obtain the equality:

$$c_1 c_2 = c_1 p.$$

TABLE 10.5　Cost of the steps in Miller's algorithm for elliptic curves admitting a twist of degree d.

Coordinates	$f_{-,P}(Q)$ Doubling	Addition
Affine	$2s(1k/d)m1i1S1M$	$(1k/d)m1s1i1M_k$
Projective	$7s(2k/d2)m1S_k1M_k$	$(2k/d12)m3s1S_k1M_k$
Jacobian	$11s(2k/d1)m1a1S_k1M_k$	$6s(k/d6)m1M_k$

Coordinates	$f_{-,Q}(P)$ Doubling	Addition
Affine	$2S_{k/d}2M_{k/d}1I_{k/d}1S_k1M_k$	$2M_{k/d}1S_{k/d}1I_{k/d}1M_k$
Projective	$2k/dm6S_{k/d}2M_{k/d}1S_k1M_k$	$2k/dm2S_{k/d}12M_{k/d}1S_k1M_k$
Jacobian	$2k/dm11S_{k/d}1M_{k/d}1S_kM_k$	$k/dm6S_{k/d}1S_k6M_{k/d}1M_k$

As we know the value of p, for a fixed c_1 only one value is possible for c_2. The easiest possibility is $c_1 = 1$ and $c_2 = p - 1$.

Considering the explanations above, when constructing $\mathbb{G}_2$, we also construct $\mathbb{G}_1$.

- We construct a point Q' of order r with coordinates in $\mathbb{F}_{p^k}$.
- We compute $P = \pi_p(Q') - pQ'$, the order of this point is r, and if $P \in E(\mathbb{F}_p)$, then P is a generator of $\mathbb{G}_1$.
- We compute $Q = pP(p-1)pQ'$, which is a point of order r in $E(\mathbb{F}_{p^k})$ and verify that $\pi(Q) = pQ$.

Once we have generators of $\mathbb{G}_1$ and $\mathbb{G}_2$, all the possible points in $\mathbb{G}_1$ and $\mathbb{G}_2$ are multiples of the generators.

10.6　Construction of the Pairing

Let us make a summary. We have the security level, the embedding degree, the elliptic curve. We have now to choose which pairing should be implemented. In practice the most efficient pairing belongs to the set of optimal Ate, optimal twisted Ate, and pairing lattices. As presented in Chapter 3, the optimal pairing and pairing lattices are powers of the Tate pairing. It is possible to calculate the algebraic complexity of the step of Miller's algorithm according to the position of $P \in \mathbb{G}_1$ and $Q \in \mathbb{G}_2$, for a given system of coordinates.

Table 10.5 recalls the algebraic cost of the doubling and addition step during Miller's algorithm for a computation of pairing, considering an even embedding degree and a twist of degree d. Considering an even value for k allows us to use the denominator elimination. We do not consider any other optimization as they are related to the equation of the elliptic curves.

The choice of the pairing optimal Ate or twisted Ate depends on the number of iterations and the number of addition steps that must be executed. In Chapter 3, methods to compute the number of iterations are more detailed. Theorem 10.1 presents the formulae for the construction of the optimal Ate in Equation 10.1 and optimal twisted ate in Equation 10.2.

THEOREM 10.1　*[36, 21] Let E be an elliptic curve defined over $\mathbb{F}_p$, r a large prime with $r|\#E(\mathbb{F}_p)$, k the embedding degree, and t the trace of the Frobenius.*

(a) *Let $\lambda = \sum_{i=0}^{\phi(k)-1} c_i p^i$ such that $\lambda = mr$, for some integer m. Then $a_{[c_0,...,c_l]} : \mathbb{G}_2 \times \mathbb{G}_1 \to \mu_r$ defined as*

$$(Q,P) \to \left(\prod_{i=0}^{\phi(k)-1} f_{c_i,Q}(P) \cdot \prod_{i=0}^{\phi(k)-1} \frac{l_{s_{i1}Q,c_ip^iQ}(P)}{v_{s_iQ}(P)} \right)^{(p^k-1)/r} \tag{10.1}$$

TABLE 10.6 Existing pairings

Pairing	Weil	Tate	Ate	Twisted Ate
Definition	$\dfrac{f_{r,P}(Q)}{f_{r,Q}(P)}$	$(f_{r,P}(Q))^{(p^k-1)/r}$	$(f_{t-1,Q}(P))^{(p^k-1)/r}$	$\left(f_{(t-1)^{k/d},P}(Q)\right)^{(p^k-1)/r}$

with $s_i = \sum_{j=i}^{\phi(k)-1} c_j p^j$ *defines a bilinear pairing. This pairing is non-degenerate if and only if* $mkp^{k-1} \neq ((p^k - 1)/r) \sum_{i=0}^{\phi(k)-1} i c_i p^{i-1} \pmod{r}$.

(b) *Assume that E has a twist of degree d and set $n = gcd(k, d)$ and $e = k/n$. Let $\lambda = \sum_{i=0}^{\phi(k)/e-1} c_i p^{ie}$ such that $\lambda = mr$, for some integer m. Then $a_{[c_0,...,c_l]} : \mathbb{G}_1 \times \mathbb{G}_2 \to \mu_r$ defined as*

$$(P,Q) \to \left(\prod_{i=0}^{\phi(k)/e-1} f_{c_i,P}(Q) \cdot \prod_{i=0}^{\phi(k)/e-1} \frac{l_{s_{i1}P,c_i p^i P}(Q)}{v_{s_i P}(Q)} \right)^{(p^k-1)/r} \qquad (10.2)$$

with $s_i = \sum_{j=i}^{\phi(k)/e-1} c_j p^j$ *defines a bilinear pairing. This pairing is non-degenerate if and only if* $mkp^{k-1} \neq ((q^k - 1)/r) \sum_{i=0}^{\phi(k)/e-1} i c_i p^{e(i-1)} \pmod{r}$.

The number of iterations for the optimal Ate pairing over $E(\mathbb{F}_p)$ is computed through the LLL reduction of the lattice Λ [36]. The reduction of this lattices provides a short vector from which we can extract the polynomial s_i.

$$\Lambda = \begin{bmatrix} r & 0 & \cdots & \cdots & 0 \\ -p & 1 & 0 & \cdots & 0 \\ -p^2 & 0 & 1 & 0 & 0 \\ \vdots & & 0 & 0 & \ddots & 0 \\ -p^{(\varphi(k)-1)} & 0 & \cdots & 0 & 1 \end{bmatrix}.$$

Choosing the right pairing

Assume that we have defined an elliptic curve $E(\mathbb{F}_p)$, with r a prime divisor of Card(E) and with embedding degree k relative to r. Let $P \in \mathbb{G}_1$ and $Q \in \mathbb{G}_2$. Then one can compute the Weil pairing $e_{r,W}(P,Q)$, the Tate pairing $e_{r,T}(P,Q)$, the Ate (or Optimal Ate) pairing $a_{s,T}(P,Q)$, the (optimal) twisted Ate pairing $a'_{s,T}(P,Q)$, for s a given integer. The Weil pairing requires us to compute both $f_{r,P}(Q)$ and $f_{r,Q}(P)$ so the Miller loop will be much more expensive than for the Tate pairing, which only requires $f_{r,P}(Q)$ (especially since $Q \in \mathbb{G}_2$ constructing $f_{r,Q}$ is expensive). Furthermore, when k is even we can apply denominator elimination for the Tate pairing because of the final exponentiation, so even the computation of $f_{r,P}(Q)$ is faster than the one for the Weil pairing. However, one should remember that for the Tate or Ate pairing on elliptic curves, the final exponentiation may be expensive, too. Indeed, the loop length of the final exponentiation is around $k \log q$ compared to $\log q$ for the Miller step. So, the implementation of the final exponentiation step should not be neglected, as described in Chapter 7. We recall the definition of the Weil, Tate, Ate, and twisted Ate pairing in Table 10.6. The integer t is the trace of the elliptic curve $E(\mathbb{F}_p)$. The optimal Ate and twisted Ate are constructed on the same scheme as the Ate and twisted Ate pairings. The difference is the number of iterations for the Miller algorithm.

It is harder to choose between the Tate and Ate (or optimal Ate) pairing. Usually for the Ate pairing, the cost of computing the Miller functions over $\mathbb{F}_{q^k}$ does not compensate the shortened Miller loop, but this can change when a high degree twist is available. For instance, Barreto-Naehrig curves have a twist of degree 6, so the Optimal Ate pairing is well suited for these curves.

For higher security than 128 bits, one will need curves with a higher embedding degree, so the Tate or twisted Ate may be faster for these curves. In this book we will explore several families and give optimized algorithms for each of them. See also [16], which compares the different versions of the pairings.

Remark 10.6 The choice of k is expected to give the most efficient arithmetic. For the AES 128-bits security level, the implementations of pairings is indeed the most efficient using the BN curves with $k = 12$. But, according to the work [3], the most efficient implementation is not obtained for the expected value of k. In the work [3], the security level is AES 192. According to Table 10.1, k should be chosen between 20 and 26 for $\rho \approx 1$. Among the possible choices, the value $k = 24$ seems to be the most suitable. Indeed, $\rho \approx 1$, 6 divides 24 and the denominator elimination is possible. However, experimentations surprisingly show that it would be more efficient to consider a family of curves with $k = 12$ and $\rho \approx 2$ than $k = 24$ with $\rho \approx 1$. That is to say that when considering pairing implementation, one should compare all the possible options before concluding. It is not sufficient to choose what seems to be the most adapted value of k minimizing ρ. All the pairing computation aspects must be taken into consideration.

10.7 Example of the BN-Curve

In the following, we made the choice for the security level AES 128 bits. For this security level, the BN curves have been demonstrated to allow the most efficient implementation of pairings. We describe step by step how we generate parameters for the computation of a pairing over BN curves. The method we describe can easily be adapted for any other family of pairing-friendly elliptic curves that can be constructed. See Chapter 4 for other families of curves.

10.7.1 BN Curves

A Barreto-Naehrig (BN) curve [8] is an elliptic curve E over a finite field $\mathbb{F}_p$, $p \geq 5$, with order $r = \#E(\mathbb{F}_p)$, such that p and r are prime numbers given by

$$
\begin{aligned}
p(x_0) &= 36x_0^4 36x_0^3 24x_0^2 6x_0 1, \\
r(x_0) &= 36x_0^4 36x_0^3 18x_0^2 6x_0 1,
\end{aligned}
$$

for some x_0 in $\mathbb{Z}$. It has an equation of the form

$$y^2 = x^3 b,$$

where $b \in \mathbb{F}_p^*$. Its neutral element is denoted by 0_E.

BN curves have been designed to have an embedding degree equal to 12. This makes them particularly appropriate for the 128-bit security level. Indeed, a prime p of size 256 bits leads to a BN curve whose group order is roughly 256 bits together with pairings taking values in $\mathbb{F}_{p^{12}}^*$, which is a 3072-bit multiplicative group. According to the NIST recommendations [1], both groups involved are matching the 128-bit security level. As a consequence, BN curves at this security level have been the object of numerous recent publications ([13, 4, 9, 33, 31, 19, 35]). They prove that the most efficient pairing in this case is the optimal Ate pairing.

10.7.2 Construction of the Twist

Finally, BN curves always have degree 6 twists. If ξ is an element that is neither a square nor a cube in $\mathbb{F}_{p^2}$, the twisted curve E' of E is defined over $\mathbb{F}_{p^2}$ by the equation

$$E' : y^2 = x^3 b',$$

with $b' = b/\xi$ or $b' = b\xi$. The choice of b' is related to the equation of the elliptic curve, as described in [33, Theorem 1].

In order to simplify the computations, the element ξ should also be used to represent $\mathbb{F}_{p^{12}}$ as a degree 6 extension of $\mathbb{F}_{p^2}$ ($\mathbb{F}_{p^{12}} = \mathbb{F}_{p^2}[\gamma]$ with $\gamma^6 = \xi$) [13], [27]. In this paper, we deal only with the case $b' = b/\xi$ as is usually done in the literature, but $b' = b/\xi^5$ can also be used with a very small additional cost [19].

As BN curves have twists of order 6, the twisted version of the optimal Ate pairing allows us to take Q in $E'\left(\mathbb{F}_{p^2}\right)$. Using the isomorphism between the curve and its twist, the point Q in our definition can then be chosen in the form $\left(x_Q\gamma^2, y_Q\gamma^3\right) \in E\left(\mathbb{F}_{p^{12}}\right)$ where $x_Q, y_Q \in \mathbb{F}_{p^2}$ ($(x_Q, y_Q) \in E'\left(\mathbb{F}_{p^2}\right)$). This means that the elliptic curve operations lie in $\mathbb{F}_{p^2}$ instead of $\mathbb{F}_{p^{12}}$ (but the result remains in $\mathbb{F}_{p^{12}}$), during the computation of the Ate pairing or any optimization of the Ate pairing. This makes computations of course easier but this also allows denominator elimination as in [7] because all the factors lying in a proper subfield of $\mathbb{F}_{p^{12}}$ (as $\mathbb{F}_{p^2}$) are wiped out by the final exponentiation. See Chapter 3 for the details of the denominator elimination.

10.7.3 Optimal Ate Pairing

Considering the BN curves, it has been proven in [36] that the shortest possible loop has length $r/\varphi(12) = r/4$ and that this length is reached by the so-called optimal Ate pairing.

Let $\pi(x, y) = (x^p, y^p)$ be the Frobenius map on the curve. If P is a rational point on E and Q is a point in $E\left(\mathbb{F}_{p^{12}}\right)$ that is in the p-eigenspace of π, the optimal Ate pairing [31] can be defined by

$$a_{opt}(Q, P) = \left(f_{v,Q}(P).\ell_{vQ,\pi(Q)}(P).\ell_{vQ\pi(Q),-\pi^2(Q)}(P)\right)^{\frac{p^{12}-1}{r}},$$

where $v = 6x_0 2$ and $\ell_{A,B}$ is the normalized line function arising in the sum of the points A and B.

In this study, we are only considering this pairing because it leaves no doubt that it is currently the most efficient for BN curves, but the same work can easily be done with other pairings. The computation of the optimal Ate pairing is done in four steps:

1. A Miller loop to compute $f_{|v|,Q}(P)$. The algorithmic choices for this step are discussed in Section 10.7.4.

2. If $v < 0$, the result f of the Miller loop must be inverted to recover $f_{v,Q}(P)$. Such an inversion is potentially expensive, but thanks to the final exponentiation, f^{-1} can be replaced by f^{p^6} [4], which is nothing but the conjugation in $\mathbb{F}_{p^{12}}/\mathbb{F}_{p^6}$; thus it is done for free.

3. Two line computations, $\ell_{vQ,\pi(Q)}(P)$ and $\ell_{vQ\pi(Q),-\pi^2(Q)}(P)$, which are nothing but extra addition steps of the Miller loop.

4. A final exponentiation to the power of $\frac{p^{12}-1}{r}$. The algorithmic choices for this step are discussed in Chapter 7.

10.7.4 Miller Algorithm

The first step of the pairing computation evaluates $f_{|v|,Q}(P)$ thanks to the Miller algorithm presented in Algorithm 3.3, which was presented in Chapter 3 and introduced in [29]. It is based on the double-and-add scheme used for the computation of $|v|Q$ by evaluating at P the lines occurring in the doubling and addition steps of this computation.

Several choices are possible for the system of coordinates in order to perform the operations over the elliptic curve during the Miller loop. We discuss them in Section 10.7.9.

Since the Miller algorithm is based on the double-and-add algorithm, it is natural to try to improve it by using advanced exponentiation techniques like the sliding window method [10,

Algorithm 9.10] or the NAF representation [10, Algorithm 9.14]. However, the interest is limited in practice for two reasons:

- In the context of pairing-based cryptography, the exponent is not a secret. Then it is usually chosen sparse, so these advanced exponentiation methods are useless.
- Such methods involve operations like $T \leftarrow T3Q$. We need to compute $f \leftarrow f \times f_{3,Q} \times \ell_{T,3Q}$ to obtain the corresponding function. Of course, $f_{3,Q}$ can be precomputed but such a step requires an additional $\mathbb{F}_{p^{12}}$ multiplication, which is the most consuming operation in Algorithm 3.3.

The only interesting case is a signed binary representation of the exponent (i.e., a 2-NAF) because it can help to find a sparse exponent. In this case, the subtraction step of Algorithm 3.3 involves an additional division by the vertical line passing through Q and $-Q$, which could be expensive, but fortunately it is wiped out by the final exponentiation if Q comes from the twisted curve.

10.7.5   Final Exponentiation

We refer to Chapter 7 for the explanation, of the simplification of the final exponentiation, which corresponds to raising the result of Miller's algorithm at the power $\frac{p^k-1}{r}$. When $k = 12$, the final exponentiation can be decomposed by $\frac{p^{12}-1}{r} = (p^6-1)(p^2 1)\frac{p^4-p^2 1}{r}$. The most popular way to perform the hard part $\frac{p^4-p^2 1}{r}$ uses the following addition chain [34]:

$$f^{\frac{p^4-p^2 1}{r}} = y_0 y_1^2 y_2^6 y_3^{12} y_4^{18} y_5^{30} y_6^{36},$$

where
$$y_0 = f^p f^{p^2} f^{p^3}, \quad y_1 = \frac{1}{f}, \quad y_2 = \left(f^{x_0^2}\right)^{p^2}, \quad y_3 = (f^u)^p,$$

$$y_4 = \frac{\left(f^{u^2}\right)^p}{f^u}, \quad y_5 = \frac{1}{f^{u^2}}, \quad y_6 = \frac{\left(f^{u^3}\right)^p}{f^{u^3}}.$$

The cost of this method is $13\mathbf{M}_{12}, 4\mathbf{S}_{12}$ and 7 Frobenius maps, in addition to the cost of 3 exponentiations by x_0. The method given in [18] is slightly more efficient but computes a power of the optimal Ate pairing. The main drawback of these methods is that they are memory consuming (up to 4Ko), which can be annoying in restricted environments. Some variants of these methods optimized in terms of memory consumption are given in [14].

10.7.6   Arithmetic for Optimal Ate Pairing over BN-Curve

The computation over BN-curves involves the arithmetic of finite fields $\mathbb{F}_p$, $\mathbb{F}_{p^2}$, $\mathbb{F}_{p^4}$, $\mathbb{F}_{p^6}$, and $\mathbb{F}_{p^{12}}$. The construction of the extension fields influences the algebraic complexity of arithmetical operations. We consider that the arithmetical operations over $\mathbb{F}_p$ are natively implemented, which is often the case in practice. In [15], the authors give advice on the construction of the tower field extensions and on the finite fields arithmetic. We only recall their results in the following Tables 10.7–10.12.

The following notations for $\mathbb{F}_{p^i}$ arithmetic will be used:

- An addition is denoted by $\mathbf{a}_i$ and a multiplication by 2 by $\mathbf{A}'_i$.
- A multiplication is denoted by $\mathbf{M}_i$ (and $\mathbf{M}_i^M$ if the method M is used).
- A sparse multiplication is denoted by $\mathbf{sM}_i$.
- A multiplication by the constant c is denoted by $\mathbf{m}_{i,c}$.

TABLE 10.7   Cost of the arithmetic of extension of degree 2.

Operation	Cost
Addition	$\mathbf{a}_{2i} = 2\mathbf{a}_i$
Doubling	$\mathbf{A'}_{2i} = 2\mathbf{A'}_i$
Multiplication, schoolbook	$\mathbf{M}_{2i}^{SB} = 4\mathbf{M}_i\,\mathbf{m}_{i,\mu}2\mathbf{a}_i$
Multiplication, Karatsuba	$\mathbf{M}_{2i}^{K} = 3\mathbf{M}_i\,\mathbf{m}_{i,\mu}5\mathbf{a}_i$
Squaring, schoolbook	$\mathbf{S}_{2i}^{SB} = \mathbf{M}_i\,2\mathbf{S}_i\,\mathbf{m}_{i,\mu}\mathbf{a}_i\,\mathbf{A'}_i$
Squaring, Karatsuba	$\mathbf{S}_{2i}^{K} = 3\mathbf{S}_i\,\mathbf{m}_{i,\mu}4\mathbf{a}_i$
Inversion, norm	$\mathbf{I}_{2i} = \mathbf{I}_i\,2\mathbf{M}_i\,2\mathbf{S}_i\,\mathbf{a}_i\,\mathbf{m}_{i,\mu}$

- A squaring is denoted by $\mathbf{S}_i$ (and $\mathbf{S}_i^M$ if the method M is used).
- An inversion is denoted by $\mathbf{I}_i$.

Extension of degree 2

In theory, any irreducible polynomial can be used to build $\mathbb{F}_{p^{2i}}$ over $\mathbb{F}_{p^i}$ but non-zero coefficients of this polynomial imply extra operations for $\mathbb{F}_{p^{2i}}$ arithmetic. So $\mathbb{F}_{p^{2i}}$ is usually built with a polynomial in the form $X^2 - \mu$ where μ is not a square in $\mathbb{F}_{p^i}$.

$$\mathbb{F}_{p^{2i}} = \mathbb{F}_{p^i}[\alpha] \text{ with } \alpha^2 = \mu.$$

Table 10.7 gives recommendations for the arithmetic of degree 2 extensions.

Remark 10.7

- The schoolbook method should be preferred to the Karatsuba one while $3\mathbf{a}_i > \mathbf{M}_i$.
- Determining which is the best method is not so easy as for the multiplication in the general case because it depends on both the relative cost of $\mathbf{M}_i$ and $\mathbf{a}_i$ and of $\mathbf{M}_i$ and $\mathbf{S}_i$.

Extension of degree 3

As in Section 10.7.6, it is preferable to choose a sparse polynomial to minimize the cost of $\mathbb{F}_{p^{3i}}$ arithmetic. Thus, $\mathbb{F}_{p^{3i}}$ is built as $\mathbb{F}_{p^i}[\alpha]$ where $\alpha^3 = \xi$ for some ξ in $\mathbb{F}_{p^i}$, which is not a cube. Of course $\mathbb{F}_{p^{3i}}$ arithmetic will involve some multiplications by ξ so that ξ must be chosen carefully.
Table 10.8 resumes the cost of the arithmetic of an extension of degree 3.

Building $\mathbb{F}_{p^{12}}$

In this section, we discuss the ways to build the extension tower $\mathbb{F}_{p^{12}}$ for pairings on BN curves. All the ways to build $\mathbb{F}_{p^{12}}$ are mathematically equivalent. However, we will use this extension in the specific case of pairings on BN curves, which implies some constraints in order to be compatible with other improvements of pairing computations.

- In order to use the sextic twist, $\mathbb{F}_{p^{12}}$ must be built as an extension of $\mathbb{F}_{p^2}$.
- $\mathbb{F}_{p^{12}}$ must be built over $\mathbb{F}_{p^2}$ thanks to a polynomial $X^6 - \xi$ where ξ, which is neither a square nor a cube, is the element used to defined the twisted curve. This allows the line involved in the Miller algorithm to be a sparse element of $\mathbb{F}_{p^{12}}$ (see Section 10.7.9 for more details).

Then, $\mathbb{F}_{p^{12}}$ should be built:

- Case $2, 2, 3$: as a cubic extension of a quadratic extension of $\mathbb{F}_{p^2}$,
- Case $2, 3, 2$: as a quadratic extension of a cubic extension of $\mathbb{F}_{p^2}$,

TABLE 10.8   Cost of the arithmetic of extension of degree 3.

Operation	Cost
Addition	$\mathbf{a}_{3i} = 3\mathbf{a}_i$
Doubling	$\mathbf{A'}_{3i} = 3\mathbf{A'}_i$
Multiplication, schoolbook	$\mathbf{M}_{3i}^{SB} = 9\mathbf{M}_i\,2\mathbf{m}_{i,\xi}6\mathbf{a}_i$
Multiplication, Karatsuba	$\mathbf{M}_{3i}^{K} = 6\mathbf{M}_i\,15\mathbf{a}_i\,2\mathbf{m}_{i,\xi}$
Squaring, schoolbook	$\mathbf{S}_{3i}^{SB} = 3\mathbf{M}_i\,3\mathbf{S}_i\,3\mathbf{a}_i\,2\mathbf{A'}_i\,2\mathbf{m}_{i,\xi}$
Squaring, Karatsuba	$\mathbf{S}_{3i}^{K} = 6\mathbf{S}_i\,12\mathbf{a}_i\,2\mathbf{m}_{i,\xi}$
Squaring, Chung-Hasan	$\mathbf{S}_{3i}^{CH} = 2\mathbf{M}_i\,3\mathbf{S}_i\,8\mathbf{a}_i\,\mathbf{A'}_i\,2\mathbf{m}_{i,\xi}$

TABLE 10.9   Cost of the arithmetic of extension of degree 6.

Operation	Addition	Doubling	Multiplication, Karatsuba	Squaring, Chung-Hasan
Cost	$\mathbf{a}_6 = 6\mathbf{a}_1$	$\mathbf{A'}_6 = 6\mathbf{A'}_1$	$\mathbf{M}_6 = 6\mathbf{M}_2\,15\mathbf{a}_2\,2\mathbf{m}_{2,\xi}$	$\mathbf{S}_6 = 2\mathbf{M}_2\,3\mathbf{S}_2\,8\mathbf{a}_2\,\mathbf{A'}_2\,2\mathbf{m}_{2,\xi}$

- Case $2, 6$: as a sextic extension of $\mathbb{F}_{p^2}$.

The latter case is proved to be less efficient in [12], so in [15], the authors only consider the first two. In any case, we have

$$\mathbb{F}_{p^{12}} = \mathbb{F}_{p^2}[\gamma] \text{ with } \gamma^6 = \xi \in \mathbb{F}_{p^2}.$$

In the case $2, 2, 3$, we will use $\beta = \gamma^3$ to define $\mathbb{F}_{p^4}$, and in the case $2, 3, 2$, we will use $\beta = \gamma^2$ to define $\mathbb{F}_{p^6}$. Of course $\mathbb{F}_{p^{12}}$ arithmetic will involve some multiplications by ξ or β so that ξ must be chosen carefully.

Case 2, 3, 2

We assume in this case that $\mathbb{F}_{p^{12}}$ is built over $\mathbb{F}_{p^2}$ via $\mathbb{F}_{p^6}$, and thanks to some ξ that is neither a square nor a cube in $\mathbb{F}_{p^2}$.

$$\mathbb{F}_{p^6} = \mathbb{F}_{p^2}[\beta] \text{ where } \beta^3 = \xi \text{ and } \mathbb{F}_{p^{12}} = \mathbb{F}_{p^6}[\gamma] \text{ with } \gamma^2 = \beta.$$

The different costs for the arithmetic over $\mathbb{F}_{p^6}$ are resumed in Table 10.9.
Table 10.10 resumes the cost of the arithmetic over $\mathbb{F}_{p^{12}}$ in the case of the tower field $2, 3, 2$.

Case 2, 2, 3

We assume in this case that $\mathbb{F}_{p^{12}}$ is built over $\mathbb{F}_{p^2}$ via $\mathbb{F}_{p^4}$, and thanks to some ξ that is neither a square nor a cube in $\mathbb{F}_{p^2}$.

$$\mathbb{F}_{p^4} = \mathbb{F}_{p^2}[\beta] \text{ where } \beta^2 = \xi \text{ and } \mathbb{F}_{p^{12}} = \mathbb{F}_{p^4}[\gamma] \text{ with } \gamma^3 = \beta.$$

Table 10.11 gives the complexity of arithmetical operations over $\mathbb{F}_{p^4}$.

TABLE 10.10 Cost of the arithmetic of extension of degree 12, case 2, 3, 2.

Operation	Cost
Addition	$a_{12} = 2a_6 = 12a_1$
Doubling	$A'_{12} = 2A'_6 = 12A'_1$
Multiplication, Karatsuba	$M_{12} = 18M_2 60a_2 7m_{2,\xi}$
Sparse multiplication (Miller line)	$sM_{12} = 13M_2 25a_2 3m_{2,\xi}$
Squaring, Karatsuba	$S_{12} = 6M_2 9S_2 36a_2 3A'_2 7m_{2,\xi}$
Squaring, complex method	$S_{12} = 12M_2 42a_2 3A'_2 6m_{2,\xi}$

TABLE 10.11 Complexities of S_4 depending on the context.

μ	condition	method	complexity
	assuming $a_1 \leq 0.33M_1$		
$-1, -2$ or -5		K	$3S_2 m_{2,\xi} 4a_2$
any		$\mathbb{C}$	$2M_2 m_{2,\xi} m_{2,\xi} 1 3a_2 A'_2$
	assuming $a_1 > 0.33M_1$		
-1 or -5		K	$3S_2 m_{2,\xi} 4a_2$
-2	$S_1 = 0.8M_1$		
any		SB	$M_2 2S_2 m_{2,\xi} a_2 A'_2$
-2 or any	$S_1 = M_1$	$\mathbb{C}$	$2M_2 m_{2,\xi} m_{2,\xi} 1 3a_2 A'_2$

$\mathbb{F}_{p^{12}}$ arithmetic

Table 10.12 presents the cost of the arithmetic of $\mathbb{F}_{p^{12}}$ in the case of the tower field 2, 2, 3.

10.7.7 Choosing x_0

The parameter x_0 is involved at several levels of the pairing computation, so that the best choice is not trivial to do. Let us summarize the constraints on x_0 that we have to deal with in order to make a good choice.

- The parameter x_0 defines the security level. Indeed, it is both parametrizing the size of the elliptic curve (whose prime order is $36x_0^4 36x_0^3 18x_0^2 6x_0 1$) and the number of elements of the target finite field (which is $(36x_0^4 36x_0^3 24x_0^2 6x_0 1)^{12}$).

- It is involved as an exponent in the Miller loop. More precisely, in the case of an optimal Ate pairing, the exponent of the Miller loop is $6x_0 2$. In order to optimize this step, x_0 should be chosen such that $6x_0 2$ is sparse.

- It is involved as an exponent in the final exponentiation; see Chapter 7. If the addition chain given in Section 10.7.5 is used, x_0 is directly used (three times) as an exponent, so it should be sparse to ensure a fast final exponentiation. Other final exponentiation methods may involve exponentiations by $6x_0 5$ and $6x_0^2 1$ [13, 14] or $6x_0 4$ [14], but these quantities are usually sparse at the same time as x_0.

- The sign of x_0 has no consequence in terms of complexities of the algorithms involved. Indeed, changing x_0 in $-x_0$ costs an $\mathbb{F}_{p^{12}}$ inversion, but this inversion can be replaced by a conjugation in $\mathbb{F}_{p^{12}}/\mathbb{F}_{p^6}$ thanks to the final exponentiation.

- Choosing x_0 with a signed binary representation (to facilitate the research of a sparse x_0) is possible if the exponentiation algorithms are adapted.

TABLE 10.12 Cost of the arithmetic of extension of degree 12, case 2, 3, 2.

Operation	Cost
Addition	$a_{12} = 3a_4 = 12a_1$
Doubling	$A'_{12} = 3A'_4 = 12A'_1$
Multiplication, Karatsuba	$M_{12} = 18M_2 60a_2 8m_{2,\xi}$
Sparse multiplication (Miller line)	$sM_{12} = 13M_2 26a_2 4m_{2,\xi}$
Squaring, Chung-Hasan	$S_{12} = 3S_4 6M_2 26a_2 2A'_2 4m_{2,\xi}$

- The choice of x_0 has a great impact on $\mathbb{F}_{p^{12}}$ arithmetic. The best choice is $x_0 = 7$ or 11 modulo 12, so that we can use $\mu = -1$ and $\xi = 1\mathbf{i}$. Depending on the situation, it could be better to choose a sparser $x_0 \neq 7, 11$ modulo 12 or reciprocally a x_0 of higher Hamming weight, but congruent to 7 or 11 modulo 12.

Hence, according to Table 10.1, x_0 should be chosen as sparse as possible and with the best possible way to build $\mathbb{F}_{p^{12}}$. Moreover, its size must ensure the right security level. For example, at the 128-bit security level, x_0 should be a 63-bit integer. A 95-bit integer provides a 192-bit security level on the elliptic curve but not in $\mathbb{F}_{p^{12}}$. To get this level of security, a 169- or 170-bit integer x_0 should be chosen.

With these constraints, finding an appropriate value of x_0 can easily be done by an exhaustive search with any software that is able to check integer's primality. Unfortunately, only a few values of x_0 with very low Hamming weight can be found. The best choice at the 128-bit security level is given by $x_0 = -2^{61} - 2^{55} - 1$ [32], even if it is ensuring a slightly smaller security level than 128. It has weight 3 and is congruent to 11 modulo 12, so that $\mu = -1$ and $\xi = 1\mathbf{i}$ can be used to build $\mathbb{F}_{p^{12}}$ (and to twist the curve). For these reasons, it is widely used in the literature. However, relaxing the constraint on the weight of x_0 allows us to generate many good values of x_0 (of weight 4, 5, or 6, for example) that can be used in a database of pairing-friendly parameters or for higher (or smaller) security levels.

Remark 10.8 Since, for BN curves, $E(\mathbb{F}_p)$ has prime order, it is naturally protected against subgroup attacks that exploit small prime divisors of the cofactor [28]. However, this is not the case of $E'(\mathbb{F}_{p^2})$ whose order equals $r(2p-r)$. For example, the value of x_0 given in Section 10.7.7, and usually used in the literature for the 128-bits security level, is not naturally protected against subgroup attacks. This can be prevented by using (possibly expensive) membership tests. If we want to avoid these tests, the parameter x_0 should be chosen such that both r and $2p - r$ are prime numbers [6].

10.7.8   Generation of the Group $\mathbb{G}_1$ and $\mathbb{G}_2$

In this section, we present an explicit method to construct the points P and Q potentially involved in the optimal Ate pairing computations. Since $E(\mathbb{F}_p)$ has prime order r, it is trivial to find a suitable candidate for P: any point $P \neq 0_E$ has order r.

However, generating a suitable point Q seems less easy because it must be of order r in $E\left(\mathbb{F}_{p^{12}}\right)$, come from the twisted curve, and be an eigenvector for the eigenvalue p of the Frobenius map. In fact, it is not so difficult because the last condition is a consequence of the other ones [22]. Finding a suitable point Q is then done in the following way:

1. Choose a random point Q' in $E'(\mathbb{F}_{p^2})$.
2. If Q' does not have order r (which is statistically always the case), replace it with $(p - 2r)Q'$, which has order r since $\#E'(\mathbb{F}_{p^2}) = (p - 2r)r$.
3. If $Q' = 0_E$, then repeat steps 1 and 2.
4. Map Q' to $E\left(\mathbb{F}_{p^{12}}\right)$ thanks to the twist isomorphism between E and E' over $\mathbb{F}_{p^{12}}$. If $b' = b/\xi$, this is nothing but

$$Q = \left(x_{Q'}\gamma^2, y_{Q'}\gamma^3\right) \text{ if } \gamma^6 = \xi.$$

Then Q is a valuable candidate, in particular, it lies in the p-eigenspace of π.

Example 10.3 For $b = 15$, the point $P = (1, 4)$ is a generator of $E(\mathbb{F}_p)$. We also need a point

Q of order r in the big field, independent of P. Since the order of $E(\mathbb{F}_{p^{12}})$ is about p^{12} and r is close to p, there seems to be no reason for such a point to have simple coordinates. So the simplest way seems to be to choose a random point Q_0 in $E(\mathbb{F}_{p^{12}})$ and then calculate $Q = sQ_0$, where $r^2 s = \sharp\big(E(\mathbb{F}_{p^{12}})\big)$. In PariGP, for example, this can be done as follows.

We find by trial and error that $X^{12}X8$ is irreducible over $\mathbb{F}_p$. To initialize $E : y^2 = x^3 15$ over the field $\mathbb{F}_{p^{12}}$, we can, for instance type

```
{p12 = Mod(1, p)*Y^12    Mod(1, p)*Y    Mod(8,p);  /* so Y is a generator of F_{p^12}
                    and satisfies Y^12 Y  8=0 */
 F12 = ffgen(p12); /* creates the field F_{p^12} */
 E12 = ellinit([0,0,0,0,15]*(F12^0)); /* creates E over F_{p^12}, since F12^0 is the
                    element 1 of F_{p^12} */
 Q0 = random(E12);}
```

Now we need to know $\sharp\big(E(\mathbb{F}_{p^{12}})\big)$, which can be calculated from $\sharp\big(E(\mathbb{F}_p)\big)$ as follows: If π is the Frobenius element over $\mathbb{F}_p$, then $\sharp\big(E(\mathbb{F}_p)\big) = p1 - \pi - \bar{\pi}$ and $\pi\bar{\pi} = p$, so if $t = \pi\bar{\pi}$, then

$$\sharp\big(E(\mathbb{F}_{p^{12}})\big) = p^{12}1 - \pi^{12} - \bar{\pi}^{12}$$
$$= p^{12}1 - (\pi^6\bar{\pi}^6)^2 2(\pi\bar{\pi})^6$$
$$= p^{12}2p^6 1 - ((\pi^3\bar{\pi}^3)^2 - 2(\pi^3\bar{\pi}^3))^2$$
$$= (p^6 1)^2 - ((\pi^3\bar{\pi}^3)^2 - 2p^3)^2$$
$$= (p^6 1)^2 - (((\pi\bar{\pi})^3 - 3(\pi\bar{\pi})\pi\bar{\pi})^2 - 2p^3)^2$$
$$= (p^6 1)^2 - ((t^3 - 3tp)^2 - 2p^3)^2$$
$$= (p^6 1(t^3 - 3tp)^2 - 2p^3)(p^6 1 - (t^3 - 3tp)^2 2p^3).$$

Thus we can define s to be $\sharp\big(E(\mathbb{F}_{p^{12}})\big)/r^2$ (which should be an integer), and type

```
Q1=ellpow(E12,Q0,s);
```

We now have a point of order r that is independent of P. If we want a point belonging to the q-eigenspace of Frobenius $\mathbb{G}_2$, we can finally type

```
Q = ellsub(E12,[Q1[1]^q,Q1[2]^q],Q1);
```

since $\pi(Q1) - Q1$ has trace 0, and so belongs to $\mathbb{G}_2$.

10.7.9   System of Coordinates over the BN-Curve

There exist several systems of coordinates over the elliptic curve. The affine, projective, and Jacobian are the classical ones for Weiestrass elliptic curves.

In this section, we give the formulas for adding and doubling points on BN curves (with the line computation) and their complexities in the affine and the projective cases (it is now well known that Jacobian coordinates are always less efficient than projective coordinates for pairing computations [11]). This allows us to determine which system of coordinates should be chosen, depending on the context. Assuming the previous choices, the two operations involved in the Miller loop are

The doubling step

In this step, we have to

- double a temporary point $T = \big(x_T\gamma^2, y_T\gamma^3\big) \in E\big(\mathbb{F}_{p^{12}}\big)$ with $x_T, y_T \in \mathbb{F}_{p^2}$,
- compute the tangent line to E at T,
- evaluate it at $P = (x_P, y_P) \in E(\mathbb{F}_p)$.

The addition step

In this step, we have to

- add $Q = (x_Q\gamma^2, y_Q\gamma^3)$ and $T = (x_T\gamma^2, y_T\gamma^3)$ in $E(\mathbb{F}_{p^{12}})$ with $x_Q, y_Q, x_T, y_T \in \mathbb{F}_{p^2}$,
- compute the line passing through T and Q,
- evaluate it at $P = (x_P, y_P) \in E(\mathbb{F}_p)$.

10.7.10 Affine Coordinates

The slope of the line passing through T and Q (or the tangent line at T if $T = Q$) is $\lambda\gamma$, with

$$\lambda = \frac{y_T - y_Q}{x_T - x_Q} \quad \left(\text{or} \quad \lambda = \frac{3x_T^2}{2y_T} \right).$$

Then TQ (or $2T$) can be written in the form $(x_{TQ}\gamma^2, y_{TQ}\gamma^3)$ with

$$x_{TQ} = \lambda^2 - x_T - x_Q \quad \text{and} \quad y_{TQ} = \lambda(x_T - x_{TQ}) - y_T.$$

The equation of the line involved in the operation is $y = \lambda\gamma(x - x_T\gamma^2) - y_T\gamma^3$, thus the $\mathbb{F}_{p^{12}}$ element involved in the update of f in Algorithm 3.3 is

$$\ell = y_P - \lambda x_P\gamma (\lambda x_T - y_T)\gamma^3.$$

Assuming that $-x_P$ is precomputed, the cost of the addition step (including the line computation) is then $\mathbf{I}_2 3\mathbf{M}_2 \mathbf{S}_2 2\mathbf{M}_1 7\mathbf{a}_2$ and the cost of the doubling step is $\mathbf{I}_2 3\mathbf{M}_2 2\mathbf{S}_2 2\mathbf{M}_1 5\mathbf{a}_2 2\mathbf{A}'_2$.

Remark 10.9 Since λ is used three times in $\mathbb{F}_{p^2}$ operations ($\lambda^2, \lambda(x_T - x_{TQ})$ and λx_T), $2\mathbf{a}_1$ can be saved using our idea of precomputing its trace if the Karatsuba/complex methods are used for $\mathbb{F}_{p^2}$ arithmetic. In the same way, x_T is used twice in the doubling step so that an additional $\mathbf{a}_1$ can be saved in this case.

10.7.11 Projective Coordinates

In order to avoid inversions in $\mathbb{F}_{p^2}$, projective coordinates are used for the point T, so that $T = (X_T\gamma^2, Y_T\gamma^3, Z_T)$ with X_T, Y_T and $Z_T \in \mathbb{F}_{p^2}$. However, the point Q is kept in affine coordinates (mixed addition method). According to [11], $2T = (X_{2T}\gamma^2, Y_{2T}\gamma^3, Z_{2T})$ with

$$\begin{aligned}
X_{2T} &= 2X_TY_T(Y_T^2 - 9bZ_T^2) \\
Y_{2T} &= (Y_T^2 9bZ_T^2)^2 - 12(3bZ_T^2)^2 \\
Z_{2T} &= 8Y_T^3 Z_T
\end{aligned}$$

and the equation of the tangent to the curve at T is (up to some subfield multiple)

$$\ell = 2y_P Y_T Z_T - 3x_P X_T^2\gamma (Y_T^2 - 3bZ_T^2)\gamma^3.$$

Assuming that $-3x_P$ is precomputed, the doubling step (including the line computation) then requires $2\mathbf{M}_2 7\mathbf{S}_2 4\mathbf{M}_1 13\mathbf{a}_2 5\mathbf{A}'_2 2\mathbf{m}_{1,b}$. In order to obtain this complexity, the double products like $2X_TY_T$ are computed by $(X_TY_T)^2 - X_T^2 - Y_T^2$. This trick is not always interesting over $\mathbb{F}_p$ (e.g., if $\mathbf{M}_1 = \mathbf{S}_1$) but it is always interesting over $\mathbb{F}_{p^2}$ because $\mathbf{S}_2$ is clearly cheaper than $\mathbf{M}_2$ according to Section 10.7.6.

In the same way, if

$$
\begin{aligned}
N &= Y_T - y_Q Z_T, \\
D &= X_T - x_Q Z_T \quad \text{(so that } \lambda = \tfrac{N}{D}\text{)}, \\
X &= N^2 Z_T - X_T D^2 - x_Q D^2 Z_T,
\end{aligned}
$$

we compute the addition step with

$$
\begin{aligned}
X_{TQ} &= DX \\
Y_{TQ} &= N(x_Q D^2 Z_T - X) - y_Q D^3 Z_T \\
Z_{TQ} &= D^3 Z_T \\
\ell &= y_P D - N x_P \gamma (N x_Q - D y_Q) \gamma^3.
\end{aligned}
$$

Assuming that $-x_P$ is precomputed, this requires $12\mathbf{M}_2 2\mathbf{S}_2 4\mathbf{M}_1 7\mathbf{a}_2$.

Remark 10.10   Again, many $\mathbb{F}_{p^2}$ operands are used several times during the computation, so that precomputing the traces saves additions in $\mathbb{F}_p$. We do not give details here because the addition step is rarely used in the Miller loop, but it is not difficult to see that $16\mathbf{a}_1$ can be saved if Karatsuba/complex arithmetic is used for $\mathbb{F}_{p^2}$ arithmetic.

10.7.12   Consequences of Formulas

Several remarks can be made looking at these formulas.

The first one is that the influence of b is small since it is just involved in two multiplications by $\mathbb{F}_p$ elements. Hence a sparse x_0 or a value of x_0 enabling a nice choice of μ (and ξ) should be preferred, even if a very small value of b is not available.

The second one is that, as mentioned in Section 10.7.6, the line ℓ is of the form $b_0 b_1 \gamma b_3 \gamma^3$ with $b_i \in \mathbb{F}_{p^2}$, thus it is sparse in $\mathbb{F}_{p^{12}}$ and a multiplication by ℓ is faster than a full multiplication in $\mathbb{F}_{p^{12}}$.

The third one is that, as already mentioned in [25, 2, 26, 19], it can be better to use affine coordinates than projective coordinates, depending on the context. Indeed, using the complexity formula for $\mathbf{I}_2$ given in Section 10.7.6 and the complexities obtained for the doubling step in affine and projective coordinates, it is easy to verify that affine coordinates become interesting for this step (and then for the full Miller loop) as soon as

$$
\mathbf{I}_1 < 5\mathbf{S}_2 - \mathbf{M}_2 - 2\mathbf{S}_1 15\mathbf{a}_1 6\mathbf{A}'_1 2\mathbf{m}_{1,b} - \mathbf{m}_{1,\mu}.
$$

For example, in the case $\mu = -1, \mathbf{a}_1 \leq 0.33\mathbf{M}_1$, it is shown in [15] that affine coordinates are interesting as soon as

$$
\mathbf{I}_1 < 7\mathbf{M}_1 - 2\mathbf{S}_1 20\mathbf{a}_1 11\mathbf{A}'_1 2\mathbf{m}_{1,b}. \tag{10.3}
$$

Depending on the way to implement $\mathbb{F}_p$ inversion, this inequality may hold in practice, especially if $\mathbb{F}_p$ addition are not negligible. In Table 10.13, we give the maximum cost of $\mathbf{I}_1$ for which affine coordinates should be chosen, depending on the context. To make the results more readable, we assumed that $\mathbf{S}_1 = \mathbf{M}_1, \mathbf{a}_1 = \mathbf{A}'_1$ and $b = 2$ (which is the least advantageous value for affine coordinates). In any case, the non-simplified result is very similar to Equation (10.3).

TABLE 10.13 Affine coordinates versus projective ones.

μ	Use affine coordinates if
-1	$\mathbf{I}_1 < 5\mathbf{M}_1 33\mathbf{a}_1$
-2	$\mathbf{I}_1 < 5\mathbf{M}_1 41\mathbf{a}_1$
-5	$\mathbf{I}_1 < 5\mathbf{M}_1 42\mathbf{a}_1$
any	$\mathbf{I}_1 < 13\mathbf{M}_1 28\mathbf{a}_1$

References

[1] SP-800-57. *Recommendation for Key Management – Part 1: General.* National Institute of Standards and Technology, U.S. Department of Commerce, July 2012.

[2] Tolga Acar, Kristin Lauter, Michael Naehrig, and Daniel Shumow. Affine pairings on ARM. In M. Abdalla and T. Lange, editors, *Pairing-Based Cryptography – Pairing 2012*, volume 7708 of *Lecture Notes in Computer Science*, pp. 203–209. Springer, Heidelberg, 2013.

[3] Diego F. Aranha, Laura Fuentes-Castañeda, Edward Knapp, Alfred Menezes, and Francisco Rodríguez-Henríquez. Implementing pairings at the 192-bit security level. In M. Abdalla and T. Lange, editors, *Pairing-Based Cryptography – Pairing 2012*, volume 7708 of *Lecture Notes in Computer Science*, pp. 177–195. Springer, Heidelberg, 2013.

[4] Diego F. Aranha, Koray Karabina, Patrick Longa, Catherine H. Gebotys, and Julio López. Faster explicit formulas for computing pairings over ordinary curves. In K. G. Paterson, editor, *Advances in Cryptology – EUROCRYPT 2011*, volume 6632 of *Lecture Notes in Computer Science*, pp. 48–68. Springer, Heidelberg, 2011.

[5] Razvan Barbulescu, Pierrick Gaudry, Antoine Joux, and Emmanuel Thomé. A heuristic quasi-polynomial algorithm for discrete logarithm in finite fields of small characteristic. In P. Q. Nguyen and E. Oswald, editors, *Advances in Cryptology – EUROCRYPT 2014*, volume 8441 of *Lecture Notes in Computer Science*, pp. 1–16. Springer, Heidelberg, 2014.

[6] Paulo S. L. M. Barreto, Craig Costello, Rafael Misoczki, Michael Naehrig, Geovandro C. C. F. Pereira, and Gustavo Zanon. Subgroup security in pairing-based cryptography. In K. E. Lauter and F. Rodríguez-Henríquez, editors, *Progress in Cryptology – LATINCRYPT 2015*, volume 9230 of *Lecture Notes in Computer Science*, pp. 245–265. Springer, Heidelberg, 2015.

[7] Paulo S. L. M. Barreto, Hae Yong Kim, Ben Lynn, and Michael Scott. Efficient algorithms for pairing-based cryptosystems. In M. Yung, editor, *Advances in Cryptology – CRYPTO 2002*, volume 2442 of *Lecture Notes in Computer Science*, pp. 354–368. Springer, Heidelberg, 2002.

[8] Paulo S. L. M. Barreto and Michael Naehrig. Pairing-friendly elliptic curves of prime order. In B. Preneel and S. Tavares, editors, *Selected Areas in Cryptography (SAC 2005)*, volume 3897 of *Lecture Notes in Computer Science*, pp. 319–331. Springer, Heidelberg, 2006.

[9] Ray C. C. Cheung, Sylvain Duquesne, Junfeng Fan, Nicolas Guillermin, Ingrid Verbauwhede, and Gavin Xiaoxu Yao. FPGA implementation of pairings using residue number system and lazy reduction. In B. Preneel and T. Takagi, editors, *Cryptographic Hardware and Embedded Systems – CHES 2011*, volume 6917 of *Lecture Notes in Computer Science*, pp. 421–441. Springer, Heidelberg, 2011.

[10] Henri Cohen and Gerhard Frey, editors. *Handbook of Elliptic and Hyperelliptic Curve Cryptography*, volume 34 of *Discrete Mathematics and Its Applications*. Chapman & Hall/CRC, 2006.

[11] Craig Costello, Tanja Lange, and Michael Naehrig. Faster pairing computations on curves with high-degree twists. In P. Q. Nguyen and D. Pointcheval, editors, *Public Key Cryptography – PKC 2010*, volume 6056 of *Lecture Notes in Computer Science*, pp. 224–242. Springer, Heidelberg, 2010.

[12] Augusto Jun Devegili, Colm Ó hÉigeartaigh, Michael Scott, and Ricardo Dahab. Multiplication and squaring on pairing-friendly fields. Cryptology ePrint Archive, Report 2006/471, 2006. http://eprint.iacr.org/2006/471.

[13] Augusto Jun Devegili, Michael Scott, and Ricardo Dahab. Implementing cryptographic pairings over Barreto-Naehrig curves (invited talk). In T. Takagi et al., editors, *Pairing-Based Cryptography – Pairing 2007*, volume 4575 of *Lecture Notes in Computer Science*, pp. 197–207. Springer, Heidelberg, 2007.

[14] Sylvain Duquesne and Loubna Ghammam. Memory-saving computation of the pairing final exponentiation on BN curves. Cryptology ePrint Archive, Report 2015/192, 2015. http://eprint.iacr.org/2015/192.

[15] Sylvain Duquesne, Nadia El Mrabet, Safia Haloui, and Franck Rondepierre. Choosing and generating parameters for low level pairing implementation on BN curves. *IACR Cryptology ePrint Archive*, 2015:1212, 2015.

[16] Andreas Enge and Jérôme Milan. Implementing cryptographic pairings at standard security levels. In R. S. Chakraborty, V. Matyas, and P. Schaumont, editors, *Security, Privacy, and Applied Cryptography Engineering (SPACE 2014)*, volume 8804 of *Lecture Notes in Computer Science*, pp. 28–46. Springer, 2014.

[17] David Freeman, Michael Scott, and Edlyn Teske. A taxonomy of pairing-friendly elliptic curves. *Journal of Cryptology*, 23(2):224–280, 2010.

[18] Laura Fuentes-Castañeda, Edward Knapp, and Francisco Rodríguez-Henríquez. Faster hashing to $\mathbb{G}_2$. In A. Miri and S. Vaudenay, editors, *Selected Areas in Cryptography – SAC 2011*, volume 7118 of *Lecture Notes in Computer Science*, pp. 412–430. Springer, Heidelberg, 2012.

[19] Gurleen Grewal, Reza Azarderakhsh, Patrick Longa, Shi Hu, and David Jao. Efficient implementation of bilinear pairings on ARM processors. In L. R. Knudsen and H. Wu, editors, *Selected Areas in Cryptography – SAC 2012*, volume 7707 of *Lecture Notes in Computer Science*, pp. 149–165. Springer, Heidelberg, 2013.

[20] Aurore Guillevic. Kim–Barbulescu variant of the number field sieve to compute discrete logarithms in finite fields. Research Report Elliptic News 2016/05/02, 2016.

[21] Florian Hess. Pairing lattices (invited talk). In S. D. Galbraith and K. G. Paterson, editors, *Pairing-Based Cryptography – Pairing 2008*, volume 5209 of *Lecture Notes in Computer Science*, pp. 18–38. Springer, Heidelberg, 2008.

[22] Florian Hess, Nigel P. Smart, and Frederik Vercauteren. The Eta pairing revisited. *IEEE Transactions on Information Theory*, 52(10):4595–4602, 2006.

[23] Marc Joye and Gregory Neven, editors. *Identity-Based Cryptography*, volume 2 of *Cryptology and Information Security Series*. IOS press, 2009.

[24] Taechan Kim and Razvan Barbulescu. Extended tower number field sieve: A new complexity for medium prime case. *IACR Cryptology ePrint Archive*, 2015:1027, 2015.

[25] Kristin Lauter, Peter L. Montgomery, and Michael Naehrig. An analysis of affine coordinates for pairing computation. In M. Joye, A. Miyaji, and A. Otsuka, editors, *Pairing-Based Cryptography – Pairing 2010*, volume 6487 of *Lecture Notes in Computer Science*, pp. 1–20. Springer, Heidelberg, 2010.

[26] Duc-Phong Le and Chik How Tan. Speeding up ate pairing computation in affine coordinates. In T. Kwon, M. Lee, and D. Kwon, editors, *Information Security and*

Cryptology – ICISC 2012, volume 7839 of *Lecture Notes in Computer Science*, pp. 262–277. Springer, Heidelberg, 2013.

[27] Rudolf Lidl and Harald Niederreiter. *Finite Fields*, volume 20 of *Encyclopedia of Mathematics and Its Applications*. Cambridge University Press, 2nd edition, 1997.

[28] Chae Hoon Lim and Pil Joong Lee. A key recovery attack on discrete log-based schemes using a prime order subgroup. In B. S. Kaliski Jr., editor, *Advances in Cryptology – CRYPTO'97*, volume 1294 of *Lecture Notes in Computer Science*, pp. 249–263. Springer, Heidelberg, 1997.

[29] Victor S. Miller. The Weil pairing, and its efficient calculation. *Journal of Cryptology*, 17(4):235–261, 2004.

[30] Nadia El Mrabet, Nicolas Guillermin, and Sorina Ionica. A study of pairing computation for elliptic curves with embedding degree 15. Cryptology ePrint Archive, Report 2009/370, 2009. http://eprint.iacr.org/2009/370.

[31] Michael Naehrig, Ruben Niederhagen, and Peter Schwabe. New software speed records for cryptographic pairings. In M. Abdalla and P. S. L. M. Barreto, editors, *Progress in Cryptology – LATINCRYPT 2010*, volume 6212 of *Lecture Notes in Computer Science*, pp. 109–123. Springer, Heidelberg, 2010.

[32] Yasuyuki Nogami, Masataka Akane, Yumi Sakemi, Hidehiro Katou, and Yoshitaka Morikawa. Integer variable chi-based Ate pairing. In S. D. Galbraith and K. G. Paterson, editors, *Pairing-Based Cryptography – Pairing 2008*, volume 5209 of *Lecture Notes in Computer Science*, pp. 178–191. Springer, Heidelberg, 2008.

[33] Geovandro C. C. F. Pereira, Marcos A. Simplício, Jr., Michael Naehrig, and Paulo S. L. M. Barreto. A family of implementation-friendly BN elliptic curves. *Journal of Systems and Software*, 84(8):1319–1326, 2011.

[34] Michael Scott, Naomi Benger, Manuel Charlemagne, Luis J. Dominguez Perez, and Ezekiel J. Kachisa. On the final exponentiation for calculating pairings on ordinary elliptic curves. In H. Shacham and B. Waters, editors, *Pairing-Based Cryptography – Pairing 2009*, volume 5671 of *Lecture Notes in Computer Science*, pp. 78–88. Springer, Heidelberg, 2009.

[35] Thomas Unterluggauer and Erich Wenger. Efficient pairings and ECC for embedded systems. In L. Batina and M. Robshaw, editors, *Cryptographic Hardware and Embedded Systems – CHES 2014*, volume 8731 of *Lecture Notes in Computer Science*, pp. 298–315. Springer, Heidelberg, 2014.

[36] Frederik Vercauteren. Optimal pairings. *IEEE Transactions on Information Theory*, 56(1):455–461, 2009.

11

Software Implementation

Diego F. Aranha
University of Campinas

Luis J. Dominguez Perez
CONACyT / CIMAT-ZAC

Amine Mrabet
Université Paris 8

Peter Schwabe
Raboud University

Since the introduction of cryptographic pairings as a constructive cryptographic primitive by Sakai, Ohgishi, and Kasahara in [31, 33, 34], and by Joux in [22, 23], the efficient implementation of pairings has become an increasingly important research topic. Early works still mainly considered the Weil pairing [43], whose computation essentially consists of two so-called *Miller loops*, but soon it became clear that variants of the Tate pairing [40, 41, 27] are more efficient. All those variants have in common that they consist of the computation of one Miller loop and one *final exponentiation*. Both the computation of the Miller loop and the computation of the final exponentiation ultimately break down into operations in large finite fields, further into arithmetic on large integers (or polynomials), and finally into machine instructions. Optimizing software for cryptographic pairings consists of

1. reducing the amount of finite-field operations, and
2. implementing finite-field and big-integer operations as efficiently as possible given the machine instructions of a certain target architecture.

In particular the first direction of optimization is largely influenced by the choice of a suitable pairing-friendly elliptic curve (see Chapter 10). As we will see, the two directions of optimizations are not entirely orthogonal, yet throughout this chapter we will refer to the first direction as *high-level optimization* and to the second one as *low-level optimization*.

This chapter will start off by reviewing common curve choices for high-performance software in Section 11.1. It will then present high-level optimization techniques for the Miller loop in Section 11.2 and for the final exponentiation in Section 11.3. Section 11.4 explains high-level optimizations for the computation of *multiple* pairings that are relevant in some cryptographic protocols. Section 11.5 explains low-level optimization techniques to efficiently map finite-field arithmetic to machine instructions. Section 11.6 gives a brief introduction to *timing attacks* and implementation techniques to protect against them. Section 11.7 gives an overview of existing software libraries and frameworks for pairings and related operations. Finally, Section 11.8 gives a full implementation of an optimal ate pairing on a Barreto-Naehrig curve in the open computer-algebra system Sage [39].

11.1   Curves for Fast Pairing Software

Pairings can be instantiated over different elliptic curves, producing maps of form $e : \mathbb{G}_2 \times \mathbb{G}_1 \to \mathbb{G}_T$. We commonly distinguish Type-1 (symmetric) pairings with $\mathbb{G}_1 = \mathbb{G}_2$), and Type-3 (asymmetric) settings.

11.1.1   Curves for Type-1 Pairings

In the symmetric setting, a *distortion map* is used to map elements from $\mathbb{G}_2$ to a linearly independent subgroup. This is required for pairing computation to satisfy non-degeneracy, so elements from $\mathbb{G}_1$ and $\mathbb{G}_2$ are mapped to non-trivial elements in $\mathbb{G}_T$. The most efficient proposals for curves in this setting were supersingular binary and ternary curves, but recent advances in the discrete logarithm computation in small characteristic derived a quasi-polynomial time that makes these instantiations insecure [4]. As far as the research literature has discovered, the only supersingular curves allowing secure instantiations of Type-1 pairings are supersingular curves with embedding degree $k = 2, 3$ defined over prime fields [9, 45, 42] (also see Chapter 4 for details).

11.1.2   Barreto-Naehrig Curves

The Barreto-Naehrig (BN) family of prime-order elliptic curves is ideal from an implementation point of view, under several different aspects. These curves have embedding degree $k = 12$, which makes them perfectly suited for 128 bits of security and a strong contender at the 192-bit security level as well [2]. The family of curves is large enough to facilitate generation and tweaking of curve parameters for optimal performance, allowing for customization of software implementations to very different platforms in the computer architecture spectrum [32]. Concrete efficient parameters for such curves can be found in [3], but parameters that attain higher security for instantiating some protocols are proposed in [5]. A recent IEFT draft proposes additional parameters.* Refer to Chapter 10 for further details on parameters selection.

11.1.3   Curves at Higher Security Levels

At higher security levels, other parameterized families of curves become more efficient. Kachisa-Schaefer-Scott (KSS) curves [24] with embedding degree $k = 18$ and Barreto-Lynn-Scott elliptic curves [6], or BLS24 curves, are well suited for the task. Another family of curves of embedding

---

*https://datatracker.ietf.org/doc/draft-kasamatsu-bncurves/

degree $k = 12$, called BLS12 [6] (also see [12]), have composite group order and provide competitive performance. In real applications, the best choice of curve to implement may depend on several factors, including protocol-level operaitons and number of required pairing computations [11]. Efficient parameters for the aforementioned families can be found in [2], while concrete curves with additional security properties are proposed in [5].

In the following, we illustrate the concepts involved in pairing computation with Barreto-Naehrig curves because of their implementation-friendliness.

11.2 The Miller Loop

The optimal ate pairing construction applied to general BN curves also provides a rather simple formulation:

$$a_{opt} : \mathbb{G}_2 \times \mathbb{G}_1 \rightarrow \mathbb{G}_T$$

$$(Q, P) \mapsto \left(f_{\ell,Q}(P) \cdot g_{[\ell]Q, \psi(Q)}(P) \cdot g_{[\ell]Q+\psi(Q), -\psi^2(Q)}(P) \right)^{\frac{p^{12}-1}{n}},$$

with $\ell = 6u+2$, ψ the homomorphism on $\mathbb{G}_2$, line functions $g_{T,Q}$ passing through points T, Q; and groups $\mathbb{G}_2, \mathbb{G}_1, \mathbb{G}_T$ as previously defined. A specialization of Miller's algorithm for computing the optimal ate pairing can be found in Algorithm 11.1 below. In this version, both negative and positive parameterizations are supported and the first iteration of the Miller loop is unrolled to avoid trivial computations.

ALGORITHM 11.1 Optimized version of optimal ate pairing on general BN curves.

Input : $P \in \mathbb{G}_1, Q \in \mathbb{G}_2, \ell = |6u + 2| = \sum_{i=0}^{\log_2(\ell)} \ell_i 2^i$
Output: $a_{opt}(Q, P)$

$d \leftarrow g_{Q,Q}(P), e \leftarrow 1, T \leftarrow 2Q$
if $\ell_{\lfloor \log_2(\ell) \rfloor - 1} = 1$ **then**
$\quad | \quad e \leftarrow g_{T,Q}(P), T \leftarrow T + Q$
end
$f \leftarrow d \cdot e$
for $i = \lfloor \log_2(\ell) \rfloor - 2$ **downto** 0 **do**
$\quad | \quad f \leftarrow f^2 \cdot g_{T,T}(P), T \leftarrow [2]T$
$\quad | \quad$ **if** $\ell_i = 1$ **then**
$\quad | \quad \quad | \quad f \leftarrow f \cdot g_{T,Q}(P), T \leftarrow T + Q$
$\quad | \quad$ **end**
end
$Q_1 \leftarrow \psi(Q), Q_2 \leftarrow \psi^2(Q)$
if $u < 0$ **then**
$\quad | \quad T \leftarrow -T, f \leftarrow f^{p^6}$
end
$d \leftarrow g_{T,Q_1}(P), T \leftarrow T + Q_1$
$e \leftarrow g_{T,-Q_2}(P), T \leftarrow T - Q_2$
$f \leftarrow f \cdot (d \cdot e)$
$f \leftarrow f^{(p^6-1)(p^2+1)(p^4-p^2+1)/n}$
return f

The main building block of pairing computation is extension-field arithmetic. Hence, its efficient implementation is crucial. A popular choice consists of implementing the extension field through a tower of extensions, built with appropriate choices of irreducible polynomials:

$$\mathbb{F}_{p^2} = \mathbb{F}_p[i]/(i^2 - \beta), \text{ with } \beta \text{ a non-square,} \tag{11.1}$$

$$\mathbb{F}_{p^4} = \mathbb{F}_{p^2}[s]/(s^2 - \xi), \text{ with } \xi \text{ a non-square,} \tag{11.2}$$

$$\mathbb{F}_{p^6} = \mathbb{F}_{p^2}[v]/(v^3 - \xi), \text{ with } \xi \text{ a non-cube,} \tag{11.3}$$

$$\mathbb{F}_{p^{12}} = \mathbb{F}_{p^4}[t]/(t^3 - s) \tag{11.4}$$

$$\text{or } \mathbb{F}_{p^6}[w]/(w^2 - v) \tag{11.5}$$

$$\text{or } \mathbb{F}_{p^{12}}[w]/(w^6 - \xi), \text{ with } \xi \text{ a non-square and non-cube.} \tag{11.6}$$

Notice that converting from one towering scheme to another is possible by simply reordering coefficients. As previously stated, a remarkably efficient set of parameters arising from the curve choice $E(\mathbb{F}_p) : y^2 = x^3 + 2$, with $p \equiv 3 \pmod 4$, is $\beta = -1$, $\xi = (1+i)$, simultaneously optimizing finite field and curve arithmetic.

Line function evaluations computed inside the Miller loop (lines 7, 9, 16, and 17 of the algorithm) generally have a rather sparse format, which motivates the implementation of dedicated multiplication routines for accumulating the line evaluations into the Miller variable f (*sparse* multiplication) or for multiplying line functions together (*sparser* multiplication). The choice of tower representation can also change the number of multiplication and addition operations for such routines, thus a careful performance analysis must be performed in the target architecture to inform what is the best decision.

11.2.1   Point Representation

Pairings can be computed over elliptic curves represented in any coordinate system, but homogeneous projective and affine coordinates are the most common, depending on the ratio between inversion and multiplication of a specific implementation. Due to the low Hamming weight of the curve parameter and its effect on reducing the number of additions, the cost of the Miller loop is usually dominated by point doubling and the corresponding line evaluations. Elements from $\mathbb{G}_2$ can be defined over both divisive twists (*D*-type) or multiplicative twists (*M*-type). Below, we review and slightly refine the formulas presented in previous chapters for the curve arithmetic involved in pairing computation on affine and homogeneous projective coordinates. When evaluating the costs of each formula, we use bold typeface for representing operation counts in $\mathbb{F}_{p^2}$ and normal typeface for operations in the base field. These operations can be multiplication (m), squaring (s), additions (a), and inverses (i). Complete and detailed operation counts can be found in [1].

Affine coordinates

Affine coordinates have limited application in pairing computation, but they have proven useful at higher security levels and embedding degrees due to simpler computation of inverses at higher extensions. Another possible use case for affine coordinates is for evaluating protocols requiring multiple pairing computations, when inversion can be batched together. The main advantages of affine coordinates are the simplicity of implementation and format of the line functions, allowing faster accumulation inside the Miller loop if the additional sparsity is exploited. If $T = (x_1, y_1)$ is a point in $E'(\mathbb{F}_{p^2})$, one can compute the point $2T := T + T$ with the following formula:

$$\lambda = \frac{3x_1^2}{2y_1}, \quad x_3 = \lambda^2 - 2x_1, \quad y_3 = (\lambda x_1 - y_1) - \lambda x_3. \tag{11.7}$$

When the twist curve E' is of *D*-type and given by the twisting isomorphism ψ, the tangent line evaluated at $P = (x_P, y_P)$ has the format $g_{2\psi(T)}(P) = y_P - \lambda x_P w + (\lambda x_1 - y_1)w^3$, using the tower representation given by Equation (11.6). This function can be evaluated at a cost of $3\mathbf{m_2} + 2\mathbf{s_2} + 7\mathbf{a_2} + \imath_2 + 2m$ with the precomputation cost of $1a$ to compute $\overline{x}_P = -x_P$. By performing more precomputation as $y'_P = 1/y_P$ and $x'_P = \overline{x}_P/y_P$, we can simplify the tangent line further:

$$y'_P \cdot g_{2\psi(T)}(P) = 1 + \lambda x'_P w + y'_P(\lambda x_1 - y_1)w^3.$$

Recall that the final exponentiation eliminates any subfield element multiplying the pairing result. Hence, this modification does not change the pairing value and computing the simpler line function now requires $3\mathbf{m_2} + 2\mathbf{s_2} + 7\mathbf{a_2} + \imath_2 + 4m$, with an additional precomputation cost of $(i + m + a)$:

$$A = \frac{1}{2y_1}, \quad B = 3x_1^2, \quad C = AB, \quad D = 2x_1, \quad x_3 = C^2 - D,$$
$$E = Cx_1 - y_1, \quad y_3 = E - Cx_3, \quad F = Cx'_P, \quad G = Ey'_P,$$
$$y'_P \cdot g_{2\psi(T)}(P) = 1 + Fw + Gw^3.$$

This trick does not have any operations compared to the previous equation and increases the cost by $2m$. However, the simpler format allows the faster accumulation $f^2 \cdot g_{2\psi(T)}(P) = (f_0 + f_1 w)(1 + g_1 w)$, where $f_0, f_1, g_1 \in \mathbb{F}_{p^6}$, by saving 6 base field multiplications required to multiply y_P by each subfield element of f_0. The performance trade-off compared to the previous formula is thus $4m$ per Miller doubling step.

When different points $T = (x_1, y_1)$ and $Q = (x_2, y_2)$ are considered, the point $T + Q$ can be computed with the following formula:

$$\lambda = \frac{y_2 - y_1}{x_2 - x_1}, \quad x_3 = \lambda^2 - x_2 - x_1, \quad y_3 = \lambda(x_1 - x_3) - y_1. \tag{11.8}$$

Applying the same trick described above gives the same performance trade-off, with a cost of $3\mathbf{m} + \mathbf{s} + 6\mathbf{a} + \imath + 4m$:

$$A = \frac{1}{x_2 - x_1}, \quad B = y_2 - y_1, \quad C = AB, \quad D = x_1 + x_2, \quad x_3 = C^2 - D,$$
$$E = Cx_1 - y_1, \quad y_3 = E - Cx_3, \quad F = Cx'_P, \quad G = Ey'_P,$$
$$y'_P \cdot g_{\psi(T),\psi(Q)}(P) = 1 + Fw + Gw^3.$$

The same technique can be further employed in M-type twists, conserving their equivalent performance to D-type twists, with some slight changes in the formula format and accumulation multiplier. A generalization for other pairing-friendly curves with degree-d twists and even embedding degree k would provide a performance trade-off of $(k/2 - k/d)$ multiplications per step in Miller's Algorithm.

Homogeneous projective coordinates

Projective coordinates improve performance in single pairing computation compared to affine coordinates due to the typically large inversion/multiplication ratio in this setting. A particular choice of projective coordinates that optimizes pairing computation is the homogeneous projective coordinate system. Below, we present optimizations proposed in [19, 3, 1] for the formulas initially presented in [14]. If $T = (X_1, Y_1, Z_1) \in E'(\mathbb{F}_{p^2})$ is a point in homogeneous coordinates, one can compute the point doubling $2T = (X_3, Y_3, Z_3)$ with the following formula [3]:

$$X_3 = \frac{X_1 Y_1}{2}(Y_1^2 - 9b'Z_1^2),$$
$$Y_3 = \left(\frac{1}{2}(Y_1^2 + 9b'Z_1^2)\right)^2 - 27b'^2 Z_1^4, \tag{11.9}$$
$$Z_3 = 2Y_1^3 Z_1.$$

The twisting point P can be represented by $(x_P w, y_P)$. When E' is a D-type twist given by the twisting isomorphism ψ, the tangent line evaluated at $P = (x_P, y_P)$ can be computed with the following formula:

$$g_{2\psi(T)}(P) = -2Y_1 Z_1 y_P + 3X_1^2 x_P w + (3b' Z_1^2 - Y_1^2) w^3. \tag{11.10}$$

This formula has additional optimization tricks, such as division by 4 to save additions and a modification of the format to reduce the number of additions required for the accumulation in the Miller variable.

The complete formula costs $2\mathbf{m_2} + 7\mathbf{s_2} + 23\mathbf{a_2} + 4m + \mathbf{d_{b'}}$ and can be optimized further by saving additions, by precomputing $\bar{y}_P = -y_P$ and $x'_P = 3x_P$. Note that all these costs consider the computation of $X_1 \cdot Y_1$ using the equivalence $2XY = (X+Y)^2 - X^2 - Y^2$. In platforms where it is more efficient to compute such terms with a direct multiplication because of $\mathbf{m_2} - \mathbf{s_2} < 3\mathbf{a_2}$, the cost would then be given by $3\mathbf{m_2} + 6\mathbf{s_2} + 15\mathbf{a_2} + 4m + \mathbf{d_{b'}}$. Finally, further improvements are possible if b is cleverly selected. For instance, if $b = 2$ then $b' = 2/(1+i) = 1 - i$ using our tower representation, which minimizes the number of additions and subtractions:

$$
\begin{aligned}
A &= X_1 \cdot Y_1/2, \quad B = Y_1^2, \quad C = Z_1^2, \quad D = 3C, \quad E_0 = D_0 + D_1, \\
E_1 &= D_1 - D_0, \quad F = 3E, \quad X_3 = A \cdot (B - F), \quad G = (B + F)/2, \\
Y_3 &= G^2 - 3E^2, \quad H = (Y_1 + Z_1)^2 - (B + C), \quad Z_3 = B \cdot H, \\
&g_{2\psi(T)}(P) = H\bar{y}_P + X_1^2 x'_P w + (E - B) w^3.
\end{aligned}
\tag{11.11}
$$

Similarly, if $T = (X_1, Y_1, Z_1)$ and $Q = (x_2, y_2) \in E'(\mathbb{F}_{p^2})$ are points in homogeneous and affine coordinates, respectively, one can compute the mixed point addition $T + Q = (X_3, Y_3, Z_3)$ with the following formula:

$$
\begin{aligned}
X_3 &= \lambda(\lambda^3 + Z_1\theta^2 - 2X_1\lambda^2), \\
Y_3 &= \theta(3X_1\lambda^2 - \lambda^3 - Z_1\theta^2) - Y_1\lambda^3, \\
Z_3 &= Z_1\lambda^3,
\end{aligned}
\tag{11.12}
$$

where $\theta = Y_1 - y_2 Z_1$ and $\lambda = X_1 - x_2 Z_1$. In the case of a D-type twist, the line evaluated at $P = (x_P, y_P)$ can be computed with the following formula:

$$g_{\psi(T+Q)}(P) = -\lambda y_P - \theta x_P w + (\theta X_2 - \lambda Y_2) w^3. \tag{11.13}$$

Analogously to point doubling, Equation (11.13) is basically the same line evaluation formula presented in [14] with a repositioning of terms to obtain faster sparse multiplication. The complete formula can be evaluated at a cost of $11\mathbf{m} + 2\mathbf{s} + 8\mathbf{a} + 4m$ with the precomputation cost of $2a$ to compute $\bar{x}_P = -x_P$ and $\bar{y}_P = -y_P$:

$$
\begin{aligned}
A &= Y_2 Z_1, \quad B = X_2 Z_1, \quad \theta = Y_1 - A, \quad \lambda = X_1 - B, \quad C = \theta^2, \\
D &= \lambda^2, \quad E = \lambda^3, \quad F = Z_1 C, \quad G = X_1 D, \quad H = E + F - 2G, \\
X_3 &= \lambda H, \quad I = Y_1 E, \quad Y_3 = \theta(G - H) - I, \quad Z_3 = Z_1 E, \quad J = \theta X_2 - \lambda Y_2, \\
&g_{2\psi(T)}(P) = \lambda \bar{y}_P + \theta \bar{x}_P w + J w^3.
\end{aligned}
$$

In the case of an M-type twist, the line function evaluated at $\psi(P) = (x_P w^2, y_P w^3)$ can be computed with the same sequence of operations shown above.

11.3 The Final Exponentiation

In the context of BN curves, the final exponentiation can be computed through the state-of-the-art approach by [16]. As initially proposed in [37], the power $\frac{p^{12}-1}{r}$ can be factored into the easy exponent $(p^6 - 1)(p^2 + 1)$ and the hard exponent $\frac{p^4 - p^2 + 1}{n}$. The easy power is computed by a short sequence of multiplications, conjugations, fast applications of the Frobenius map [8], and a single inversion in $\mathbb{F}_{p^{12}}$. The hard power is computed in the cyclotomic subgroup, where additional algebraic structure allows elements to be compressed and squared consecutively in their compressed form, with decompression required only when performing multiplications [3]. Lattice reduction is able to obtain parameterized multiples of the hard exponent and significantly reduce the length of the addition chain involved in that exponentiation. In total, the hard part of the final exponentiation requires 3 exponentiations by parameter u, 3 squarings in the cyclotomic subgroup, 10 full extension field multiplications, and 3 applications of the Frobenius maps with increasing pth-powers. Chapter 7 discusses theses methods in finer detail.

11.4 Computing Multiple and Fixed-Argument Pairings

In some protocols, part of the parameters and variables can be known in advance. If the storage area is not a constraint, one can precompute some of the operations beforehand. Not only the pairing function could be precomputed, but also some of the ancillary functions around it. This way, the pairing function could be speeded up by 15% in the case of the BN curves at 128-bit level of security, and between 53 to 68% for the exponentiation in the pairing groups. [44]

11.4.1 Pairing Computations in the Fixed Argument

One should note that the line functions of the pairing algorithm can be partially computed in advance using the argument from the $\mathbb{G}_2$ group if it is fixed (indeed, almost all of the function). One can store these partial values, and complete them at running time by evaluating the remainder of the function on the argument from the $\mathbb{G}_1$ group. Hence, one can gain a sensitive speed-up of the pairing computation in exchange for storage space if the parameter of the $\mathbb{G}_2$ group is fixed, as it was discussed in [36]. The optimized line functions from the Miller loop are described in Equations 11.11, and 11.10.

For example, for the case of the BN curves at a security level of 128-bits, with the Aranha et al. [3] parameter $x = -(2^{62} + 2^{55} + 1)$, the Miller loop of the optimal ate pairing (Algorithm 11.1) requires 70 evaluations of line functions (64 correspond to point doubling, and 6 to point additions). Recalling that these functions use values in $\mathbb{F}_{p^6}$, then, one needs to store 70 elements in $\mathbb{F}_{p^6}$ in order to gain a speed-up of around 15% [44].

To do the precomputation, we need the values of the line function for each stage in the Miller loop, which requires a modified pairing function that saves each line function result without the coordinates of the point P;* we need to store m line function states: $\lfloor log_2(6x+2) \rfloor$ line doublings, and $HammingWeight(6x + 2) + 2$ line additions for each fixed Q in the case of the BN curves; we would need to change the Miller loop length $(6x + 2)$ in the case of other families of curves. Taking Algorithm 11.1 as a base, instead of accumulating into f, d, or e the results from the line functions $(g_{T,T}(P), g_{T,Q}(P), g_{T,Q_1}, $ or $g_{T,-Q_2}(P))$, we can store the output values as $\tilde{g}_{j,i}$ with $0 \le j < $ *number of fixed points*, and $0 \le i < m$-line function states.

---

*One can use the multiplicative identity to use the same, unmodified line functions.

11.4.2   Product of Pairings

When computing products of pairings in a pairing-based protocol, one has two lines of optimization: reducing the number of pairings using the bilinearity property, and sharing operations.

For the first line of optimization, one can recall the bilinearity property of the pairing function: for $P_1, P_2, Q \in E[n]$,

$$e_n(P_1 + P_2, Q) = e_n(P_1, Q) \cdot e_n(P_2, Q)$$
$$e_n(Q, P_1 + P_2) = e_n(Q, P_1) \cdot e_n(Q, P_2).$$

One can group pairings that share one of the input parameters. If all arguments of the pairing function share the same element from the pairing group, then one can reduce the n pairing products by $n - 1$ point additions with a single pairing function as:

$$\prod_{i=0}^{n-1} e(Q, P_i) = e(Q, \sum_{i=0}^{n-1} P_i),$$

which saves a very significant amount of operations, namely, $n - 1$ pairings, and $n - 1$ multiplications in $\mathbb{G}_T$. It may be possible to apply this optimization several times, depending on the protocol.

The second line of optimization is by sharing the computation between several pairing instances.

For example, one can calculate the product of pairings by performing a simultaneous product of pairings (or multipairing). Essentially, one can apply the same techniques used when dealing with multiple point multiplications; i.e., *Shamir's trick*.

The use of the Shamir's trick has been discussed in [38], [35], and [18]. Essentially, in a product of pairings one can share the pairing accumulator f from Algorithm 11.1. This way, one is simultaneously performing the several pairing functions, and the multiplication. Additionally, in the case of the Tate pairing, one must recall that the final exponentiation is, broadly speaking, used to map the result of the pairing into the desired pairing subgroup; since this operation is independent from the Miller loop, one can apply a single final exponentiation at the end of the product of pairings, and get an element in the desired subgroup.

Algorithm 11.2 presents an explicit multi-pairing version of Algorithm 11.1 that computes the product of n optimal BN pairings.

One should recall that optimized versions of the line functions will also compute the point doubling, and point addition; however, Algorithm 11.2 is explicit in the point doubling, and point addition after the line functions in the Miller loop; this was written for completeness.

As a final remark, a mixed environment can be presented where there are fixed arguments, and unknown parameters; see Algorithm 11.3 for a version with both fixed and unknown parameters.

11.4.3   Simultaneous Normalization

An important implementation note on the pairing function is that both line functions (Equations 11.11 and 11.10) expect the parameters of the pairing to be in affine coordinates. Most pairing-based protocols involve either the computation of point additions, or scalar-point multiplications before a pairing — for example, when using the bilinearity of the pairing, or when needing a product of pairings. In both cases, the result would probably be a point, or a set of points in Jacobian coordinates; however, in the case of the scalar-point multiplication, it is customary to normalize the point into affine coordinates, but this could be inefficient in the case where multiple pairings are involved in a protocol. For implementation purposes, it is recommended to use Montgomery's trick, and do a *simultaneous multi-normalization* of the parameters; this

---

ALGORITHM 11.2 Explicit multipairing version of Algorithm 11.1.

Input : $P_1, P_2, \ldots P_n \in \mathbb{G}_1$,
$\quad\quad\quad Q_1, Q_2, \ldots Q_n \in \mathbb{G}_2$

Output: $f = \displaystyle\prod_{i=1}^{n} e_\ell(Q_i, P_i)$

Write ℓ in binary form, $\ell = \sum_{i=0}^{m-1}$
$f \leftarrow 1, \ell \leftarrow \mathbf{abs}(6x + 2)$
for $j \leftarrow 1$ *to* n **do**
$\quad\mid\ T_j \leftarrow Q_j$
end

for $i = m - 2$ ***down to*** 0 **do**
$\quad\mid\ f \leftarrow f^2$
$\quad\mid$ **for** $j \leftarrow 1$ *to* n **do**
$\quad\mid\quad\mid\ f \leftarrow f \cdot g_{T_j, T_j}(P_j),\ T_j \leftarrow [2]T_j$
$\quad\mid\quad\mid$ **if** $\ell_i = 1$ **then**
$\quad\mid\quad\mid\quad\mid\ f \leftarrow f \cdot g_{T_j, Q_j}(P_j),\ T_j \leftarrow T_j + Q_j$
$\quad\mid\quad\mid$ **end**
$\quad\mid$ **end**

end

if $(6x + 2) < 0$ **then**
$\quad\mid\ f \leftarrow f^{p^6}$
end

for $j \leftarrow 1$ *to* n **do**
$\quad\mid$ **if** $(6x + 2) < 0$ **then**
$\quad\mid\quad\mid\ T = -T$
$\quad\mid$ **end**
$\quad\mid\ R \leftarrow \psi(Q_j);\ f \leftarrow f \cdot g_{T_j, R}(P_j);$
$\quad\mid\ T_j \leftarrow T_j + R$
$\quad\mid\ R \leftarrow \psi^2(Q_j);\ f \leftarrow f \cdot g_{T_j, -R}(P_j);$
$\quad\mid\ T_j \leftarrow T_j - R$
end

$f^{(p^k - 1)/r}$
return f

---

has to be done for elements in both $\mathbb{G}_1$ and $\mathbb{G}_2$, and taking care that the Montgomery's trick is still worth the number of multiplications in the recovering part, if there are sufficient inversions involved.

11.5   Low-Level Optimization

All pairing computations and computations in the related groups eventually break down to operations in the underlying finite field. The task of low-level optimization is to map these finite-field operations to machine instructions of a given target architecture as efficiently as possible. The main challenge in this mapping is that the finite-field elements are typically much larger than the size of machine words. For example, in the 128-bit security setup using BN curves considered in this chapter, elements of the finite field have ≈ 256 bits, whereas the size of machine words ranges somewhere between 8 bits (for example, on small microcontrollers) through 16 bits and 32 bits up to 64 bits (for example, on recent Intel and AMD processors). One way to accomplish this task is to use a software library for "big-integer arithmetic" or "multiprecision arithmetic." There are at least three potential downsides to this approach: First, various big-integer libraries were not originally written for use in cryptography, and the arithmetic may leak timing information (cf. Section 11.6). One of the most widely used free multiprecision libraries, the GNU multiprecision library (GMP), has acknowledged this problem since Version 5 and includes low-level `mpn_sec_` functions for timing-attack-protected use in cryptography. Second, general-purpose software libraries for finite-field arithmetic have to support arbitrary finite fields and don't include optimizations for reduction modulo a special-shape integer. This is not very much of an issue in pairing-based cryptography (where moduli typically do not have a "nice" shape), but it is a critical disadvantage for non-pairing elliptic-curve software. Finally, software libraries typically optimize separate operations (like addition, subtraction, multiplication) and not sequences of operations. This may make it harder to include lazy-reduction techniques such as, for example, the ones extensively used in [3].

ALGORITHM 11.3   Multipairing version of Algorithm 11.1 with mixed unknown and fixed arguments.

Input   : $P_1, \ldots, P_n, P_{n+1}, \ldots P_\nu \in \mathbb{G}_1, \tilde{g}_1, \ldots, \tilde{g}_n$ (the m precomputed line functions for each known $Q_1, \ldots, Q_n$), $Q_{n+1}, \ldots Q_\nu \in \mathbb{G}_2$

Output: $f = \displaystyle\prod_{i=1}^{\nu} e_\ell(Q_i, P_i)$

Simultaneously normalize the $n + \nu$ parameters in $\mathbb{G}_1$, and the ν parameters in $\mathbb{G}_2$

Write ℓ in binary form, $\ell = \sum_{i=0}^{s-1}$
$f \leftarrow 1, \ell \leftarrow \mathbf{abs}(6x + 2)$
for $j \leftarrow 1$ *to* ν **do**
  | $T_j \leftarrow Q_j$
end
for $i = s - 2$ **down to** 0 **do**
  | $f \leftarrow f^2$
  | **for** $j \leftarrow 1$ *to* n **do**
  | | $l \leftarrow \tilde{g}_{j,m}$ (The m-th precomputed value of line function of the j-th fixed Q)
  | | Include P_j in l (See Equations 11.11 and 11.10)
  | | $f \leftarrow f \cdot l$
  | **end**
  | $m \leftarrow m + 1$
  | **for** $j \leftarrow 1$ *to* ν **do**
  | | $f \leftarrow f \cdot g_{T_j,T_j}(P_{n+j}), T_j \leftarrow [2]T_j$
  | **end**
  | **if** $\ell_i = 1$ **then**
  | | **for** $j \leftarrow 1$ *to* n **do**
  | | | $l \leftarrow \tilde{g}_{j,m}$
  | | | Include P_j in l (See Equations 11.11 and 11.10)
  | | | $f \leftarrow f \cdot l$
  | | **end**
  | | $m \leftarrow m + 1$
  | | **for** $j \leftarrow 1$ *to* ν **do**
  | | | $f \leftarrow f \cdot g_{T_j,Q_j}(P_{n+j}), T_j \leftarrow T_j + Q_j$
  | | **end**
  | **end**
end

end
if $(6x + 2) < 0$ **then**
  | $f \leftarrow f^{p^6}$
end
for $j \leftarrow 1$ *to* n **do**
  | $l \leftarrow \tilde{g}_{j,m}$
  | $l_2 \leftarrow \tilde{g}_{j,m+1}$
  | Include P_j in l, and l_2 (See Equations 11.11 and 11.10)
  | $f \leftarrow f \cdot l \cdot l_2$
end
for $j \leftarrow 1$ *to* ν **do**
  | $T_j \leftarrow -T_j$
  | $R \leftarrow \psi(Q_j); f \leftarrow f \cdot g_{T_j,R}(P_{n+j});$
  | $T_j \leftarrow T_j + R$
  | $R \leftarrow \psi^2(Q_j); f \leftarrow f \cdot g_{T_j,-R}(P_{n+j});$
  | $T_j \leftarrow T_j - R$
end
$f^{(p^k-1)/r}$

return f

All speed-record setting software for pairing (and more generally elliptic-curve) software therefore optimizes low-level arithmetic in hand-written assembly. There are three main choices to make in this low-level optimization, which we will briefly describe in the following.

- **Choice of radix.** The typical approach to split an n-bit integer A into w-bit machine words is to use $m = \lceil n/w \rceil$ unsigned w-bit integers $(a_0, \ldots, a_{m-1})$ such that $A = \sum_{i=0}^{m-1} a_i 2^{iw}$. This approach is called *unsigned radix-2^w* representation. On many architectures this approach turns out to be the most efficient not just in terms of space but also in terms of speed of the arithmetic operations performed on big integers in this representation. On many other architectures, it turns out that a "redundant" (or "unsaturated") approach yields better performance. Essentially, the idea is to represent an integer A as $\sum_{i=0}^{\ell} a_i 2^{ki}$, where k is smaller than the word length of the target architecture and ℓ is larger than $\lceil n/w \rceil$. The advantage is that the *limbs* a_i do not need all bits of a machine word and arithmetic can thus simplify carry handling.

- **Choice of multiplication algorithm.** Most of the computation time of pairings and group arithmetic eventually boils down to modular multiplications and squarings in the underlying prime field. These in turn break down into multiplications and additions of saturated or unsaturated limbs, i.e., machine words. Many approaches for multiprecision multiplication have quadratic complexity: Multiplying two n-limb numbers takes n^2 multiplications and $(n-1)^2$ additions of (two-limb) partial results. Many of those operations are independent, which gives a large degree of freedom for instruction scheduling. The most common approaches to instruction scheduling are operand-scanning and product scanning, but more involved approaches like hybrid multiplication [20] or operand caching [21] have been shown to yield better performance on some architectures. As the number of limbs in the representation of bit integers increases, algorithms with sub-quadratic complexity, in particular Karatsuba multiplication [25] become faster than any of the quadratic-complexity algorithms.

- **Choice of reduction algorithm.** Non-pairing elliptic-curve cryptography typically chooses curves over prime fields with particularly "nice" (i.e., sparse) primes that come with very efficient modular-reduction algorithms. In pairing-based cryptography this is not possible, because the field of definition of the pairing-friendly elliptic curves falls out of the parameterized constructions for pairing-friendly curves. This is why the standard approach for implementing modular reduction is to represent field elements in the Montgomery domain and use the efficient Montgomery reduction algorithm [29]. Some implementations attempted to exploit the structure of the prime (which comes from the parameterized construction) for efficient reduction [15, 30], but at least in software, those approaches have been outperformed by Montgomery arithmetic.

Note that the above choices do not have independent influence on performance; there are typically various subtle interactions. For example, the choice of a different radix may favor a different multiplication algorithm or the choice of a specific reduction algorithm may *require* a certain radix. Consequently, there is a large body of literature on multiprecision arithmetic on a broad variety of computer architectures. A good starting point to get an idea of the general algorithmic approaches is [26, Section 4.3] and [28, Section 14.2].

11.6   Constant-Time Software

Many modern cryptographic primitives, including suitably chosen cryptographic pairings, are considered infeasible to break in the so-called *black-box* setting. This setting allows the attacker to choose inputs and observe outputs of cryptographic primitives and protocols, but not to make any observations about the computation. Unfortunately, this model is often too weak to describe the actual power of an attacker. Real-world attackers could focus on the ancillary functions around the pairing function. With the help of the discrete logarithm problem, one is able to hide a secret into an element of $\mathbb{G}_1$, $\mathbb{G}_2$, or $\mathbb{G}_T$, before or after the pairing function. The exponentiation in these groups may include functions leaking information about the secret, the so-called side-channel information that leaks during the computation; this is discussed in much more detail in Chapter 12 since the pairing itself could be eventually protected against these attacks.

In this section we will briefly discuss one specific class of side-channel attacks, namely timing attacks, and describe how to inherently protect software against those attacks by implementations that follow the *constant-time-software* paradigm.

The basic idea of timing attacks is that the execution time of (cryptographic) software depends on secret data. An attacker measures the execution time and deduces information about the secret data. This basic idea also shows how to fully protect software against timing attacks: The software needs to be written in such a way that it takes the same amount for any values of secret data. This may sound like a very harsh statement, but luckily there are essentially two sources for timing variability in software: secret branch conditions and secret memory addresses.

11.6.1   Eliminating Secret Branch Conditions

The general structure of a secretly conditioned branch is "if $(s = 1)$ do $R \leftarrow A()$; else do $R \leftarrow B()$," where s is a secret bit of information, $A()$ and $B()$ are some computations, and R is the data that is manipulated by $A()$ or $B()$. This kind of code will obviously leak information about s through timing if $A()$ and $B()$ take a different amount of time. Maybe less obvious is that it typically also leaks information about s if $A()$ and $B()$ take the same amount of time. One reason is that most modern CPUs try to predict whether a branch is taken or not to avoid pipeline flushes. If this prediction is correct, the computation is going to be faster than if the prediction is wrong. Another reason is instruction caches. If $A()$ uses code from different memory locations than $B()$, it can be the case that the code of only one of the two branches is in cache and then this code runs faster. The generic approach to avoid this kind of timing leak is to replace branches by arithmetic that replaces the above sequence of operations by something like $R \leftarrow s \cdot A() + (1 - s) \cdot B()$. Note that this means that both branches, $A()$ and $B()$, have to be computed. Also note that it is not necessary to use multiplication and addition for this arithmetic approach; it is very common to expand s to an all-one or all-zero mask and then use a bit-logical AND instead of multiplication and a bit-logical XOR or OR instead of addition. To illustrate this, consider the following function written in C that conditionally copies n bytes from a to r if c is set to 1:

```
/* c has to be either 1 or 0 */
void cmov(unsigned char *r, const unsigned char *b, size_t n, int c) {
  size_t i;
  int nc;
  c  = -c; /* assume 2s-complement for negative ints */
  nc = ~c;
  for(i=0;i<n;i++) {
    r[i] = (nc & r[i]) | (c & a[i]);
  }
}
```

11.6.2   Eliminating Secret Memory Addresses

The second main sources of timing variability are secret memory addresses, i.e., loads of the form "$a \leftarrow \text{mem}[s]$" or stores of the form "$\text{mem}[s] \leftarrow a$", where the address s depends on secret data. The most obvious reason that a load from a secret address leaks information about the address is caching: If the cache line containing the address is in cache, the load is going to be fast ("cache hit"); if the cache line is not in cache, the load needs to retrieve data from main memory and the load is much slower ("cache miss"). This explanation suggests that loads can only leak information about cache lines; however, there are multiple other effects that potentially leak information about the least-significant bits of the address. Examples are cache-bank conflicts and store-to-load forwarding. Constant-time software needs to avoid all memory access at addresses that depend on secret data. The general approach to eliminate loads from secret positions is to load all possible values, e.g., from a lookup table, and then use conditional copies (such as the one listed in the `cmov` function above) to copy data to the result. A general approach to performing a constant-time lookup of the datastructure of type `elem` at position `pos` from a table with `TABLE_SIZE` entries looks like this:

```
elem lookup(const elem *table, size_t pos) {
  size_t i;
  int d;
  elem r = table[0];
  for(i=0;i<TABLE_SIZE;i++) {
    d = int_isequal(i, pos);
    cmov{(unsigned char *)&r, (unsigned char *)table+i, sizeof(elem), d);
  }
  return r;
}
```

Note that this code does not use `d = (i == pos)`, because this might cause a compiler to re-introduce a timing leak. It instead relies on a safe implementation of the equality check called `int_isequal`. This comparison can be implemented as follows (under the reasonable assumptions that a `sizeof(int)` is strictly smaller than `sizeof(long long)`, that a `long long` has 64 bits, and that negative numbers are represented in the 2's complement):

```
int int_isequal(int a, int b) {
  long long t = a ^ b;
  t = (-t) >> 63;
  return 1 - (t & 1);
}
```

11.7   Software Libraries

There exist various libraries and software frameworks to compute cryptographic pairings and related arithmetic operations in the groups. In this section, we briefly introduce two of them and for each of them give example code that implements the BLS signature scheme [10] following the eBACS API [7] for cryptographic signatures. Other well-known software implementations are contained in the Pairing-Based Cryptography library* and MIRACL Cryptographic SDK.[†] The former can be considered outdated and does not include modern instantiations of pairings, while the latter can also be found in a flavor tailored to typical embedded devices.[‡] We start by reviewing the BLS signature scheme.

---

*https://crypto.stanford.edu/pbc/
[†]https://github.com/CertiVox/MIRACL
[‡]https://github.com/CertiVox/MiotCL

11.7.1   The BLS Signature Scheme

Signature schemes are an important cryptographic primitive. A signer possessing a secret signing key can sign messages using a signing algorithm; for a matching public verification key, the message-signature pair can be checked with a verification algorithm.

It is possible to construct a signature scheme from an Identity-Based Encryption (IBE) scheme. Suppose there is an IBE scheme secure against adaptive-identity attacks; here, the adversary submits identities, and receives the corresponding decryption keys. The analogy from an IBE scheme could be the message submissions (identities): In order to get valid signatures, the signatures would play the role of the decryption keys, if the adversary can generate valid message-signature pairs, then the scheme has failed. Verification of the decryption key (signature) can be done by encrypting a random string under the identity (message), and then using the decryption algorithm with the decryption key to see if the string is recovered properly. [13]

Based on this construction, Boneh, Lynn, and Shacham in 2001 [10, 2004 version] introduced a short signature scheme based on the computational Diffie-Hellman assumption on certain elliptic and hyperelliptic curves. Their scheme is particularly interesting since it produces a significantly smaller signature than traditional elliptic curve-based schemes.

DEFINITION 11.1   Let $E(\mathbb{F}_p)$ be an elliptic curve, and let $P, Q \in E(\mathbb{F}_p)$ have prime order r. The **co-DDH problem** is: given $(P, [a]P, Q, [b]Q)$ to determine if $a \stackrel{?}{\equiv} b \bmod r$. [17]

The computational variant is: given $(P, Q, [a]Q)$, compute $[a]P$.

The BLS short signature scheme makes use of a hash function $H : \{0,1\}^* \to \mathbb{G}_1$, and an asymmetric pairing $e : \mathbb{G}_1 \times \mathbb{G}_2 \to \mathbb{G}_T$. Let $\mathbb{G}_1 = <P>$, and $\mathbb{G}_2 = <Q>$, viewing H as a random oracle. The scheme is as follows:

- **Key Generation.** Choose $x \in_R \mathbb{Z}_r$, set $R \leftarrow [x]Q$. The public key is Q, R, whereas the private key is x.
- **Sign.** Map-to-Point the message to sign as $P_M = H(M) \in <P>$, set $S_M \leftarrow [x]P_M$. The signature is the x-coordinate of S_M.
- **Verify.** Given the x-coordinate of S_M, find $\pm S_M$. Decide: $e(Q, S_M) \stackrel{?}{=} e(R, H(M))$, and (P_M, S_M, Q, R) is a valid Computational Diffie-Hellman tuple.

For implementation purposes, the sign of the y-coordinate of S_M can also be included as part of the signature.

11.7.2   RELIC is an Efficient LIbrary for Cryptography

RELIC is a modern cryptographic meta-toolkit with emphasis on efficiency and flexibility, and can be used to build efficient and usable cryptographic toolkits tailored for specific security levels and algorithmic choices. The focus is to provide portability, easy support to architecture-dependent code, flexible configuration, and maximum efficiency. In terms of pairing-based cryptographics, RELIC implements several types of pairings and pairing-based protocols, including pairings over BN curves and other parameterized curves at different security levels; the Sakai-Ohgishi-Kasahara ID-based authenticated key agreement, Boneh-Lynn-Schacham and Boneh-Boyen short signatures, and a version of the Boneh-Go-Nissin homomorphic encryption system adapted to asymmetric pairings. Algorithms 11.4, 11.5, and 11.6 present code portions implementing the BLS signature scheme (as included in the library), and signature/verification operations with an eBACS-compatible interface, together with illustrative test code.

ALGORITHM 11.4   BLS implementation with RELIC.

```
//Function to generate signing, and verification keys
//Input: memory space to store the signing, and verification keys
//Output: pub - verification key, priv - signing key
int cp_bls_gen(bn_t priv, g2_t pub) {
    bn_t ord;
    g2_get_ord(ord);
    bn_rand_mod(priv, ord);
    g2_mul_gen(pub, priv);
    return STS_OK;
}


//Function to sign a message
//Input: msg - message to sign, len - length of message in bytes,
//priv - signing key
//Output: sig - signature of message msg with signing key priv
int cp_bls_sig(g1_t sig, uint8_t *msg, int len, bn_t priv) {
    g1_t p;
    g1_new(p);
    g1_map(p, msg, len);
    g1_mul(sig, p, priv);
    return STS_OK;
}


//Function to verify a message
//Input: sig - signature of message msg, msg - message to verify,
//len - length of message msg, q - verification key
//Output: 1 - True, signature verifies,
// 0 - False, signature verification fails
int cp_bls_ver(g1_t sig, uint8_t *msg, int len, g2_t q) {
    g1_t p;
    g2_t g;
    gt_t e1, e2;

    g2_get_gen(g);

    g1_map(p, msg, len);
    pc_map(e1, p, q);
    pc_map(e2, sig, g);

    if (gt_cmp(e1, e2) == CMP_EQ) {
        return 1;
    }
    return 0;
}
```

ALGORITHM 11.5   eBACs interface for BLS signatures in RELIC.

```
#include "relic.h"

#define CRYPTO_SECRETKEYBYTES    PC_BYTES
#define CRYPTO_PUBLICKEYBYTES    (2 * PC_BYTES + 1)
#define CRYPTO_BYTES             (PC_BYTES + 1)
#define TEST_BYTES               20
#define COMPRESS                 1

//eBats function to generate signing, and verification keys
//Input: memory space to store the signing, and verification keys
//Output: pk - verification key, sk - signing key
int crypto_sign_keypair(unsigned char *pk, unsigned char *sk) {
    bn_t k;
    g2_t pub;

    if (cp_bls_gen(k, pub) == STS_OK) {
        bn_write_bin(sk, CRYPTO_SECRETKEYBYTES, k);
        g2_write_bin(pk, CRYPTO_PUBLICKEYBYTES, pub, COMPRESS);
        return 0;
    }
    return 1;
}

//eBats function to sign a message
//Input: m - message to sign, mlen - length of message in bytes,
// sk - signing key
//Output: sm - signature of message m with signing key sk,
// smlen - length of signature
int crypto_sign(unsigned char *sm, unsigned long long *smlen, const unsigned
        char *m, unsigned long long mlen, const unsigned char *sk) {
    g1_t sig;
    bn_t k;
    int i, len = CRYPTO_BYTES;

    if (*smlen < CRYPTO_BYTES + mlen)
            return 1;
    bn_init(k, BN_DIGS);
    bn_read_bin(k, sk, CRYPTO_SECRETKEYBYTES);

    if (cp_bls_sig(sig, m, mlen, k) == STS_OK) {
            g1_write_bin(sm, &len, sig, COMPRESS);
            for (i = 0; i < mlen; i++)
                    sm[i + CRYPTO_BYTES] = m[i];
            *smlen = mlen + CRYPTO_BYTES;
            return 0;
    }
    return 1;
}
```

ALGORITHM 11.6 Verification and test code for BLS signatures in RELIC.

```
//eBats function to verify a message
//Input: m - message to verify, mlen - length of message m,
// sm - signature of message m, smlen - length of signature sm,
// and pk - verification key
//Output: 0 signature verifies, -1 - signature verification fails
int crypto_sign_open(unsigned char *m, unsigned long long *mlen, const
        unsigned char *sm, unsigned long long smlen, const unsigned
        char *pk) {
    g1_t sig;
    g2_t pub;

    g2_read_bin(pub, pk, CRYPTO_PUBLICKEYBYTES);
    g1_read_bin(sig, sm, CRYPTO_BYTES);
    if (cp_bls_ver(sig, sm + CRYPTO_BYTES, smlen - CRYPTO_BYTES, pub) == 1) {
            for (int i = 0; i < smlen - CRYPTO_BYTES; i++)
                    m[i] = sm[i + CRYPTO_BYTES];
            *mlen = smlen - CRYPTO_BYTES;
            return 0;
    } else {
            for (int i = 0; i < smlen - CRYPTO_BYTES; i++)
                    m[i] = 0;
            *mlen = (unsigned long long)(-1);
            return -1;
    }
}

int main(int argc, char *arv[]) {
    unsigned char m[TEST_BYTES], sig[CRYPTO_BYTES + TEST_BYTES];
    unsigned char sk[CRYPTO_SECRETKEYBYTES], pk[CRYPTO_PUBLICKEYBYTES];
    unsigned long long len = CRYPTO_BYTES + TEST_BYTES;

    core_init();
    rand_bytes(m, TEST_BYTES);
    if (pc_param_set_any() == STS_OK) {
        if (crypto_sign_keypair(pk, sk) != 0)
            return 1;
        if (crypto_sign(sig, &len, m, TEST_BYTES, sk) != 0)
            return 1;
        /* Check signature. */
        len = TEST_BYTES;
        if (crypto_sign_open(m, &len, sig, CRYPTO_BYTES + TEST_BYTES, pk) =
            = -1) return 1;
        len = TEST_BYTES;
        /* Make signature invalid and check if it fails. */
        sig[CRYPTO_BYTES] ^= 1;
        if (crypto_sign_open(m, &len, sig, CRYPTO_BYTES + TEST_BYTES, pk) =
          = 0) return 1;
    }
    core_clean();
}
```

11.7.3   PandA – Pairings and Arithmetic

PandA is a framework for pairing computations and arithmetic in the related groups. The main idea of PandA is to provide the definition of an API together with tests and benchmarks; implementors of pairing software can implement this API and protocol implementors can simply include a different header file and recompile their protocol to use a different implemention in PandA.

A particular emphasis in the API design of PandA is to distinguish between computations that operate on secret data and those that only involve public inputs. By default, all functions assume to possibly receive secret input and the implementor of the pairing software must ensure that the software does not leak (timing) information about those secret values (cf. Section 11.6). If the protocol implementor knows that inputs to a certain function are always public, he can choose to use a _publicinputs version of the respective function, and the library implementor is free to provide a faster (but unprotected) implementation for that function.

Algorithms 11.7, 11.9, and 11.10 present code portions implementing the BLS signature scheme following the eBACS API for cryptographic signatures.

Implementing BLS with PandA

ALGORITHM 11.7   The file `api.h` for the eBACS API.

```
#include "panda.h"

#define CRYPTO_SECRETKEYBYTES BGROUP_SCALAR_BYTES
#define CRYPTO_PUBLICKEYBYTES BGROUP_G2E_PACKEDBYTES
#define CRYPTO_BYTES BGROUP_G1E_PACKEDBYTES
```

ALGORITHM 11.8   BLS keypair generation with PandA.

```
//Function to generate signing, and verification keys
//Input: memory space to store the signing, and verification keys
//Output: pk - verification key, sk - signing key
  int crypto_sign_keypair(
    unsigned char *pk,
    unsigned char *sk
  )
  {
    // private key //
    bgroup_scalar x;
    bgroup_scalar_setrandom(&x);
    bgroup_scalar_pack(sk, &x);

    // public key //
    bgroup_g2e r;
    bgroup_g2_scalarmult_base(&r, &x);
    bgroup_g2_pack(pk, &r);

    return 0;
  }
```

ALGORITHM 11.9   BLS signing with PandA.

```
//Function to sign a message
//Input: m - message to sign, mlen - length of message in bytes,
// sk - signing key
//Output: sm - signature of message m with signing key sk,
// smlen - length of signature
  int crypto_sign(
    unsigned char *sm, unsigned long long *smlen,
    const unsigned char *m, unsigned long long mlen,
    const unsigned char *sk)
  {

    bgroup_g1e p, p1;
    bgroup_scalar x;
    int i,r;

    bgroup_g1e_hashfromstr_publicinputs(&p, m, mlen);
    r = bgroup_scalar_unpack(&x, sk);
    bgroup_g1e_scalarmult(&p1, &p, &x);
    bgroup_g1e_pack(sm, &p1);

    for (i = 0; i < mlen; i++)
    sm[i + CRYPTO_BYTES] = m[i];
    *smlen = mlen + CRYPTO_BYTES;

    return -r;
  }
```

ALGORITHM 11.10   BLS verification with PandA.

```
//Function to verify a message
//Input: m - message to verify, mlen - length of message m,
// sm - signature of message m, smlen - length of signature sm,
// and pk - verification key
//Output: 0 signature verifies, -1 - signature verification fails
int crypto_sign_open(
    unsigned char *m, unsigned long long *mlen,
    const unsigned char *sm, unsigned long long smlen,
    const unsigned char *pk)
  {

    bgroup_g1e p[2];
    bgroup_g2e q[2];
    bgroup_g3e r;
    unsigned long long i;
    int ok;

    ok  = !bgroup_g1e_unpack(p, sm);
    bgroup_g1e_negate_publicinputs(p, p);
    q[0] = bgroup_g2e_base;
    bgroup_g1e_hashfromstr_publicinputs(p+1, sm + CRYPTO_BYTES, smlen -
     CRYPTO_BYTES); ok &= !bgroup_g2e_unpack(q+1, pk);
    bgroup_pairing_product(&r, p, q, 2);

    ok &= bgroup_g3e_equals(&r, &bgroup_g3e_neutral);

    if (ok)
    {
      for (i = 0; i < smlen - CRYPTO_BYTES; i++)
        m[i] = sm[i + CRYPTO_BYTES];
      *mlen = smlen - CRYPTO_BYTES;
      return 0;
    }
    else
    {
      for (i = 0; i < smlen - CRYPTO_BYTES; i++)
        m[i] = 0;
      *mlen = (unsigned long long) (-1);
      return -1;
    }
  }
```

11.8   A Pairing in SAGE

Listing 50 File `parameters-negative.sage`. Definition of parameters and computation of derived parameters for the optimal ate pairing using the Barreto-Naehrig curve $E : Y^2 = X^3 + 2$ over $\mathbb{F}_p$, where $p = 36 * t^4 + 36 * t^3 + 24 * t^2 + 6 * t + 1$ and $t = -(2^{62} + 2^{55} + 1)$. The group order $r = 36 * t^4 + 36 * t^3 + 18 * t^2 + 6 * t + 1$ is a 254-bit prime.

```
1   t   = -(2**62 + 2**55 + 1)
2   p   = 36*t^4+36*t^3+24*t^2+6*t+1
3   r   = 36*t^4+36*t^3+18*t^2+6*t+1
4   tr  = 6*t^2+1
5
6   # MSB-first signed binary representation of abs(6*t+2)
7   L = [1,0,0,0,0,0,1,1,0,0,0,0,0,0,0,0,0,0,0,0,0,0,
8        0,0,0,0,0,0,0,0,0,0,0,0,0,0,0,0,0,0,0,0,0,0,
9        0,0,0,0,0,0,0,0,0,0,0,0,0,0,0,0,1,0,0]
10
11  # Definition of finite fields
12  F       = GF(p)
13  K2.<x>  = PolynomialRing(F)
14  F2.<u>  = F.extension(x^2+1)
15  K6.<y>  = PolynomialRing(F2)
16  F6.<v>  = F2.extension(y^3-(u+1))
17  K12.<z> = PolynomialRing(F6)
18  F12.<w> = F6.extension(z^2-v)
19  # Required to work around limitations of Sage
20  F12.is_field = lambda:True
21
22  # Constants required in Frobenius computation
23  c1 = v**(p-1)
24  c2 = (v*w)**(p-1)
25
26  # Definition of curve E and twist EE
27  b   = 2
28  E   = EllipticCurve(F,[0,b])
29  EE  = EllipticCurve(F2,[0,b/(u+1)])
30  # cofactor of the twist
31  h   = EE.order()//r
```

Listing 51 File `parameters-positive.sage`. Definition of parameters and computation of derived parameters for the optimal ate pairing using the Barreto-Naehrig curve $E : Y^2 = X^3 + 5$ over $\mathbb{F}_p$, where $p = 36 * t^4 + 36 * t^3 + 24 * t^2 + 6 * t + 1$ and $t = 2^{62} - 2^{54} + 2^{44}$. The group order $r = 36 * t^4 + 36 * t^3 + 18 * t^2 + 6 * t + 1$ is a 254-bit prime.

```
1   t   = 2^62-2^54+2^44
2   p   = 36*t^4+36*t^3+24*t^2+6*t+1
3   r   = 36*t^4+36*t^3+18*t^2+6*t+1
4   tr  = 6*t^2+1
5
6   # MSB-first signed binary representation of abs(6*t+2)
7   L = [1,0,0,0,0,0,0,-1,-1,0,0,0,0,0,0,0,0,1,1,0,0,
8        0,0,0,0,0,0,0,0,0,0,0,0,0,0,0,0,0,0,0,0,0,0,
9        0,0,0,0,0,0,0,0,0,0,0,0,0,0,0,0,0,0,1,0]
10
11  # Definition of finite fields
12  F       = GF(p)
13  K2.<x>  = PolynomialRing(F)
14  F2.<u>  = F.extension(x^2+5)
15  K6.<y>  = PolynomialRing(F2)
16  F6.<v>  = F2.extension(y^3-u)
17  K12.<z> = PolynomialRing(F6)
18  F12.<w> = F6.extension(z^2-v)
19  # Required to work around limitations of Sage
20  F12.is_field = lambda:True
21
22  # Constants required in Frobenius computation
23  c1 = v**(p-1)
24  c2 = (v*w)**(p-1)
25
26  # Definition of curve E and twist EE
27  b   = 5
28  E   = EllipticCurve(F,[0,b])
29  EE  = EllipticCurve(F2,[0,b/u])
30  # cofactor of the twist
31  h   = EE.order()//r
32  OA
```

Listing 52 File `linefunctions.sage`. Point addition and doubling and evaluation of the corresponding line functions using Jacobian Coordinates. See Algorithms 11.13 and 11.10.

```
1    load("parameters-negative.sage")
2    #load("parameters-positive.sage")
3
4    #Line function
5    #Input: P=(xp, yp) \in G1, T=(XR, YR, ZR), Q=(XQ, YQ, ZQ) \in G2
6    #Output: T(XR, YR, ZR) \gets [2]T(XR, YR, ZR),
7    # f \gets \g2_{\pi_E(T+Q)}(P)
8    def add_eval(XR,YR,ZR,XQ,YQ,ZQ,xp,yp):
9        ZR2 = ZR**2
10       t0  = XQ*ZR2
11       t1  = (YQ+ZR)**2 - YQ*YQ - ZR2
12       t1  = t1*ZR2
13       t2  = t0 - XR
14       t3  = t2**2
15       t4  = 4*t3
16       t5  = t4*t2
17       t6  = t1-2*YR
18       t9  = t6*XQ
19       t7  = XR*t4
20       XT  = t6**2-t5-2*t7
21       ZT  = (ZR+t2)**2 - ZR2 - t3
22       t10 = YQ + ZT
23       t8  = (t7-XT)*t6
24       t0  = 2*(YR*t5)
25       YT  = t8-t0
26       t10 = t10**2 - YQ**2 - ZT**2
27       t9  = 2*t9-t10
28       t10 = 2*(ZT*yp)
29       t6  = -t6
30       t1  = 2*(t6*xp)
31       l00 = t10
32       l10 = t1
33       l11 = t9
34       f   = l00 + l10*w + l11*v*w
35       return (XT,YT,ZT,f)
36
37
38   #Line function
39   #Input: P=(xp, yp) \in G1, T=(XR, YR, ZR) \in G2
40   #Output: T(XR, YR, ZR) \gets [2]T(XR, YR, ZR),
41   # f \gets \g2_{\pi_E(T)}(P)
42   def dbl_eval(XR,YR,ZR,xp,yp):
43       t0  = XR**2
44       t1  = YR**2
45       ZR2 = ZR**2
46       t2  = t1**2
47       t3  = (t1+XR)**2 - t0 -t2
48       t3  = 2*t3
49       t4  = 3*t0
50       t6  = XR+t4
51       t5  = t4**2
52       XT  = t5 - 2*t3
53       ZT  = (YR+ZR)**2 - t1 - ZR2
54       YT  = (t3 - XT)*t4 - 8*t2
55       t3  = -2*(t4 *ZR2)
56       l10 = t3*xp
57       l11 = t6**2-t0-t5-4*t1
58       t0  = 2*(ZT*ZR2)
59       l00 = t0*yp
60       f   = l00 + l10*w + l11*v*w
61       return (XT,YT,ZT,f)
```

Listing 53 File `millerloop.sage`. The Miller loop computation of the optimal ate pairing using the signed binary representation of $6t + 2$ is defined in file `parameters.sage` and calls to the line-function evaluations defined in file `linefunctions.sage`.

```
1    load("linefunctions.sage")
2
3    #Miller loop e(.,.)
4    #Input: P \in G1, Q \in G2
5    #Output: f \in \F_{p^k}
6    def millerloop(P,Q):
7        def frob_point_prime(XP,YP,ZP):
8            XR = c1*(XP**p) #XXX: Frobenius operator
9            YR = c2*(YP**p) #XXX: Frobenius operator
10           ZR = ZP**p      #XXX: Frobenius operator
11           return (XR, YR, ZR)
12
13       def conjugation(f):
14           (a0,a1) = vector(f)
15           return a0-w*a1
16
17       xq,yq,zq = Q
18       xt,yt,zt = Q
19       f=1
20
21       for i in L:
22           xt,yt,zt,d = dbl_eval(xt,yt,zt,P[0],P[1])
23           f = f**2
24           f = f*d #XXX: sparse mul
25           if (i==-1):
26               xt,yt,zt,d = add_eval(xt,yt,zt,xq,-yq,zq,P[0],P[1])
27               f = f*d #XXX: sparse mul
28           elif (i==1):
29               xt,yt,zt,d = add_eval(xt,yt,zt,xq,yq,zq,P[0],P[1])
30               f = f*d #XXX: sparse mul
31
32       xq1,yq1,zq1 = frob_point_prime(xq,yq,zq)
33       xq2,yq2,zq2 = frob_point_prime(xq1,yq1,zq1)
34
35       if t < 0:
36           f = conjugation(f)
37           yt = -yt
38
39       xt,yt,zt,d  = add_eval(xt,yt,zt,xq1,yq1,zq1,P[0],P[1])
40       f = f*d #XXX: sparse mul
41
42       yq_2 = -yq2
43       xt,yt,zt,d = add_eval(xt,yt,zt,xq2,yq_2,zq2,P[0],P[1])
44       f = f*d #XXX: sparse mul
45
46       return f
```

Listing 54 File `finalexponentiation.sage`. Optimized final exponentiation with exponent $e = (p^{12} - 1)/r$.

```
1   #load("fp12.sage")
2
3   #Function for the final exponentiation
4   #Input: f \in \F_{p^k}
5   #Output: f^{(p^{12} - 1)/r}
6   def final_expo(f):
7       def conjugation(f):
8           (a0,a1) = vector(f)
9           return a0-w*a1
10      f = conjugation(f) * f^(-1)
11      f = f^(p^2)*f
12      if t < 0:
13          ft1 = conjugation(f)
14          ft1 = ft1^abs(t)
15          ft2 = conjugation(ft1)
16          ft2 = ft2^abs(t)
17          ft3 = conjugation(ft2)
18          ft3 = ft3^abs(t)
19      else:
20          ft1 = f^abs(t)
21          ft2 = ft1^abs(t)
22          ft3 = ft2^abs(t)
23      fp1 = f^p
24      fp2 = fp1^p
25      fp3 = fp2^p
26      y0 = fp1*fp2*fp3
27      y1 = conjugation(f)
28      y2 = ft2^(p^2)
29      y3 = ft1^p
30      y3 = conjugation(y3)
31      y4 = ft2^p
32      y4 = y4*ft1
33      y4 = conjugation(y4)
34      y5 = conjugation(ft2)
35      y6 = ft3^p
36      y6 = y6*ft3
37      y6 = conjugation(y6)
38      t0 = y6^2*y4*y5
39      t1 = y3*y5*t0
40      t0 = t0*y2
41      t1 = (t1^2*t0)^2
42      t0 = t1*y1
43      t1 = t1*y0
44      t0 = t0^2
45      f = t1*t0
46      return f
```

Listing 55 File `optimal_ate.sage`. Optimal ate pairing consisting of the Miller loop in file `millerloop.sage` and the final exponentation in file `finalexponentiation.sage`.

```
1   load ("millerloop.sage")
2   load ("finalexponentiation.sage")
3
4   #Function to compute the ate pairing
5   #Input: P \in G1, Q \in G2
6   #Output: f \in \mu_r
7   def Optimal_Ate (P,Q):
8       f = millerloop (P,Q)
9       f = final_expo (f)
10      return f
```

Listing 56 File `test.sage`. Minimal test of the optimal ate pairing defined in file `optimal_ate.sage`.

```
1   load ("optimal_ate.sage")
2
3   ntests = 1
4
5   for n in range (ntests):
6       D1=E.random_point ()
7       D2=h*EE.random_point ()
8       s = randint (0,r)
9       r1 = Optimal_Ate (D1,s*D2)
10      r2 = Optimal_Ate (s*D1,D2)
11      r3 = Optimal_Ate (D1,D2)^s
12      print r1 == r2 == r3
13      print r1 != 0 and r2 != 0 and r3 != 0
14      if r1 == 1:
15          print "Warning: pairing computes 1"
```

References

[1] Diego F. Aranha, Paulo S. L. M. Barreto, Patrick Longa, and Jefferson E. Ricardini. The realm of the pairings. In T. Lange, K. Lauter, and P. Lisonek, editors, *Selected Areas in Cryptography – SAC 2013*, volume 8282 of *Lecture Notes in Computer Science*, pp. 3–25. Springer, Heidelberg, 2014.

[2] Diego F. Aranha, Laura Fuentes-Castañeda, Edward Knapp, Alfred Menezes, and Francisco Rodríguez-Henríquez. Implementing pairings at the 192-bit security level. In M. Abdalla and T. Lange, editors, *Pairing-Based Cryptography – Pairing 2012*, volume 7708 of *Lecture Notes in Computer Science*, pp. 177–195. Springer, Heidelberg, 2013.

[3] Diego F. Aranha, Koray Karabina, Patrick Longa, Catherine H. Gebotys, and Julio López. Faster explicit formulas for computing pairings over ordinary curves. In K. G. Paterson, editor, *Advances in Cryptology – EUROCRYPT 2011*, volume 6632 of *Lecture Notes in Computer Science*, pp. 48–68. Springer, Heidelberg, 2011.

[4] Razvan Barbulescu, Pierrick Gaudry, Aurore Guillevic, and François Morain. Improving NFS for the discrete logarithm problem in non-prime finite fields. In E. Oswald and M. Fischlin, editors, *Advances in Cryptology – EUROCRYPT 2015, Part I*, volume 9056 of *Lecture Notes in Computer Science*, pp. 129–155. Springer, Heidelberg, 2015.

[5] Paulo S. L. M. Barreto, Craig Costello, Rafael Misoczki, Michael Naehrig, Geovandro C. C. F. Pereira, and Gustavo Zanon. Subgroup security in pairing-based cryptography. In K. E. Lauter and F. Rodríguez-Henríquez, editors, *Progress in Cryptology – LATINCRYPT 2015*, volume 9230 of *Lecture Notes in Computer Science*, pp. 245–265. Springer, Heidelberg, 2015.

[6] Paulo S. L. M. Barreto, Ben Lynn, and Michael Scott. Constructing elliptic curves with prescribed embedding degrees. In S. Cimato, C. Galdi, and G. Persiano, editors, *Security in Communication Networks (SCN 2002)*, volume 2576 of *Lecture Notes in Computer Science*, pp. 257–267. Springer, Heidelberg, 2003.

[7] Daniel J. Bernstein and Tanja Lange (editors). eBACS: ECRYPT benchmarking of cryptographic systems. http://bench.cr.yp.to (accessed 2014-05-17).

[8] Jean-Luc Beuchat, Jorge E. González-Díaz, Shigeo Mitsunari, Eiji Okamoto, Francisco Rodríguez-Henríquez, and Tadanori Teruya. High-speed software implementation of the optimal Ate pairing over Barreto-Naehrig curves. In M. Joye, A. Miyaji, and A. Otsuka, editors, *Pairing-Based Cryptography – Pairing 2010*, volume 6487 of *Lecture Notes in Computer Science*, pp. 21–39. Springer, Heidelberg, 2010.

[9] Dan Boneh and Matthew K. Franklin. Identity-based encryption from the Weil pairing. In J. Kilian, editor, *Advances in Cryptology – CRYPTO 2001*, volume 2139 of *Lecture Notes in Computer Science*, pp. 213–229. Springer, Heidelberg, 2001.

[10] Dan Boneh, Ben Lynn, and Hovav Shacham. Short signatures from the Weil pairing. *Journal of Cryptology*, 17(4):297–319, 2004.

[11] Joppe W. Bos, Craig Costello, and Michael Naehrig. Exponentiating in pairing groups. In T. Lange, K. Lauter, and P. Lisonek, editors, *Selected Areas in Cryptography – SAC 2013*, volume 8282 of *Lecture Notes in Computer Science*, pp. 438–455. Springer, Heidelberg, 2014.

[12] Friederike Brezing and Annegret Weng. Elliptic curves suitable for pairing based cryptography. *Designs, Codes and Cryptography*, 37(1):133–141, 2005.

[13] Sanjit Chatterjee and Palash Sarkar. *Identity-Based Encryption*. Springer, 2011.

[14] Craig Costello, Tanja Lange, and Michael Naehrig. Faster pairing computations on curves with high-degree twists. In P. Q. Nguyen and D. Pointcheval, editors, *Public Key Cryptography – PKC 2010*, volume 6056 of *Lecture Notes in Computer Science*, pp. 224–242. Springer, Heidelberg, 2010.

[15] Junfeng Fan, Frederik Vercauteren, and Ingrid Verbauwhede. Efficient hardware implementation of $\mathbb{F}_p$-arithmetic for pairing-friendly curves. *IEEE Transactions on Computers*, 61(5):676–685, 2012.

[16] Laura Fuentes-Castañeda, Edward Knapp, and Francisco Rodríguez-Henríquez. Faster hashing to $\mathbb{G}_2$. In A. Miri and S. Vaudenay, editors, *Selected Areas in Cryptography – SAC 2011*, volume 7118 of *Lecture Notes in Computer Science*, pp. 412–430. Springer, Heidelberg, 2012.

[17] Steven D. Galbraith. *Mathematics of Public Key Cryptography*. Cambridge University Press, 2012.

[18] Robert Granger and Nigel P. Smart. On computing products of pairings. Cryptology ePrint Archive, Report 2006/172, 2006. http://eprint.iacr.org/2006/172.

[19] Gurleen Grewal, Reza Azarderakhsh, Patrick Longa, Shi Hu, and David Jao. Efficient implementation of bilinear pairings on ARM processors. In L. R. Knudsen and H. Wu, editors, *Selected Areas in Cryptography – SAC 2012*, volume 7707 of *Lecture Notes in Computer Science*, pp. 149–165. Springer, Heidelberg, 2013.

[20] Nils Gura, Arun Patel, Arvinderpal Wander, Hans Eberle, and Sheueling Chang Shantz. Comparing elliptic curve cryptography and rsa on 8-bit cpus. In *Cryptographic Hardware and Embedded Systems - CHES 2004*, pp. 119–132. Springer Berlin Heidelberg, 2004.

[21] Michael Hutter and Erich Wenger. Fast multi-precision multiplication for public-key cryptography on embedded microprocessors. In B. Preneel and T. Takagi, editors, *Cryptographic Hardware and Embedded Systems – CHES 2011*, volume 6917 of *Lecture Notes in Computer Science*, pp. 459–474. Springer, Heidelberg, 2011.

[22] Antoine Joux. A one round protocol for tripartite Diffie-Hellman. In W. Bosma, editor, *Algorithmic Number Theory (ANTS-IV)*, volume 1838 of *Lecture Notes in Computer Science*, pp. 385–393. Springer, 2000.

[23] Antoine Joux. A one round protocol for tripartite Diffie-Hellman. *Journal of Cryptology*, 17(4):263–276, 2004.

[24] Ezekiel J. Kachisa, Edward F. Schaefer, and Michael Scott. Constructing Brezing-Weng pairing-friendly elliptic curves using elements in the cyclotomic field. In S. D. Galbraith and K. G. Paterson, editors, *Pairing-Based Cryptography – Pairing 2008*, volume 5209 of *Lecture Notes in Computer Science*, pp. 126–135. Springer, Heidelberg, 2008.

[25] Antolii A. Karatusba and Yuri P. Ofman. Multiplication of many-digital numbers by automatic computers. *Proceedings of the USSR Academy of Sciences 145: 293–294. Translation in the academic journal Physics-Doklady*, pp. 595–596, 1963.

[26] Donald E. Knuth. *The Art of Computer Programming, Third Edition*. Addison-Wesley, 1997.

[27] Stephen Lichtenbaum. Duality theorems for curves over p-adic fields. *Inventiones mathematicae*, 7(2):120–136, 1969.

[28] Alfred J. Menezes, Paul C. van Oorschot, and Scott A. Vanstone. *Handbook of Applied Cryptography*. The CRC Press series on discrete mathematics and its applications. CRC Press, 2000 N.W. Corporate Blvd., Boca Raton, FL 33431-9868, USA, 1997.

[29] Peter L. Montgomery. Modular multiplication without trial division. *Mathematics of Computation*, 44(170):519–521, 1985.

[30] Michael Naehrig, Ruben Niederhagen, and Peter Schwabe. New software speed records for cryptographic pairings. In M. Abdalla and P. S. L. M. Barreto, editors, *Progress in Cryptology – LATINCRYPT 2010*, volume 6212 of *Lecture Notes in Computer Science*, pp. 109–123. Springer, Heidelberg, 2010.

[31] Kiyoshi Ohgishi, Ryuichi Sakai, and Masao Kasahara. Notes on ID-based key sharing systems over elliptic curve (in Japanese). Technical Report ISEC99-57, IEICE, 1999.

[32] Geovandro C. C. F. Pereira, Marcos A. Simplício, Jr., Michael Naehrig, and Paulo S. L. M. Barreto. A family of implementation-friendly BN elliptic curves. *Journal of Systems and Software*, 84(8):1319–1326, 2011.

[33] Ryuichi Sakai, Kiyoshi Ohgishi, and Masao Kasahara. Cryptosystems based on pairing. In *2000 Symposium on Cryptography and Information Security (SCIS 2000)*, pp. 135–148, 2000.

[34] Ryuichi Sakai, Kiyoshi Ohgishi, and Masao Kasahara. Cryptosystems based on pairing over elliptic curve (in Japanese). In *2001 Symposium on Cryptography and Information Security (SCIS 2001)*, pp. 23–26, 2001.

[35] Michael Scott. Computing the Tate pairing. In A. Menezes, editor, *Topics in Cryptology – CT-RSA 2005*, volume 3376 of *Lecture Notes in Computer Science*, pp. 293–304. Springer, Heidelberg, 2005.

[36] Michael Scott. Implementing cryptographic pairings. In T. Takagi et al., editors, *Pairing-Based Cryptography – Pairing 2007*, volume 4575 of *Lecture Notes in Computer Science*, pp. 177–196. Springer, Heidelberg, 2007.

[37] Michael Scott, Naomi Benger, Manuel Charlemagne, Luis J. Dominguez Perez, and Ezekiel J. Kachisa. On the final exponentiation for calculating pairings on ordinary elliptic curves. In H. Shacham and B. Waters, editors, *Pairing-Based Cryptography – Pairing 2009*, volume 5671 of *Lecture Notes in Computer Science*, pp. 78–88. Springer, Heidelberg, 2009.

[38] Jerome Solinas. ID-based digital signature algorithms. Invited talk, *7th Workshop on Elliptic Curve Cryptography (ECC 2003)*, August 11–13, 2003. Slides available at http://www.cacr.math.uwaterloo.ca/conferences/2003/ecc2003/solinas.pdf.

[39] William Stein. *SAGE: Software for Algebra and Geometry Experimentation*. http://www.sagemath.org/.

[40] John Tate. WC-groups over p-adic fields. Exposé 156, Séminaire Bourbaki, 1957/58.

[41] John Tate. Duality theorems in Galois cohomology over number fields. In *Proc. Internat. Congr. Mathematicians (Stockholm, 1962)*, pp. 288–295. Inst. Mittag-Leffler, Djursholm, 1963.

[42] Tadanori Teruya, Kazutaka Saito, Naoki Kanayama, Yuto Kawahara, Tetsutaro Kobayashi, and Eiji Okamoto. Constructing symmetric pairings over supersingular elliptic curves with embedding degree three. In Z. Cao and F. Zhang, editors, *Pairing-Based Cryptography – Pairing 2013*, volume 8365 of *Lecture Notes in Computer Science*, pp. 97–112. Springer, Heidelberg, 2014.

[43] André Weil. Sur les fonctions algébriques à corps de constantes fini. *Comptes rendus de l'Académie des sciences*, 210:592–594, 1940.

[44] Eric Zavattoni, Luis J. Dominguez Perez, Shigeo Mitsunari, Ana H. Sánchez-Ramírez, Tadanori Teruya, and Francisco Rodríguez-Henríquez. Software implementation of an attribute-based encryption scheme. *IEEE Transactions on Computers*, 64(5):1429–1441, 2015.

[45] Xusheng Zhang and Kunpeng Wang. Fast symmetric pairing revisited. In Z. Cao and F. Zhang, editors, *Pairing-Based Cryptography – Pairing 2013*, volume 8365 of *Lecture Notes in Computer Science*, pp. 131–148. Springer, Heidelberg, 2014.

12

Physical Attacks

Nadia El Mrabet
EMSE

Louis Goubin
UVSQ

Sylvain Guilley
Telecom ParisTech & Secure-IC

Jacques Fournier
CEA Tech

Damien Jauvart
UVSQ/CEA Tech

Martin Moreau
Telecom ParisTech & Secure-IC

Pablo Rauzy
Université Paris 8

Franck Rondepierre
Oberthur Technologies

The security of modern cryptography is based on the impossibility of breaking the implemented algorithms in practice. In order to reach such a goal, the algorithms are built in such a way that breaking them in theory is as expensive as doing an exhaustive search on the whole key. The cryptosystems are made public to ensure a high knowledge of potential attacks and the key length is selected according to the existing computation power available to prevent any brute force of the key. The strength of a given algorithm grows exponentially with the length of the key used.

However, if by any means someone can access some part of the key and test whether he guessed the correct value of the key, independently from the rest of the key, he would be able to brute force the key. Indeed, if the size of such parts is small enough, the exhaustive search for each part will be practically feasible and by repeating the attack on the different parts, the cost of finding the whole key will grow linearly with the length of the key.

The study of whether it is possible to access small parts of the key or not has been a new field in cryptographic engineering since the middle of the nineties. This has been made possible thanks to a class of attacks called *Physical Attacks* against the implementations of cryptographic algorithms.

Physical attacks exploit the underlying intrinsic weaknesses of the integrated circuits used to run the cryptographic algorithms. In the context of cryptographic engineering, two types of physical attacks are of special interest. The first one, called *Side-Channel Analysis*, is based on the non-invasive measurement of side-channel information (power, electromagnetic, timing, temperature...) leaked during cryptographic computations. The second one, called *Fault Attacks*, consists of semi-invasively stressing the integrated circuit running the cryptographic algorithm (using lasers, clock or power glitches, or electromagnetic pulses, for example) to corrupt the calculations.

In this chapter, we shall see to what extent physical attacks have been successful so far in attacking implementations of pairing calculations. Both side-channel and fault attacks are covered. We also look at the countermeasures that have to be added to pairing calculations to increase their robustness against such attacks.

12.1   Side-Channel Attacks

The integrated circuits running the cryptographic algorithms are mostly made of transistors whose switching is directly correlated to the data being manipulated by the circuit. The difference in the switching activities of transistors when manipulating, say, a '0' or a '1' gives rise to measurable physical characteristics that provide the so-called side-channel information leakage. Those measurable physical characteristics can be, for example, timing information, power consumption, or electromagnetic emissions.

In order to capture the power consumption during the execution of algorithms, a small resistor is placed in series with the power ground input. The measured power traces can have a shape, like the time of duration, that depends on the program's inputs (Figure 12.1).

Kocher, Jaffe, and Jun introduced the power analysis as a means of side-channel attacks against cryptographic algorithms [39]. The main assumption of a power analysis attack is based on the fact that the power traces are correlated to the instructions performed by the device. Thus, studying the power traces can allow us to recover information about the instructions and data registers, and then about the involved operands.

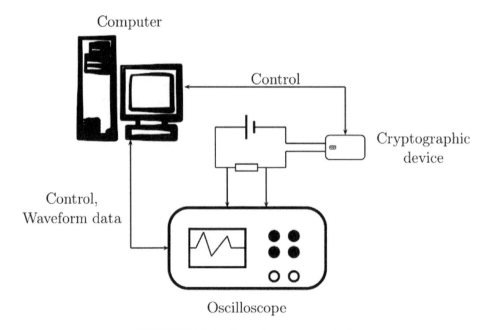

FIGURE 12.1   Setup for power analysis.

Physical side-channel leakage is not restricted to power consumption. The electric current also generates an electromagnetic (EM) field that can be captured with a small probe placed close to the part of the targeted circuit. Such a technique has the advantage of allowing an access restricted to some module (AES or big integer coprocessor, for instance) that limits the noise induced by uncorrelated operations.

In order to mount such attacks, the device shall be mounted on a dedicated board that has to be adapted to the form factor of the original target: bank chip card, sim card, secure element of a smartphone, rfid tag, ... Besides, especially for electromagnetic analysis, some chip preparation may be required, such as removing the black epoxy and the glue.

12.1.1 Simple Side-Channel Analysis

Timing attacks

Timing attacks were introduced by P. Kocher in 1996 and are the first known example of side-channel analysis. They are based on the fact that — in a given implementation — some predictable variation of the computational time may depend on the inputs (and in particular the secret key).

All 'basic' or 'straight-forward' implementations of cryptosystems can potentially succumb to such an attack. For instance, for a scalar ECC multiplication using the "double-and-add" method, the "add" operation, implemented with the Montgomery technique, may have different durations (according to the presence — or not — of a final subtraction). Therefore, by partitioning the inputs into two sets, the adversary can learn — thanks to the timing attack — whether this "add" operation is executed or not, and thus deduce the value of a secret key bit.

Note that symmetric algorithms can also be threatened by timing attacks, as illustrated in [10]. This type of attack is considered to be very powerful, mainly due to its low cost.

In the case of pairing-based cryptography, the Miller loop usually does not contain any conditional operations that depend on the secret data. However, at a lower level, the implementation may involve a basic operation (for instance, multiplication on a finite field) whose computational time depends on a secret bit, as highlighted, for instance in [37]. Therefore, timing attacks have also to be taken into account in the context of pairing-based cryptography. However, the required countermeasures appear to be the same as those developed to resist DPA-like attacks; we therefore refer to the sections about DPA.

Simple power analysis

Following Kerckhoff's principles, one may assume that the implemented cryptographic algorithms are publicly known. This is a legitimate assumption if we take into account the following facts: the publication of new algorithms within the cryptography community and the possibility of analyzing side-channel emissions coming from the algorithms' executions. Indeed, by looking at the power or execution trace of an algorithm, one can quickly recognize some patterns, and figure out which operations are being executed; it is especially true in public key cryptography where expensive modular operations are usually needed.

But the study of traces of execution is more powerful than simply giving access to the implemented algorithms. Such analyses, referred to as Simple Power Analysis (SPA) (since the first acquisitions were power consumption traces), are also a threat against weak implementations whose traces are dependant on the secret value. And this is an actual threat, since straightforward and fastest implementations won't thwart this dependency. A classic example is the study of a scalar multiplication $[n]P$ based on the double-and-add algorithm; see Algorithm 12.1. In this algorithm, we can see that an addition $Q + P$ is only performed if the secret bit n_i is equal to 1. So, if the trace of a doubling is different from the trace of the addition — and this is the case for the Short-Weierstrass formulae — the attacker will have direct access to the secret scalar n, which otherwise would have required him or her to solve the discrete logarithm problem.

ALGORITHM 12.1   Double-and-add algorithm.

Input   : $n = (n_{t-1}, \ldots, n_0)_2$ and $P \in E$
Output : $[n]P$

$Q \leftarrow P$
for $i = t - 1$ **downto** 0 **do**
    $Q \leftarrow [2]Q$ `// double`
    if $n_i == 1$ **then**
       $Q \leftarrow Q + P$ `// add`
    end
end
return R_0

Several countermeasures have been proposed in the literature to thwart such attacks. For the case of scalar multiplications, which is pretty similar to the exponentiation case, the flaw is twofold: the branch conditionally selected from a secret and also a different implementation of the doubling and addition operations. This last point also applies to exponentiations that use optimized modular squaring. Existing countermeasures consist of either solving the flaw or in changing the algorithm in order to perform a regular flow of operations, independent from the secret. For the first solution we can cite the atomicity principle [14] and for the second one we can cite the Montgomery ladder [35] (see Algorithm 12.2). Other solutions exist, but have larger impacts on the performances (timing) of implemented algorithms.

ALGORITHM 12.2   Montgomery ladder.

Input   : $n = (n_{t-1}, \ldots, n_0)_2$ and $P \in E$
Output : $[n]P$

$R_0 \leftarrow P$
$R_1 \leftarrow [2]P$
for $i = t - 1$ **downto** 0 **do**
    $b \leftarrow n_i$
    $R_{1-b} \leftarrow R_0 + R_1$
    $R_b \leftarrow [2]R_b$
end
return R_0

Besides, the whole security does not rely only on software countermeasures. At the hardware level, techniques such as clock jitters, additional power noise, dummy cycles, or power filtering help to increase the resistance against (simple) side-channel analysis.

12.1.2   Advanced Side-Channel Analysis

Differential power analysis

Differential Power Analysis (DPA) was initially defined by Kocher, Jaffe, and Jun [39] to target the Data Encryption Standard (DES). In the family of differential analysis attacks, we include, for example, the differential power and electromagnetic attacks. Differential power analysis works on several power/EM traces that are analyzed using statistical tools, which helps in getting rid of variations due to data manipulated, and some noise, which are embarrassing problems in the case of a single trace.

The principle is to build, for the system under attack, a 'function' parameterized by a small part of the algorithm that we want to attack. The aim is to recover the set 'function' corresponding to the secret. For this we acquire a large number of images of the 'function.' Furthermore, we construct a theoretical series of corresponding images for each set function. Then we choose a distinguisher to compare theoretical series and series from the acquisition. There are many such distinguishers, the main ones being difference of means, correlation coefficient, and mutual information.

In the case of public key cryptography, most classical DPA attacks target a scalar multiplication operation with the aim of recovering the scalar bits one by one. The description of a DPA attack against ECC is well introduced in [32].

Assume that the double-and-add method is implemented with one of the regular variants given in Algorithm 12.1. Let $n = (n_{t-1}, \ldots, n_0)_2$ be the scalar multiplier. Suppose that an attacker already knows the most significant bits, $n_{t-1}, \ldots, n_{j+1}$. Then, the attacker has to make a guess on the next bit n_j, which is equal to 1. He randomly chooses several points $P_1, \ldots, P_r$ and computes $Q_s = \left[\sum_{i=j}^{t-1} n_i 2^i \right] P_s$ for $1 \le s \le r$.

Using a boolean selection function g, the attacker prepares two sets: the first set, S_{true}, contains the points P_s such that $g(Q_s) = \texttt{true}$ and the second set, S_{false}, contains those such that $g(Q_s) = \texttt{false}$. Then, a candidate for the selection function may, for example, be the value of a given bit in the representation of Q_s.

Let $C^{(s)}$ denote the side-channel information associated to the computation of $[n]P_s$ by the cryptographic device (e.g., the power consumption). If the guess $n_j == 1$ is incorrect then the difference obtained in Equation 12.1 will be $\simeq 0$.

$$\left\langle C^{(s)} \right\rangle_{\substack{1 \le s \le r \\ P_s \in S_{\text{true}}}} - \left\langle C^{(s)} \right\rangle_{\substack{1 \le s \le r \\ P_s \in S_{\text{false}}}} . \tag{12.1}$$

If the guess is wrong, both sets appear as two random sets, otherwise the guess is correct. After revealing n_j, the remaining bits $n_{j-1}, \ldots, n_0$ are recovered recursively by the same method.

Correlation power analysis

In DPA, the classification of power traces is based on comparing the differences between the measured traces. Brier, Clavier, and Olivier in 2004 at CHES proposed an improvement of DPA based on the use of Pearson's correlation for comparing the measured side-channel traces and a leakage model based on the Hamming Weight (HW) of the manipulated data.

The side-channel information of the device is supposed to be linear in $H(D \oplus R)$, the Hamming distance of the data manipulated D, with respect to a *reference state* R. The linear correlation factor is used to correlate the side-channel curves with this value $H(D \oplus R)$. The maximum correlation factor is obtained for the right guess of secret key bits.

Let C be the side channel (power consumption for instance) of the chip; its consumption model is:

$$W = \mu H(D \oplus R) + \nu . \tag{12.2}$$

The correlation factor $\rho_{C,H}$ between the set of power curves C and values $H(D \oplus R)$ is defined as: $\rho_{C,H} = \frac{cov(C,H)}{\sigma_C \sigma_H}$.

The principle of the attack is then the following:

- Perform r executions on the chip with input data $m_1, \ldots, m_r$ and collect the corresponding power curves $C^{(1)}, \ldots, C^{(r)}$.

- Predict some intermediate data D_i as a function of m_i and key hypothesis g.

- Produce the set of the r predicted Hamming distances: $\{H_{i,R} = H(D_i \oplus R), i = 1, \ldots, r\}$.
- Calculate the estimated correlation factor:

$$\widehat{\rho_{C,H}} = \frac{r \sum C^{(i)} H_{i,R} - \sum C^{(i)} \sum H_{i,R}}{\sqrt{r \sum (C^{(i)})^2 - (\sum C^{(i)})^2} \sqrt{r \sum H_{i,R}^2 - (\sum H_{i,R})^2}}. \tag{12.3}$$

When the attacker makes the right guesses for values of the reference state R and secret leading to data D, the correlation factor ρ is maximum.

This attack is more powerful than DPA in the sense that the 'leakage' peaks are generally more visible in CPA with the same conditions as in DPA.

12.1.3 Side-Channel Attacks against Pairings

In the case of pairings, side-channel attacks are relevant whenever pairings are used in schemes involving some secret data, which is typically the case when pairings are used in identity-based encryption schemes.

The fundamental idea of identity-based encryption is to allow the user's public key to be a public function of his identity. This requires a trusted authority (T_A) that sends him his private key. This trusted authority creates all the private keys related to an Identity-Based (IB) protocol. The advantage of IB is to simplify the transmission of public keys while sending the encryption of a message. Indeed, it is no longer necessary to use certificates or public-key infrastructure (PKI), since the public key used for encryption can be deterministically (and publicly) deduced from the identity of the receiver.

The important point during an IB protocol is that the decryption involves a pairing computation between the private key of the user and a public key. We call the public key the part of the message used during the pairing calculation involving the secret key. A potential attacker can know the algorithm used, the number of iterations, and the exponent. The secret is only one of the arguments of the pairing. The secret key influences neither the time execution nor the number of iterations of the algorithm, which is different from RSA or ECC protocols.

From here on, the secret will be denoted P and the public parameter (or the point used by the attacker) Q. We are going to describe a DPA attack against the Miller algorithm. We restrict this study to the case where the secret is used as the first argument of the pairing. If the secret is used as the second argument, the same attack can be applied; this assumption is shown theoretically and practically in [21] and also in [56]. We assume that the algorithm is implemented on an electronic device such as a smart card and used in a protocol involving IB cryptography. The attacker can send as many known entries Q for the decryption operation of IBC as he wants, and he can collect the power consumption curves.

Most pairing computations are based on the use of the Miller algorithm. This is in particular true for the Weil, Tate, and Ate pairings. We assume that the Miller algorithm is implemented in software running on an electronic device: for example, a smart card. The attacks are performed during the execution of a cryptographic protocol based on identity. Let Q be the public message. The private key will be represented by the point P in the computation of the pairing $e(P, Q)$. We restrict the study to the case where the secret is the first argument of the pairing. Placing the first secret of the coupling parameter is a first countermeasure against some side-channel attacks as proposed in [60]. If the secret is the second argument of the pairing, the same attack patterns may apply and allow us to recover the secret used. The attacker can compute as many times as necessary pairings between the secret P (that will not change) and inputs Q (that changes at will). He can record and store the power consumption curves for each of those computations, together with the final result of the Miller algorithm.

Description of the attack

When implementing pairings, different coordinate systems may be used. This does not have any significant impact on the feasibility of side-channel attacks. Indeed, in the Miller loop, even if the choice of the coordinate system will give rise to different implementations of the 'lines' and 'tangents' computations, the underlying internal operations will be the same modular multiplications and additions on long precision numbers.

As described in the general DPA/CPA approach, we try to identify some operations that involve a secret and a known operand; such operations are in bold in the following equations. As already explained, there are several ways of implementing the Miller loop. For example, [60] takes the case of affine coordinates; in this case the line and the tangent equations are the following.

The line equation is the formula to compute $l_{T,P}(Q)$, the line passing through T and P evaluated in Q is:

$$l_{T,P}(Q) = \boldsymbol{y_Q} - \boldsymbol{y_T} - \frac{y_P - y_T}{x_P - x_T}(\boldsymbol{x_Q} - \boldsymbol{x_T}).$$

The tangent equation is $l_{T,T}(Q)$, the tangent line through point T evaluated in Q. This equation is:

$$l_{T,T}(Q) = \boldsymbol{y_Q} - \boldsymbol{y_T} - \frac{3x_T^2 + a}{2y_T}(\boldsymbol{x_Q} - \boldsymbol{x_T}).$$

The case for Jacobian coordinates is treated by [21] and [23] with the same aim of targeting an operation in order to recover one coordinate of the secret input point. If the points are a three-tuple, then it is necessary to recover a second component. Now we use the elliptic curve equation to find the last coordinate. The secret point is recovered.

The line and tangent equation in Jacobian coordinates are the following:

$$l_{T,P}(Q) = \frac{2\boldsymbol{y_Q}\boldsymbol{y_T}z_T^4 - 2y_T^2\boldsymbol{z_T}z_Q^3 + (y_P z_T^3 - y_T z_P^3)(x_T\boldsymbol{z_T}z_Q^3 - \boldsymbol{x_Q}z_Q z_T^3)}{2y_T z_T^4 z_Q^3}$$

and

$$l_{T,T}(Q) = \frac{2y_T(\boldsymbol{y_Q}z_T^3 - \boldsymbol{y_T}z_Q^3) - z_Q(3x_T^2 + az_T^4)(\boldsymbol{x_Q}z_T^2 - \boldsymbol{x_T}z_Q^2)}{2y_T z_T^3 z_Q^3}.$$

The same approach works when in mixed coordinates as described in [21] and [11]. For optimization reasons it is also possible to mix system coordinates. The equations are available in [21]. Let $T = (X_T, Y_T, Z_T)$ be a point in Jacobian coordinates, P and Q in affine coordinates, then the line and tangent equation in mixed coordinates are the following:

$$l_{T,T}(Q) = 2\boldsymbol{y_Q}\boldsymbol{Y_T}Z_T^3 - 2Y_T^2 - (3X_T^2 + aZ_T^4)(\boldsymbol{x_Q}\boldsymbol{Z_T^2} - X_T)$$

and

$$l_{T,P}(Q) = (\boldsymbol{y_Q} - \boldsymbol{y_P})Z_T(X_T - Z_T^2 x_P) - (Y_T - Z_T^3 y_P)(\boldsymbol{x_Q} - \boldsymbol{x_P}).$$

Multiplication in $\mathbb{F}_q$

We describe the attacks as if we have the embedded degree $k = 1$, and then the coordinates of Q being elements of $\mathbb{F}_q$. This way, the targeted multiplication $Z_P^2 x_Q$ is a multiplication in $\mathbb{F}_q$. The DPA attack also works when $k > 1$. Even if the multiplication $Z_P^2 x_Q$ becomes a multiplication between an element of $\mathbb{F}_q$ and an element of $\mathbb{F}_{q^k}$, we can consider a multiplication between two $\mathbb{F}_q$ elements.

Indeed, $x_Q \in \mathbb{F}_{q^k}$ is written: $x_Q = \sum_{i=0}^{k-1} x_{Q_i}\xi^i$, with $(1, \xi, \xi^2, \ldots, \xi^{k-1})$ a basis of $\mathbb{F}_{q^k}$, and there exists a polynomial R such that $deg(R) = k$ with ξ root of R, $(R(\xi) = 0)$. Then

$Z_P^2 x_Q = \sum_{i=0}^{k-1} \left(Z_P^2 \times x_{Q_i} \right) \xi^i$, is composed of k products in $\mathbb{F}_q$. So we can focus on one of these k products in $\mathbb{F}_q$ to apply the DPA attack as described.

In the same way, to compute the difference $(Z_P^2 x_Q - X)$, we compute a difference between elements of $\mathbb{F}_q$ as in the affine case.

Indeed, if $Z_P^2 x_Q = \sum_{i=0}^{k-1} (Z_P^2 x_Q)_i \xi^i$ then

$$Z_P^2 x_Q - X = \left((Z_P^2 x_Q)_0 - X \right) + \sum_{i=1}^{k-1} (Z_P^2 x_Q)_i \xi^i.$$

Targeting the first iteration in the Miller loop

We describe the attack for the first iteration. It is the simplest case, because we know that for this iteration, $T = P$. We can provide the attack for the jth iteration. For this iteration we find $T = [j]P$, where $[j]P$ represents the scalar multiplication of point P by the integer j.

We know l, the order of the point Q (as P and Q have the same order). By counting the number of clock cycles, we can find the number d of iterations we have made before the DPA attack. Then, reading the binary decomposition of l directly gives us j. We consider that at the beginning $j = 1$, if $l_{n-1} = 0$ then $j \leftarrow 2j$, else $j \leftarrow 2j + 1$, and we go on, until we arrive at the $(n-1-d)^{th}$ bit of l.

If the attack is done during the jth iteration of the Miller algorithm, we find the coordinates of $[j]P$. In order to find P, we just have to compute j', the inverse of j modulo l, and then $P = [j'][j]P$.

Furthermore, we present the attack against the basic Miller algorithm. The attack can be straightforwardly generalized to the optimised Miller algorithm given in [38].

Description of the attack

In order to retrieve the secret key $P = (X_P, Y_P, Z_P)$, the circuit has to be used to perform some calculations while the power consumption of the physical device is monitored. In particular, the measurement of the consumed power must be done during a time slot when the circuit calculates a result that depends on both the secret key and some controllable input data.

For example, we decided to observe the power consumption when the circuit performs the multiplication between Z_P^2 (a part of the secret key) and x_Q (the input data). This operation is done during the second control step. To retrieve the second part of the key (X_P), we focused on the subtraction between the previously performed multiplication and the key.

The DPA attack against the Miller algorithm was first proposed by Page and Vercauteren [44]. Over the years, the proposed schemes have been enhanced. Reference [60] extends the attack to several other operations and proposes a scheme using CPA. Another remarkable improvement is proposed by [11], where the authors attack the modular addition and multiplication of elements in a finite field of large prime characteristics. In this chapter, we present those attacks against pairings and provide simulation results.

To implement the attack, it is necessary to target an operation in the line or tangent equation. Let $\star$ be the general targeted operator between $g \in \mathbb{F}_q$ and $U \in E(\mathbb{F}_q)$. For instance, $g \star U$ can be $g - U_x \in \mathbb{F}_q$.

The attack scheme proposed against Miller is Algorithm 12.3.

The last part of the key (Y_P) can be mathematically inferred from X_P and Z_P^2. Indeed, the elliptic curve equation $E : Y^2 = X^3 + aXZ^4 + bZ^6$ is a quadratic equation in Y_P. The square root of $\sqrt{X_P^3 + aX_P Z_P^4 + bZ_P^6}$ gives us two possibilities for the value of Y_P; testing them by an execution of the Miller algorithm will give the correct coordinates for P.

ALGORITHM 12.3 A Messerges-style DPA attack to reveal $P = (x_P, y_P)$ by guessing y_P **one bit at a time.**

Input   : n is the bitlength max of y_P
Output : A candidate for the coordinate y_P

Set g to 0
for $i = 0$ **upto** $n - 1$ **do**
  │ Set S_{hi} and S_{lo} to empty
  │ Guess the ith bit of g to one
  │ **for** $k = 0$ **upto** $r - 1$ **do**
  │ │ Select at random a point U of E
  │ │ Calculate $X = g \star U$
  │ │ Use device to execute $e(P, U)$, collect power signal S_k
  │ │ **if** *the ith bit of X is* 1 **then**
  │ │ │ add S_k to S_{hi}
  │ │ **else**
  │ │ │ add S_k to S_{lo}
  │ │ **end**
  │ **end**
  │ Average power signals to get DPA bias $D = \overline{S_{hi}} - \overline{S_{lo}}$
  │ **if** *DPA bias signal has a spike* **then**
  │ │ The guess was right: set i^{th} bit of g to 1
  │ **else**
  │ │ The guess was wrong: set i^{th} bit of g to 0
  │ **end**
end
return g

The practical feasibility of such attacks is illustrated in [56]. The target is an Ate pairing over BN curves $e(P, Q)$ with P the secret input. The targeted operation is a modular multiplication. To implement this over long integer ($\simeq 256$ bits), they use the Montgomery method. The device architecture imposes on the attacker to target 16 bits at time.

12.2   Fault Attacks

In 1984, A. Shamir challenged the cryptography community to find a protocol based on the user's identity [51]. This challenge was solved nearly twenty years later by D. Boneh and M. Franklin. In 2003, D. Boneh and M. Franklin created an identity-based encryption (IBE) scheme based on pairings [13]. The general scheme of an identity-based encryption is described in [13], and several protocols based on pairings have been developed since [33]. A feature of identity-based protocols is that a computation of a pairing involving the private key and the plaintext is performed in order to decipher a message. A pairing is a bilinear map e taking as inputs two points P and Q of an elliptic curve. The pairing computation gives the result $e(P, Q)$. Several pairings have been described in the literature. The Weil and the Tate pairing was developed [54] without any consideration for the efficiency of the computation. Once pairings were used to construct protocols, cryptographers sought more efficient algorithms. In chronological order, the Duursma and Lee algorithm [18], the Eta [8], Ate, twisted Ate [29], optimal pairings [57], and pairing lattices [28] were invented. Recently, a construction of pairing over a general abelian variety was proposed in [42]. The latest implementations results [3, 26, 49] of pairing computations are fast enough to consider the use of pairing-based protocols in embedded devices. Consequently, it seems fair to wonder if pairing-based protocols involving a secret are secure against physical attacks in general, and fault attacks in particular. We focus here on fault attacks against pairing-based cryptography.

Since 2006, several fault attacks against pairings have been proposed. Here we will present what are in our opinion the most significant ones. For each attack, we assume that the pairing is used during an identity-based protocol. The secret point is stored into an embedded electronic device that can be attacked with fault attacks. The location of the secret is not important in practice. Indeed, the equations that leak information about the secret can provide information as to whether the secret is the first or the second parameter. Often, the attack is easier when the secret is the second parameter. That is why we consider the cases where the first parameter is the secret argument.

The necessary background in order to understand pairings and IBE is presented in Chapter 1. The first fault attack against a pairing was proposed by Page and Vercauteren [45] and is presented in Section 12.2.2. Then, we describe the adaptations of the previous attack against the Miller algorithm in Section 12.2.2. Whelan and Scott [59] highlighted the fact that pairings without a final exponentiation are more sensitive to a sign-change fault attack. After that, El Mrabet [19] generalized the attack of Page and Vercauteren to the Miller algorithm used to compute all the recent optimizations of pairings. Another method is adopted in [4], based on instruction skips, and presented in Section 12.2.2. In [40], Lashermes et al. proposed a fault attack against the final exponentiation during a Tate-like pairing. Their attack is described in Section 12.2.3. Finally, we conclude the description of fault attack in Section 12.2.4.

12.2.1   What Are Fault Attacks?

The goal of a fault attack is to inject errors during the calculation of an algorithm in order to reveal sensitive data. At first these attacks required a very precise positioning and expensive equipment to be performed, but now even some cheap equipment allows us to perform them [27]. The faults can be performed using a laser, an electromagnetic pulse, and power or clock glitches [16, 17, 36].

The effect of a fault can be permanent, i.e., a modification of a value in memory, or transient, i.e., a modification of data that is not stored into memory at one precise moment.

At the bit level, a fault can be a bit-flip if the value of a bit is complemented. Or it can be stuck-at (0 or 1) if the bit modification depends on its value.

The fault cannot only modify the data manipulated but also modify a program's execution. As an example in a microcontroller, if a fault occurs on the opcode and modifies it, the executed instruction will be modified. This method gives rise what is called an instruction skip fault model where an instruction is skipped by modifying its opcode to a value representing an instruction without effect (e.g., NOP).

12.2.2   Fault Attacks against the Miller Algorithm

In this section we present the existing attacks against the Miller algorithm. We describe in Section 12.2.2 an attack against the Duursma and Lee algorithm, since it was the first attack against a pairing and, more importantly, all the following attacks are constructed on this scheme. Then, in Section 12.2.2 , we describe the attacks against the Miller algorithm.

Attacks against the Dursma and Lee algorithm

The Duursma and Lee algorithm is not constructed using the Miller algorithm. But it was the first implementation of a pairing to be attacked. The attack was developed by Page and Vercauteren in [45].

Duursma and Lee [18] define a pairing over hyperelliptic curves, and in particular, over super-singular elliptic curves over finite fields of characteristic 3. For $\mathbb{F}_q$ with $q = 3^m$ and $k = 6$, suitable curves are defined by

$$E : y^2 = x^3 - x + b$$

with $b = \pm 1 \in \mathbb{F}_3$. Let $\mathbb{F}_{q^3} = \mathbb{F}_q[\rho]/(\rho^3 - \rho - b)$ and $\mathbb{F}_{q^6} = \mathbb{F}_{q^3}[\sigma]/(\sigma^2 + 1)$. The distortion map $\phi : E(\mathbb{F}_q) \to E(\mathbb{F}_{q^6})$ is defined by $\phi(x, y) = (\rho - x, \sigma y)$. Then, with $\mathbb{G}_1 = \mathbb{G}_2 = E(\mathbb{F}_{3^m})$ and $\mathbb{G}_T = \mathbb{F}_{q^6}$, Algorithm 12.4 computes an admissible, symmetric pairing.

ALGORITHM 12.4   The Duursma-Lee pairing algorithm.

Input   : $P = (x_P, y_P) \in \mathbb{G}_1$ and $Q = (x_Q, y_Q) \in \mathbb{G}_2$.
Output: $e(P, Q) \in \mathbb{G}_3$.

$f \leftarrow 1$
for $i = 1$ **upto** m **do**
$\quad x_P \leftarrow x_P^3,\ y_P \leftarrow y_P^3$
$\quad \mu \leftarrow x_P + x_Q + b$
$\quad \lambda \leftarrow -y_P y_Q \sigma - \mu^2$
$\quad g \leftarrow \lambda - \mu \rho - \rho^2$
$\quad f \leftarrow f \cdot g$
$\quad x_Q \leftarrow x_Q^{1/3},\ y_Q \leftarrow y_Q^{1/3}$
end
return $f^{q^3 - 1}$

The attack developed by Page and Vercauteren in [45] consists of modifying the number of iterations during the Duursma and Lee algorithm. The hypotheses to perform the attack are that

- the two inputs parameters (points P and Q) are fixed, one is secret and the other public;
- the pairing implementation is public;
- two pairing computations are done, one valid and one faulty.

The analysis of the quotient of the two results gives information about the secret. Indeed, the quotient of the two results cancel terms that are not influenced by the fault. Firstly, Page and Vercauteren described how to recover the secret point if the final exponentiation is not performed (i.e., Line 9 of Algorithm 12.4). Then they explained how to reverse the final exponentiation for a complete attack.

Attack without the final exponentiation

Let $P = (x_P, y_P)$ be the secret input during the pairing computation and let $Q = (x_Q, y_Q)$ be selected by the attacker. We consider the Duursma and Lee algorithm without the final exponentiation (Line 9).

Let $\bar{e}[\Delta]$ be the execution of Algorithm 12.4 where the fault replaces the loop bound m (in Line 2) with Δ. Then the result of the Duursma and Lee algorithm without the final exponentiation, instead of being a product of polynomials of the form

$$\prod_{i=1}^{m}\left[(-y_P^{3^i}\cdot y_2^{3^{m-i+1}}\sigma - (x_P^{3^i}+x_2^{3^{m-i+1}}+b)^2) - (x_P^{3^i}+x_2^{3^{m-i+1}}+b)\rho - \rho^2\right],$$

is a product of the form

$$\prod_{i=1}^{\Delta}\left[(-y_P^{3^i}\cdot y_2^{3^{m-i+1}}\sigma - (x_P^{3^i}+x_2^{3^{m-i+1}}+b)^2) - (x_P^{3^i}+x_2^{3^{m-i+1}}+b)\rho - \rho^2\right]$$

for a random integer Δ.

If $\Delta = m+1$, then recovering the secret point P is easy. We have two results

$$\begin{aligned} R_1 &= \bar{e}[m](P,Q) \\ R_2 &= \bar{e}[m+1](P,Q) \end{aligned}$$

where R_1 is correct and R_2 is faulty. Let $g_{(i)}$ be the i-th factor of a product produced by the algorithm. The quotient of the two results produces a single factor,

$$g_{(m+1)} = (-y_P^{3^{m+1}}\cdot y_2\sigma - (x_P^{3^{m+1}}+x_2+b)^2) - (x_P^{3^{m+1}}+x_2+b)\rho - \rho^2.$$

Given that $\forall z \in \mathbb{F}_q, z^{3^m} = z$, the attacker can easily extract x_P or y_P based on the knowledge of x_Q and y_Q.

In practice, the faulty result Δ cannot be forced to $m+1$. It is more realistic to assume that the fault gives $\Delta = m \pm \tau$ for a random unknown integer τ. As a consequence, the attacker computes two results

$$\begin{aligned} R_1 &= \bar{e}[m \pm \tau](P,Q) \\ R_2 &= \bar{e}[m \pm \tau + 1](P,Q), \end{aligned}$$

and once again, considering the quotient, the attacker obtains a single term $g_{(m \pm \tau+1)}$.

In order to apply the same approach, the attacker should discover the exact value of τ. Indeed, this value is needed to correct the powers of x_P, y_P, x_Q, and y_Q. As the implementation of Duursma and Lee algorithm is supposed to be public, the number of operations performed during the faulty execution leaks the value of τ. Then the attack consists of several faulty executions of Algorithm 12.4, until we find two results R_1 and R_2 satisfying the requirements. The probability to obtain two values R_1 and R_2 after a realistic number of tests was computed in [19].

The probability to obtain two consecutive numbers after n picks among N integers is

$$P(n,N) = 1 - \frac{B(n,N)}{C_{n+N}^n},$$

where

$$\begin{cases} N \leq 0, n > 0, B(n,N) &= 0, \\ \forall N, n = 0\, B(n,N) &= 1 \\ B(n,N) &= \sum_{j=1}^{N}\sum_{k=1}^{n} B(n-k, j-2). \end{cases}$$

For instance, for an 8-bit architecture only 15 tests are needed to obtain a probability larger than one half, $P(15, 2^8) = 0.56$, and only 28 for a probability larger than 0.9.

Reversing the final exponentiation

The attack described above is efficient without the final exponentiation. But since the final exponentiation is a part of the Duursma and Lee algorithm, Page and Vercauteren present a method to reverse it. The problem is that given the result $R = e(P, Q)$ the attacker wants to recover S, the value obtained in Line 7 of Algorithm 12.4, before the final exponentiation (i.e., $R = S^{q^3-1}$). Given R, the value of S is only determined up to a non-zero factor in $\mathbb{F}_{q^3}$. Indeed, the Fermat little theorem implies that $\forall c \in \mathbb{F}_{q^3} \setminus \{0\}$, $c^{q^3-1} = 1$. Furthermore, for one solution S of the equation $X^{q^3-1} - R = 0$, all the other solutions are of the form cS, for $c \in \mathbb{F}_{q^3} \setminus \{0\}$. At first sight, the attacker would not be able to choose the correct value S among the $q^3 - 1$ possibilities. However, given the description of the attack, the attacker does not need to reverse the powering of a full factor, but only a single factor with a special form:

$$R = \frac{R_2}{R_1} = \frac{\bar{e}[m \pm \tau + 1](P, Q)}{\bar{e}[m \pm \tau](P, Q)} = g_{(m \pm \tau + 1)}^{q^3-1}.$$

We want to recover $g_{(m \pm \tau + 1)}$, in order to find the coordinates of the secret point x_P and y_P: In order to solve this problem, Page and Vercauteren split it in two:

1. a method to compute one valid root of $R = g^{q^3-1}$ for some factor g, and

2. a method to derive the correct value of g from among all possible solutions.

The first problem is solved throughout the method of Lidl and Niederreiter [41] to compute roots of the linear operator $X^{q^3} - R \cdot X$ on the vector space $\mathbb{F}_{q^6}/\mathbb{F}_{q^3}$. They use a matrix representation of the problem to find all the solutions of the equation $X^{q^3-1} - R = 0$. Then, in order to find the correct root among the $q^3 - 1$ possibilities, Page and Vercauteren use the specific form of the factors in the product. Indeed, the terms $\rho\sigma$ and $\rho^2\sigma$ do not appear in the correct value and this gives a linear system of equations providing the solution. As the method to reverse the final exponentiation is specific to the Duursma and Lee algorithm, we do not give the equations. They are presented with examples in [22, 45].

Attacks against the Miller algorithm

A specific sign-change attack

The first attack against the Miller algorithm was developed by Whelan and Scott [59]. They use the same approach as the attack against Duursma and Lee. They compute two pairing values, one correct and one faulty. However, the fault is no longer on the Miller loop bound but into the Miller variable. Whelan and Scott analyze several pairings and study the success of the attack whether the secret is the point P or Q. They consider the case of the Eta pairing [8]. This pairing is defined over super-singular curves for small characteristics. Considering the recent result on the discrete logarithm problem [31] and the fact that the attack is based on the scheme of the Page and Vercauteren attack, we do not describe it. Whelan and Scott target the Weil pairing. First they try to describe a general fault model: any fault is injected during any iteration of the Miller algorithm. The attacker needs to solve a non-linear system and they conclude that it cannot be done. So they consider a more specific attack: a sign cannot change fault attack (a single sign bit is flipped [12]). They consider that the attacker modifies the sign of one of the coordinates of the point P or Q. This attack is the most efficient when exactly the last iteration of the Miller algorithm is corrupted. They consider the ratio between a valid and a faulty execution of the Weil pairing, and, using the equations, they obtain a linear system in the coordinates of the secret point. In this case, the attack is successful. If the fault is injected earlier in the Miller algorithm, the analysis is more complex, as several square and cubic roots have to be computed, but possible. Then they consider the Tate pairing. As the Tate pairing is also constructed using the Miller algorithm, the attack described for the Weil pairing should be efficient. However, due to the complex final exponentiation, they conclude that the Tate pairing is efficiently protected against the sign-change fault they propose.

A general fault attack

In [19], El Mrabet considers a fault attack based on the Page and Vercauteren attack [45]. The fault consists of modifying the number of iterations during the execution of the Miller algorithm. As the Miller algorithm is the central step for the Weil, the Tate, Ate, twisted Ate, optimal pairings, and pairing lattices, the fault model is valuable for a wide class of pairings. However, the attack targets only the Miller algorithm, the final exponentiation is not reversed cryptanalytically, and the author assumes that another attack could annihilate it. In Section 12.2.3 we describe a recent attack that reverses the final exponentiation. We describe here the general attack against the Miller algorithm. The difficulty of the attack relates to the resolution of a non-linear system.

El Mrabet considers that the number of iterations in the Miller algorithm is modified by a fault attack and denotes τ the new number of iterations. The value of τ is random but can be determined afterwards if the attacker knows the number of iterations, by monitoring the timing of the computation, for example. The aim is to obtain two consecutive results of Miller's algorithm $F_{\tau,P}(Q)$ and $F_{\tau+1,P}(Q)$. As in the attack on the Duursma and Lee algorithm, we consider the ratio $\frac{F_{\tau+1,P}(Q)}{F_{\tau,P}(Q)^2}$. Then an identification in the basis of $\mathbb{F}_{p^k}$ leads to a system that reveals the secret point.

Without loss of generality, we describe the attack when the embedding degree of the curve is $k = 4$. This allows the description of the equation. As the important point of the method is the identification of the decomposition in the basis of $\mathbb{F}_{p^k}$, it is easily applicable when k is larger than 3. Indeed, $k = 3$ is the minimal value of the embedding degree for which the system obtained can be solved. At the τ-th step, the Miller algorithm calculates $[j]P$. During the $(\tau + 1)^{th}$ iteration, it calculates $[2j]P$, and considering the value of the $(\tau + 1)^{th}$ bit of $\log_2(r)$, it either stops at this moment, or it calculates $[2j + 1]P$.

Let $B = \{1, \xi, \sqrt{\nu}, \xi\sqrt{\nu}\}$ be the basis of $\mathbb{F}_{p^k}$; this basis is constructed using tower extensions. The point $P \in E(\mathbb{F}_p)$ is given in Jacobian coordinates, $P = (X_P, Y_P, Z_P)$, and the point $Q \in E(\mathbb{F}_{p^k})$ is in affine coordinates. As k is even, we can use a classical optimization in pairing-based cryptography, which consists of using the twisted elliptic curve to write $Q = (x, y\sqrt{\nu})$, with x, y and $\nu \in \mathbb{F}_{p^{k/2}}$ and $\sqrt{\nu} \in \mathbb{F}_{p^k}$ [6]. We will consider here only the case where $r_{\tau+1} = 0$. The case where $r_{\tau+1} = 1$ can be treated similarly is described in [19]. The non-linear system in the case $r_{\tau+1} = 1$ is a bit more complex and must be solved using the discriminant theory.

When $r_{\tau+1} = 0$, we have that $F_{\tau+1,P}(Q) = (F_{\tau,P}(Q))^2 \times h_1(Q)$, $[j]P = (X_j, Y_j, Z_j)$, where j is obtained by reading the τ first bits of r and $T = [2j]P = (X_{2j}, Y_{2j}, Z_{2j})$.

Using the equation of h_1, we obtain the following equality:

$$F_{\tau+1,P}(Q) = (F_{\tau,P}(Q))^2 \times$$

$$\left(Z_{2j} Z_j^2 y \sqrt{\nu} - 2Y_j^2 - 3(X_j - Z_j^2)(X_j + Z_j^2)(xZ_j^2 - X_j) \right).$$

Considering that the secret is the point P, we know j, τ, the coordinates of Q. The Miller algorithm gives us $F_{\tau+1,P}(Q)$ and $F_{\tau,P}(Q)$. We calculate the ratio $R = \frac{F_{\tau+1,P}(Q)}{(F_{\tau,P}(Q))^2}$. Using the theoretical form of R and its decomposition in the base B, by identification we can obtain, after simplification, the following system:

$$\begin{cases} Y_j Z_j^3 = \lambda_2, \\ Z_j^2(X_j^2 - Z_j^4) = \lambda_1, \\ 3X_j(X_j^2 - Z_j^4) + 2Y_j^2 = \lambda_0, \end{cases}$$

where we know the three values $\lambda_{0,1,2}$.

The resolution [19] of this non-linear system gives the following equation:

$$(\lambda_0^2 - 9\lambda_1^2)Z^{12} - (4\lambda_0\lambda_2^2 + 9\lambda_1^3)Z^6 + 4\lambda_1^4 = 0.$$

Solving the equation in Z_j, we find at most $24 = 12 \times 2 \times 1$ possible triplets (X_j, Y_j, Z_j) for the coordinates of the point $[j]P$. Once we have the coordinates of $[j]P$, to find the possible points P, we have to find j' the inverse of j modulo r, and then calculate $[j'][j]P = [j'j]P = P$. Using the elliptic curve equation, we eliminate triplets that do not lie on E. Then we just have to perform the Miller algorithm with the remaining points and compare it with the result obtained with the secret point P. So we recover the secret point P, in the case where $r_{\tau+1} = 0$. The case of $r_{\tau+1} = 1$ also leads to a non-linear system that can be solved using a Grobner basis.

Remark 12.1 We present the attack in Jacobian coordinates. As the attack is not dependent on the system of coordinates, it will be successful for other systems. In [19], the affine, projective, and Edwards coordinates are also treated. In the paper [58], the authors consider Hessian coordinates.

Remark 12.2 We describe the attack with the secret point being P. If the secret is the point Q, the attack is also valid — we just obtain an easier system to solve.

The attack against the Miller algorithm is efficient. A model of the attack was implemented in [46]. It is fair to wonder if this attack can be applied to a complete pairing. As the Weil pairing consists of two applications of the Miller algorithm, the Weil pairing is sensitive to this attack. For the Tate-like pairings (Ate, twisted Ate,...) the final exponentiation must be cancelled for the attack to be efficient. As the result of the Miller algorithm has no particular form, it seems difficult to cryptanalytically reverse the final exponentiation. As far as we know, it has not been done yet. El Mrabet cites several works in microelectronics that would give the result of the Miller algorithm during a Tate-like pairing computation: for example, the scan attack [61] or the under-voltage technique [2]. We describe in Section 12.2.3 a recent fault attack against the final exponentiation.

Attack against the *if* instruction

In [4], the authors propose a new fault model as well as an implementation of their fault attack.

The *if* skip fault model

In the Miller algorithm, the addition step is performed or not according to the bits of r. This decision is usually implemented with an *if* instruction. If an attacker is able to skip an *if* instruction, he can avoid the addition step if he wants to.

This fault model has several advantages. It can target the last iteration only of the Miller algorithm, and as a consequence only one fault injection is required to find the value $h_2(Q)$. This is better than when altering the counter value, where the attacker had to perform fault attacks until he finds the faulty result for two consecutive iterations. Then it is not as easy to develop a countermeasure against it as for an attack on the loop counter. In the latter case, it is enough to check the number of iterations that the chip executed. In the *if* skip case, the number of addition steps is highly dependant on the l value and can vary even if the security level of the parameters do not.

Recovery of $h_2(Q)$

Let $F_P(Q) = f^2 \cdot h_1(Q) \cdot h_2(Q)$ be the result of the (correct) Miller algorithm expressed with the variables of the last iteration.

If an attacker skips the *if* instruction in the last iteration, he obtains the value $F_P(Q)^* = f^2 \cdot h_1(Q)$.

With a faulty result and a correct one, he can then compute the ratio

$$\frac{F_P(Q)}{F_P(Q)^*} = \frac{f^2 \cdot h_1(Q) \cdot h_2(Q)}{f^2 \cdot h_1(Q)} = h_2(Q). \tag{12.4}$$

Finding the secret with $h_2(Q)$.   With the value $h_2(Q)$, the attacker still has to find the secret (the point P in our case). The following computations are done for the Tate pairing in particular. In this case the value r is the order of the groups used in the pairing. As a consequence, in the last iteration, the equation $T = -P$ holds.

In affine coordinates in the last iteration, with an embedding degree 2, $h_2(Q) = x_Q - x_P$ since $T = -P$: the line is the vertical passing through P. So, knowing the value $h_2(Q)$, the attacker can find x_P with x_Q known. Using the elliptic curve equation, two candidates are possible for the y_P value. By trying the two possible input points in the Miller algorithm, he can find y_P with the comparison of these two Miller results and the correct one.

The result in Jacobian coordinates is slightly different. The equations are computed with an embedding degree 4 and the basis $B = \{1, \xi, \sqrt{\nu}, \xi\sqrt{\nu}\}$. The point P has Jacobian coordinates (x_P, y_P, z_P) and Q has coordinates $(x_Q, y_Q\sqrt{\nu})$.

In the last iteration, the simplified value $h_2(Q)$ is $h_2(Q) = z_P^2 x_Q - x_P$. When the attacker computes the ratio $R = \frac{F_P(Q)}{F_P(Q)^*}$, he finds a value that can be decomposed on the basis B:

$$R = R_0 + R_1\xi + R_2\sqrt{\nu} + R_3\xi\sqrt{\nu}.$$

The decomposition of $h_2(Q)$ on the basis B yields the system

$$R_1 = z_P^2 x_{Q1} \tag{12.5}$$
$$R_2 = z_P^2 x_{Q0} - x_P, \tag{12.6}$$

where $x_Q = x_{Q0} + x_{Q1}\xi$.

Since Q is known to the attacker, this system can be solved to provide the values z_P^2 and then x_P. There are four possible candidates for the point P, which have to be verified by comparing them with the correct result of the Miller algorithm.

Remark 12.3   In the case of other pairings (Ate,...), the same attack can be applied. The main difference is that we find a point multiple of P: λP for a public integer λ. Indeed, we consider that except for the secret point, every detail of the implementation is public.

An implementation of the attack.   The authors of this attack [4] implemented their attack on a chip, an ATmega128L, with a laser fault injection. They demonstrated the feasibility of the *if* instruction skip on a dummy algorithm mimicking the structure of the Miller algorithm. After locating the right spot for the laser fault injection, they were able to successfully skip an *if* instruction.

The *if* instruction skip has two big advantages. It easily targets a specific iteration in the Miller algorithm. It is possible to combine it with another instruction skip in the final exponentiation in order to realize a full attack on the pairing computation algorithm. But this latter possibility is yet to be proven experimentally.

Countermeasures

Several countermeasures can be implemented to prevent a fault attack. They are referenced in [22], and we briefly recall them here. We can preventively use randomization of the inputs in order to prevent any leakage of information or detect any alteration of the circuit and then abort the pairing computation.

In order to detect any alteration of the computation we can

- duplicate the computation using bilinerarity: $R_1 = e(P, Q)$, $R_2 = e(aP, bQ)$ and check if $R_2 = R_1^{ab}$ [22];
- check intermediate results during the computation: verify that the points are still on the elliptic curve, compare the last point T with $(r - 1)P$ [22];
- use fault-resilient counters to avoid attacks focused on changing the Miller loop bound [43, Section 5.3];
- implement the algorithm to perform a random number of iterations greater than the correct one [24, Section 4].

The randomization and blinding methods are both based on the bilinearity of pairings:

- choose integers a and b such that $ab = 1 \mod (r)$ and compute $e(P, Q) = e(aP, bQ)$ [45];
- choose a random point R such that $S = e(P, R)^{-1}$ is defined and compute $e(P, Q) = e(P, Q + R)S$;
- use the homogeneity property of Jacobian and projective coordinates to represent the point P;
- use the homogeneity property of Jacobian and projective coordinates to represent the point Q (with a modification of the equations in the Miller algorithm);
- randomize the input points using a random field element and modify the pairing algorithm in order to cancel out the effects [53].

12.2.3 A Fault Attack against the Final Exponentiation

The main difficulty faced by fault attacks on the pairing is the final exponentiation. Even if efficient schemes are able to reverse the Miller algorithm, they still require the attacker to have access to the result of the Miller algorithm, correct or faulty.

Several possibilities have been proposed to access these values. First, for some exponents (e.g., $q^3 - 1$), it is possible to reverse the final exponentiation by using the structure of the Miller result as shown in [45]. A more implementation-dependent approach has been proposed in [19], where the authors propose to realize a scan chain attack or to completely override the final exponentiation, to directly read the result of the Miller algorithm.

Despite having been previously considered unrealistic, multiple fault injections during one execution of an algorithm seem to be more and more feasible, with some new results in this direction [55]. This new possibility opens the door to a new scheme, where two fault attacks are combined: one to reverse the final exponentiation, one to reverse the Miller algorithm.

Until recently, the final exponentiation was thought to be an efficient countermeasure against the fault attacks on the Miller algorithm, since it is mathematically impossible to find the unique preimage of the exponentiation and thus the result of the Miller loop. However, in [40], the authors propose a fault attack to reverse the final exponentiation.

Description of the attack

They chose the case where the embedding degree $k = 2d$ is even, and they attack the final exponentiation algorithm proposed in [50].

The exponent is $\frac{p^k-1}{r}$ and can be decomposed as $\frac{p^k-1}{r} = (p^d - 1) \cdot \frac{p^d+1}{r}$. If the result of the Miller algorithm is noted f, we choose the following notation: $f_2 = f^{p^d-1}$ and $f_3 = f_2^{\frac{p^d+1}{r}}$ (f_3 is the pairing result observed at the end of the computation). Since $f \in \mathbb{F}_{p^k}^*$, f, f_2, and f_3 satisfy the relations

$$f^{p^k-1} = 1 \; ; \; f_2^{p^d+1} = 1 \; ; \; f_3^r = 1. \qquad (12.7)$$

These relations show that these intermediary values belong to the groups noted $f_2 \in \mu_{p^d+1}$ and $f_3 \in \mu_r$.

Let $\mathbb{F}_{p^k} = \mathbb{F}_{p^d}[w]/(w^2 - v)$ be the construction rule for the $\mathbb{F}_{p^k}$ extension field. v is a quadratic nonresidue in $\mathbb{F}_{p^d}$ and is a public parameter.

Let $f_2 = g_2 + h_2 \cdot w$ with $g_2, h_2 \in \mathbb{F}_{p^d}$. Then $f_2^{p^d+1} = 1$ implies $g_2^2 - v \cdot h_2^2 = 1$.

First fault

But this equation holds because $f_2 \in \mu_{p^d+1}$. If an attacker now injects a fault of value $e \in \mathbb{F}_{p^d}$ such that the faulty value f_2^* equals

$$f_2^* = f_2 + e \notin \mu_{p^d+1}, \qquad (12.8)$$

it is possible to write the fault effect as

$$f_2^* = (g_2 + e) + h_2 \cdot w, \qquad (12.9)$$

and the value $(f_2^*)^{p^d+1}$ can be computed by the attacker, since he can measure the value f_3^* and r:

$$(f_2^*)^{p^d+1} = (f_3^*)^{\in \mathbb{F}_{p^d}}. \qquad (12.10)$$

Moreover,

$$\begin{aligned} (f_2^*)^{p^d+1} &= (g_2 + e)^2 - v \cdot h_2^2 \\ &= 1 + 2 \cdot e \cdot g_2 + e^2. \end{aligned}$$

If the attacker knows the error value e, he can compute

$$g_2 = \frac{(f_3^*)^l - 1 - e^2}{2 \cdot e}, \qquad (12.11)$$

and deduce the two candidates for h_2

$$h_2^+ = \sqrt{\frac{g_2^2 - 1}{v}} \; ; \; h_2^- = -\sqrt{\frac{g_2^2 - 1}{v}}. \qquad (12.12)$$

With one fault, the attacker found the intermediary value f_2 by checking the two candidates and comparing $(f_2^+)^{\frac{p^d+1}{r}}$ and $(f_2^-)^{\frac{p^d+1}{r}}$ with f_3.

Second fault

At this step, the attacker knows that f_3 is the correct result of the pairing computation, and that the intermediary value is f_2. Let $f = g + h \cdot w$, $f^{-1} = g' + h' \cdot w$, and $f_2 = f^{p^d-1}$. Then we note K, the ratio

$$K = \frac{g_2 - 1}{v \cdot h_2} = \frac{h'}{g'} = -\frac{h}{g}. \qquad (12.13)$$

In order to recover f, the attacker creates a new fault $e_2 \in \mathbb{F}_{p^d}$ during the inversion in the exponentiation by exponent $p^d - 1$.

Then
$$f_2 = f^{p^d-1} = \bar{f} \cdot f^{-1} \text{ and } f_2^* = \bar{f} \cdot (f^{-1} + e_2). \tag{12.14}$$

Let Δ_{f_2} be the difference $\Delta_{f_2} = f_2^* - f_2 = \bar{f} \cdot e_2$. Since $e_2 \in \mathbb{F}_{p^d}$, Δ_{f_2} can be written $\Delta_{f_2} = \Delta_{g_2} + \Delta_{h_2} \cdot w$ with $\Delta_{g_2} = e_2 \cdot g$ and $\Delta_{h_2} = -e_2 \cdot h$.

As f_2^* is not in μ_{p^d+1} with high probability, the attacker can compute $(f_2^*)^{p^d+1} = (f_3^*)^r \in \mathbb{F}_{p^d}$. Here

$$
\begin{aligned}
(f_3^*)^{p^d+1} &= (g_2 + \Delta_{g_2})^2 - v \cdot (h_2 + \Delta_{h_2})^2 \\
&= (g_2 + e_2 \cdot g)^2 - v \cdot (h_2 - e_2 \cdot h)^2.
\end{aligned}
$$

Using the relation $h = -g \cdot K$, we obtain

$$g^2 \cdot e_2^2 \cdot (1 - v \cdot K^2) + g \cdot 2 \cdot e_2 \cdot (g_2 - v \cdot K \cdot h_2) + 1 - (f_3^*)^r = 0. \tag{12.15}$$

This quadratic equation provides two solutions for g, each one giving only one possibility thanks to K. The attacker has two candidates for f if he knows e_2.

If he does not exactly know the fault values but is able to have a limited number of guesses, he can still find f_2 easily. But in order to find f he will have to inject more faults similar to the second one in order to uniquely determine f.

As a conclusion, with a minimum of two separate faults during two executions (plus one correct execution) of the pairing computation, the attacker is able to reverse the final exponentiation.

A notable fact about this fault attack is that it can be achieved with instruction skip faults. As a consequence, it is possible to combine it with a fault on the Miller algorithm, if the attacker can inject two faults in the same execution, in order to achieve a full-pairing fault attack.

A major disadvantage of this attack, making it easy to counter, is that the attacker must be able to observe $f_3^* = (f_2^*)^{\frac{p^d+1}{r}}$. But often, since $f_2 \in \mu_{p^d+1}$ is called a unitary element, it is possible to speed up the final exponentiation computation by replacing the inversions in the computation of f_3 by conjugations (which is equivalent to an inversion for unitary elements). As a consequence, the attacker cannot observe f_3^* in this case and he cannot realize the attack.

12.2.4 Conclusion

We presented the vulnerability to fault attacks of pairing algorithms when used in an identity based protocol. The first attack against Duursma and Lee algorithm targets the number of iterations. The final exponentiation in this case can be reversed using cryptanalytic equations. The most efficient pairings are constructed on the Tate model: an execution of the Miller algorithm followed by a final exponentiation. The Miller algorithm and the final exponentiation were separately submitted to fault attacks. The Miller algorithm was attacked by a modification of the number of iterations and by the corruption of the *if* condition during the last iteration. The final exponentiation was attacked using two "independent" errors in the computation.

For once, it would be interesting to validate all those fault attack schemes on practical implementations running on an embedded chip. Moreover, in order to attack a whole Tate-like pairing, further work is necessary. It would be interesting to try to attack, at the same time, the Miller algorithm and the final exponentiation. We also highlight the fact that a more general pairing constructed over an algebraic variety is sensitive to a fault attack. As a conclusion, we can say that the fault attack is a threat against an identity-based protocol, and consequently any implementation of pairings should be protected against physical attacks.

TABLE 12.1   Summary of the presented attacks.

Attack name	Target	Attack path	Fault model	Number of faults required (+ correct execution)
Page and Vercauteren [45]	Duursma and Lee algorithm	Loop counter	Data modification	$n \| P(n,N) > 0.5$ (+1)
Whelan and Scott [59]	Miller algorithm	Sign change	Bit-flip	1 (+1)
El Mrabet [19]	Miller algorithm	Loop counter	Data modification	$n \| P(n,N) > 0.5$ (+1)
Bae et al. [4]	Miller algorithm	If skip	Instruction skip	1 (+1)
Lashermes et al. [40]	Final exponentiation	Group change	Data modification	2+ (+1)
El Mrabet [20]	Pairing on Theta functions	Loop counter	Data modification	1

In Table 12.1 $P(n,N)$ is the probability to obtain two consecutive numbers after n picks among N integers (cf Section 12.2.2).

12.3   Countermeasures against Fault Attacks on Pairing-Based Cryptography

The protection scheme that we present here is based on the technique of *modular extension*, which was introduced by Shamir along with the first software countermeasure against fault injection attacks on CRT-RSA [52]. Joye, Paillier, and Yen noticed, two years later in [34], that the same protection could extend to any modular function. Since then, many countermeasures based on modular extension have been developed for CRT-RSA, and the method made its way to elliptic curve cryptography (ECC). In particular, Blömer, Otto, and Seifert [12], and Baek and Vasyltsov [5] applied this protection method to elliptic curve scalar multiplication (ECSM). More recently, Rauzy, Moreau, Guilley, and Najm [48] have formally studied the protection of ECSM computations with the modular extension method. We here extend it to pairing-based cryptography.

12.3.1   Modular Extension

The general idea of modular extension is to lift the computation into an over-structure (e.g., an overring) which allows us to quotient the result of the computation back to the original structure, as well as quotienting a "checksum" of the computation to a smaller structure. What has just been described (the original structure and the smaller structure) is the *direct product* of the underlying algebraic structures. If an equivalent computation is performed in parallel in the smaller structure, its result can be compared with the checksum of the main computation. If they are the same, we have a high confidence in the integrity of the main computation. This protection is sketched in Figure 12.2.

The confidence degree depends directly on the size of the smaller structure, which is thus a security parameter: the larger it is, the less probable it is to have an unwanted collision, but the more costly the redundancy will be. Indeed, the fault non-detection probability ($\mathbb{P}_{\text{n.d.}}$) is inversely proportional to the size of the small structure.

When the basic structure underlying the original computation is a field, as is the case in pairing-based cryptography (contrary to, e.g., RSA, which only requires a ring), a problem arises with inversions. Indeed, if we call $\mathbb{F}_p$ the original structure and $\mathbb{F}_r$ the smaller one, the nonzero elements of their direct product $\mathbb{Z}_{pr}$ do not all have an inverse. Nonetheless, this problem can be circumvented.

FIGURE 12.2   Sketch of the principle of *modular extension*.

PROPOSITION 12.1 *To get the inverse of z in $\mathbb{F}_p$ while computing in $\mathbb{Z}_{pr}$, one has:*

- $z = 0 \bmod r \implies (z^{p-2} \bmod pr) \equiv z^{-1} \mod p,$
- *otherwise* $(z^{-1} \bmod pr) \equiv z^{-1} \mod p.$

Remark 12.4 Golić and Tymen introduce in [25] a masking countermeasure of the Advanced Encryption Standard (AES [1]), called the "embedded multiplicative masking", which also requires embedding a finite field into a larger ring. In this context, the over-structure is a *polynomial extension* of some extension of $\mathbb{F}_2$, but the idea is similar to *modular extension*. In particular, the authors note in Section 5.1 of their paper [25] that inversion in the base field can be obtained in the overring as an exponentiation to the base field order minus two.

But the inversion procedure we give in Proposition 12.1 is novel, in that we allow an optimization if the number is inversible in the overring. This requires a test, which we can do safely without disclosing information in the context of fault-attacks detection. Nonetheless, such optimization would be insecure in the context of the "embedded multiplicative masking" countermeasure, since this would leak information about the value of the mask. This is a first-order flaw that would undermine the security of the "embedded multiplicative masking" protection against side-channel attacks.

In addition, it is possible to write pairing algorithms that use very few divisions (as little as a single one in our mini-pairing implementation; see hereafter in Section 12.3.3).

12.3.2   Other Existing Countermeasures

We review the three known methods to apply and/or adapt the modular extension countermeasure to ECSM (which is central to pairing-based cryptography).

In [12], Blömer, Otto, and Seifert (BOS) suggest applying the modular extension countermeasure by replacing finite fields and rings with elliptic curves over finite fields and rings. Let us denote the nominal elliptic curve as $E(\mathbb{F}_p)$. Then the protection by BOS consists in achieving the same computation, but on a larger elliptic curve $E(\mathbb{Z}_{pr})$, and on a small elliptic curve $E(\mathbb{F}_r)$. According to the authors, the reduction of the result of the ECSM on $E(\mathbb{Z}_{pr})$ modulo r should yield exactly the result of the ECSM on $E(\mathbb{F}_r)$. If not, then an error is suspected, otherwise the result of the ECSM on $E(\mathbb{Z}_{pr})$ is reduced modulo p, which should be the correct result. The rationale of BOS is illustrated in Figure 12.3. Apart from the lacunar management of inversions in $\mathbb{Z}_{pr}$, one other caveat is pinpointed in [48, § 3.1]. Due to the existence of unrelated tests (e.g., equality of intermediate points to the point at infinity) on $E(\mathbb{Z}_{pr})$ and $E(\mathbb{F}_r)$, the algorithm proposed by BOS is incorrect, meaning that it can return an error when there has been none. These *false positives* are harmful in that they leak information on the scalar.

In [5], Baek and Vasyltsov (BV) present an optimization of BOS. The idea is to avoid the computation on $E(\mathbb{F}_r)$, but to trade it for a verification that the ECSM result on $E(\mathbb{Z}_{pr})$ modulo r belongs to $E(\mathbb{F}_r)$, i.e., that it satisfies its Weierstrass equation taken modulo r. The rationale of BV is illustrated in Figure 12.4: notice that in this figure, the security parameter r is chosen to be a prime. This was not mandated in the original BV publication, but shall definitely be preferred for the countermeasure to have reasonable detection probability. The BV protection is more efficient than BOS, since the verification of BV is, computationally speaking, easier than an ECSM on $E(\mathbb{F}_r)$ (even if the scalar is reduced modulo the order of the small curve $E(\mathbb{F}_r)$). Besides, to avoid dealing with inversions in $\mathbb{Z}_{pr}$, BV is executed in projective coordinates, the projective-to-affine conversion only being carried out after the integrity verification. Still, BV runs into the problem of inconsistent tests before elliptic curves point addition and doubling. The consequence is that, depending on the scalar (and the fixed generator point), the "virtual"

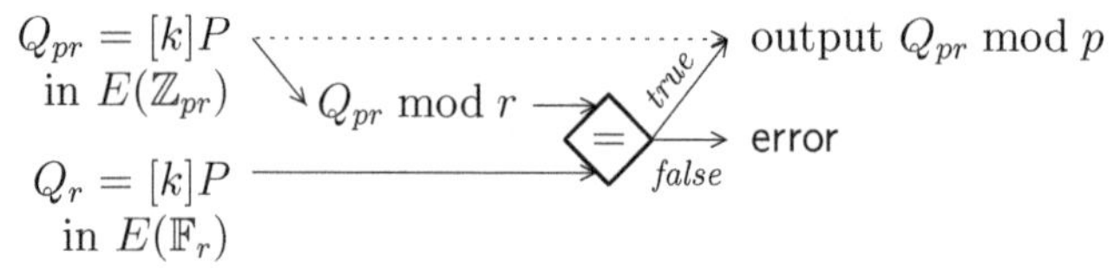

FIGURE 12.3   Principle of protection of ECSM against fault attacks by BOS [12].

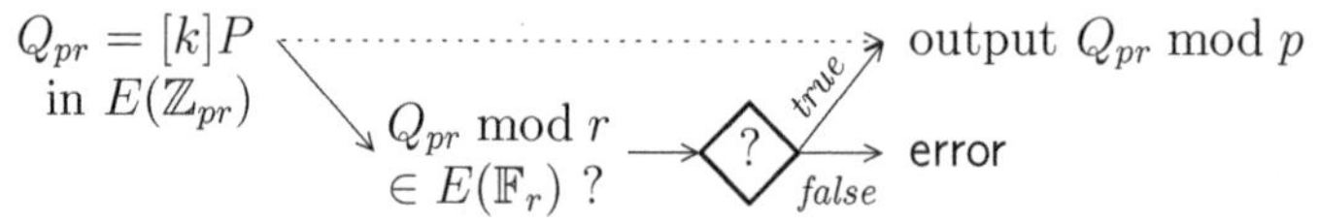

FIGURE 12.4   Principle of protection of ECSM against fault attacks by BV [5].

computation on $E(\mathbb{F}_r)$ (the modulo r of computation on the embedding elliptic curve $E(\mathbb{Z}_{pr})$) can be stuck at the point at infinity.

In [48], Rauzy, Moreau, Guilley, and Najm (RMGN) notice that the tests' inconsistencies on $E(\mathbb{F}_r)$ can also be security weaknesses. Indeed, when the mirror computation on $E(\mathbb{F}_r)$ is stuck at the point at infinity, most faults (for instance, faults touching only one of three projective coordinates) are undetected, because the computation naturally brings the intermediate point at the point at infinity on $E(\mathbb{F}_r)$ (and to a point with coordinates that are null modulo r on $E(\mathbb{Z}_{pr})$). Thus, the probability of fault non-detection is increased with respect to the expected $O(1/r)$. Consequently, RMGN propose a straightforward application of the modular extension method (as suggested by Joye, Paillier, and Yen [34]) to ECSM, where all tests on points are simply removed. From a functional point of view, this does not raise an issue, as in practice scalars are chosen to be smaller than the base point order, so that tests can be safely skipped. The pro is that this method is correct (it has no false positives), but the con is that some faults are undetected (the behavior is identical to that of BV). Indeed, even though in RMGN there is no notion of elliptic curve $E(\mathbb{F}_r)$, the values in $\mathbb{F}_r$ can be stuck at 0 (though we can still detect those faulty cases beforehand by comparing the order of $\mathbb{F}_r$ with the scalar). However, as in the case of BV, the increase of fault non-detection probability is limited, and can be tolerated with large enough values of r (e.g., 32-bit values). Indeed, as detailed in [48, Proposition 7 in Sec. 6.3], the probability of fault non-detection remains $O\left(\frac{1}{r}\right)$.

12.3.3   Walk through Mini-Pairing

As an example of the *modular extension* protection scheme that we present here, we provide both an unprotected and a protected implementations of the optimal Ate pairing that we call "mini-pairing".[*] The provided code has been implemented in C using the GMP big number library, more precisely `mini-gmp`, a portable version of GMP with a reduced number of functions. The parameters of the optimal Ate pairing we used are presented in Figure 12.5.

The protected version of the optimal Ate pairing is given in Algorithm 12.5.

Here we discuss the differences between the unprotected and the protected mini-pairing implementations. Indeed, for the sake of simplicity, we emphasize our comments on the necessary code modifications to implement the modular extension protection scheme, rather than focusing on the underlying algorithm, namely an optimal Ate pairing. For the same reasons, the implementation has not been optimized.

---

[*]The code is available here: `http://pablo.rauzy.name/research/sources/hopbc_mini-pairing.tgz`.

Field characteristic	$p =$	0x2523648240000001ba344d80000000086121000000000013a700000000000013
Curve equation coefficients	$a =$	0x0
	$b =$	0x2
Points coordinates	$Qx =$	0x1c2141648fed8ba0f2a3febe8b98509bf86398d1fd1050c88fc3d88be3d15db1 + 0x93772aa06ea4acf488ce113f4a56aeb6c23264001c1501c1c59cd47faac6d0f·u
	$Qy =$	0x1eb672f0d5335990c9b12f9839b1a8804393211b198237c5acfc4d69d51186a0 + 0x118c5d037558e51efdd3cf3530d8c5cb65c52f9cf639ed6d81ddc6c16b76eec0·u
	$Px =$	0x2523648240000001ba344d80000000086121000000000013a700000000000012
	$Py =$	0x1

FIGURE 12.5   Parameters of mini-pairing.

ALGORITHM 12.5   Optimal Ate pairing capable of detecting faults (using entanglement strategy).

---

let p,r be two primes, and $\mathbb{F}_p$, $\mathbb{F}_r$ two fields with p and r elements
let $\mathbb{Z}_{pr}$ be the direct product ofq $\mathbb{F}_p$ and $\mathbb{F}_r$
let G_1,G_2 be two additive cyclic groups of prime order p
let e be the pairing mapping G_1 and G_2
let $P \in G_1$ and $Q \in G_2$
compute $e_{\mathbb{Z}_{pr}} = e(P,Q)$ in $\mathbb{Z}_{pr}$
compute $e_{\mathbb{F}_r} = e(P,Q)$ in $\mathbb{F}_r$
if $e_{\mathbb{F}_r} = e_{\mathbb{Z}_{pr}} \bmod r$ **then**
| **return** $e(P,Q) = e_{\mathbb{Z}_{pr}} \bmod p$
else
| **return** error
end

The first modification is obviously the addition of variables that store newly needed values, such as the security parameter r (lines 1314 and 1315 at the beginning of the `main` function in `mini-pairing_protected.c`). Then, the main change induced by the protection is that the pairing algorithm is now called twice: once in $\mathbb{Z}_{pr}$, and once in $\mathbb{F}_r$. Following these computations, we need to check whether the redundancy invariant held, i.e., to test whether both outputs are equal modulo r. Two additional functions are defined for this purpose: one to compare two elements of $\mathbb{F}_{r^{12}}$ (`p12_is_eq`, lines 1241 to 1308 of `mini-pairing_protected.c`), and another to cast an element from $\mathbb{Z}_{(pr)^{12}}$ to $\mathbb{F}_{r^{12}}$ (`p12in`, lines 1180 to 1239 of `mini-pairing_protected.c`). The redundancy check and error display (if need be) are then performed at line 1365.

Another difference is in the inversion (lines 316 to 324 of `mini-pairing_protected.c`). Inversions seldom occur in this pairing algorithm; however, it will fail if the input number is a multiple of r in $\mathbb{Z}_{pr}$. In such a case, we simulate an inversion in $\mathbb{F}_p$ (which is what we actually need) by exponentiating to $p-2$, as explained in Proposition 12.1. Computing an exponentiation is more costly than computing an inversion with an extended Euclidean algorithm. But there are few enough occurrences of this case in practice that this workaround does not have a significant impact on the execution time.

12.3.4   Overhead

Here we present the cost of the countermeasure as deployed in our mini-pairing implementation. Note that for the sake of simplicity and clarity, the implementation is not optimized and is thus quite slow. However, the overhead factor is still relevant, since optimizations of the pairing algorithm would directly benefit the protected version as we constructed it (see Section 12.3.3).

TABLE 12.2   Performance for mini-pairing on an ARM Cortex-M4 μc.

Miller's loop	$\mathbb{F}_p$ time (ms) intermediate computations	final exponentiation	sum
4160	621	6519	11343

TABLE 12.3   Modular extension performance for mini-pairing on an ARM Cortex-M4 μc.

r size (bit)	$\mathbb{Z}_{pr}$ time (ms) ML	IC	FE	sum	$\mathbb{F}_r$ time (ms) ML	IC	FE	sum	total (ms)	over-head
8	4576	703	7201	12443	1186	142	1781	3105	15587	×1.37
16	4617	706	7263	12546	1185	141	1777	3097	15685	×1.38
32	4706	725	7407	12864	1042	126	1565	2726	15590	×1.37
64	5260	834	8302	14334	1370	171	2071	3618	17984	×1.59

ML = Miller's loop, IC = intermediate computations, FE = final exponentiation.

Speed

Times are measured on an ARM Cortex-M4 microcontroller. Table 12.2 gives the timing of the unprotected implementation that serves as reference to compute the overheads given in Table 12.3. Table 12.3 presents the cost of the countermeasure for different sizes of the security parameter r, using the largest prime number of each size.

Table 12.3 shows the good performance results of the modular extension protection scheme. We can see that when r is on 32 bits, the alignment with `int` makes `mini-gmp` faster, incurring a factor of only ≈ 1.37 in the total run time compared to the unprotected algorithm, similar to the cost with r on 8 bits but with a much higher resistance.

Space

Table 12.4 shows the cost of the countermeasure in terms of code size, both for the programmer (in number of lines of C code), and for the hardware (in kilobytes of executable code and in bytes of occupied memory). Note that the executable code size also accounts for embedded libraries such as `mini-gmp`.

In order to measure the RAM usage, the maximal value of the heap pointer address is monitored. This is achieved by equipping the `_sbrk()` function, located in `syscall.c`, which is called by `malloc()` and `free()`. Notice that most of the RAM is indeed used by the heap (and not the stack), because in the ECSM code, all parameters are passed by address, and there are neither recursive functions nor pre-initialized tables (but for the elliptic curve parameters).

As expected, we can see in Table 12.4 that the implementation of the modular extension countermeasure is cheap in terms of engineering: less than 150 additional lines of code (for a total of almost 1400 lines); as well as in terms of resources: the executable code is only marginally larger, and memory usage is essentially the same (probably due to the way `mini-gmp`'s and libC memory allocation works).

TABLE 12.4   Modular extension cost in terms of space for mini-pairing on an ARM Cortex-M4 μc.

implementation	code size (LoC)	executable size (B)	occupied RAM (kB)
unprotected	1404	95032	≈ 20
protected	1545 (+141)	96832 (+1800)	≈ 20

12.3.5 Security Evaluation

DEFINITION 12.1 (Fault model) We consider an attacker who is able to fault data by randomizing or zeroing any intermediate variable, and fault code by skipping any number of consecutive instructions.

DEFINITION 12.2 (Attack order) We call order of the attack the number of faults (in the sense of Definition 12.1) injected during the target execution.

Remark 12.5 In the rest of this section, we focus on the resistance to first-order attacks on data. Indeed, Rauzy and Guilley have shown in [47] that it is possible to adapt the modular extension protection scheme to resist attacks of order D for any D by chaining D repetitions of the final check in a way that forces each repetition of the modular extension invariant verification to be faulted independently, and faults on the code can be formally captured (simulated) by faults on intermediate variables.

The security provided by the modular extension protection scheme has been formally studied in [48, § 5]. Although the practical study was carried out on an ECSM algorithm, the theoretical results are still valid in the context of pairing algorithms (or actually any other modular arithmetic computations): the probability of not detecting a fault $\mathbb{P}_{\text{n.d.}}$ is inversely proportional to the security parameter r, i.e., $\mathbb{P}_{\text{n.d.}} = O(\frac{1}{r})$.

Indeed, we consider that a fault might be exploitable as soon as the algorithm outputs a value that is different from the expected result in absence of faults. In the modular extension setting, this can happen if and only if the result of the computation in $\mathbb{Z}_{(pr)^{12}}$ is equal to the result of the computation in $\mathbb{F}_{r^{12}}$ modulo r, while being different from the expected result modulo p. The probability of this happening is $\frac{1}{r}$ if we consider that values in $\mathbb{F}_r$ are uniformly distributed, which is quite reasonable given that $r \ll p$. As a matter of fact, we can quantify this distribution. Let U uniformly distributed in $\{0, \ldots, p-1\}$, then $V = U \mod r$ has a piecewise constant distribution. Let v in $\{0, \ldots, r-1\}$, we have:

$$\mathbb{P}(V = v) = \begin{cases} \frac{1}{p}(\lfloor \frac{p}{r} \rfloor + 1) & \text{if } v < (p \bmod r) \\ \frac{1}{p}(\lfloor \frac{p}{r} \rfloor) & \text{otherwise.} \end{cases}$$

There are other vulnerabilities, but they do not alter $\mathbb{P}_{\text{n.d.}}$. For instance, the final exponentiation always returns 0 in $\mathbb{F}_{r^{12}}$ for some (small) values of r. In addition, Miller's algorithm manipulates an element from an elliptic curve on $\mathbb{F}_{p^2}$, and if a fault manages to set the Y coordinate of that element to 0 mod r, its other coordinates will also become multiples of r after few iterations of the Miller's loop, thereby "infecting" the computation by being completely equal to 0 modulo r. Therefore, the exponentiation will also output 0 modulo r in $\mathbb{Z}_{(pr)^{12}}$, and the final test won't detect the fault. However, such faults are highly unlikely in practice, the probability being roughly $\frac{1}{r^2}$, which is why $\mathbb{P}_{\text{n.d.}}$ stays $O(\frac{1}{r})$. Anyway, it is advised to use large enough values for the security parameter r. In practice, 32-bits values are recommended as they are large enough to offer a good security while not being big enough for the overhead induced by the countermeasure to be prohibitive (see Table 12.3). It is also advised to use prime numbers for r as it will diminish the probability of occurrence of the inversion problem mentioned above.

Remark 12.6 One must also be careful with the choice of parameters. For instance, manipulating a P whose coordinates are multiples of r might lead to singularities in the computation in $\mathbb{F}_r$, singularities such as the $\mathbb{F}_{r^{12}}$-output being equal to 1, therefore making the pairing computation more vulnerable to fault injections.

Remark 12.7   Taking the BV fault detection as an example (recall Figure 12.4), one might be tempted to lift the computation from $\mathbb{F}_r$ to $\mathbb{Z}_{pr}$, and do a sanity check of the pairing computation instead of redoing a redundant pairing computation in $\mathbb{F}_r$. One property that could be checked is the bilinearity. Unfortunately, the elliptic curve changes modulo r, as the Weierstrass coefficients are reduced modulo r. Therefore, the bilinearity remarkable identity is not preserved in $\mathbb{F}_r$ after reducing the pairing computation from $\mathbb{Z}_{pr}$ modulo r.

The presented countermeasure can also bring a reasonable security against simple side-channel attacks. Indeed, the 32-bits r parameter can be chosen randomly, and there are 98182657 prime numbers between 2^{31} and 2^{32}, hence providing many different possible execution traces against power analysis.

12.4   Countermeasures against Side-Channel Attacks on Pairing-Based Cryptography

In order to protect pairing implementations against the side-channel attacks described in this chapter, several countermeasures have been proposed. The aim of most of those countermeasures is to avoid any predictable link between the manipulated data and the known input.

In practice, in the pairing computation context, there are different randomization levels. One category of countermeasures consists of randomizing the inputs before the pairing computation. Another one consists of adding a random mask directly into the Miller algorithm. Moreover, a method based on arithmetic randomization can be adapted for the pairing.

12.4.1   Randomization of the Inputs

Page and Vercauteren [44] proposed two countermeasures for their passive attack. The first one is based on the pairing bilinearity. Let a and b be two random values, then $e([a]P, [b]Q)^{\frac{1}{ab}} = e(P, Q)$. For each pairing computation, one can thus take different a and b and compute $e([a]P, [b]Q)^{\frac{1}{ab}}$. This method is clearly very costly in terms of computation time. Then, the random choice for a and b can be adapted to have $a = b^{-1} \mod q$, so the exponent $\frac{1}{ab}$ is equal to 1.

The same authors propose another method, for instance in the case where P is secret, consisting of adding the mask to the point Q in the following way: select a random point $R \in \mathbb{G}_2$ and compute $e(P, Q + R)e(P, R)^{-1}$ instead of $e(P, Q)$, with different values of R at every call to e.

Widely inspired by the previous protection, Blömer et al. in [11] proposed an improvement applied for the Tate pairing. In the reduced Tate pairing, they note that the set of the second argument input is the equivalence class $\frac{E(\mathbb{F}_{p^k})}{rE(\mathbb{F}_{p^k})}$. They hence choose a random point $R \in E(\mathbb{F}_{p^k})$ with order l and coprime to r. Then $Q + R \sim Q$. Hence $e(P, Q + R) = e(P, Q)$. This method avoids the second pairing computation that is used to find the same result without mask.

12.4.2   Randomization of Intermediate Variables

Kim et al. [37] use the third countermeasure proposed by Coron in [15], using random projective coordinates to protect the Eta pairing in characteristic 2. But it can be adapted to pairing algorithms based over a large prime characteristic field. At the beginning of the algorithm, they proceed with this randomization based on the homogeneity of projective or Jacobian coordinates. For non-zero integer λ, the point $P = (X_P, Y_P, Z_P)$ in projective coordinates is also the point $P = (\lambda X_P, \lambda Y_P, \lambda Z_P)$. The point $P = (X_P, Y_P, Z_P)$ in Jacobian coordinates is also the point

$P = (\lambda^2 X_P, \lambda^3 Y_P, \lambda Z_P).$

12.4.3 Arithmetic Randomization

However, all previous attacks against pairings targeted an arithmetic operation. Securing multiplications was originally studied in [30] in order to protect ECDSA against side-channel attacks. The aim is to avoid all possible predictions during a modular multiplication. A mask is randomly chosen before processing a multiplication. Then it is impossible to make any hypothesis on the output of internal modular multiplication. We find another masking technique in the paper [9], the aim being the same: avoiding any predictable link between known and secret data directly in the arithmetic.

Protected arithmetic can also be obtained with the well-known Residue Number System method [7].

Arithmetic protection seems to be a robust method against side-channel. However, it is necessary to evaluate the overhead cost. Indeed, changing permutation in randomized multiplication or refreshing RNS basis in case of RNS implementation have a significant overhead.

12.4.4 Countermeasures against Loop Reduction

The fault attacks against the Miller algorithm rely on the modification of the number of iterations performed by the algorithm. We can add a counter to the Miller algorithm.

12.4.5 Pseudo Code of Unified Existing Countermeasures

A set of these protections is relatively easy to implement. Algorithm 12.6 shows a possible combination of existing countermeasures. The arithmetic randomization is directly implemented in the arithmetic. For example, the multiplication of two long integers in $\mathbb{F}_q$ can be realized by Algorithm 2 of [9] instead of classic long integer multiplication.

ALGORITHM 12.6   Computation of pairing using Miller's loop.

Input   : $P \in \mathbb{G}_1, Q \in \mathbb{G}_2$ with Q secret, $r = (r_{w-1} \ldots r_0)_2$ radix 2 representation
Output : $e(P, Q)$

Randomly pick a and b in $\{1, \ldots, q-1\}$ such that $a = b^{-1} \mod q$
Set $P' \leftarrow [a]P$ and $Q' \leftarrow [b]Q$ // Randomization of the inputs
Randomly pick $\lambda \in \mathbb{F}'_q$
Set $T \leftarrow (\lambda x_{P'}, \lambda y_{P'}, \lambda)$ // Randomized projective coordinates
$f \leftarrow 1$
for $i = w - 2$ **downto** 0 **do**
  $\quad f \leftarrow f^2 \cdot l_{T,T}(Q')$
  $\quad T \leftarrow [2]T$
  $\quad$ **if** $r_i == 1$ **then**
    $\quad\quad f \leftarrow f \cdot l_{T,P'}(Q')$
    $\quad\quad T \leftarrow T + P'$
  $\quad$ **end**
end
return $f^{\frac{q^k - 1}{r}}$

References

[1] FIPS PUB 197. *Advanced Encryption Standard (AES)*. National Institute of Standards and Technology, U.S. Department of Commerce, November 2001.

[2] Ross Anderson and Markus Kuhn. Tamper resistance—a cautionary note. In *Proceedings of the Second USENIX Workshop on Electronic Commerce*, pp. 1–11. USENIX Association, 1996.

[3] Diego F. Aranha, Jean-Luc Beuchat, Jérémie Detrey, and Nicolas Estibals. Optimal Eta pairing on supersingular genus-2 binary hyperelliptic curves. In O. Dunkelman, editor, *Topics in Cryptology – CT-RSA 2012*, volume 7178 of *Lecture Notes in Computer Science*, pp. 98–115. Springer, Heidelberg, 2012.

[4] Kiseok Bae, Sangjae Moon, and Jaecheol Ha. Instruction fault attack on the Miller algorithm in a pairing-based cryptosystem. In *7th International Conference on Innovative Mobile and Internet Services in Ubiquitous Computing (IMIS 2013)*, pp. 167–174. IEEE Press, 2013.

[5] Yoo-Jin Baek and Ihor Vasyltsov. How to prevent DPA and fault attack in a unified way for ECC scalar multiplication - Ring extension method. In E. Dawson and D. S. Wong, editors, *Information Security Practice and Experience (ISPEC 2007)*, volume 3439 of *Lecture Notes in Computer Science*, pp. 225–237. Springer, 2007.

[6] Jean-Claude Bajard and Nadia El Mrabet. Pairing in cryptography: an arithmetic point of view. In F. T. Luk, editor, *Advanced Signal Processing Algorithms, Architectures, and Implementations XVII*, volume 6697 of *Proc. SPIE*. SPIE, 2007.

[7] Jean-Claude Bajard, Laurent Imbert, Pierre-Yvan Liardet, and Yannick Teglia. Leak resistant arithmetic. In M. Joye and J.-J. Quisquater, editors, *Cryptographic Hardware and Embedded Systems – CHES 2004*, volume 3156 of *Lecture Notes in Computer Science*, pp. 62–75. Springer, Heidelberg, 2004.

[8] Paulo S. L. M. Barreto, Steven Galbraith, Colm Ó hÉigeartaigh, and Michael Scott. Efficient pairing computation on supersingular Abelian varieties. *Designs, Codes and Cryptography*, 42(3):239–271, 2007.

[9] Aurélie Bauer, Eliane Jaulmes, Emmanuel Prouff, and Justine Wild. Horizontal and vertical side-channel attacks against secure RSA implementations. *CT-RSA*, pp. 1–17, 2013.

[10] Daniel J. Bernstein. Cache-timing attacks on AES. Unpublished manuscript, available at http://cr.yp.to/antiforgery/cachetiming-20050414.pdf, 2005.

[11] Johannes Blömer, Peter Günther, and Gennadij Liske. Improved side-channel attacks on pairing based cryptography. In E. Prouff, editor, *Constructive Side-Channel Analysis and Secure Design (COSADE 2013)*, volume 7864 of *Lecture Notes in Computer Science*, pp. 154–168. Springer, 2013.

[12] Johannes Blömer, Martin Otto, and Jean-Pierre Seifert. Sign change fault attacks on elliptic curve cryptosystems. In L. Breveglieri et al., editors, *Fault Diagnosis and Tolerance in Cryptography (FDTC 2006)*, volume 4236 of *Lecture Notes in Computer Science*, pp. 154–168. Springer, 2006.

[13] Dan Boneh and Matthew K. Franklin. Identity based encryption from the Weil pairing. *SIAM Journal on Computing*, 32(3):586–615, 2003.

[14] Benoît Chevallier-Mames, Mathieu Ciet, and Marc Joye. Low-cost solutions for preventing simple side-channel analysis: Side-channel atomicity. *IEEE Transactions on Computers*, 53(6):760–768, 2004.

[15] Jean-Sébastien Coron. Resistance against differential power analysis for elliptic curve cryptosystems. In Çetin Kaya. Koç and C. Paar, editors, *Cryptographic Hardware*

and Embedded Systems – CHES'99, volume 1717 of *Lecture Notes in Computer Science*, pp. 292–302. Springer, Heidelberg, 1999.

[16] Elke De Mulder, Siddika B. Örs, Bart Preneel, and Ingrid Verbauwhede. Differential power and electromagnetic attacks on a FPGA implementation of elliptic curve cryptosystems. *Computers & Electrical Engineering*, 33(5/6):367–382, 2007.

[17] Amine Dehbaoui, Jean-Max Dutertre, Bruno Robisson, and Assia Tria. Electromagnetic transient faults injection on a hardware and a software implementation of AES. In G. Bertoni and B. Gierlichs, editors, *2012 Workshop on Fault Diagnosis and Tolerance in Cryptography (FDTC 2012)*, pp. 7–15. IEEE Computer Society, 2012.

[18] Iwan Duursma and Hyang-Sook Lee. Tate-pairing implementations for tripartite key agreement. Cryptology ePrint Archive, Report 2003/053, 2003. http://eprint.iacr.org/2003/053.

[19] Nadia El Mrabet. Fault attack against Miller's algorithm. Cryptology ePrint Archive, Report 2011/709, 2011. http://eprint.iacr.org/2011/709.

[20] Nadia El Mrabet. Side-channel attacks against pairing over Theta functions. In T. Muntean, D. Poulakis, and R. Rolland, editors, *Algebraic Informatics (CAI 2013)*, volume 8080 of *Lecture Notes in Computer Science*, pp. 132–146. Springer, 2013.

[21] Nadia El Mrabet, Giorgio Di Natale, and Marie Lise Flottes. A practical differential power analysis attack against the Miller algorithm. In *2009 Conference on Ph.D. Research in Microelectronics and Electronics (PRIME 2009)*, pp. 308–311. IEEE Press, 2009.

[22] Nadia El Mrabet, Dan Page, and Frederik Vercauteren. Fault attacks on pairing-based cryptography. In M. Joye and M. Tunstall, editors, *Fault Analysis in Cryptography*, Information Security and Cryptography, pp. 221–236. Springer, 2012.

[23] Santosh Ghosh and Dipanwita Roy Chowdhury. Security of prime field pairing cryptoprocessor against differential power attack. In M. Joye, D. Mukhopadhyay, and M. Tunstall, editors, *Security Aspects in Information Technology (InfoSecHiComNet 2011)*, volume 7011 of *Lecture Notes in Computer Science*, pp. 16–29. Springer, 2011.

[24] Santosh Ghosh, Debdeep Mukhopadhyay, and Dipanwita Roy Chowdhury. Fault attack, countermeasures on pairing based cryptography. *International Journal of Network Security*, 12(1):21–28, 2011.

[25] Jovan Dj. Golic and Christophe Tymen. Multiplicative masking and power analysis of AES. In B. S. Kaliski Jr., Ç. K. Koç, and C. Paar, editors, *Cryptographic Hardware and Embedded Systems – CHES 2002*, volume 2523 of *Lecture Notes in Computer Science*, pp. 198–212. Springer, Heidelberg, 2003.

[26] Gurleen Grewal, Reza Azarderakhsh, Patrick Longa, Shi Hu, and David Jao. Efficient implementation of bilinear pairings on ARM processors. In L. R. Knudsen and H. Wu, editors, *Selected Areas in Cryptography – SAC 2012*, volume 7707 of *Lecture Notes in Computer Science*, pp. 149–165. Springer, Heidelberg, 2013.

[27] D. H. Habing. The use of lasers to simulate radiation-induced transients in semiconductor devices and circuits. *IEEE Transactions on Nuclear Science*, 12(5):91–100, 1965.

[28] Florian Hess. Pairing lattices (invited talk). In S. D. Galbraith and K. G. Paterson, editors, *Pairing-Based Cryptography – Pairing 2008*, volume 5209 of *Lecture Notes in Computer Science*, pp. 18–38. Springer, Heidelberg, 2008.

[29] Florian Hess, Nigel P. Smart, and Frederik Vercauteren. The Eta pairing revisited. *IEEE Transactions on Information Theory*, 52(10):4595–4602, 2006.

[30] Michael Hutter, Marcel Medwed, Daniel Hein, and Johannes Wolkerstorfer. Attacking ECDSA-Enabled RFID devices. *Applied Cryptography and Network Security*, pp. 519–534, 2009.

[31] Antoine Joux. A new index calculus algorithm with complexity $L(1/4 + o(1))$ in small characteristic. In T. Lange, K. Lauter, and P. Lisonek, editors, *Selected Areas in Cryptography – SAC 2013*, volume 8282 of *Lecture Notes in Computer Science*, pp. 355–379. Springer, Heidelberg, 2014.

[32] Marc Joye. Elliptic curves and side-channel analysis. *ST Journal of System Research*, 4(1):17–21, 2003.

[33] Marc Joye and Gregory Neven, editors. *Identity-Based Cryptography*, volume 2 of *Cryptology and Information Security Series*. IOS press, 2009.

[34] Marc Joye, Pascal Paillier, and Sung-Ming Yen. Secure evaluation of modular functions. In R. J. Hwang and C. K. Wu, editors, *2001 International Workshop on Cryptology and Network Security*, pp. 227–229, 2001.

[35] Marc Joye and Sung-Ming Yen. The Montgomery powering ladder. In B. S. Kaliski Jr., Ç. K. Koç, and C. Paar, editors, *Cryptographic Hardware and Embedded Systems – CHES 2002*, volume 2523 of *Lecture Notes in Computer Science*, pp. 291–302. Springer, Heidelberg, 2003.

[36] Chong Hee Kim and Jean-Jacques Quisquater. Faults, injection methods, and fault attacks. *Design Test of Computers, IEEE*, 24(6):544–545, 2007.

[37] Tae Hyun Kim, Tsuyoshi Takagi, Dong-Guk Han, Ho Won Kim, and Jongin Lim. *Cryptology and Network Security: 5th International Conference, CANS 2006, Suzhou, China, December 8-10, 2006. Proceedings*, chapter Side-Channel Attacks and Countermeasures on Pairing Based Cryptosystems over Binary Fields, pp. 168–181. Springer Berlin Heidelberg, Berlin, Heidelberg, 2006.

[38] Neal Koblitz and Alfred Menezes. Pairing-based cryptography at high security levels (invited paper). In N. P. Smart, editor, *Cryptography and Coding*, volume 3796 of *Lecture Notes in Computer Science*, pp. 13–36. Springer, Heidelberg, 2005.

[39] Paul C. Kocher, Joshua Jaffe, and Benjamin Jun. Differential power analysis. In M. J. Wiener, editor, *Advances in Cryptology – CRYPTO '99*, volume 1666 of *Lecture Notes in Computer Science*, pp. 388–397. Springer, Heidelberg, 1999.

[40] Ronan Lashermes, Jacques Fournier, and Louis Goubin. Inverting the final exponentiation of Tate pairings on ordinary elliptic curves using faults. In G. Bertoni and J.-S. Coron, editors, *Cryptographic Hardware and Embedded Systems – CHES 2013*, volume 8086 of *Lecture Notes in Computer Science*, pp. 365–382. Springer, Heidelberg, 2013.

[41] Rudolf Lidl and Harald Niederreiter. *Finite Fields*, volume 20 of *Encyclopedia of Mathematics and Its Applications*. Cambridge University Press, 2nd edition, 1997.

[42] David Lubicz and Damien Robert. Efficient pairing computation with theta functions. In G. Hanrot, F. Morain, and E. Thomé, editors, *Algorithmic Number Theory (ANTS-IX)*, volume 6197 of *Lecture Notes in Computer Science*, pp. 251–269. Springer, 2010.

[43] Erdinç Öztürk, Gunnar Gaubatz, and Berk Sunar. Tate pairing with strong fault resiliency. In L. Breveglieri et al., editors, *Fourth Workshop on Fault Diagnosis and Tolerance in Cryptography (FDTC 2007)*, pp. 103–111. IEEE Computer Society, 2007.

[44] Dan Page and Frederik Vercauteren. Fault and side-channel attacks on pairing based cryptography. Cryptology ePrint Archive, Report 2004/283, 2004. http://eprint.iacr.org/2004/283.

[45] Dan Page and Frederik Vercauteren. A fault attack on pairing-based cryptography. *IEEE Transactions on Computers*, 55(9):1075–1080, 2006.

[46] Jea-Hoon Park, Gyo Yong-Sohn, and Sang-Jae Moon. A simplifying method of fault attacks on pairing computations. *IEICE Transactions on Fundamentals of Electronics, Communications and Computer Sciences*, 94(6):1473–1475, 2011.

[47] Pablo Rauzy and Sylvain Guilley. Countermeasures against high-order fault-injection attacks on CRT-RSA. In A. Tria and D. Choi, editors, *2014 Workshop on Fault Diagnosis and Tolerance in Cryptography (FDTC 2014)*, pp. 68–82. IEEE Computer Society, 2014.

[48] Pablo Rauzy, Martin Moreau, Sylvain Guilley, and Zakaria Najm. Using modular extension to provably protect ECC against fault attacks. Cryptology ePrint Archive, Report 2015/882, 2015. http://eprint.iacr.org/2015/882.

[49] Michael Scott. On the efficient implementation of pairing-based protocols. In L. Chen, editor, *Cryptography and Coding*, volume 7089 of *Lecture Notes in Computer Science*, pp. 296–308. Springer, Heidelberg, 2011.

[50] Michael Scott, Naomi Benger, Manuel Charlemagne, Luis J. Dominguez Perez, and Ezekiel J. Kachisa. On the final exponentiation for calculating pairings on ordinary elliptic curves. In H. Shacham and B. Waters, editors, *Pairing-Based Cryptography – Pairing 2009*, volume 5671 of *Lecture Notes in Computer Science*, pp. 78–88. Springer, Heidelberg, 2009.

[51] Adi Shamir. Identity-based cryptosystems and signature schemes. In G. R. Blakley and D. Chaum, editors, *Advances in Cryptology, Proceedings of CRYPTO '84*, volume 196 of *Lecture Notes in Computer Science*, pp. 47–53. Springer, Heidelberg, 1984.

[52] Adi Shamir. Method and apparatus for protecting public key schemes from timing and fault attacks. US Patent # 5,991,415, 1999. Presented at the rump session of EUROCRYPT '97.

[53] Masaaki Shirase, Tsuyoshi Takagi, and Eiji Okamoto. An efficient countermeasure against side-channel attacks for pairing computation. In L. Chen, Y. Mu, and W. Susilo, editors, *Information Security Practice and Experience (ISPEC 2008)*, volume 4991 of *Lecture Notes in Computer Science*, pp. 290–303. Springer, 2008.

[54] Joseph H. Silverman. *The Arithmetic of Elliptic Curves*, volume 106 of *Graduate Texts in Mathematics*. Springer-Verlag, 2nd edition, 2009.

[55] Elena Trichina and Roman Korkikyan. Multi fault laser attacks on protected CRT-RSA. In L. Breveglieri et al., editors, *2010 Workshop on Fault Diagnosis and Tolerance in Cryptography (FDTC 2010)*, pp. 75–86. IEEE Computer Society, 2010.

[56] Thomas Unterluggauer and Erich Wenger. Practical attack on bilinear pairings to disclose the secrets of embedded devices. In *9th International Conference on Availability, Reliability and Security (ARES 2014)*, pp. 69–77. IEEE Computer Society, 2014.

[57] Frederik Vercauteren. Optimal pairings. *IEEE Transactions on Information Theory*, 56(1):455–461, 2009.

[58] Jiang Weng, Yunqi Dou, and Chuangui Ma. Fault attacks against the Miller algorithm in Hessian coordinates. In C.-K. Wu, M. Yung, and D. Lin, editors, *Information Security and Cryptology (Inscrypt 2011)*, volume 7537 of *Lecture Notes in Computer Science*, pp. 102–112. Springer, 2012.

[59] Claire Whelan and Michael Scott. The importance of the final exponentiation in pairings when considering fault attacks. In T. Takagi et al., editors, *Pairing-Based Cryptography – Pairing 2007*, volume 4575 of *Lecture Notes in Computer Science*, pp. 225–246. Springer, Heidelberg, 2007.

[60] Claire Whelan and Mike Scott. Side-channel analysis of practical pairing implementa-
      tions: Which path is more secure? In P. Q. Nguyen, editor, *Progress in Cryptology –
      VIETCRYPT 2006*, volume 4341 of *Lecture Notes in Computer Science*, pp. 99–114.
      Springer, Heidelberg, 2006.

[61] Bo Yang, Kaijie Wu, and Ramesh Karri. Scan based side-channel attack on dedi-
      cated hardware implementations of data encryption standard. In *International Test
      Conference (ITC 2004)*, pp. 339–344. IEEE Press, 2004.

Bibliography

D.SPA.20. *ECRYPT2 Yearly Report on Algorithms and Keysizes (2011-2012)*. European Network of Excellence in Cryptology II, September 2012.

ETSI TS 102 225. *Smart Cards; Secured packet structure for UICC based applications (Release 11)*. European Telecommunications Standards Institute, March 2012.

FIPS PUB 186-3. *Digital Signature Standard (DSS)*. National Institute of Standards and Technology, U.S. Department of Commerce, June 2009.

FIPS PUB 197. *Advanced Encryption Standard (AES)*. National Institute of Standards and Technology, U.S. Department of Commerce, November 2001.

ISO/IEC FDIS 15946-1. *Information technology – Security techniques – Cryptographic techniques based on elliptic curves*. International Organization for Standardization/International Electrotechnical Commission, July 2015.

NSA Suite B. *Fact Sheet Suite B Cryptography*. National Security Agency, U.S.A., September 2014.

RGS-B1. *Mécanismes cryptographiques - Règles et recommandations concernant le choix et le dimensionnement des mécanismes cryptographiques*. Agence Nationale de la Sécurité des Systèmes d'Information, France, February 2014. version 2.03.

SP-800-57. *Recommendation for Key Management – Part 1: General*. National Institute of Standards and Technology, U.S. Department of Commerce, July 2012.

TEE White Paper. *The Trusted Execution Environment: Delivering Enhanced Security at a Lower Cost to the Mobile Market*. GlobalPlatform, June 2015. Revised from February 2011.

Tolga Acar, Kristin Lauter, Michael Naehrig, and Daniel Shumow. Affine pairings on ARM. In M. Abdalla and T. Lange, editors, *Pairing-Based Cryptography – Pairing 2012*, volume 7708 of *Lecture Notes in Computer Science*, pp. 203–209. Springer, Heidelberg, 2013.

Ben Adida. Helios: Web-based open-audit voting. In *17th USENIX Security Symposium*, pp. 335–348. USENIX Association, 2008.

Gora Adj, Alfred Menezes, Thomaz Oliveira, and Francisco Rodríguez-Henríquez. Computing discrete logarithms in $\mathbb{F}_{3^{6\cdot137}}$ and $\mathbb{F}_{3^{6\cdot163}}$ using Magma. In Ç. K. Koç, S. Mesnager, and E. Savas, editors, *Arithmetic of Finite Fields (WAIFI 2014)*, volume 9061 of *Lecture Notes in Computer Science*, pp. 3–22. Springer, 2014.

Gora Adj and Francisco Rodríguez-Henríquez. Square root computation over even extension fields. *IEEE Transactions on Computers*, 63(11):2829–2841, 2014.

Leonard Adleman. A subexponential algorithm for the discrete logarithm problem with applications to cryptography. In *20th Annual Symposium on Foundations of Computer Science*, pp. 55–60. IEEE Computer Society Press, 1979.

Leonard Adleman. The function field sieve. In L. M. Adleman and M.-D. Huang, editors, *Algorithmic Number Theory (ANTS-I)*, volume 877 of *Lecture Notes in Computer Science*, pp. 141–154. Springer, 1994.

Leonard M. Adleman and Ming-Deh A. Huang. Function field sieve method for discrete logarithms over finite fields. *Information and Computation*, 151(1/2):5–16, 1999.

David Adrian, Karthikeyan Bhargavan, Zakir Durumeric, Pierrick Gaudry, Matthew Green, J. Alex Halderman, Nadia Heninger, Drew Springall, Emmanuel Thomé, Luke Valenta, Benjamin VanderSloot, Eric Wustrow, Santiago Zanella-Béguelin, and Paul Zimmermann. Imperfect forward secrecy: How Diffie-Hellman fails in practice. In I. Ray, N. Li, and C. Kruegel, editors, *22nd ACM Conference on Computer and Communications Security*, pp. 5–17. ACM Press, 2015.

M. Albrecht, S. Bai, D. Cadé, X. Pujol, and D. Stehlé. fplll-4.0, a floating-point LLL implementation. Available at http://perso.ens-lyon.fr/damien.stehle.

Ross Anderson and Markus Kuhn. Tamper resistance—a cautionary note. In *Proceedings of the Second USENIX Workshop on Electronic Commerce*, pp. 1–11. USENIX Association, 1996.

Diego F. Aranha, Paulo S. L. M. Barreto, Patrick Longa, and Jefferson E. Ricardini. The realm of the pairings. In T. Lange, K. Lauter, and P. Lisonek, editors, *Selected Areas in Cryptography – SAC 2013*, volume 8282 of *Lecture Notes in Computer Science*, pp. 3–25. Springer, Heidelberg, 2014.

Diego F. Aranha, Jean-Luc Beuchat, Jérémie Detrey, and Nicolas Estibals. Optimal Eta pairing on supersingular genus-2 binary hyperelliptic curves. In O. Dunkelman, editor, *Topics in Cryptology – CT-RSA 2012*, volume 7178 of *Lecture Notes in Computer Science*, pp. 98–115. Springer, Heidelberg, 2012.

Diego F. Aranha, Laura Fuentes-Castañeda, Edward Knapp, Alfred Menezes, and Francisco Rodríguez-Henríquez. Implementing pairings at the 192-bit security level. In M. Abdalla and T. Lange, editors, *Pairing-Based Cryptography – Pairing 2012*, volume 7708 of *Lecture Notes in Computer Science*, pp. 177–195. Springer, Heidelberg, 2013.

Diego F. Aranha, Koray Karabina, Patrick Longa, Catherine H. Gebotys, and Julio López. Faster explicit formulas for computing pairings over ordinary curves. In K. G. Paterson, editor, *Advances in Cryptology – EUROCRYPT 2011*, volume 6632 of *Lecture Notes in Computer Science*, pp. 48–68. Springer, Heidelberg, 2011.

Christophe Arène, Tanja Lange, Michael Naehrig, and Christophe Ritzenthaler. Faster computation of the Tate pairing. *Journal of Number Theory*, 131(5):842–857, 2011.

Giuseppe Ateniese, Jan Camenisch, Marc Joye, and Gene Tsudik. A practical and provably secure coalition-resistant group signature scheme. In M. Bellare, editor, *Advances in Cryptology – CRYPTO 2000*, volume 1880 of *Lecture Notes in Computer Science*, pp. 255–270. Springer, Heidelberg, 2000.

A.O.L. Atkin. Probabilistic primality testing, summary by F. Morain. Research Report 1779, INRIA, 1992.

Arthur O. L. Atkin and François Morain. Elliptic curves and primality proving. *Mathematics of Computation*, 61(203):29–68, 1993.

László Babai. On Lovász' lattice reduction and the nearest lattice point problem. *Combinatorica*, 6(1):1–13, 1986.

Eric Bach and Klaus Huber. Note on taking square-roots modulo N. *IEEE Transactions on Information Theory*, 45(2):807–809, 1999.

Kiseok Bae, Sangjae Moon, and Jaecheol Ha. Instruction fault attack on the Miller algorithm in a pairing-based cryptosystem. In *7th International Conference on Innovative Mobile and Internet Services in Ubiquitous Computing (IMIS 2013)*, pp. 167–174. IEEE Press, 2013.

Joonsang Baek and Yuliang Zheng. Identity-based threshold decryption. In F. Bao, R. Deng, and J. Zhou, editors, *Public Key Cryptography – PKC 2004*, volume 2947 of *Lecture Notes in Computer Science*, pp. 262–276. Springer, Heidelberg, 2004.

Yoo-Jin Baek and Ihor Vasyltsov. How to prevent DPA and fault attack in a unified way for ECC scalar multiplication - Ring extension method. In E. Dawson and D. S. Wong, editors, *Information Security Practice and Experience (ISPEC 2007)*, volume 3439 of *Lecture Notes in Computer Science*, pp. 225–237. Springer, 2007.

Shi Bai. *Polynomial Selection for the Number Field Sieve*. PhD thesis, Australian National University, 2011. http://maths.anu.edu.au/~brent/pd/Bai-thesis.pdf.

Shi Bai, Richard Brent, and Emmanuel Thomé. Root optimization of polynomials in the number field sieve. *Mathematics of Computation*, 84(295):2447–2457, 2015.

Daniel V. Bailey and Christof Paar. Optimal extension fields for fast arithmetic in public-key algorithms. In H. Krawczyk, editor, *Advances in Cryptology – CRYPTO '98*, volume 1462 of *Lecture Notes in Computer Science*, pp. 472–485. Springer, Heidelberg, 1998.

Jean-Claude Bajard, Laurent-Stéphane Didier, and Peter Kornerup. An RNS montgomery modular multiplication algorithm. *IEEE Trans. Computers*, 47(7):766–776, 1998.

Jean-Claude Bajard, Laurent-Stéphane Didier, and Peter Kornerup. Modular multiplication and base extensions in residue number systems. In *IEEE Symposium on Computer Arithmetic*, pp. 59–65. IEEE Computer Society, 2001.

Jean-Claude Bajard, Sylvain Duquesne, and Milos D. Ercegovac. Combining leak-resistant arithmetic for elliptic curves defined over f_p and RNS representation. *IACR Cryptology ePrint Archive*, 2010:311, 2010.

Jean-Claude Bajard and Nadia El Mrabet. Pairing in cryptography: an arithmetic point of view. In F. T. Luk, editor, *Advanced Signal Processing Algorithms, Architectures, and Implementations XVII*, volume 6697 of *Proc. SPIE*. SPIE, 2007.

Jean-Claude Bajard and Laurent Imbert. A full RNS implementation of RSA. *IEEE Trans. Computers*, 53(6):769–774, 2004.

Jean-Claude Bajard, Laurent Imbert, Pierre-Yvan Liardet, and Yannick Teglia. Leak resistant arithmetic. In M. Joye and J.-J. Quisquater, editors, *Cryptographic Hardware and Embedded Systems – CHES 2004*, volume 3156 of *Lecture Notes in Computer Science*, pp. 62–75. Springer, Heidelberg, 2004.

Jean-Claude Bajard, Marcelo E. Kaihara, and Thomas Plantard. Selected RNS bases for modular multiplication. In *IEEE Symposium on Computer Arithmetic*, pp. 25–32. IEEE Computer Society, 2009.

Selçuk Baktir and Berk Sunar. Optimal tower fields. *IEEE Transactions on Computers*, 53(10):1231–1243, 2004.

R. Balasubramanian and Neal Koblitz. The improbability that an elliptic curve has subexponential discrete log problem under the Menezes - Okamoto - Vanstone algorithm. *Journal of Cryptology*, 11(2):141–145, 1998.

Razvan Barbulescu. *Algorithmes de logarithmes discrets dans les corps finis*. PhD thesis, Université de Lorraine, 2013. https://tel.archives-ouvertes.fr/tel-00925228.

Razvan Barbulescu, Cyril Bouvier, Jérémie Detrey, Pierrick Gaudry, Hamza Jeljeli, Emmanuel Thomé, Marion Videau, and Paul Zimmermann. Discrete logarithm in GF(2^{809}) with ffs, April 2013. Announcement available at the NMBRTHRY archives, item 004534.

Razvan Barbulescu, Cyril Bouvier, Jérémie Detrey, Pierrick Gaudry, Hamza Jeljeli, Emmanuel Thomé, Marion Videau, and Paul Zimmermann. Discrete logarithm in GF(2809) with FFS. In H. Krawczyk, editor, *PKC 2014: 17th International Conference on Theory and Practice of Public Key Cryptography*, volume 8383 of *Lecture Notes in Computer Science*, pp. 221–238. Springer, Heidelberg, 2014.

Razvan Barbulescu, Pierrick Gaudry, Aurore Guillevic, and François Morain. Improving NFS for the discrete logarithm problem in non-prime finite fields. In E. Oswald and M. Fischlin, editors, *Advances in Cryptology – EUROCRYPT 2015, Part I*, volume 9056 of *Lecture Notes in Computer Science*, pp. 129–155. Springer, Heidelberg, 2015.

Razvan Barbulescu, Pierrick Gaudry, Antoine Joux, and Emmanuel Thomé. A heuristic quasi-polynomial algorithm for discrete logarithm in finite fields of small characteristic.

In P. Q. Nguyen and E. Oswald, editors, *Advances in Cryptology – EUROCRYPT 2014*, volume 8441 of *Lecture Notes in Computer Science*, pp. 1–16. Springer, Heidelberg, 2014.

Razvan Barbulescu, Pierrick Gaudry, and Thorsten Kleinjung. The tower number field sieve. In T. Iwata and J. H. Cheon, editors, *Advances in Cryptology – ASIACRYPT 2015, Part II*, volume 9453 of *Lecture Notes in Computer Science*, pp. 31–55. Springer, Heidelberg, 2015.

Razvan Barbulescu and Cécile Pierrot. The Multiple Number Field Sieve for Medium and High Characteristic Finite Fields. *LMS Journal of Computation and Mathematics*, 17:230–246, 2014.

Paulo S. L. M. Barreto, Craig Costello, Rafael Misoczki, Michael Naehrig, Geovandro C. C. F. Pereira, and Gustavo Zanon. Subgroup security in pairing-based cryptography. In K. E. Lauter and F. Rodríguez-Henríquez, editors, *Progress in Cryptology – LATIN-CRYPT 2015*, volume 9230 of *Lecture Notes in Computer Science*, pp. 245–265. Springer, Heidelberg, 2015.

Paulo S. L. M. Barreto, Steven Galbraith, Colm Ó hÉigeartaigh, and Michael Scott. Efficient pairing computation on supersingular Abelian varieties. *Designs, Codes and Cryptography*, 42(3):239–271, 2007.

Paulo S. L. M. Barreto and Hae Yong Kim. Fast hashing onto elliptic curves over fields of characteristic 3. Cryptology ePrint Archive, Report 2001/098, 2001. http://eprint.iacr.org/2001/098.

Paulo S. L. M. Barreto, Hae Yong Kim, Ben Lynn, and Michael Scott. Efficient algorithms for pairing-based cryptosystems. In M. Yung, editor, *Advances in Cryptology – CRYPTO 2002*, volume 2442 of *Lecture Notes in Computer Science*, pp. 354–368. Springer, Heidelberg, 2002.

Paulo S. L. M. Barreto, Ben Lynn, and Michael Scott. Constructing elliptic curves with prescribed embedding degrees. In S. Cimato, C. Galdi, and G. Persiano, editors, *Security in Communication Networks (SCN 2002)*, volume 2576 of *Lecture Notes in Computer Science*, pp. 257–267. Springer, Heidelberg, 2003.

Paulo S. L. M. Barreto, Ben Lynn, and Michael Scott. On the selection of pairing-friendly groups. In M. Matsui and R. J. Zuccherato, editors, *Selected Areas in Cryptography (SAC 2003)*, volume 3006 of *Lecture Notes in Computer Science*, pp. 17–25. Springer, Heidelberg, 2004.

Paulo S. L. M. Barreto and Michael Naehrig. Pairing-friendly elliptic curves of prime order. In B. Preneel and S. Tavares, editors, *Selected Areas in Cryptography (SAC 2005)*, volume 3897 of *Lecture Notes in Computer Science*, pp. 319–331. Springer, Heidelberg, 2006.

Paul Barrett. Implementing the Rivest Shamir and Adleman public key encryption algorithm on a standard digital signal processor. In A. M. Odlyzko, editor, *Advances in Cryptology – CRYPTO '86*, volume 263 of *Lecture Notes in Computer Science*, pp. 311–323. Springer, Heidelberg, 1987.

Paul T. Bateman and Roger A. Horn. A heuristic asymptotic formula concerning the distribution of prime numbers. *Mathematics of Computation*, 16(79):363–367, 1962.

Aurélie Bauer, Eliane Jaulmes, Emmanuel Prouff, and Justine Wild. Horizontal and vertical side-channel attacks against secure RSA implementations. *CT-RSA*, pp. 1–17, 2013.

Mihir Bellare, Anand Desai, David Pointcheval, and Phillip Rogaway. Relations among notions of security for public-key encryption schemes. In H. Krawczyk, editor, *Advances in Cryptology – CRYPTO '98*, volume 1462 of *Lecture Notes in Computer Science*, pp. 26–45. Springer, Heidelberg, 1998.

Mihir Bellare and Phillip Rogaway. Random oracles are practical: A paradigm for designing efficient protocols. In V. Ashby, editor, *1st ACM Conference on Computer and Communications Security*, pp. 62–73. ACM Press, 1993.

Mihir Bellare and Phillip Rogaway. The exact security of digital signatures: How to sign with RSA and Rabin. In U. M. Maurer, editor, *Advances in Cryptology – EURO-CRYPT '96*, volume 1070 of *Lecture Notes in Computer Science*, pp. 399–416. Springer, Heidelberg, 1996.

Mihir Bellare, Haixia Shi, and Chong Zhang. Foundations of group signatures: The case of dynamic groups. In A. Menezes, editor, *Topics in Cryptology – CT-RSA 2005*, volume 3376 of *Lecture Notes in Computer Science*, pp. 136–153. Springer, Heidelberg, 2005.

Naomi Benger. *Cryptographic Pairings: Efficiency and DLP Security*. PhD thesis, Dublin City University, 2010.

Naomi Benger, Manuel Charlemagne, and David Mandell Freeman. On the security of pairing-friendly abelian varieties over non-prime fields. In H. Shacham and B. Waters, editors, *Pairing-Based Cryptography – Pairing 2009*, volume 5671 of *Lecture Notes in Computer Science*, pp. 52–65. Springer, Heidelberg, 2009.

Naomi Benger and Michael Scott. Constructing tower extensions of finite fields for implementation of pairing-based cryptography. In M. A. Hasan and T. Helleseth, editors, *Arithmetic of Finite Fields (WAIFI 2010)*, volume 6087 of *Lecture Notes in Computer Science*, pp. 180–195. Springer, 2010.

Daniel J. Bernstein. Pippenger's exponentiation algorithm. Unpublished manuscript, available at http://cr.yp.to/papers.html#pippenger, 2001.

Daniel J. Bernstein. Cache-timing attacks on AES. Unpublished manuscript, available at http://cr.yp.to/antiforgery/cachetiming-20050414.pdf, 2005.

Daniel J. Bernstein, Peter Birkner, Marc Joye, Tanja Lange, and Christiane Peters. Twisted Edwards curves. In S. Vaudenay, editor, *Progress in Cryptology – AFRICACRYPT 2008*, volume 5023 of *Lecture Notes in Computer Science*, pp. 389–405. Springer, Heidelberg, 2008.

Daniel J. Bernstein, Chitchanok Chuengsatiansup, David Kohel, and Tanja Lange. Twisted hessian curves. In K. E. Lauter and F. Rodríguez-Henríquez, editors, *Progress in Cryptology – LATINCRYPT 2015*, volume 9230 of *Lecture Notes in Computer Science*, pp. 269–294. Springer, Heidelberg, 2015.

Daniel J. Bernstein and Tanja Lange (editors). eBACS: ECRYPT benchmarking of cryptographic systems. http://bench.cr.yp.to (accessed 2014-05-17).

Daniel J. Bernstein and Tanja Lange. Explicit-formulas database. http://www.hyperelliptic.org/EFD.

Daniel J. Bernstein and Tanja Lange. Faster addition and doubling on elliptic curves. In K. Kurosawa, editor, *Advances in Cryptology – ASIACRYPT 2007*, volume 4833 of *Lecture Notes in Computer Science*, pp. 29–50. Springer, Heidelberg, 2007.

Daniel J. Bernstein and Tanja Lange. Computing small discrete logarithms faster. In S. D. Galbraith and M. Nandi, editors, *Progress in Cryptology – INDOCRYPT 2012*, volume 7668 of *Lecture Notes in Computer Science*, pp. 317–338. Springer, Heidelberg, 2012.

Daniel J. Bernstein and Tanja Lange. Non-uniform cracks in the concrete: The power of free precomputation. In K. Sako and P. Sarkar, editors, *Advances in Cryptology – ASIACRYPT 2013, Part II*, volume 8270 of *Lecture Notes in Computer Science*, pp. 321–340. Springer, Heidelberg, 2013.

Daniel J. Bernstein, Tanja Lange, and Peter Schwabe. On the correct use of the negation map in the Pollard rho method. In D. Catalano et al., editors, *Public Key Cryptography – PKC 2011*, volume 6571 of *Lecture Notes in Computer Science*, pp. 128–146. Springer, Heidelberg, 2011.

Jean-Luc Beuchat, Jorge E. González-Díaz, Shigeo Mitsunari, Eiji Okamoto, Francisco Rodríguez-Henríquez, and Tadanori Teruya. High-speed software implementation of the optimal Ate pairing over Barreto-Naehrig curves. In M. Joye, A. Miyaji, and A. Otsuka, editors, *Pairing-Based Cryptography – Pairing 2010*, volume 6487 of *Lecture Notes in Computer Science*, pp. 21–39. Springer, Heidelberg, 2010.

Jean-Luc Beuchat, Emmanuel López-Trejo, Luis Martínez-Ramos, Shigeo Mitsunari, and Francisco Rodríguez-Henríquez. Multi-core implementation of the Tate pairing over supersingular elliptic curves. In J. A. Garay, A. Miyaji, and A. Otsuka, editors, *Cryptology and Network Security (CANS 2009)*, volume 5888 of *Lecture Notes in Computer Science*, pp. 413–432. Springer, Heidelberg, 2009.

Olivier Billet and Marc Joye. The Jacobi model of an elliptic curve and side-channel analysis. In M. Fossorier, T. Høholdt, and A. Poli, editors, *Applied Algebra, Algebraic Algorithms and Error-Correcting Codes (AAECC 2003)*, volume 2643 of *Lecture Notes in Computer Science*, pp. 34–42. Springer, 2003.

Yuval Bistritz and Alexander Lifshitz. Bounds for resultants of univariate and bivariate polynomials. *Linear Algebra and its Applications*, 432(8):1995–2005, 2009.

Simon R. Blackburn and Edlyn Teske. Baby-step giant-step algorithms for non-uniform distributions. In W. Bosma, editor, *Algorithmic Number Theory (ANTS-IV)*, volume 1838 of *Lecture Notes in Computer Science*, pp. 153–168. Springer, 2000.

Ian F. Blake, Ryoh Fuji-Hara, Ronald C. Mullin, and Scott A. Vanstone. Computing logarithms in finite fields of characteristic two. *SIAM Journal on Algebraic Discrete Methods*, 5(2):276–285, 1984.

Ian F. Blake, Ronald C. Mullin, and Scott A. Vanstone. Computing logarithms in gf(2^n). In G. R. Blakley and D. Chaum, editors, *Advances in Cryptology, Proceedings of CRYPTO '84*, volume 196 of *Lecture Notes in Computer Science*, pp. 73–82. Springer, Heidelberg, 1984.

Ian F. Blake, Gadiel Seroussi, and Nigel P. Smart. *Elliptic Curves in Cryptography*, volume 265 of *London Mathematical Society Lecture Notes Series*. Cambridge University Press, 1999.

Ian F. Blake, Gadiel Seroussi, and Nigel P. Smart, editors. *Advances in Elliptic Curve Cryptography*, volume 317 of *London Mathematical Society Lecture Notes Series*. Cambridge University Press, 2004.

Johannes Blömer, Peter Günther, and Gennadij Liske. Improved side-channel attacks on pairing based cryptography. In E. Prouff, editor, *Constructive Side-Channel Analysis and Secure Design (COSADE 2013)*, volume 7864 of *Lecture Notes in Computer Science*, pp. 154–168. Springer, 2013.

Johannes Blömer, Martin Otto, and Jean-Pierre Seifert. Sign change fault attacks on elliptic curve cryptosystems. In L. Breveglieri et al., editors, *Fault Diagnosis and Tolerance in Cryptography (FDTC 2006)*, volume 4236 of *Lecture Notes in Computer Science*, pp. 154–168. Springer, 2006.

Alexandra Boldyreva. Threshold signatures, multisignatures and blind signatures based on the gap-Diffie-Hellman-group signature scheme. In Y. Desmedt, editor, *Public Key Cryptography – PKC 2003*, volume 2567 of *Lecture Notes in Computer Science*, pp. 31–46. Springer, Heidelberg, 2003.

Dan Boneh and Xavier Boyen. Short signatures without random oracles and the SDH assumption in bilinear groups. *Journal of Cryptology*, 21(2):149–177, 2008.

Dan Boneh, Xavier Boyen, and Eu-Jin Goh. Hierarchical identity based encryption with constant size ciphertext. In R. Cramer, editor, *Advances in Cryptology – EUROCRYPT 2005*, volume 3494 of *Lecture Notes in Computer Science*, pp. 440–456. Springer, Heidelberg, 2005.

Dan Boneh, Xavier Boyen, and Hovav Shacham. Short group signatures. In M. Franklin, editor, *Advances in Cryptology – CRYPTO 2004*, volume 3152 of *Lecture Notes in Computer Science*, pp. 41–55. Springer, Heidelberg, 2004.

Dan Boneh and Matthew K. Franklin. An efficient public key traitor tracing scheme. In M. J. Wiener, editor, *Advances in Cryptology – CRYPTO '99*, volume 1666 of *Lecture Notes in Computer Science*, pp. 338–353. Springer, Heidelberg, 1999.

Dan Boneh and Matthew K. Franklin. Identity-based encryption from the Weil pairing. In J. Kilian, editor, *Advances in Cryptology – CRYPTO 2001*, volume 2139 of *Lecture Notes in Computer Science*, pp. 213–229. Springer, Heidelberg, 2001.

Dan Boneh and Matthew K. Franklin. Identity based encryption from the Weil pairing. *SIAM Journal on Computing*, 32(3):586–615, 2003.

Dan Boneh, Craig Gentry, Ben Lynn, and Hovav Shacham. Aggregate and verifiably encrypted signatures from bilinear maps. In E. Biham, editor, *Advances in Cryptology – EUROCRYPT 2003*, volume 2656 of *Lecture Notes in Computer Science*, pp. 416–432. Springer, Heidelberg, 2003.

Dan Boneh, Ben Lynn, and Hovav Shacham. Short signatures from the Weil pairing. In C. Boyd, editor, *Advances in Cryptology – ASIACRYPT 2001*, volume 2248 of *Lecture Notes in Computer Science*, pp. 514–532. Springer, Heidelberg, 2001.

Dan Boneh, Ben Lynn, and Hovav Shacham. Short signatures from the Weil pairing. *Journal of Cryptology*, 17(4):297–319, 2004.

Joppe W. Bos, Craig Costello, Hüseyin Hisil, and Kristin Lauter. High-performance scalar multiplication using 8-dimensional GLV/GLS decomposition. In G. Bertoni and J.-S. Coron, editors, *Cryptographic Hardware and Embedded Systems – CHES 2013*, volume 8086 of *Lecture Notes in Computer Science*, pp. 331–348. Springer, Heidelberg, 2013.

Joppe W. Bos, Craig Costello, and Michael Naehrig. Exponentiating in pairing groups. In T. Lange, K. Lauter, and P. Lisonek, editors, *Selected Areas in Cryptography – SAC 2013*, volume 8282 of *Lecture Notes in Computer Science*, pp. 438–455. Springer, Heidelberg, 2014.

Viktor I. Bouniakowsky. Sur les diviseurs numériques invariables des fonctions rationnelles entières. *Mém. Acad. Sc. St-Petersbourg*, VI:305–329, 1857.

Cyril Bouvier. *Algorithmes pour la factorisation d'entiers et le calcul de logarithme discret*. PhD thesis, Université de Lorraine, 2015. https://tel.archives-ouvertes.fr/tel-01167281.

Colin Boyd, Paul Montague, and Khanh Quoc Nguyen. Elliptic curve based password authenticated key exchange protocols. In V. Varadharajan and Y. Mu, editors, *Information Security and Privacy (ACISP 2001)*, volume 2119 of *Lecture Notes in Computer Science*, pp. 487–501. Springer, Heidelberg, 2001.

Xavier Boyen. Multipurpose identity-based signcryption (a swiss army knife for identity-based cryptography). In D. Boneh, editor, *Advances in Cryptology – CRYPTO 2003*, volume 2729 of *Lecture Notes in Computer Science*, pp. 383–399. Springer, Heidelberg, 2003.

R. P. Brent and H. T. Kung. The area-time complexity of binary multiplication. *J. ACM*, 28(3):521–534, 1981.

Richard P. Brent. An improved Monte Carlo factorization algorithm. *BIT*, 20:176–184, 1980.

Friederike Brezing and Annegret Weng. Elliptic curves suitable for pairing based cryptography. *Designs, Codes and Cryptography*, 37(1):133–141, 2005.

Eric Brier, Jean-Sébastien Coron, Thomas Icart, David Madore, Hugues Randriam, and Mehdi Tibouchi. Efficient indifferentiable hashing into ordinary elliptic curves. In T. Rabin, editor, *Advances in Cryptology – CRYPTO 2010*, volume 6223 of *Lecture Notes in Computer Science*, pp. 237–254. Springer, Heidelberg, 2010.

Reinier M. Bröker. *Constructing elliptic curves of prescribed order*. PhD thesis, Leiden University, 2006.

Peter Bruin. The tate pairing for abelian varieties over finite fields. *J. de theorie des nombres de Bordeaux*, 23(2):323–328, 2011.

Joe P. Buhler, Hendrik W. Lenstra Jr., and Carl Pomerance. Factoring integers with the number field sieve. In A. K. Lenstra and H. W. Lenstra Jr., editors, *The Development of the Number Field Sieve*, volume 1554 of *Lecture Notes in Mathematics*, pp. 50–94. Springer, 1993.

The CADO-NFS Development Team. CADO-NFS, an implementation of the number field sieve algorithm, 2015. Release 2.2.0.

Jan Camenisch, Susan Hohenberger, and Anna Lysyanskaya. Compact e-cash. In R. Cramer, editor, *Advances in Cryptology – EUROCRYPT 2005*, volume 3494 of *Lecture Notes in Computer Science*, pp. 302–321. Springer, Heidelberg, 2005.

Sébastien Canard, Aline Gouget, and Jacques Traoré. Improvement of efficiency in (unconditional) anonymous transferable e-cash. In G. Tsudik, editor, *Financial Cryptography and Data Security (FC 2008)*, volume 5143 of *Lecture Notes in Computer Science*, pp. 202–214. Springer, Heidelberg, 2008.

Sébastien Canard, David Pointcheval, Olivier Sanders, and Jacques Traoré. Divisible E-cash made practical. In J. Katz, editor, *Public Key Cryptography – PKC 2015*, volume 9020 of *Lecture Notes in Computer Science*, pp. 77–100. Springer, Heidelberg, 2015.

Sébastien Canard, Berry Schoenmakers, Martijn Stam, and Jacques Traoré. List signature schemes. *Discrete Applied Mathematics*, 154(2):189–201, 2006.

Earl R. Canfield, Paul Erdős, and Carl Pomerance. On a problem of Oppenheim concerning "factorisatio numerorum". *Journal of Number Theory*, 17(1):1–28, 1983.

Jae Choon Cha and Jung Hee Cheon. An identity-based signature from gap Diffie-Hellman groups. In Y. Desmedt, editor, *Public Key Cryptography – PKC 2003*, volume 2567 of *Lecture Notes in Computer Science*, pp. 18–30. Springer, Heidelberg, 2003.

Sanjit Chatterjee, Darrel Hankerson, and Alfred Menezes. On the efficiency and security of pairing-based protocols in the Type 1 and Type 4 settings. In M. A. Hasan and T. Helleseth, editors, *Arithmetic of Finite Fields (WAIFI 2010)*, volume 6087 of *Lecture Notes in Computer Science*, pp. 114–134. Springer, 2010.

Sanjit Chatterjee and Alfred Menezes. On cryptographic protocols employing asymmetric pairings – The role of Ψ revisited. *Discrete Applied Mathematics*, 159(13):1311–1322, 2011.

Sanjit Chatterjee and Palash Sarkar. *Identity-Based Encryption*. Springer, 2011.

David Chaum. Showing credentials without identification: Signatures transferred between unconditionally unlinkable pseudonyms. In F. Pichler, editor, *Advances in Cryptology – EUROCRYPT '85*, volume 219 of *Lecture Notes in Computer Science*, pp. 241–244. Springer, Heidelberg, 1986.

Liqun Chen, Zhaohui Cheng, and Nigel P. Smart. Identity-based key agreement protocols from pairings. *International Journal of Information Security*, 6(4):213–241, 2007.

Yuanmi Chen. *Réduction de réseau et sécurité concrète du chiffrement complètement homomorphe*. PhD thesis, UniversitÃl Paris 7 Denis Diderot, 2013. http://www.di.ens.fr/~ychen/research/these.pdf.

Jung Hee Cheon, Jin Hong, and Minkyu Kim. Speeding up the pollard rho method on prime fields. In J. Pieprzyk, editor, *Advances in Cryptology – ASIACRYPT 2008*, volume 5350 of *Lecture Notes in Computer Science*, pp. 471–488. Springer, Heidelberg, 2008.

Ray C. C. Cheung, Sylvain Duquesne, Junfeng Fan, Nicolas Guillermin, Ingrid Verbauwhede, and Gavin Xiaoxu Yao. FPGA implementation of pairings using residue number system and lazy reduction. In *CHES*, volume 6917 of *Lecture Notes in Computer Science*, pp. 421–441. Springer, 2011.

Ray C. C. Cheung, Sylvain Duquesne, Junfeng Fan, Nicolas Guillermin, Ingrid Verbauwhede, and Gavin Xiaoxu Yao. FPGA implementation of pairings using residue number system and lazy reduction. In B. Preneel and T. Takagi, editors, *Cryptographic Hardware and Embedded Systems – CHES 2011*, volume 6917 of *Lecture Notes in Computer Science*, pp. 421–441. Springer, Heidelberg, 2011.

Benoît Chevallier-Mames, Mathieu Ciet, and Marc Joye. Low-cost solutions for preventing simple side-channel analysis: Side-channel atomicity. *IEEE Transactions on Computers*, 53(6):760–768, 2004.

Benoît Chevallier-Mames, Pascal Paillier, and David Pointcheval. Encoding-free ElGamal encryption without random oracles. In M. Yung, Y. Dodis, A. Kiayias, and T. Malkin, editors, *Public Key Cryptography – PKC 2006*, volume 3958 of *Lecture Notes in Computer Science*, pp. 91–104. Springer, Heidelberg, 2006.

Young Ju Choie, Eun Kyung Jeong, and Eun Jeong Lee. Supersingular hyperelliptic curves of genus 2 over finite fields. *Journal of Applied Mathematics and Computation*, 163(2):565–576, 2005.

Benny Chor, Amos Fiat, and Moni Naor. Tracing traitors. In Y. Desmedt, editor, *Advances in Cryptology – CRYPTO '94*, volume 839 of *Lecture Notes in Computer Science*, pp. 257–270. Springer, Heidelberg, 1994.

David V. Chudnovsky and Gregory V. Chudnovsky. Sequences of numbers generated by addition in formal groups and new primality and factorization tests. *Advances in Applied Mathematics*, 7(4):385–434, 1986.

Chitchanok Chuengsatiansup, Michael Naehrig, Pance Ribarski, and Peter Schwabe. PandA: Pairings and arithmetic. In Z. Cao and F. Zhang, editors, *Pairing-Based Cryptography – Pairing 2013*, volume 8365 of *Lecture Notes in Computer Science*, pp. 229–250. Springer, Heidelberg, 2014.

Jaewook Chung and M. Anwar Hasan. Asymmetric squaring formulæ. In *18th IEEE Symposium on Computer Arithmetic (ARITH 2007)*, pp. 113–122. IEEE Computer Society, 2007.

M. Cipolla. Un metodo per la risoluzione della congruenza di secondo grado. *Rend. Accad. Sci. Fis. Mat. Napoli*, vol. 9:154–163, 1903.

Clifford Cocks. An identity based encryption scheme based on quadratic residues. In B. Honary, editor, *Cryptography and Coding*, volume 2260 of *Lecture Notes in Computer Science*, pp. 360–363. Springer, Heidelberg, 2001.

Clifford Cocks and Richard G. E. Pinch. Identity-based cryptosystems based on the Weil pairing. Unpublished manuscript, 2001.

Henri Cohen. *A Course in Computational Algebraic Number Theory*, volume 138 of *Graduate Texts in Mathematics*. Springer-Verlag, 4th printing, 2000.

Henri Cohen and Gerhard Frey, editors. *Handbook of Elliptic and Hyperelliptic Curve Cryptography*, volume 34 of *Discrete Mathematics and Its Applications*. Chapman & Hall/CRC, 2006.

An Commeine and Igor Semaev. An algorithm to solve the discrete logarithm problem with the number field sieve. In M. Yung, Y. Dodis, A. Kiayias, and T. Malkin, editors, *Public Key Cryptography – PKC 2006*, volume 3958 of *Lecture Notes in Computer Science*, pp. 174–190. Springer, Heidelberg, 2006.

D. Coppersmith. Solving linear equations over GF(2) via block Wiedemann algorithm. *Mathematics of Computation*, 62(205):333–350, 1994.

Don Coppersmith. Fast evaluation of logarithms in fields of characteristic two. *IEEE Transactions on Information Theory*, 30(4):587–594, 1984.

Don Coppersmith. Modifications to the number field sieve. *Journal of Cryptology*, 6(3):169–180, 1993.

Don Coppersmith, Andrew M. Odlyzko, and Richard Schroeppel. Discrete logarithms in GF(p). *Algorithmica*, 1(1):1–15, 1986.

Jean-Sébastien Coron. Resistance against differential power analysis for elliptic curve cryptosystems. In Çetin Kaya. Koç and C. Paar, editors, *Cryptographic Hardware and Embedded Systems – CHES'99*, volume 1717 of *Lecture Notes in Computer Science*, pp. 292–302. Springer, Heidelberg, 1999.

Craig Costello, Tanja Lange, and Michael Naehrig. Faster pairing computations on curves with high-degree twists. In P. Q. Nguyen and D. Pointcheval, editors, *Public Key Cryptography – PKC 2010*, volume 6056 of *Lecture Notes in Computer Science*, pp. 224–242. Springer, Heidelberg, 2010.

Craig Costello, Kristin Lauter, and Michael Naehrig. Attractive subfamilies of BLS curves for implementing high-security pairings. In D. J. Bernstein and S. Chatterjee, editors, *Progress in Cryptology – INDOCRYPT 2011*, volume 7107 of *Lecture Notes in Computer Science*, pp. 320–342. Springer, Heidelberg, 2011.

Craig Costello and Patrick Longa. FourQ: Four-dimensional decompositions on a Q-curve over the Mersenne prime. In T. Iwata and J. H. Cheon, editors, *Advances in Cryptology – ASIACRYPT 2015, Part I*, volume 9452 of *Lecture Notes in Computer Science*, pp. 214–235. Springer, Heidelberg, 2015.

M. Prem Laxman Das and Palash Sarkar. Pairing computation on twisted Edwards form elliptic curves. In S. D. Galbraith and K. G. Paterson, editors, *Pairing-Based Cryptography – Pairing 2008*, volume 5209 of *Lecture Notes in Computer Science*, pp. 192–210. Springer, Heidelberg, 2008.

Elke De Mulder, Siddika B. Örs, Bart Preneel, and Ingrid Verbauwhede. Differential power and electromagnetic attacks on a FPGA implementation of elliptic curve cryptosystems. *Computers & Electrical Engineering*, 33(5/6):367–382, 2007.

Amine Dehbaoui, Jean-Max Dutertre, Bruno Robisson, and Assia Tria. Electromagnetic transient faults injection on a hardware and a software implementation of AES. In G. Bertoni and B. Gierlichs, editors, *2012 Workshop on Fault Diagnosis and Tolerance in Cryptography (FDTC 2012)*, pp. 7–15. IEEE Computer Society, 2012.

Max Deuring. Die Typen der Multiplikatorenringe elliptischer Funktionenkörper. *Abh. Math. Sem. Hansischen Univ.*, 14:197–272, 1941.

Augusto Jun Devegili, Colm Ó hÉigeartaigh, Michael Scott, and Ricardo Dahab. Multiplication and squaring on pairing-friendly fields. Cryptology ePrint Archive, Report 2006/471, 2006. http://eprint.iacr.org/2006/471.

Augusto Jun Devegili, Michael Scott, and Ricardo Dahab. Implementing cryptographic pairings over Barreto-Naehrig curves (invited talk). In T. Takagi et al., editors, *Pairing-Based Cryptography – Pairing 2007*, volume 4575 of *Lecture Notes in Computer Science*, pp. 197–207. Springer, Heidelberg, 2007.

Whitfield Diffie and Martin E. Hellman. New directions in cryptography. *IEEE Transactions on Information Theory*, 22(6):644–654, 1976.

Luis J. Dominguez Perez, Ezekiel J. Kachisa, and Michael Scott. Implementing cryptographic pairings: A Magma tutorial. Cryptology ePrint Archive, Report 2009/072, 2009. http://eprint.iacr.org/2009/072.

Régis Dupont, Andreas Enge, and François Morain. Building curves with arbitrary small MOV degree over finite prime fields. *Journal of Cryptology*, 18(2):79–89, 2005.

Sylvain Duquesne. RNS arithmetic in $\mathbb{F}_{p^k}$ and application to fast pairing computation. *Journal of Mathematical Cryptology*, 5(1):51–88, 2011.

Sylvain Duquesne and Emmanuel Fouotsa. Tate pairing computation on Jacobi's elliptic curves. In M. Abdalla and T. Lange, editors, *Pairing-Based Cryptography – Pairing 2012*, volume 7708 of *Lecture Notes in Computer Science*, pp. 254–269. Springer, Heidelberg, 2013.

Sylvain Duquesne and Loubna Ghammam. Memory-saving computation of the pairing final exponentiation on BN curves. Cryptology ePrint Archive, Report 2015/192, 2015. http://eprint.iacr.org/2015/192.

Sylvain Duquesne, Nadia El Mrabet, and Emmanuel Fouotsa. Efficient computation of pairings on jacobi quartic elliptic curves. *J. Mathematical Cryptology*, 8(4):331–362, 2014.

Sylvain Duquesne, Nadia El Mrabet, Safia Haloui, and Franck Rondepierre. Choosing and generating parameters for low level pairing implementation on BN curves. *IACR Cryptology ePrint Archive*, 2015:1212, 2015.

Stephen R. Dussé and Burton S. Kaliski Jr. A cryptographic library for the Motorola DSP56000. In I. Damgard, editor, *Advances in Cryptology – EUROCRYPT'90*, volume 473 of *Lecture Notes in Computer Science*, pp. 230–244. Springer, Heidelberg, 1991.

Iwan Duursma and Hyang-Sook Lee. Tate-pairing implementations for tripartite key agreement. Cryptology ePrint Archive, Report 2003/053, 2003. http://eprint.iacr.org/2003/053.

Iwan M. Duursma, Pierrick Gaudry, and François Morain. Speeding up the discrete log computation on curves with automorphisms. In K.-Y. Lam, E. Okamoto, and C. Xing, editors, *Advances in Cryptology – ASIACRYPT'99*, volume 1716 of *Lecture Notes in Computer Science*, pp. 103–121. Springer, Heidelberg, 1999.

Harold. M. Edwards. A normal form for elliptic curves. *Bulletin of the American Mathematical Society*, 44(3):393–422, 2007.

Nadia El Mrabet, Nicolas Guillermin, and Sorina Ionica. A study of pairing computation for elliptic curves with embedding degree 15. Cryptology ePrint Archive, Report 2009/370, 2009. http://eprint.iacr.org/2009/370.

Nadia El Mrabet. Fault attack against Miller's algorithm. Cryptology ePrint Archive, Report 2011/709, 2011. http://eprint.iacr.org/2011/709.

Nadia El Mrabet. Side-channel attacks against pairing over Theta functions. In T. Muntean, D. Poulakis, and R. Rolland, editors, *Algebraic Informatics (CAI 2013)*, volume 8080 of *Lecture Notes in Computer Science*, pp. 132–146. Springer, 2013.

Nadia El Mrabet, Giorgio Di Natale, and Marie Lise Flottes. A practical differential power analysis attack against the Miller algorithm. In *2009 Conference on Ph.D. Research in Microelectronics and Electronics (PRIME 2009)*, pp. 308–311. IEEE Press, 2009.

Nadia El Mrabet, Dan Page, and Frederik Vercauteren. Fault attacks on pairing-based cryptography. In M. Joye and M. Tunstall, editors, *Fault Analysis in Cryptography*, Information Security and Cryptography, pp. 221–236. Springer, 2012.

Taher ElGamal. A public key cryptosystem and a signature scheme based on discrete logarithms. In G. R. Blakley and D. Chaum, editors, *Advances in Cryptology, Proceedings of CRYPTO '84*, volume 196 of *Lecture Notes in Computer Science*, pp. 10–18. Springer, Heidelberg, 1984.

Andreas Enge and Jérôme Milan. Implementing cryptographic pairings at standard security levels. In R. S. Chakraborty, V. Matyas, and P. Schaumont, editors, *Security, Privacy, and Applied Cryptography Engineering (SPACE 2014)*, volume 8804 of *Lecture Notes in Computer Science*, pp. 28–46. Springer, 2014.

A. E. Escott, J. C. Sager, A. P. L. Selkirk, and D. Tsapakidis. Attacking elliptic curve cryptosystems using the parallel Pollard rho method. *CryptoBytes*, 4, 1999.

Junfeng Fan, Frederik Vercauteren, and Ingrid Verbauwhede. Faster arithmetic for cryptographic pairings on Barreto-Naehrig curves. In C. Clavier and K. Gaj, editors, *Cryptographic Hardware and Embedded Systems – CHES 2009*, volume 5747 of *Lecture Notes in Computer Science*, pp. 240–253. Springer, Heidelberg, 2009.

Junfeng Fan, Frederik Vercauteren, and Ingrid Verbauwhede. Efficient hardware implementation of $\mathbb{F}_p$-arithmetic for pairing-friendly curves. *IEEE Transactions on Computers*, 61(5):676–685, 2012.

Reza Rezaeian Farashahi, Pierre-Alain Fouque, Igor Shparlinski, Mehdi Tibouchi, and José Felipe Voloch. Indifferentiable deterministic hashing to elliptic and hyperelliptic curves. *Mathematics of Computation*, 82(281):491–512, 2013.

Reza Rezaeian Farashahi and Marc Joye. Efficient arithmetic on hessian curves. In P. Q. Nguyen and D. Pointcheval, editors, *Public Key Cryptography – PKC 2010*, volume 6056 of *Lecture Notes in Computer Science*, pp. 243–260. Springer, Heidelberg, 2010.

Reza Rezaeian Farashahi, Igor E. Shparlinski, and José Felipe Voloch. On hashing into elliptic curves. *Journal of Mathematical Cryptology*, 3(4):353–360, 2009.

Armando Faz-Hernández, Patrick Longa, and Ana H. Sánchez. Efficient and secure algorithms for GLV-based scalar multiplication and their implementation on GLV-GLS curves. *Journal of Cryptographic Engineering*, 5(1):31–52, 2015.

Min Feng, Bin B. Zhu, Maozhi Xu, and Shipeng Li. Efficient comb elliptic curve multiplication methods resistant to power analysis. Cryptology ePrint Archive, Report 2005/222, 2005. http://eprint.iacr.org/2005/222.

Amos Fiat and Moni Naor. Broadcast encryption. In D. R. Stinson, editor, *Advances in Cryptology – CRYPTO '93*, volume 773 of *Lecture Notes in Computer Science*, pp. 480–491. Springer, Heidelberg, 1994.

Amos Fiat and Adi Shamir. How to prove yourself: Practical solutions to identification and signature problems. In A. M. Odlyzko, editor, *Advances in Cryptology – CRYPTO '86*, volume 263 of *Lecture Notes in Computer Science*, pp. 186–194. Springer, Heidelberg, 1987.

Philippe Flajolet and Andrew M. Odlyzko. Random mapping statistics. In J.-J. Quisquater and J. Vandewalle, editors, *Advances in Cryptology – EUROCRYPT '89*, volume 434 of *Lecture Notes in Computer Science*, pp. 329–354. Springer, Heidelberg, 1990.

Pierre-Alain Fouque and Mehdi Tibouchi. Estimating the size of the image of deterministic hash functions to elliptic curves. In M. Abdalla and P. S. L. M. Barreto, editors, *Progress in Cryptology – LATINCRYPT 2010*, volume 6212 of *Lecture Notes in Computer Science*, pp. 81–91. Springer, Heidelberg, 2010.

Pierre-Alain Fouque and Mehdi Tibouchi. Indifferentiable hashing to Barreto-Naehrig curves. In A. Hevia and G. Neven, editors, *Progress in Cryptology – LATINCRYPT 2012*, volume 7533 of *Lecture Notes in Computer Science*, pp. 1–17. Springer, Heidelberg, 2012.

David Freeman. Constructing pairing-friendly elliptic curves with embedding degree 10. In F. Hess, S. Pauli, and M. E. Pohst, editors, *Algorithmic Number Theory (ANTS-VII)*, volume 4076 of *Lecture Notes in Computer Science*, pp. 452–465. Springer, 2006.

David Freeman, Michael Scott, and Edlyn Teske. A taxonomy of pairing-friendly elliptic curves. *Journal of Cryptology*, 23(2):224–280, 2010.

Gerhard Frey and Hans-Georg Rück. A remark concerning m-divisibility and the discrete logarithm in the divisor class group of curves. *Mathematics of Computation*, 62(206):865–874, 1994.

Laura Fuentes-Castañeda, Edward Knapp, and Francisco Rodríguez-Henríquez. Faster hashing to $\mathbb{G}_2$. In A. Miri and S. Vaudenay, editors, *Selected Areas in Cryptography*

– *SAC 2011*, volume 7118 of *Lecture Notes in Computer Science*, pp. 412–430. Springer, Heidelberg, 2012.

Atsushi Fujioka, Tatsuaki Okamoto, and Kazuo Ohta. A practical secret voting scheme for large scale elections. In J. Seberry and Y. Zheng, editors, *Advances in Cryptology – AUSCRYPT '92*, volume 718 of *Lecture Notes in Computer Science*, pp. 244–251. Springer, Heidelberg, 1993.

Fujitsu Laboratories, NICT, and Kyushu University. DL record in $\mathbb{F}_{3^{6 \cdot 97}}$ of 923 bits (278 dd). NICT press release, June 18, 2012. http://www.nict.go.jp/en/press/2012/06/18en-1.html.

Jun Furukawa, Kengo Mori, and Kazue Sako. An implementation of a mix-net based network voting scheme and its use in a private organization. In D. Chaum et al., editors, *Towards Trustworthy Elections, New Directions in Electronic Voting*, volume 6000 of *Lecture Notes in Computer Science*, pp. 141–154. Springer, 2010.

Steven Galbraith. Quasi-polynomial-time algorithm for discrete logarithm in finite fields of small/medium characteristic. The Elliptic Curve Cryptography blog, June 2013. https://ellipticnews.wordpress.com/2013/06/21.

Steven D. Galbraith. Supersingular curves in cryptography. In C. Boyd, editor, *Advances in Cryptology – ASIACRYPT 2001*, volume 2248 of *Lecture Notes in Computer Science*, pp. 495–513. Springer, Heidelberg, 2001.

Steven D. Galbraith. *Mathematics of Public Key Cryptography*. Cambridge University Press, 2012.

Steven D. Galbraith and Pierrick Gaudry. Recent progress on the elliptic curve discrete logarithm problem. Cryptology ePrint Archive, Report 2015/1022, 2015. http://eprint.iacr.org/.

Steven D. Galbraith, Xibin Lin, and Michael Scott. Endomorphisms for faster elliptic curve cryptography on a large class of curves. *Journal of Cryptology*, 24(3):446–469, 2011.

Steven D. Galbraith, James F. McKee, and P. C. Valença. Ordinary abelian varieties having small embedding degree. *Finite Fields and Their Applications*, 13(4):800–814, 2007.

Steven D. Galbraith, Kenneth G. Paterson, and Nigel P. Smart. Pairings for cryptographers. *Discrete Applied Mathematics*, 156(16):3113–3121, 2008.

Steven D. Galbraith and Victor Rotger. Easy decision Diffie-Hellman groups. *LMS Journal of Computation and Mathematics*, 7:201–218, 2004.

Steven D. Galbraith and Raminder S. Ruprai. An improvement to the Gaudry-Schost algorithm for multidimensional discrete logarithm problems. In M. G. Parker, editor, *Cryptography and Coding*, volume 5921 of *Lecture Notes in Computer Science*, pp. 368–382. Springer, Heidelberg, 2009.

Steven D. Galbraith and Raminder S. Ruprai. Using equivalence classes to accelerate solving the discrete logarithm problem in a short interval. In P. Q. Nguyen and D. Pointcheval, editors, *Public Key Cryptography – PKC 2010*, volume 6056 of *Lecture Notes in Computer Science*, pp. 368–383. Springer, Heidelberg, 2010.

Steven D. Galbraith and Michael Scott. Exponentiation in pairing-friendly groups using homomorphisms. In S. D. Galbraith and K. G. Paterson, editors, *Pairing-Based Cryptography – Pairing 2008*, volume 5209 of *Lecture Notes in Computer Science*, pp. 211–224. Springer, Heidelberg, 2008.

Steven D. Galbraith, Ping Wang, and Fangguo Zhang. Computing elliptic curve discrete logarithms with improved baby-step giant-step algorithm. Cryptology ePrint Archive, Report 2015/605, 2015. http://eprint.iacr.org/2015/605.

Robert P. Gallant, Robert J. Lambert, and Scott A. Vanstone. Faster point multiplication on elliptic curves with efficient endomorphisms. In J. Kilian, editor, *Advances in*

Cryptology – CRYPTO 2001, volume 2139 of *Lecture Notes in Computer Science*, pp. 190–200. Springer, Heidelberg, 2001.

Pierrick Gaudry and Éric Schost. A low-memory parallel version of Matsuo, Chao, and Tsujii's algorithm. In D. A. Buell, editor, *Algorithmic Number Theory (ANTS-VI)*, volume 3076 of *Lecture Notes in Computer Science*, pp. 208–222. Springer, 2004.

Craig Gentry and Alice Silverberg. Hierarchical ID-based cryptography. In Y. Zheng, editor, *Advances in Cryptology – ASIACRYPT 2002*, volume 2501 of *Lecture Notes in Computer Science*, pp. 548–566. Springer, Heidelberg, 2002.

Santosh Ghosh and Dipanwita Roy Chowdhury. Security of prime field pairing cryptoprocessor against differential power attack. In M. Joye, D. Mukhopadhyay, and M. Tunstall, editors, *Security Aspects in Information Technology (InfoSecHiComNet 2011)*, volume 7011 of *Lecture Notes in Computer Science*, pp. 16–29. Springer, 2011.

Santosh Ghosh, Debdeep Mukhopadhyay, and Dipanwita Roy Chowdhury. Fault attack, countermeasures on pairing based cryptography. *International Journal of Network Security*, 12(1):21–28, 2011.

Shafi Goldwasser, Silvio Micali, and Ronald L. Rivest. A digital signature scheme secure against adaptive chosen-message attacks. *SIAM Journal on Computing*, 17(2):281–308, 1988.

Jovan Dj. Golic and Christophe Tymen. Multiplicative masking and power analysis of AES. In B. S. Kaliski Jr., Ç. K. Koç, and C. Paar, editors, *Cryptographic Hardware and Embedded Systems – CHES 2002*, volume 2523 of *Lecture Notes in Computer Science*, pp. 198–212. Springer, Heidelberg, 2003.

Daniel M. Gordon. Discrete logarithms in GF(p) using the number field sieve. *SIAM Journal on Discrete Mathematics*, 6(1):124–138, 1993.

Robert Granger, Thorsten Kleinjung, and Jens Zumbrägel. Breaking '128-bit secure' supersingular binary curves - (or how to solve discrete logarithms in $F_{2^{4 \cdot 1223}}$ and $F_{2^{12 \cdot 367}}$). In J. A. Garay and R. Gennaro, editors, *Advances in Cryptology – CRYPTO 2014, Part II*, volume 8617 of *Lecture Notes in Computer Science*, pp. 126–145. Springer, Heidelberg, 2014.

Robert Granger, Thorsten Kleinjung, and Jens Zumbragel. Discrete logarithms in GF(2^{9234}), 2014. Announcement available at the NMBRTHRY archives, item 004666.

Robert Granger, Dan Page, and Nigel P. Smart. High security pairing-based cryptography revisited. In F. Hess, S. Pauli, and M. E. Pohst, editors, *Algorithmic Number Theory (ANTS-VII)*, volume 4076 of *Lecture Notes in Computer Science*, pp. 480–494. Springer, 2006.

Robert Granger, Dan Page, and Martijn Stam. On small characteristic algebraic tori in pairing-based cryptography. *LMS Journal of Computation and Mathematics*, 9:64–85, 2006.

Robert Granger and Michael Scott. Faster squaring in the cyclotomic subgroup of sixth degree extensions. In P. Q. Nguyen and D. Pointcheval, editors, *Public Key Cryptography – PKC 2010*, volume 6056 of *Lecture Notes in Computer Science*, pp. 209–223. Springer, Heidelberg, 2010.

Robert Granger and Nigel P. Smart. On computing products of pairings. Cryptology ePrint Archive, Report 2006/172, 2006. http://eprint.iacr.org/2006/172.

Gurleen Grewal, Reza Azarderakhsh, Patrick Longa, Shi Hu, and David Jao. Efficient implementation of bilinear pairings on ARM processors. In L. R. Knudsen and H. Wu, editors, *Selected Areas in Cryptography – SAC 2012*, volume 7707 of *Lecture Notes in Computer Science*, pp. 149–165. Springer, Heidelberg, 2013.

Jens Groth. Fully anonymous group signatures without random oracles. In K. Kurosawa, editor, *Advances in Cryptology – ASIACRYPT 2007*, volume 4833 of *Lecture Notes in Computer Science*, pp. 164–180. Springer, Heidelberg, 2007.

Jens Groth. Short pairing-based non-interactive zero-knowledge arguments. In M. Abe, editor, *Advances in Cryptology – ASIACRYPT 2010*, volume 6477 of *Lecture Notes in Computer Science*, pp. 321–340. Springer, Heidelberg, 2010.

Jens Groth and Amit Sahai. Efficient non-interactive proof systems for bilinear groups. In N. P. Smart, editor, *Advances in Cryptology – EUROCRYPT 2008*, volume 4965 of *Lecture Notes in Computer Science*, pp. 415–432. Springer, Heidelberg, 2008.

Nicolas Guillermin. A high speed coprocessor for elliptic curve scalar multiplications over \mathbb$F$$p$\mathbb{F}_p. In *CHES*, volume 6225 of *Lecture Notes in Computer Science*, pp. 48–64. Springer, 2010.

Aurore Guillevic. Kim–Barbulescu variant of the number field sieve to compute discrete logarithms in finite fields. Research Report Elliptic News 2016/05/02, 2016.

Philippe Guillot, Abdelkrim Nimour, Duong Hieu Phan, and Viet Cuong Trinh. Optimal public key traitor tracing scheme in non-black box model. In A. Youssef, A. Nitaj, and A. E. Hassanien, editors, *Progress in Cryptology – AFRICACRYPT 2013*, volume 7918 of *Lecture Notes in Computer Science*, pp. 140–155. Springer, Heidelberg, 2013.

Nils Gura, Arun Patel, Arvinderpal Wander, Hans Eberle, and Sheueling Chang Shantz. Comparing elliptic curve cryptography and rsa on 8-bit cpus. In *Cryptographic Hardware and Embedded Systems - CHES 2004*, pp. 119–132. Springer Berlin Heidelberg, 2004.

Juan E. Guzmán-Trampe, Nareli Cruz Cortés, Luis J. Dominguez Perez, Daniel Ortiz Arroyo, and Francisco Rodríguez-Henríquez. Low-cost addition-subtraction sequences for the final exponentiation in pairings. *Finite Fields and Their Applications*, 29:1–17, 2014.

D. H. Habing. The use of lasers to simulate radiation-induced transients in semiconductor devices and circuits. *IEEE Transactions on Nuclear Science*, 12(5):91–100, 1965.

Mike Hamburg. Fast and compact elliptic-curve cryptography. Cryptology ePrint Archive, Report 2012/309, 2012. http://eprint.iacr.org/2012/309.

D. Hankerson, A. Menezes, and M. Scott. Software implementation of pairings. In M. Joye and G. Neven, editors, *Identity-based Cryptography*, Cryptology and Information Security Series, chapter 12, pp. 188–206. IOS Press, 2009.

D. Hankerson, A. Menezes, and S. Vanstone. *Guide to Elliptic Curve Cryptography*. Springer-Verlag New York, Inc., Secaucus, NJ, 2003.

Helmut Hasse. Zur Theorie der abstrakten elliptischen Funktionenkörper III. Die Struktur des Meromorphismenrings; die Riemannsche Vermutung. *Journal für die reine und angewandte Mathematik*, 175:193–208, 1936.

Takuya Hayashi, Takeshi Shimoyama, Naoyuki Shinohara, and Tsuyoshi Takagi. Breaking pairing-based cryptosystems using η_T pairing over GF(3^{97}). In X. Wang and K. Sako, editors, *Advances in Cryptology – ASIACRYPT 2012*, volume 7658 of *Lecture Notes in Computer Science*, pp. 43–60. Springer, Heidelberg, 2012.

Takuya Hayashi, Naoyuki Shinohara, Lihua Wang, Shin'ichiro Matsuo, Masaaki Shirase, and Tsuyoshi Takagi. Solving a 676-bit discrete logarithm problem in GF(3^{6n}). In P. Q. Nguyen and D. Pointcheval, editors, *Public Key Cryptography – PKC 2010*, volume 6056 of *Lecture Notes in Computer Science*, pp. 351–367. Springer, Heidelberg, 2010.

Mustapha Hedabou, Pierre Pinel, and Lucien Bénéteau. Countermeasures for preventing comb method against SCA attacks. In R. H. Deng et al., editors, *Information Security Practice and Experience (ISPEC 2005)*, volume 3439 of *Lecture Notes in Computer Science*, pp. 85–96. Springer, 2005.

F. Hess, N.P. Smart, and F. Vercauteren. The eta pairing revisited. Cryptology ePrint Archive, Report 2006/110, 2006. http://eprint.iacr.org/2006/110.

Florian Hess. A note on the tate pairing of curves over finite fields. *Archiv der Mathematik*, 82(1):28–32, 2004.

Florian Hess. Pairing lattices (invited talk). In S. D. Galbraith and K. G. Paterson, editors, *Pairing-Based Cryptography – Pairing 2008*, volume 5209 of *Lecture Notes in Computer Science*, pp. 18–38. Springer, Heidelberg, 2008.

Florian Hess, Nigel P. Smart, and Frederik Vercauteren. The Eta pairing revisited. *IEEE Transactions on Information Theory*, 52(10):4595–4602, 2006.

Otto Hesse. Über die Elimination der Variabeln aus drei algebraischen Gleichungen vom zweiten Grade mit zwei Variabeln. *Journal für die reine und angewandte Mathematik*, 10:68–96, 1844.

Hüseyin Hişil. *Elliptic curves, group law, and efficient computation*. PhD thesis, Queensland University of Technology, 2010.

Hüseyin Hisil, Gary Carter, and Ed Dawson. New formulae for efficient elliptic curve arithmetic. In K. Srinathan, C. P. Rangan, and M. Yung, editors, *Progress in Cryptology – INDOCRYPT 2007*, volume 4859 of *Lecture Notes in Computer Science*, pp. 138–151. Springer, Heidelberg, 2007.

Hüseyin Hisil, Kenneth Koon-Ho Wong, Gary Carter, and Ed Dawson. Twisted Edwards curves revisited. In J. Pieprzyk, editor, *Advances in Cryptology – ASIACRYPT 2008*, volume 5350 of *Lecture Notes in Computer Science*, pp. 326–343. Springer, Heidelberg, 2008.

Hüseyin Hisil, Kenneth Koon-Ho Wong, Gary Carter, and Ed Dawson. Jacobi quartic curves revisited. In C. Boyd and J. M. G. Nieto, editors, *Information Security and Privacy (ACISP 2009)*, volume 5594 of *Lecture Notes in Computer Science*, pp. 452–468. Springer, Heidelberg, 2009.

Yvonne Hitchcock, Paul Montague, Gary Carter, and Ed Dawson. The efficiency of solving multiple discrete logarithm problems and the implications for the security of fixed elliptic curves. *International Journal of Information Security*, 3(2):86–98, 2004.

Laura Hitt. On the minimal embedding field. In T. Takagi et al., editors, *Pairing-Based Cryptography – Pairing 2007*, volume 4575 of *Lecture Notes in Computer Science*, pp. 294–301. Springer, Heidelberg, 2007.

Jeremy Horwitz and Ben Lynn. Toward hierarchical identity-based encryption. In L. R. Knudsen, editor, *Advances in Cryptology – EUROCRYPT 2002*, volume 2332 of *Lecture Notes in Computer Science*, pp. 466–481. Springer, Heidelberg, 2002.

Michael Hutter, Marcel Medwed, Daniel Hein, and Johannes Wolkerstorfer. Attacking ECDSA-Enabled RFID devices. *Applied Cryptography and Network Security*, pp. 519–534, 2009.

Michael Hutter and Erich Wenger. Fast multi-precision multiplication for public-key cryptography on embedded microprocessors. In B. Preneel and T. Takagi, editors, *Cryptographic Hardware and Embedded Systems – CHES 2011*, volume 6917 of *Lecture Notes in Computer Science*, pp. 459–474. Springer, Heidelberg, 2011.

Thomas Icart. How to hash into elliptic curves. In S. Halevi, editor, *Advances in Cryptology – CRYPTO 2009*, volume 5677 of *Lecture Notes in Computer Science*, pp. 303–316. Springer, Heidelberg, 2009.

Sorina Ionica and Antoine Joux. Another approach to pairing computation in Edwards coordinates. In D. R. Chowdhury, V. Rijmen, and A. Das, editors, *Progress in Cryptology – INDOCRYPT 2008*, volume 5365 of *Lecture Notes in Computer Science*, pp. 400–413. Springer, Heidelberg, 2008.

Sorina Ionica and Antoine Joux. Pairing computation on elliptic curves with efficiently computable endomorphism and small embedding degree. In M. Joye, A. Miyaji, and

A. Otsuka, editors, *Pairing-Based Cryptography – Pairing 2010*, volume 6487 of *Lecture Notes in Computer Science*, pp. 435–449. Springer, Heidelberg, 2010.

Sorina Ionica and Antoine Joux. Pairing the volcano. *Mathematics of Computation*, 82(281):581–603, 2013.

David P. Jablon. Strong password-only authenticated key exchange. *SIGCOMM Computer Communication Review*, 26(5):5–26, 1996.

David P. Jablon. Extended password key exchange protocols immune to dictionary attacks. In *6th IEEE International Workshops on Enabling Technologies: Infrastructure for Collaborative Enterprises (WETICE 1997)*, pp. 248–255. IEEE Computer Society, 1997.

N. Jacobson. *Basic Algebra I: Second Edition*. Dover Books on Mathematics. Dover Publications, 2012.

Jinhyuck Jeong and Taechan Kim. Extended tower number field sieve with application to finite fields of arbitrary composite extension degree. Cryptology ePrint Archive, Report 2016/526, 2016. http://eprint.iacr.org/.

Antoine Joux. A one round protocol for tripartite Diffie-Hellman. In W. Bosma, editor, *Algorithmic Number Theory (ANTS-IV)*, volume 1838 of *Lecture Notes in Computer Science*, pp. 385–393. Springer, 2000.

Antoine Joux. A one round protocol for tripartite Diffie-Hellman. *Journal of Cryptology*, 17(4):263–276, 2004.

Antoine Joux. Faster index calculus for the medium prime case application to 1175-bit and 1425-bit finite fields. In T. Johansson and P. Q. Nguyen, editors, *Advances in Cryptology – EUROCRYPT 2013*, volume 7881 of *Lecture Notes in Computer Science*, pp. 177–193. Springer, Heidelberg, 2013.

Antoine Joux. A new index calculus algorithm with complexity $L(1/4 + o(1))$ in small characteristic. In T. Lange, K. Lauter, and P. Lisonek, editors, *Selected Areas in Cryptography – SAC 2013*, volume 8282 of *Lecture Notes in Computer Science*, pp. 355–379. Springer, Heidelberg, 2014.

Antoine Joux and Reynald Lercier. The function field sieve is quite special. In C. Fieker and D. R. Kohel, editors, *Algorithmic Number Theory (ANTS-V)*, volume 2369 of *Lecture Notes in Computer Science*, pp. 431–445. Springer, 2002.

Antoine Joux and Reynald Lercier. Improvements to the general number field sieve for discrete logarithms in prime fields. A comparison with the Gaussian integer method. *Mathematics of Computation*, 72(242):953–967, 2003.

Antoine Joux and Reynald Lercier. The function field sieve in the medium prime case. In S. Vaudenay, editor, *Advances in Cryptology – EUROCRYPT 2006*, volume 4004 of *Lecture Notes in Computer Science*, pp. 254–270. Springer, Heidelberg, 2006.

Antoine Joux, Reynald Lercier, Nigel Smart, and Frederik Vercauteren. The number field sieve in the medium prime case. In C. Dwork, editor, *Advances in Cryptology – CRYPTO 2006*, volume 4117 of *Lecture Notes in Computer Science*, pp. 326–344. Springer, Heidelberg, 2006.

Antoine Joux and Cécile Pierrot. Improving the polynomial time precomputation of frobenius representation discrete logarithm algorithms - simplified setting for small characteristic finite fields. In P. Sarkar and T. Iwata, editors, *Advances in Cryptology – ASIACRYPT 2014, Part I*, volume 8873 of *Lecture Notes in Computer Science*, pp. 378–397. Springer, Heidelberg, 2014.

Antoine Joux and Cécile Pierrot. The special number field sieve in $\mathbb{F}_{p^n}$ - application to pairing-friendly constructions. In Z. Cao and F. Zhang, editors, *Pairing-Based Cryptography – Pairing 2013*, volume 8365 of *Lecture Notes in Computer Science*, pp. 45–61. Springer, Heidelberg, 2014.

Antoine Joux and Cécile Pierrot. Nearly sparse linear algebra. Cryptology ePrint Archive, Report 2015/930, 2015. http://eprint.iacr.org/.

Antoine Joux and Vanessa Vitse. Cover and decomposition index calculus on elliptic curves made practical - application to a previously unreachable curve over $\mathbb{F}_{p^6}$. In D. Pointcheval and T. Johansson, editors, *Advances in Cryptology – EUROCRYPT 2012*, volume 7237 of *Lecture Notes in Computer Science*, pp. 9–26. Springer, Heidelberg, 2012.

Antoine Joux and Vanessa Vitse. Elliptic curve discrete logarithm problem over small degree extension fields - application to the static Diffie-Hellman problem on $E(\mathbb{F}_{q^5})$. *Journal of Cryptology*, 26(1):119–143, 2013.

Marc Joye. Elliptic curves and side-channel analysis. *ST Journal of System Research*, 4(1):17–21, 2003.

Marc Joye and Gregory Neven, editors. *Identity-Based Cryptography*, volume 2 of *Cryptology and Information Security Series*. IOS press, 2009.

Marc Joye, Pascal Paillier, and Sung-Ming Yen. Secure evaluation of modular functions. In R. J. Hwang and C. K. Wu, editors, *2001 International Workshop on Cryptology and Network Security*, pp. 227–229, 2001.

Marc Joye and Jean-Jacques Quisquater. Hessian elliptic curves and side-channel attacks. In Ç. K. Koç, D. Naccache, and C. Paar, editors, *Cryptographic Hardware and Embedded Systems – CHES 2001*, volume 2162 of *Lecture Notes in Computer Science*, pp. 402–410. Springer, Heidelberg, 2001.

Marc Joye, Mehdi Tibouchi, and Damien Vergnaud. Huff's model for elliptic curves. In G. Hanrot, F. Morain, and E. Thomé, editors, *Algorithmic Number Theory (ANTS-IX)*, volume 6197 of *Lecture Notes in Computer Science*, pp. 234–250. Springer, 2010.

Marc Joye and Sung-Ming Yen. The Montgomery powering ladder. In B. S. Kaliski Jr., Ç. K. Koç, and C. Paar, editors, *Cryptographic Hardware and Embedded Systems – CHES 2002*, volume 2523 of *Lecture Notes in Computer Science*, pp. 291–302. Springer, Heidelberg, 2003.

Ezekiel J. Kachisa, Edward F. Schaefer, and Michael Scott. Constructing Brezing-Weng pairing-friendly elliptic curves using elements in the cyclotomic field. In S. D. Galbraith and K. G. Paterson, editors, *Pairing-Based Cryptography – Pairing 2008*, volume 5209 of *Lecture Notes in Computer Science*, pp. 126–135. Springer, Heidelberg, 2008.

Michael Kalkbrener. An upper bound on the number of monomials in determinants of sparse matrices with symbolic entries. *Mathematica Pannonica*, 8:73–82, 1997.

David Kammler, Diandian Zhang, Peter Schwabe, Hanno Scharwächter, Markus Langenberg, Dominik Auras, Gerd Ascheid, and Rudolf Mathar. Designing an ASIP for cryptographic pairings over Barreto-Naehrig curves. In C. Clavier and K. Gaj, editors, *Cryptographic Hardware and Embedded Systems – CHES 2009*, volume 5747 of *Lecture Notes in Computer Science*, pp. 254–271. Springer, Heidelberg, 2009.

Koray Karabina. Squaring in cyclotomic subgroups. *Mathematics of Computation*, 82(281):555–579, 2013.

Koray Karabina and Edlyn Teske. On prime-order elliptic curves with embedding degrees $k = 3$, 4, and 6. In A. J. van der Poorten and A. Stein, editors, *Algorithmic Number Theory (ANTS-VIII)*, volume 5011 of *Lecture Notes in Computer Science*, pp. 102–117. Springer, 2008.

Antolii A. Karatusba and Yuri P. Ofman. Multiplication of many-digital numbers by automatic computers. *Proceedings of the USSR Academy of Sciences 145: 293–294. Translation in the academic journal Physics-Doklady*, pp. 595–596, 1963.

Emilia Käsper. Fast elliptic curve cryptography in OpenSSL. In G. Danezis, S. Dietrich, and K. Sako, editors, *Financial Cryptography and Data Security (FC 2011 Workshops,*

RLCPS and WECSR 2011), volume 7126 of *Lecture Notes in Computer Science*, pp. 27–39. Springer, Heidelberg, 2012.

Shin-ichi Kawamura, Masanobu Koike, Fumihiko Sano, and Atsushi Shimbo. Cox-rower architecture for fast parallel montgomery multiplication. In *EUROCRYPT*, volume 1807 of *Lecture Notes in Computer Science*, pp. 523–538. Springer, 2000.

Chong Hee Kim and Jean-Jacques Quisquater. Faults, injection methods, and fault attacks. *Design Test of Computers, IEEE*, 24(6):544–545, 2007.

Minkyu Kim, Jung Hee Cheon, and Jin Hong. Subset-restricted random walks for Pollard rho method on $\mathbb{F}_{p^m}$. In S. Jarecki and G. Tsudik, editors, *Public Key Cryptography – PKC 2009*, volume 5443 of *Lecture Notes in Computer Science*, pp. 54–67. Springer, Heidelberg, 2009.

Tae Hyun Kim, Tsuyoshi Takagi, Dong-Guk Han, Ho Won Kim, and Jongin Lim. *Cryptology and Network Security: 5th International Conference, CANS 2006, Suzhou, China, December 8-10, 2006. Proceedings*, chapter Side-Channel Attacks and Countermeasures on Pairing Based Cryptosystems over Binary Fields, pp. 168–181. Springer Berlin Heidelberg, Berlin, Heidelberg, 2006.

Taechan Kim and Razvan Barbulescu. Extended tower number field sieve: A new complexity for medium prime case. *IACR Cryptology ePrint Archive*, 2015:1027, 2015.

Taechan Kim and Razvan Barbulescu. Extended Tower Number Field Sieve: A New Complexity for Medium Prime Case. In M. Robshaw and J. Katz, editors, *CRYPTO 2016*, LNCS. Springer, 2016. to appear, preprint available at `http://eprint.iacr.org/2015/1027`.

Taechan Kim, Sungwook Kim, and Jung Hee Cheon. Accelerating the final exponentiation in the computation of the tate pairings. Cryptology ePrint Archive, Report 2012/119, 2012. `http://eprint.iacr.org/2012/119`.

Taechan Kim, Sungwook Kim, and Jung Hee Cheon. On the final exponentiation in Tate pairing computations. *IEEE Transactions on Information Theory*, 59(6):4033–4041, 2013.

Thorsten Kleinjung. On polynomial selection for the general number field sieve. *Mathematics of Computation*, 75(256):2037–2047, 2006.

Thorsten Kleinjung. Polynomial selection. Invited talk at the CADO-NFS workshop, Nancy, France, October 2008. slides available at `http://cado.gforge.inria.fr/workshop/slides/kleinjung.pdf`.

Thorsten Kleinjung. Discrete logarithms in GF(2^{1279}), October 2014. Announcement available at the NMBRTHRY archives, item 004751.

Edward Knapp. *On the Efficiency and Security of Cryptographic Pairings*. PhD thesis, University of Waterloo, 2010.

Donald E. Knuth. *The Art of Computer Programming, Third Edition*. Addison-Wesley, 1997.

Neal Koblitz and Alfred Menezes. Pairing-based cryptography at high security levels (invited paper). In N. P. Smart, editor, *Cryptography and Coding*, volume 3796 of *Lecture Notes in Computer Science*, pp. 13–36. Springer, Heidelberg, 2005.

Neal Koblitz and Alfred Menezes. Another look at non-standard discrete log and Diffie-Hellman problems. *Journal of Mathematical Cryptology*, 2(4):311–326, 2008.

Çetin Kaya Koç, Tolga Acar, and Burton S. Kaliski Jr. Analyzing and comparing montgomery multiplication algorithms. *IEEE Micro*, 16(3):26–33, 1996.

Paul C. Kocher, Joshua Jaffe, and Benjamin Jun. Differential power analysis. In M. J. Wiener, editor, *Advances in Cryptology – CRYPTO '99*, volume 1666 of *Lecture Notes in Computer Science*, pp. 388–397. Springer, Heidelberg, 1999.

Maurice Kraitchik. *Théorie des Nombres*. Gauthier–Villars, 1922.

Maurice Kraitchik. *Recherches sur la Théorie des Nombres.* Gauthier–Villars, 1924.

Fabian Kuhn and René Struik. Random walks revisited: Extensions of Pollard's rho algorithm for computing multiple discrete logarithms. In S. Vaudenay and A. M. Youssef, editors, *Selected Areas in Cryptography (SAC 2001)*, volume 2259 of *Lecture Notes in Computer Science*, pp. 212–229. Springer, Heidelberg, 2001.

S. Lang. *Algebra.* Graduate Texts in Mathematics. Springer New York, 2005.

Serge Lang. *Elliptic Functions*, volume 112 of *Graduate Texts in Mathematics.* Springer-Verlag, 2nd edition, 1987.

Ronan Lashermes, Jacques Fournier, and Louis Goubin. Inverting the final exponentiation of Tate pairings on ordinary elliptic curves using faults. In G. Bertoni and J.-S. Coron, editors, *Cryptographic Hardware and Embedded Systems – CHES 2013*, volume 8086 of *Lecture Notes in Computer Science*, pp. 365–382. Springer, Heidelberg, 2013.

Kristin Lauter, Peter L. Montgomery, and Michael Naehrig. An analysis of affine coordinates for pairing computation. In M. Joye, A. Miyaji, and A. Otsuka, editors, *Pairing-Based Cryptography – Pairing 2010*, volume 6487 of *Lecture Notes in Computer Science*, pp. 1–20. Springer, Heidelberg, 2010.

Duc-Phong Le and Chik How Tan. Speeding up ate pairing computation in affine coordinates. In T. Kwon, M. Lee, and D. Kwon, editors, *Information Security and Cryptology – ICISC 2012*, volume 7839 of *Lecture Notes in Computer Science*, pp. 262–277. Springer, Heidelberg, 2013.

Arjen K. Lenstra. Unbelievable security: Matching AES security using public key systems (invited talk). In C. Boyd, editor, *Advances in Cryptology – ASIACRYPT 2001*, volume 2248 of *Lecture Notes in Computer Science*, pp. 67–86. Springer, Heidelberg, 2001.

Arjen K. Lenstra. Key lengths. In H. Bidgoli, editor, *Handbook of Information Security*, volume 3, pp. 617–635. John Wiley & Sons, 2006.

Arjen K. Lenstra, Hendrik W. Lenstra Jr., and László Lovász. Factoring polynomials with rational coefficients. *Mathematische Annalen*, 261(4):515–534, 1982.

Arjen K. Lenstra and Eric R. Verheul. The XTR public key system. In M. Bellare, editor, *Advances in Cryptology – CRYPTO 2000*, volume 1880 of *Lecture Notes in Computer Science*, pp. 1–19. Springer, Heidelberg, 2000.

Arjen K. Lenstra and Eric R. Verheul. Selecting cryptographic key sizes. *Journal of Cryptology*, 14(4):255–293, 2001.

Benoît Libert and Jean-Jacques Quisquater. Efficient signcryption with key privacy from gap Diffie-Hellman groups. In F. Bao, R. Deng, and J. Zhou, editors, *Public Key Cryptography – PKC 2004*, volume 2947 of *Lecture Notes in Computer Science*, pp. 187–200. Springer, Heidelberg, 2004.

Stephen Lichtenbaum. Duality theorems for curves over p-adic fields. *Inventiones mathematicae*, 7(2):120–136, 1969.

Rudolf Lidl and Harald Niederreiter. *Finite Fields*, volume 20 of *Encyclopedia of Mathematics and Its Applications.* Cambridge University Press, 2nd edition, 1997.

Chae Hoon Lim and Hyo Sun Hwang. Fast implementation of elliptic curve arithmetic in GF(p^n). In H. Imai and Y. Zheng, editors, *PKC 2000: 3rd International Workshop on Theory and Practice in Public Key Cryptography*, volume 1751 of *Lecture Notes in Computer Science*, pp. 405–421. Springer, Heidelberg, 2000.

Chae Hoon Lim and Pil Joong Lee. More flexible exponentiation with precomputation. In Y. Desmedt, editor, *Advances in Cryptology – CRYPTO '94*, volume 839 of *Lecture Notes in Computer Science*, pp. 95–107. Springer, Heidelberg, 1994.

Chae Hoon Lim and Pil Joong Lee. A key recovery attack on discrete log-based schemes using a prime order subgroup. In B. S. Kaliski, Jr., editor, *Advances in Cryptology –*

CRYPTO'97, volume 1294 of *Lecture Notes in Computer Science*, pp. 249–263. Springer, Heidelberg, 1997.

Seongan Lim, Seungjoo Kim, Ikkwon Yie, Jaemoon Kim, and Hongsub Lee. XTR extended to GF(p^{6m}). In S. Vaudenay and A. M. Youssef, editors, *Selected Areas in Cryptography (SAC 2001)*, volume 2259 of *Lecture Notes in Computer Science*, pp. 301–312. Springer, Heidelberg, 2001.

S. Lindhurst. An analysis of Shanks's algorithm for computing square roots in finite fields. *CRM Proc. and Lecture Notes*, Vol. 19:231–242, 1999.

David Lubicz and Damien Robert. Efficient pairing computation with theta functions. In G. Hanrot, F. Morain, and E. Thomé, editors, *Algorithmic Number Theory (ANTS-IX)*, volume 6197 of *Lecture Notes in Computer Science*, pp. 251–269. Springer, 2010.

Florian Luca and Igor E. Shparlinski. Elliptic curves with low embedding degree. *Journal of Cryptology*, 19(4):553–562, 2006.

Keith Matthews. The diophantine equation $x^2 - Dy^2 = N, D > 0$. *Expositiones Mathemeticae*, 18(4):323–331, 2000.

D. V. Matyukhin. Effective version of the number field sieve for discrete logarithms in the field GF(p^k) (in Russian). *Trudy po Discretnoi Matematike*, 9:121–151, 2006.

Ueli M. Maurer, Renato Renner, and Clemens Holenstein. Indifferentiability, impossibility results on reductions, and applications to the random oracle methodology. In M. Naor, editor, *Theory of Cryptography Conference (TCC 2004)*, volume 2951 of *Lecture Notes in Computer Science*, pp. 21–39. Springer, Heidelberg, 2004.

Ueli M. Maurer and Stefan Wolf. The relationship between breaking the Diffie-Hellman protocol and computing discrete logarithms. *SIAM Journal on Computing*, 28(5):1689–1721, 1999.

Ueli M. Maurer and Stefan Wolf. The Diffie-Hellman protocol. *Designs, Codes and Cryptography*, 19(2/3):147–171, 2000.

Kevin S. McCurley. The discrete logarithm problem. In C. Pomerance, editor, *Cryptology and Computational Number Theory*, volume 42 of *Proceedings of Symposia in Applied Mathematics*, pp. 49–74. AMS, 1990.

Alfred Menezes. *Elliptic Curve Public Key Cryptosystems*. Kluwer Academic Publishers, 1993.

Alfred Menezes. An introduction to pairing-based cryptography. In I. Luengo, editor, *Recent Trends in Cryptography*, volume 477 of *Contemporary Mathematics*, pp. 47–65. AMS-RMSE, 2009.

Alfred Menezes, Scott A. Vanstone, and Tatsuaki Okamoto. Reducing elliptic curve logarithms to logarithms in a finite field. In *23rd Annual ACM Symposium on Theory of Computing*, pp. 80–89. ACM Press, 1991.

Alfred J. Menezes, Tatsuaki Okamoto, and Scott A. Vanstone. Reducing elliptic curves logarithms to logarithms in a finite field. *IEEE Transactions on Information Theory*, 39(5):1639–1646, 1993.

Alfred J. Menezes, Paul C. van Oorschot, and Scott A. Vanstone. *Handbook of Applied Cryptography*. The CRC Press series on discrete mathematics and its applications. CRC Press, Boca Raton, FL, 1997.

Victor S. Miller. The Weil pairing, and its efficient calculation. *Journal of Cryptology*, 17(4):235–261, 2004.

Hermann Minkowski. *Geometrie der Zahlen*. Leipzig und Berlin, Druck ung Verlag von B.G. Teubner, 1910.

Shigeo Mitsunari. A fast implementation of the optimal Ate pairing over BN curve on Intel Haswell processor. Cryptology ePrint Archive, Report 2013/362, 2013. http://eprint.iacr.org/2013/362.

Atsuko Miyaji, Masaki Nakabayashi, and Shunzo Takano. Characterization of elliptic curve traces under FR-reduction. In D. Won, editor, *Information Security and Cryptology – ICISC 2000*, volume 2015 of *Lecture Notes in Computer Science*, pp. 90–108. Springer, Heidelberg, 2001.

Peter L. Montgomery. Modular multiplication without trial division. *Mathematics of Computation*, 44(170):519–521, 1985.

Peter L. Montgomery. Speeding the Pollard and elliptic curve methods of factorization. *Mathematics of Computation*, 48(177):243–264, 1987.

François Morain. Classes d'isomorphismes des courbes elliptiques supersingulières en caractéristique ≥ 3. *Utilitas Mathematica*, 52:241–253, 1997.

S. Müller. On the computation of square roots in finite fields. *J. Design, Codes and Cryptography*, vol. 31:301–312, 2004.

Angela Murphy and Noel Fitzpatrick. Elliptic curves for pairing applications. Cryptology ePrint Archive, Report 2005/302, 2005. http://eprint.iacr.org/2005/302.

B. A. Murphy. *Polynomial Selection for the Number Field Sieve Integer Factorisation Algorithm*. PhD thesis, Australian National University, 1999. http://maths-people.anu.edu.au/~brent/pd/Murphy-thesis.pdf.

Brian A. Murphy. Modelling the yield of number field sieve polynomials. In J. P. Buhler, editor, *Algorithmic Number Theory: Third International Symposiun, ANTS-III Portland, Oregon, USA, June 21–25, 1998 Proceedings*, Lecture Notes in Computer Science, pp. 137–150. Springer Berlin Heidelberg, 1998.

Michael Naehrig, Paulo S. L. M. Barreto, and Peter Schwabe. On compressible pairings and their computation. In S. Vaudenay, editor, *Progress in Cryptology – AFRICACRYPT 2008*, volume 5023 of *Lecture Notes in Computer Science*, pp. 371–388. Springer, Heidelberg, 2008.

Michael Naehrig, Ruben Niederhagen, and Peter Schwabe. New software speed records for cryptographic pairings. In M. Abdalla and P. S. L. M. Barreto, editors, *Progress in Cryptology – LATINCRYPT 2010*, volume 6212 of *Lecture Notes in Computer Science*, pp. 109–123. Springer, Heidelberg, 2010.

Yasuyuki Nogami, Masataka Akane, Yumi Sakemi, Hidehiro Katou, and Yoshitaka Morikawa. Integer variable chi-based Ate pairing. In S. D. Galbraith and K. G. Paterson, editors, *Pairing-Based Cryptography – Pairing 2008*, volume 5209 of *Lecture Notes in Computer Science*, pp. 178–191. Springer, Heidelberg, 2008.

Andrew M. Odlyzko. Discrete logarithms in finite fields and their cryptographic significance. In T. Beth, N. Cot, and I. Ingemarsson, editors, *Advances in Cryptology – EUROCRYPT '84*, volume 209 of *Lecture Notes in Computer Science*, pp. 224–314. Springer, Heidelberg, 1985.

Kiyoshi Ohgishi, Ryuichi Sakai, and Masao Kasahara. Notes on ID-based key sharing systems over elliptic curve (in Japanese). Technical Report ISEC99-57, IEICE, 1999.

Jorge Olivos. On vectorial addition chains. *Journal of Algorithms*, 2(1):13–21, 1981.

Hilarie Orman and Paul Hoffman. Determining strengths for public keys used for exchanging symmetric keys. Request for Comments RFC 3766, Internet Engineering Task Force (IETF), 2004.

Erdinç Öztürk, Gunnar Gaubatz, and Berk Sunar. Tate pairing with strong fault resiliency. In L. Breveglieri et al., editors, *Fourth Workshop on Fault Diagnosis and Tolerance in Cryptography (FDTC 2007)*, pp. 103–111. IEEE Computer Society, 2007.

Dan Page, Nigel P. Smart, and Frederik Vercauteren. A comparison of MNT curves and supersingular curves. *Applicable Algebra in Engineering, Communication and Computing*, 17(5):379–392, 2006.

Dan Page and Frederik Vercauteren. Fault and side-channel attacks on pairing based cryptography. Cryptology ePrint Archive, Report 2004/283, 2004. http://eprint.iacr.org/2004/283.

Dan Page and Frederik Vercauteren. A fault attack on pairing-based cryptography. *IEEE Transactions on Computers*, 55(9):1075–1080, 2006.

Jea-Hoon Park, Gyo Yong-Sohn, and Sang-Jae Moon. A simplifying method of fault attacks on pairing computations. *IEICE Transactions on Fundamentals of Electronics, Communications and Computer Sciences*, 94(6):1473–1475, 2011.

Bryan Parno, Jon Howell, Craig Gentry, and Mariana Raykova. Pinocchio: Nearly practical verifiable computation. In *2013 IEEE Symposium on Security and Privacy*, pp. 238–252. IEEE Computer Society Press, 2013.

Geovandro C. C. F. Pereira, Marcos A. Simplício, Jr., Michael Naehrig, and Paulo S. L. M. Barreto. A family of implementation-friendly BN elliptic curves. *Journal of Systems and Software*, 84(8):1319–1326, 2011.

Cécile Pierrot. The multiple number field sieve with conjugation and generalized joux-lercier methods. In E. Oswald and M. Fischlin, editors, *Advances in Cryptology – EUROCRYPT 2015, Part I*, volume 9056 of *Lecture Notes in Computer Science*, pp. 156–170. Springer, Heidelberg, 2015.

John M. Pollard. Monte Carlo methods for index computation (mod p). *Mathematics of Computation*, 32(143):918–924, 1978.

John M. Pollard. Kangaroos, monopoly and discrete logarithms. *Journal of Cryptology*, 13(4):437–447, 2000.

Karl C. Posch and Reinhard Posch. Modulo reduction in residue number systems. *IEEE Trans. Parallel Distrib. Syst.*, 6(5):449–454, 1995.

Pablo Rauzy and Sylvain Guilley. Countermeasures against high-order fault-injection attacks on CRT-RSA. In A. Tria and D. Choi, editors, *2014 Workshop on Fault Diagnosis and Tolerance in Cryptography (FDTC 2014)*, pp. 68–82. IEEE Computer Society, 2014.

Pablo Rauzy, Martin Moreau, Sylvain Guilley, and Zakaria Najm. Using modular extension to provably protect ECC against fault attacks. Cryptology ePrint Archive, Report 2015/882, 2015. http://eprint.iacr.org/2015/882.

Thomas Ristenpart, Hovav Shacham, and Thomas Shrimpton. Careful with composition: Limitations of the indifferentiability framework. In K. G. Paterson, editor, *Advances in Cryptology – EUROCRYPT 2011*, volume 6632 of *Lecture Notes in Computer Science*, pp. 487–506. Springer, Heidelberg, 2011.

Ronald L. Rivest, Adi Shamir, and Leonard M. Adleman. A method for obtaining digital signature and public-key cryptosystems. *Communications of the Association for Computing Machinery*, 21(2):120–126, 1978.

John P. Robertson. Solving the generalized Pell equation $x^2 - Dy^2 = N$. Unpublished manuscript, available at http://www.jpr2718.org/pell.pdf, 2004.

Francisco Rodríguez-Henríquez and Çetin K. Koç. On fully parallel Karatsuba multipliers for $GF(2^m)$. In A. Tria and D. Choi, editors, *International Conference on Computer Science and Technology (CST 2003)*, pp. 405–410. ACTA Press, 2003.

Karl Rubin and Alice Silverberg. Choosing the correct elliptic curve in the CM method. *Mathematics of Computation*, 79(269):545–561, 2010.

Hans-Georg Rück. On the discrete logarithm in the divisor class group of curves. *Mathematics of Computation*, 68(226):805–806, 1999.

Amit Sahai. Non-malleable non-interactive zero knowledge and adaptive chosen-ciphertext security. In *40th Annual Symposium on Foundations of Computer Science*, pp. 543–553. IEEE Computer Society Press, 1999.

Ryuichi Sakai, Kiyoshi Ohgishi, and Masao Kasahara. Cryptosystems based on pairing. In *2000 Symposium on Cryptography and Information Security (SCIS 2000)*, pp. 135–148, 2000.

Ryuichi Sakai, Kiyoshi Ohgishi, and Masao Kasahara. Cryptosystems based on pairing over elliptic curve (in Japanese). In *2001 Symposium on Cryptography and Information Security (SCIS 2001)*, pp. 23–26, 2001.

Palash Sarkar and Shashank Singh. A simple method for obtaining relations among factor basis elements for special hyperelliptic curves. Cryptology ePrint Archive, Report 2015/179, 2015. http://eprint.iacr.org/2015/179.

Palash Sarkar and Shashank Singh. A general polynomial selection method and new asymptotic complexities for the tower number field sieve algorithm. Cryptology ePrint Archive, Report 2016/485, 2016. http://eprint.iacr.org/.

Palash Sarkar and Shashank Singh. A generalisation of the conjugation method for polynomial selection for the extended tower number field sieve algorithm. Cryptology ePrint Archive, Report 2016/537, 2016. http://eprint.iacr.org/.

Palash Sarkar and Shashank Singh. Tower number field sieve variant of a recent polynomial selection method. Cryptology ePrint Archive, Report 2016/401, 2016. http://eprint.iacr.org/.

Takakazu Satoh and Kiyomichi Araki. Fermat quotients and the polynomial time discrete log algorithm for anomalous elliptic curves. *Commentarii Math. Univ. St. Pauli*, 47(1):81–92, 1998.

Jürgen Sattler and Claus-Peter Schnorr. Generating random walks in groups. *Ann. Univ. Sci. Budapest. Sect. Comput.*, 6:65–79, 1985.

Edward F. Schaefer. A new proof for the non-degeneracy of the Frey-Rück pairing and a connection to isogenies over the base field. *Computational aspects of algebraic curves*, 13:1–12, 2005.

Andrzej Schinzel and Wacław Sierpiński. Sur certaines hypothèses concernant les nombres premiers. *Acta Arithmetica*, 4(3):185–208, 1958. Erratum 5 (1958), 259.

Oliver Schirokauer. Discrete logarithms and local units. *Philosophical Transactions of the Royal Society*, 345(1676):409–423, 1993.

René Schoof. Elliptic curves over finite fields and the computation of square roots mod p. *Mathematics of Computation*, 44(170):483–494, 1985.

Michael Scott. Computing the Tate pairing. In A. Menezes, editor, *Topics in Cryptology – CT-RSA 2005*, volume 3376 of *Lecture Notes in Computer Science*, pp. 293–304. Springer, Heidelberg, 2005.

Michael Scott. Implementing cryptographic pairings. In T. Takagi et al., editors, *Pairing-Based Cryptography – Pairing 2007*, volume 4575 of *Lecture Notes in Computer Science*, pp. 177–196. Springer, Heidelberg, 2007.

Michael Scott. On the efficient implementation of pairing-based protocols. In L. Chen, editor, *Cryptography and Coding*, volume 7089 of *Lecture Notes in Computer Science*, pp. 296–308. Springer, Heidelberg, 2011.

Michael Scott. Unbalancing pairing-based key exchange protocols. Cryptology ePrint Archive, Report 2013/688, 2013. http://eprint.iacr.org/2013/688.

Michael Scott and Paulo S. L. M. Barreto. Compressed pairings. In M. Franklin, editor, *Advances in Cryptology – CRYPTO 2004*, volume 3152 of *Lecture Notes in Computer Science*, pp. 140–156. Springer, Heidelberg, 2004.

Michael Scott and Paulo S. L. M. Barreto. Generating more MNT elliptic curves. *Designs, Codes and Cryptography*, 38(2):209–217, 2006.

Michael Scott, Naomi Benger, Manuel Charlemagne, Luis J. Dominguez Perez, and Ezekiel J. Kachisa. Fast hashing to G_2 on pairing-friendly curves. In H. Shacham and B. Waters, editors, *Pairing-Based Cryptography – Pairing 2009*, Volume 5671 of *Lecture Notes in Computer Science*, pp. 102–113. Springer, Heidelberg, 2009.

Michael Scott, Naomi Benger, Manuel Charlemagne, Luis J. Dominguez Perez, and Ezekiel J. Kachisa. On the final exponentiation for calculating pairings on ordinary elliptic curves. In H. Shacham and B. Waters, editors, *Pairing-Based Cryptography – Pairing 2009*, volume 5671 of *Lecture Notes in Computer Science*, pp. 78–88. Springer, Heidelberg, 2009.

Mike Scott. Missing a trick: Karatsuba revisited. Cryptology ePrint Archive, Report 2015/1247, 2015. http://eprint.iacr.org/.

Jean-Pierre Serre. *Groupes algébriques et corps de classes*, volume 7 of *Publications de l'Institut de mathématique de l'Université de Nancago*. Hermann, 2nd edition, 1975.

Andrew Shallue and Christiaan van de Woestijne. Construction of rational points on elliptic curves over finite fields. In F. Hess, S. Pauli, and M. E. Pohst, editors, *Algorithmic Number Theory (ANTS-VII)*, volume 4076 of *Lecture Notes in Computer Science*, pp. 510–524. Springer, 2006.

Adi Shamir. Identity-based cryptosystems and signature schemes. In G. R. Blakley and D. Chaum, editors, *Advances in Cryptology, Proceedings of CRYPTO '84*, volume 196 of *Lecture Notes in Computer Science*, pp. 47–53. Springer, Heidelberg, 1984.

Adi Shamir. Method and apparatus for protecting public key schemes from timing and fault attacks. US Patent # 5,991,415, 1999. Presented at the rump session of EURO-CRYPT '97.

Daniel Shanks. Class number, a theory of factorization, and genera. In D. J. Lewis, editor, *1969 Number Theory Institute*, volume 20 of *Proceedings of Symposia in Applied Mathematics*, pp. 415–440. AMS, 1971.

Daniel Shanks. Five number-theoretic algorithms. In R. S. D. Thomas and H. C. Williams, editors, *Proceedings of the Second Manitoba Conference on Numerical Mathematics*, pp. 51–70. Utilitas Mathematica, 1972.

Masaaki Shirase, Tsuyoshi Takagi, and Eiji Okamoto. An efficient countermeasure against side-channel attacks for pairing computation. In L. Chen, Y. Mu, and W. Susilo, editors, *Information Security Practice and Experience (ISPEC 2008)*, volume 4991 of *Lecture Notes in Computer Science*, pp. 290–303. Springer, 2008.

Peter W. Shor. Polynomial-time algorithms for prime factorization and discrete logarithms on a quantum computer. *SIAM Journal on Computing*, 26(5):1484–1509, 1997.

Victor Shoup. *A Computational Introduction to Number Theory and Algebra*. Cambridge University Press, 2nd edition, 2009.

Joseph H. Silverman. *Advanced Topics in the Arithmetic of Elliptic Curves*, volume 151 of *Graduate Texts in Mathematics*. Springer-Verlag, 1994.

Joseph H. Silverman. *The Arithmetic of Elliptic Curves*, volume 106 of *Graduate Texts in Mathematics*. Springer-Verlag, 2nd edition, 2009.

Nigel P. Smart. The discrete logarithm problem on elliptic curves of trace one. *Journal of Cryptology*, 12(3):193–196, 1999.

Nigel P. Smart. The hessian form of an elliptic curve. In Ç. K. Koç, D. Naccache, and C. Paar, editors, *Cryptographic Hardware and Embedded Systems – CHES 2001*, volume 2162 of *Lecture Notes in Computer Science*, pp. 118–125. Springer, Heidelberg, 2001.

Benjamin Smith. Easy scalar decompositions for efficient scalar multiplication on elliptic curves and genus 2 Jacobians. In S. Ballet, M. Perret, and A. Zaytsev, editors, *Algorithmic Arithmetic, Geometry, and Coding Theory*, volume 637 of *Contemporary Mathematics*, pp. 127–145. AMS, 2015.

Jerome Solinas. ID-based digital signature algorithms. Invited talk, *7th Workshop on Elliptic Curve Cryptography (ECC 2003)*, August 11–13, 2003. Slides available at http://www.cacr.math.uwaterloo.ca/conferences/2003/ecc2003/solinas.pdf.

Martijn Stam and Arjen K. Lenstra. Speeding up XTR. In C. Boyd, editor, *Advances in Cryptology – ASIACRYPT 2001*, volume 2248 of *Lecture Notes in Computer Science*, pp. 125–143. Springer, Heidelberg, 2001.

Andreas Stein and Edlyn Teske. Optimized baby step–giant step methods. *J. Ramanujan Math. Soc.*, 20(1):27–58, 2005.

William Stein. *SAGE: Software for Algebra and Geometry Experimentation*. http://www.sagemath.org/.

Douglas R. Stinson. *Cryptography: Theory and Practice*. Discrete Mathematics and Its Applications. Chapman & Hall/CRC, Boca Raton, FL, 3rd edition, 2006.

Andrew V. Sutherland. Computing Hilbert class polynomials with the Chinese remainder theorem. *Mathematics of Computation*, 80(273):501–538, 2011.

Robert Szerwinski and Tim Güneysu. Exploiting the power of gpus for asymmetric cryptography. In *CHES*, volume 5154 of *Lecture Notes in Computer Science*, pp. 79–99. Springer, 2008.

Satoru Tanaka and Ken Nakamula. Constructing pairing-friendly elliptic curves using factorization of cyclotomic polynomials. In S. D. Galbraith and K. G. Paterson, editors, *Pairing-Based Cryptography – Pairing 2008*, volume 5209 of *Lecture Notes in Computer Science*, pp. 136–145. Springer, Heidelberg, 2008.

John Tate. WC-groups over p-adic fields. Exposé 156, Séminaire Bourbaki, 1957/58.

John Tate. Duality theorems in Galois cohomology over number fields. In *Proc. Internat. Congr. Mathematicians (Stockholm, 1962)*, pp. 288–295. Inst. Mittag-Leffler, Djursholm, 1963.

David C. Terr. A modification of Shanks' baby-step giant-step algorithm. *Mathematics of Computation*, 69(230):767–773, 2000.

Tadanori Teruya, Kazutaka Saito, Naoki Kanayama, Yuto Kawahara, Tetsutaro Kobayashi, and Eiji Okamoto. Constructing symmetric pairings over supersingular elliptic curves with embedding degree three. In Z. Cao and F. Zhang, editors, *Pairing-Based Cryptography – Pairing 2013*, volume 8365 of *Lecture Notes in Computer Science*, pp. 97–112. Springer, Heidelberg, 2014.

Edlyn Teske. Speeding up Pollard's rho method for computing discrete logarithms. In J. P. Buhler, editor, *Algorithmic Number Theory (ANTS-III)*, volume 1423 of *Lecture Notes in Computer Science*, pp. 541–554. Springer, 1998.

Emmanuel Thomé. Computation of discrete logarithms in $F_{2^6 07}$. In C. Boyd, editor, *Advances in Cryptology – ASIACRYPT 2001*, volume 2248 of *Lecture Notes in Computer Science*, pp. 107–124. Springer, Heidelberg, 2001.

Emmanuel Thomé. Discrete logarithms in GF(2^{607}), February 2002. Announcement available at the NMBRTHRY archives, item 001894.

Emmanuel Thomé. *Algorithmes de calcul de logarithme discret dans les corps finis*. Thèse, École polytechnique, 2003. https://tel.archives-ouvertes.fr/tel-00007532.

A. Tonelli. Bemerkung uber die auflosung quadratischer congruenzen. *Götinger Nachrichten*, pp. 344–346, 1891.

Elena Trichina and Roman Korkikyan. Multi fault laser attacks on protected CRT-RSA. In L. Breveglieri et al., editors, *2010 Workshop on Fault Diagnosis and Tolerance in Cryptography (FDTC 2010)*, pp. 75–86. IEEE Computer Society, 2010.

Thomas Unterluggauer and Erich Wenger. Efficient pairings and ECC for embedded systems. In L. Batina and M. Robshaw, editors, *Cryptographic Hardware and Embedded Systems – CHES 2014*, volume 8731 of *Lecture Notes in Computer Science*, pp. 298–315. Springer, Heidelberg, 2014.

Thomas Unterluggauer and Erich Wenger. Practical attack on bilinear pairings to disclose the secrets of embedded devices. In *9th International Conference on Availability, Reliability and Security (ARES 2014)*, pp. 69–77. IEEE Computer Society, 2014.

Jorge Jiménez Urroz, Florian Luca, and Igor E. Shparlinski. On the number of isogeny classes of pairing-friendly elliptic curves and statistics of MNT curves. *Mathematics of Computation*, 81(278):1093–1110, 2012.

Paul C. van Oorschot and Michael J. Wiener. Parallel collision search with cryptanalytic applications. *Journal of Cryptology*, 12(1):1–28, 1999.

F. Vercauteren. Optimal pairings. Cryptology ePrint Archive, Report 2008/096, 2008. http://eprint.iacr.org/2008/096.

Frederik Vercauteren. Optimal pairings. *IEEE Transactions on Information Theory*, 56(1):455–461, 2009.

Eric R. Verheul. Evidence that XTR is more secure than supersingular elliptic curve cryptosystems. *Journal of Cryptology*, 17(4):277–296, 2004.

Lawrence C. Washington. *Elliptic Curves: Number Theory and Cryptography*. Discrete Mathematics and Its Applications. Chapman & Hall/CRC, second edition, 2008.

William C. Waterhouse. Abelian varieties over finite fields. *Ann. Sci. École Norm. Sup. (4)*, 2:521–560, 1969.

Damian Weber and Thomas F. Denny. The solution of McCurley's discrete log challenge. In H. Krawczyk, editor, *Advances in Cryptology – CRYPTO '98*, volume 1462 of *Lecture Notes in Computer Science*, pp. 458–471. Springer, Heidelberg, 1998.

André Weil. Sur les fonctions algébriques à corps de constantes fini. *Comptes rendus de l'Académie des sciences*, 210:592–594, 1940.

André Weimerskirch and Christof Paar. Generalizations of the karatsuba algorithm for efficient implementations. Cryptology ePrint Archive, Report 2006/224, 2006. http://eprint.iacr.org/2006/224.

Jiang Weng, Yunqi Dou, and Chuangui Ma. Fault attacks against the Miller algorithm in Hessian coordinates. In C.-K. Wu, M. Yung, and D. Lin, editors, *Information Security and Cryptology (Inscrypt 2011)*, volume 7537 of *Lecture Notes in Computer Science*, pp. 102–112. Springer, 2012.

A. E. Western and J. C. P. Miller. *Tables of Indices and Primitive Roots*, volume 9 of *Royal Society Mathematical Tables*. Cambridge University Press, 1968.

Claire Whelan and Michael Scott. The importance of the final exponentiation in pairings when considering fault attacks. In T. Takagi et al., editors, *Pairing-Based Cryptography – Pairing 2007*, volume 4575 of *Lecture Notes in Computer Science*, pp. 225–246. Springer, Heidelberg, 2007.

Claire Whelan and Mike Scott. Side-channel analysis of practical pairing implementations: Which path is more secure? In P. Q. Nguyen, editor, *Progress in Cryptology – VIETCRYPT 2006*, volume 4341 of *Lecture Notes in Computer Science*, pp. 99–114. Springer, Heidelberg, 2006.

D. H. Wiedemann. Solving sparse linear equations over finite fields. *IEEE Transactions on Information Theory*, IT–32(1):54–62, 1986.

Bo Yang, Kaijie Wu, and Ramesh Karri. Scan based side-channel attack on dedicated hardware implementations of data encryption standard. In *International Test Conference (ITC 2004)*, pp. 339–344. IEEE Press, 2004.

Gavin Xiaoxu Yao, Junfeng Fan, Ray C. C. Cheung, and Ingrid Verbauwhede. Faster pairing coprocessor architecture. In M. Abdalla and T. Lange, editors, *Pairing-Based Cryptography – Pairing 2012*, volume 7708 of *Lecture Notes in Computer Science*, pp. 160–176. Springer, Heidelberg, 2013.

Takanori Yasuda, Tsuyoshi Takagi, and Kouichi Sakurai. Application of scalar multiplication of Edwards curves to pairing-based cryptography. In G. Hanaoka and T. Yamauchi, editors, *IWSEC 12: 7th International Workshop on Security, Advances in Information and Computer Security*, volume 7631 of *Lecture Notes in Computer Science*, pp. 19–36. Springer, Heidelberg, 2012.

Peter Yee. Updates to the internet X.509 public key infrastructure certificate and certificate revocation list (CRL) profile. Request for Comments RFC 6818, Internet Engineering Task Force (IETF), 2013.

Gideon Yuval. How to swindle rabin. *Cryptologia, Volume 3, Issue 3*, p. 187-190, 1979.

Pavol Zajac. *Discrete Logarithm Problem in Degree Six Finite Fields*. PhD thesis, Slovak University of Technology, 2008. http://www.kaivt.elf.stuba.sk/kaivt/Vyskum/XTRDL.

Eric Zavattoni, Luis J. Dominguez Perez, Shigeo Mitsunari, Ana H. Sánchez-Ramírez, Tadanori Teruya, and Francisco Rodríguez-Henríquez. Software implementation of an attribute-based encryption scheme. *IEEE Transactions on Computers*, 64(5):1429–1441, 2015.

Fangguo Zhang and Kwangjo Kim. ID-based blind signature and ring signature from pairings. In Y. Zheng, editor, *Advances in Cryptology – ASIACRYPT 2002*, volume 2501 of *Lecture Notes in Computer Science*, pp. 533–547. Springer, Heidelberg, 2002.

Xusheng Zhang and Kunpeng Wang. Fast symmetric pairing revisited. In Z. Cao and F. Zhang, editors, *Pairing-Based Cryptography – Pairing 2013*, volume 8365 of *Lecture Notes in Computer Science*, pp. 131–148. Springer, Heidelberg, 2014.

Chang-An Zhao, Fangguo Zhang, and Jiwu Huang. A note on the Ate pairing. *International Journal of Information Security*, 7(6):379–382, 2008.

Index